Forestry: Sustenance and Ecology

Forestry: Sustenance and Ecology

Edited by Lee Zieger

SYRAWOOD
PUBLISHING HOUSE
New York

Published by Syrawood Publishing House,
750 Third Avenue, 9th Floor,
New York, NY 10017, USA
www.syrawoodpublishinghouse.com

Forestry: Sustenance and Ecology
Edited by Lee Zieger

International Standard Book Number: 978-1-68286-426-5 (Hardback)

Cataloging-in-publication Data

Forestry : sustenance and ecology / edited by Lee Zieger.
p. cm.
Includes bibliographical references and index.
ISBN 978-1-68286-426-5
1. Forests and forestry. 2. Forest management. 3. Forest ecology. 4. Forest restoration. I. Zieger, Lee.
SD121 .F67 2017
634.9--dc23

Printed in China.

TABLE OF CONTENTS

PREFACE

Forestry is defined as the management and conservation of forests for human and environmental benefits. Forests are vital for ecosystem functioning. This book on forestry focuses on practices of conservation, landscape management and other practices related to forestry. Urbanization, waste management and pollution are other concerns that are faced by forestry. This book is a valuable compilation of topics, ranging from the basic to the most complex advancement in this field. Some of the diverse topics covered herein address the varied branches that fall under this category. This text studies, analysis, upholds the pillars of forestry and its utmost significance in modern times. Experts and researchers in the fields of agricultural soil sciences, forest ecology and ecological risk assessment will find this book to be especially useful.

This book has been a concerted effort by a group of academicians, researchers and scientists, who have contributed their research works for the realization of the book. This book has materialized in the wake of emerging advancements and innovations in this field. Therefore, the need of the hour was to compile all the required researches and disseminate the knowledge to a broad spectrum of people comprising of students, researchers and specialists of the field.

At the end of the preface, I would like to thank the authors for their brilliant chapters and the publisher for guiding us all-through the making of the book till its final stage. Also, I would like to thank my family for providing the support and encouragement throughout my academic career and research projects.

Editor

Dynamics of Two Multi-Stemmed Understory Shrubs in Two Temperate Forests

Xuejiao Bai[1,2], Tania Brenes-Arguedas[3], Ji Ye[2], Xugao Wang[2], Fei Lin[2,4], Zuoqiang Yuan[2], Shuai Shi[2,4], Dingliang Xing[2,4], Zhanqing Hao[2]*

1 College of Forestry, Shenyang Agricultural University, Shenyang, Liaoning Province, China, **2** State Key Laboratory of Forest and Soil Ecology, Institute of Applied Ecology, Chinese Academy of Sciences, Shenyang, Liaoning Province, China, **3** Centro Cientifico Tropical, San Jose, Costa Rica, **4** University of Chinese Academy of Sciences, Beijing, China

Abstract

A multi-stemmed growth form may be an important trait enabling the persistence of individual shrubs in the forest understory. With the aim of evaluating the role of multiple stems, neighbor competition and soil nutrients in shrub performance, we study the dynamics of two temperate multi-stemmed shrub species. We modeled stem growth and survival of *Corylus mandshurica* and *Acer barbinerve* in two temperate forests with differing structure in northeastern China. One forest was an old growth broad-leaved Korean pine (*Pinus koraiensis*) mixed forest; the other was a secondary poplar-birch forest. Growth of the two species and survival of *C. mandshurica* increased with stem number in the old growth forest, but not the secondary forest, suggesting the benefits of a multi-stemmed growth form are facultative. *C. mandshurica* also suffered more from overstory neighbor competition in the old growth forest, which may suggest that this species is less shade-tolerant than *A. barbinerve*. Moreover, the performance of the two species were clearly influenced by understory neighbors and soil variables in the old growth forest relative to the secondary forest, which may be due to different forest structure. We conclude that multiple stems are not always important for the persistence of shrub species. Even within the same species, the multi-stemmed benefits might be facultative, differing among forests and neighborhood compositions.

Editor: Ben Bond-Lamberty, DOE Pacific Northwest National Laboratory, United States of America

Funding: This work is supported by the National Natural Science Foundation of China (No. 31300378 and No. 31300517). The funders had no role in study design, data collection and analysis, decision to publish, or preparation of the manuscript.

Competing Interests: The authors have declared that no competing interests exist.

* E-mail: hzq@iae.ac.cn

Introduction

While forest studies often focus on tree dynamics, shrubs are also ubiquitous components of the forest community, and their dynamics differ from dominant trees in important ways. Shrubs influence ecosystem properties by maintaining plant diversity, influencing tree seedling regeneration [1,2], and affecting soil nutrient flow [3–5]. However, the demographic traits of understory shrub species are fundamentally different from overstory tall tree species. The dynamics of understory shrubs are mostly driven by vegetative growth and mechanical damage to stems [6]. Also, unlike overstory tall trees, shrubs can spend all their life in the forest understory.

Many shrub species produce multiple stems through vegetative reproduction. The replacement of stems within a shrub has been thought of as an adaptation enabling long persistence of individual shrubs in their environments, as it may allow for increased acquisition of resources in heterogeneous environments [7]. In multi-stemmed shrubs, each stem may experience recruitment, growth, and death semi-independently. Thus, such shrubs usually consist of stems of different ages and sizes [8]. However, interactions among stems in a multi-stemmed plant may be complex. For example, competition among stems within a shrub might restrict height growth relative to single-stemmed trees [9], while more stems can increase overall plant survival compared with individuals of the same species with fewer stems [10].

To understand the dynamics of multi-stemmed understory shrubs, it is important to account for the biotic environment in the understory. Understory shrubs are usually limited by tree competition. Denser neighborhoods may reduce shrub performance due to resource competition. One study that investigated shrub dynamics as a function of the understory neighbor structure suggested that the neighboring understory trees negatively affected the absolute diameter growth and survival of large ramets [11]. Also, large individuals can suppress the performance of smaller individuals [12], as canopy trees usually dominate forest light and root environments [13]. Some researchers have investigated the influence of overstory characteristics on diversity and abundance of shrub species [14,15]. Soil conditions may also affect the performance of understory shrub species. For example, the reproductive modes of an understory shrub was related to soil environment, and the multi-stemmed growth form was preferred in places with abundant litter substrate [16]. Similarly, the dominance and frequency patterns of multi-stemmed plants were driven by soil nutrient status [17,18]. In addition, size strongly affected the absolute diameter growth and survival of large ramets for shrub species [11]. Therefore, soil variables and stem size should be also considered to examine the dynamics of multi-stemmed understory shrubs.

During forest development, floristic composition, diversity, richness, and density of plant communities may change [19]. Hence, the existence or strength of biotic processes, such as neighborhood interactions, can vary at different successional stages, particularly if large changes occur in the vertical structure of plant communities through time [15]. Moreover, soil nutrients are more available in old growth forests owing to less efficiency in nutrient retention than secondary stage in some cases [20]. To our knowledge, no ecological research has been conducted in dynamics of understory shrubs in old growth compared with secondary forests.

The goal of this study is to evaluate the role of stem size, multiple stems, neighbor competition and soil conditions in stem demographic variables of two shrub species in two temperate forests at different successional stages. We investigated the effects of diameter at breast height (dbh), the number of stems in a shrub, overstory and understory neighbors and soil variables. We specifically ask the following questions: (1) How does the number of stems in a shrub affect stem growth and survival of these two shrub species? (2) How do basal area of overstory and understory neighbors and soil variables affect stem growth and survival of these two species? And (3) do the effects of multiple stems, neighbors and soil variables differ between the two study forests? We predict that competition from local neighbors is going to be stronger, and have a more negative impact on growth and survival of these two species in the old growth forest, where there are more larger and older trees [21].

Materials and Methods

Ethics statement

The study forests represent neither privately-owned field, nor endangered or protected species. No specific permits were required for the described field studies.

Study site

We studied two forests in northeastern China, both located in Changbai Mountain Nature Reserve (42°23′N, 128°05′E), which is one of the largest biosphere reserves in China and has been spared from logging and other severe human disturbances. The reserve was established in 1960 and joined the World Biosphere Reserve Network under the Man and the Biosphere Project in 1980 [22,23]. The reserve is about 200,000 ha with elevation ranging from 740 m to 2,691 m [24].

One of our study sites is in broad-leaved Korean pine (*Pinus koraiensis*) mixed temperate forest, which is the dominant vegetation type in northeastern China and well known for high species richness and particular species composition [23,25]. The other is a secondary poplar-birch forest, which is one of the main secondary forest types on Changbai Mountain, resulting from natural regeneration after clear-cutting or fire. The poplar-birch forest is an important stage in the secondary succession of the broad-leaved Korean pine forest [26]. For brevity, we will hereafter refer to these two forests as the 'old growth forest' and the 'secondary forest' respectively. The climate of the study region is characterized by low temperature and high precipitation [24]. Mean annual temperature is 3.3°C (−16.5°C in January and 20.5°C in August). Mean annual precipitation is 672 mm, most of which occurs between June and September (480–500 mm) [24].

Plot censuses

Both these sites have large forest dynamic plots where all individuals with diameter at breast height ≥1 cm have been measured, mapped, tagged, and identified to species. Trees and shrubs are re-measured every five years. In the old forest, the forest dynamics plot is 25 hectare (ha) (500×500 m). It was established in 2004 and recensused in 2009. The elevation in the plot ranges from 791.8 m to 809.5 m. In the 2004 census, there were 59,138 living stems, belonging to 52 species, 34 genera, and 18 families. Mean stand density of living stems was 2,366 stems·ha^{-1} and mean basal area of living trees was 43.75 $m^2 \cdot ha^{-1}$.

In the secondary forest, the forest dynamic plot is 5 ha (250×200 m). It was established in 2005 and recensused in 2009. The elevation ranges from 796.3 m to 800.4 m. In the 2005 census, there were 20,107 living stems, belonging to 50 species, 32 genera, and 18 families. Mean stand density of living stems was 4,021 stems·ha^{-1} and mean basal area of living trees was 28.79 $m^2 \cdot ha^{-1}$. Hence, relative to the old growth forest, in this plot there were much more stems, but these were considerably smaller, suggesting lower canopy height and a denser understory.

Study species

Here we focus on the stem growth and survival dynamics of two shrub species that are most abundant in the understory of both forest types. The first is *Corylus mandshurica* (English common name Manchurian Filbert), an androgynous shrub with edible fruits that are dispersed by gravity. The second species is *Acer barbinerve* (common name Barbed Vein Maple), a dioecious species with wind dispersed seeds. The distributions of these two shrub species in the two forest plot are shown in Figure 1. A large proportion of individuals of these two species have multiple stems and form clumps [27]. Here, we evaluated the dynamics of stems in these clumps.

Statistical analysis

To examine the explanatory variables affecting the dynamics of shrub stems, we used generalized linear mixed models (GLMMs), with gamma errors and log-link function for the growth rate, and with binomial errors and logit-link function for the survival probability. The variables in the models are described below.

Demographic rates. Growth of each shrub stem was calculated as basal area increment ($cm^2\ yr^{-1}$):

$$g = \frac{BA_{t2} - BA_{t1}}{t}$$

where BA_{ti} is the basal area of shrub stems in census i, and t is the inter-census interval. We excluded extremes from the calculation of all shrub stems decreasing in size by >5% yr^{-1} [28], and altered remaining negative growth rates to be zero. To log transform growth, we added 0.01.

Biotic variables. For each shrub, we calculated the following variables: (1) the number of stems in the shrub, (2) basal area of overstory neighbors (dbh≥10 cm) between 5–20 m radii, (3) basal area of understory neighbors (dbh<10 cm) between 5–20 m radii. To eliminate boundary effects, only those shrubs with a distance greater than or equal to 20 m from the forest plot edges were included.

Soil variables. We sampled soils within the two forest plots using a regular grid of points every 30 m. At each grid point, two additional sample points at 2, 5 or 15 m were selected in a random compass direction from the grid point. In total, 967 and 215 soil samples were sampled in the old growth and secondary forest plots respectively. Soil properties (organic matter, total nitrogen, total phosphorus and total potassium) showed significant differences among the two forests (organic matter: t = 8.703, P<2.2e-16; total nitrogen: t = −4.551, P= 7.331e-06; total phosphorus: t = 4.725, P= 3.28e-06; total potassium: t = 16.254, P<2.2e-16). There were

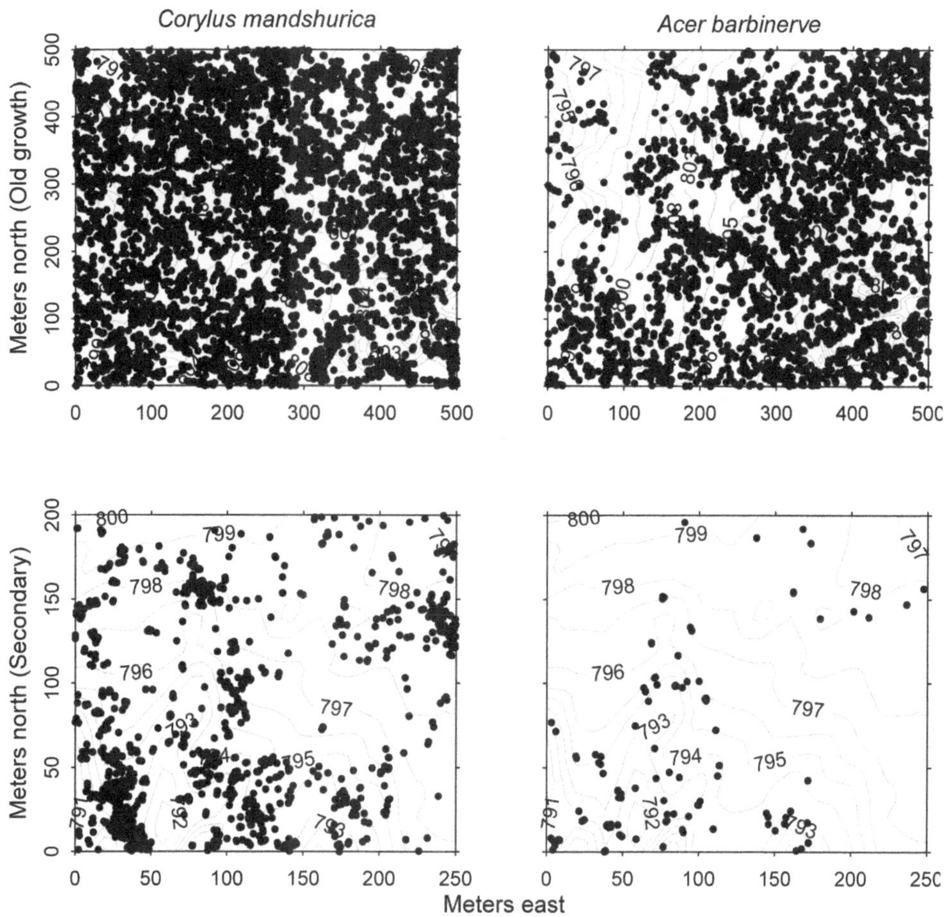

Figure 1. The distribution of *C. mandshurica* and *A. barbinerve* in the old growth and secondary forest plots. The solid points represent the individuals of the two species. The contour lines refer to the topography of the two forest plots.

higher content of organic matter, total phosphorus and total potassium, but lower content of total nitrogen in the old growth forest (Mean±S.E.: 165.284±2.646 g·kg^{-1} vs. 129.324±3.173 g·kg^{-1}; 1.255±0.015 g·kg^{-1} vs. 1.111±0.027 g·kg^{-1}; 16.472±0.086 g·kg^{-1} vs. 13.661±0.150 g·kg^{-1}; 6.370±0.087 g·kg^{-1} vs. 7.190±0.158 g·kg^{-1}). Soil organic matter did not show spatial autocorrelation in the secondary forest. We interpolated total nitrogen, total phosphorus and total potassium of the two forest plots to 5×5 m grid by kriging. Then, we converted the calculated values for each grid to z scores [29]. To reduce possibility of overfitting the models [29], we performed a principal components analysis (PCA) on these three soil variables. The first two principle components accounted for 98.8% of the total variance in the three soil variables in the old growth forest and 91.8% in the secondary forest (Table S1 in File S1). In both forests, the first component (PC1) explained variation in total nitrogen and phosphorous, while the second (PC2) represented variation in total potassium.

In all models we included dbh (log-transformed), biotic variables and soil variables (Table 1) as explanatory variables. The values of the explanatory variables were standardized by subtracting the mean value of the variable and dividing by the standard deviation. Because shrub stems from the same individual were not independent replicates of stem number, we included individual as a random effect.

Here we present only results for understory neighbors at 5 m radius and overstory neighbors at 20 m radius because GLMMs under these scales were the most likely models based on AIC and BIC for most cases when testing separately (Table S2 and S3 in File S1). We evaluated spatial autocorrelation in the residuals of the GLMMs (Figure S1 in File S1), and variograms revealed that the models adequately accounted for spatial autocorrelation.

All analyses were carried out in the statistical environment R (version 2.14.2), using the packages 'geoR' [30], 'languageR' [31], 'lme4' [32] and 'sqldf' [33].

Results

Effects of dbh and stem number in a shrub on stem growth and survival

Stem dbh had a strong negative influence on the survival of *C. mandshurica* in the two forests (Figure 2 and 3), such that larger stems had significantly higher mortality in the two forests. For example, stems of *C. mandshurica* of 1 cm dbh had probability of surviving of 0.736±0.002 in the old growth forest and 0.800±0.009 in the secondary forest. This probability declined to 0.696±0.014 and 0.522±0.064 for 3 cm dbh respectively. The effect of stem dbh was small but positive on the growth of *C. mandshurica* in the old growth forest. Stems of 1 cm dbh grew on average 0.200±0.005 cm^2·yr^{-1} in the old growth forest, and this increased to 0.256±0.051 cm^2·yr^{-1} for stems of 3 cm dbh.

Table 1. Parameters included in models.

Variables	Data		
	Range	Mean	Median
Old growth forest			
***Corylus mandshurica* (13,498)**			
DBH (cm)	1–10.9	1.580	1.5
Number of stems in a shrub	1–18	3.318	3
Basal area (m^2) of overstory neighbors within 20 m	2.217–10.447	5.340	5.284
Basal area (m^2) of understory neighbors within 5 m	0–0.044	0.010	0.009
Soil PC1	−2.024–3.279	−0.008	−0.497
Soil PC2	−1.015–1.377	0.140	0.119
***Acer barbinerve* (10,216)**			
DBH (cm)	1–13.2	1.931	1.7
Number of stems in a shrub	1–21	5.394	5
Basal area (m^2) of overstory neighbors within 20 m	3.066–9.592	5.523	5.470
Basal area (m^2) of understory neighbors within 5 m	0–0.042	0.011	0.010
Soil PC1	−2.020–3.299	−0.633	−1.066
Soil PC2	−0.946–1.391	0.034	0.013
Secondary forest			
***Corylus mandshurica* (1,269)**			
DBH (cm)	1–4.9	1.551	1.5
Number of stems in a shrub	1–13	2.940	2
Basal area (m^2) of overstory neighbors within 20 m	1.705–4.249	3.098	3.180
Basal area (m^2) of understory neighbors within 5 m	0.000–0.073	0.033	0.032
Soil PC1	−2.576–3.525	0.579	0.538
Soil PC2	−3.042–2.836	0.133	0.060
***Acer barbinerve* (227)**			
DBH (cm)	1–3.7	1.640	1.5
Number of stems in a shrub	1–14	6.366	6
Basal area (m^2) of overstory neighbors within 20 m	2.298–4.140	3.247	3.247
Basal area (m^2) of understory neighbors within 5 m	0.012–0.062	0.031	0.029
Soil PC1	−1.991–3.340	0.781	1.068
Soil PC2	−1.595–2.430	0.400	0.208

Similarly, larger stems of *A. barbinerve* had faster growth in the two forests (for example, growth increased from 0.204±0.011 and 0.240±0.036 $cm^2 \cdot yr^{-1}$ at 1 cm dbh to 0.421±0.035 and 1.099 (only one stem) $cm^2 \cdot yr^{-1}$ at 3 cm dbh) (Figure 2 and 3). While in the secondary forest larger stems had a higher survival (increased from 0.821±0.050 to 0.939±0.000, at 1 and 3 cm dbh, respectively) (Figure 3).

The number of stems in an individual positively influenced growth and survival of *C. mandshurica* in the old growth forest (Growth: Effect±S.E. = 0.070±0.011, P = 5.45e-10; Survival: Effect±S.E. = 0.069±0.026, P = 0.008), but did not influence in the secondary forest (Figure 2 and 3). For example, average growth of stems of *C. mandshurica* in the old growth forest increased from 0.180±0.003 $cm^2 \cdot yr^{-1}$ in shrubs with only one stem, to 0.209±0.003 $cm^2 \cdot yr^{-1}$ in shrubs with two or more stems. Similarly, probability of survival of *C. mandshurica* in the old growth forest increased a very small but significant amount, going from 0.711±0.001 in shrubs with only one stem to 0.724±0.001 in shrubs with two or more stems. Moreover, in the old growth forest, there was a significant interaction between stem number and basal area of overstory neighbors in the survival of *C. mandshurica* (Effect±S.E. = 0.066±0.027, P = 0.014; not shown). This interaction suggests that the positive effect of multiple stems on survival was stronger when there were larger trees in the vicinity. The number of stems in a shrub also positively influenced the growth of *A. barbinerve* in the old growth forest (Effect±S.E. = 0.051±0.014, P = 0.0003), but not in the secondary forest, nor did it influence the survival of *A. barbinerve* in either of the two forest plots.

Effects of neighborhood variables and soil variables in old growth and secondary forests

Consistent with our expectations, the effects of neighborhood variables and soil variables in the old growth forest were more apparent. For instance, in the old growth forest, basal area of overstory trees showed significant negative effects on growth and

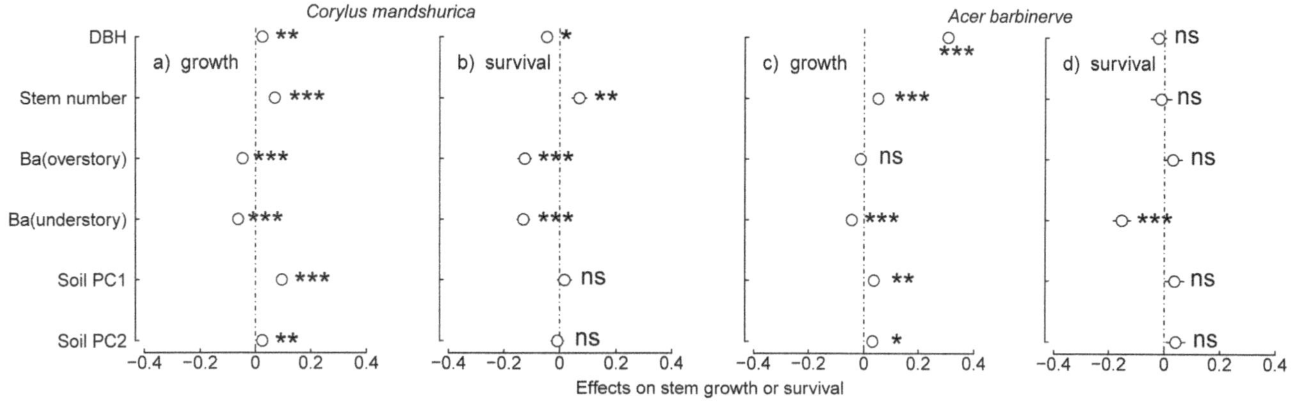

Figure 2. Effects of dbh, stem number in a shrub, neighborhood and soil variables on stem growth or survival of *C. mandshurica* and *A. barbinerve* in the old growth forest plot. *Circles* show the coefficient estimate for each parameter, with 2S.E. indicated by *horizontal lines*. "Ba" represents basal area. Asterisks represent the probability that the estimates are not different from zero: $^*P<0.05$, $^{**}P<0.01$, $^{***}P<0.001$, ns not significant ($P>0.05$).

survival of *C. mandshurica*, such that both decreased when there were more, larger neighbors (Figure 2). This effect of overstory neighbors was not seen for *A. barbinerve* (Figure 2). However, basal area of understory neighbors had significant negative effects for both species (Figure 2). In contrast, in the secondary forest neighborhood variables had little effect on growth or survival of both species, with the exception of a negative effect of basal area of overstory trees for *C. mandshurica* (Figure 3). Growth of the two species in the old growth forest was influenced by soil variables (PC1 and PC2, Figure 2), suggesting that better soil nutrients lead to better growth. In contrast, in the secondary forest only soil PC2 had a positive effect on stem growth of *C. mandshurica* (Figure 3).

Discussion

Our study showed differences in shrub dynamics between the two species and among the two forest sites.

We found that stems of *C. mandshurica* generally had decreased survival as they grew bigger. As older, larger stems probably have better resistance to physical damage and better access to light, decreasing survival can only be explained by stem senescence. We interpret this to mean that established shrubs of this species might shift their investment from surviving long lived stems to producing new stems. This same behavior was not found for *A. barvinerve*. On the contrary, larger stems of *A. barvinerve* in the secondary forest showed higher, not lower survival. While we cannot explain this with certainty, one possibility is that this population in the secondary forest is much younger. Indeed this species was much less abundant in this plot (45.4 stems·ha^{-1} vs. 408.64 stems·ha^{-1} in the old forest; Table 1), and stems were much smaller than in the old forest (max dbh of 3.7 cm vs. 13.2 cm in the old forest; Table 1). Hence, it is possible that most of the stems we captured were still growing and not senescing. *C. mandshurica* also had lower population density and smaller stems in the secondary forest, but this was not as marked as for *A. barvinerve* (Table 1).

Larger stems had higher basal area growth for *A. barvinerve* in the two forests, and also for *C. mandshurica* in the old growth forest. Similar results were found in recent studies of tree growth [34,35], suggesting that larger stems had more biomass production than smaller ones. This result does not contradict the survival trends. As noted by previous research [12], size affects resource acquisition, as large individuals can obtain a disproportionately larger amount of the resources. Because stems require less diameter growth to produce the same biomass over an increasingly large area, basal area growth may be constant or increased, while diameter growth declines with stem size.

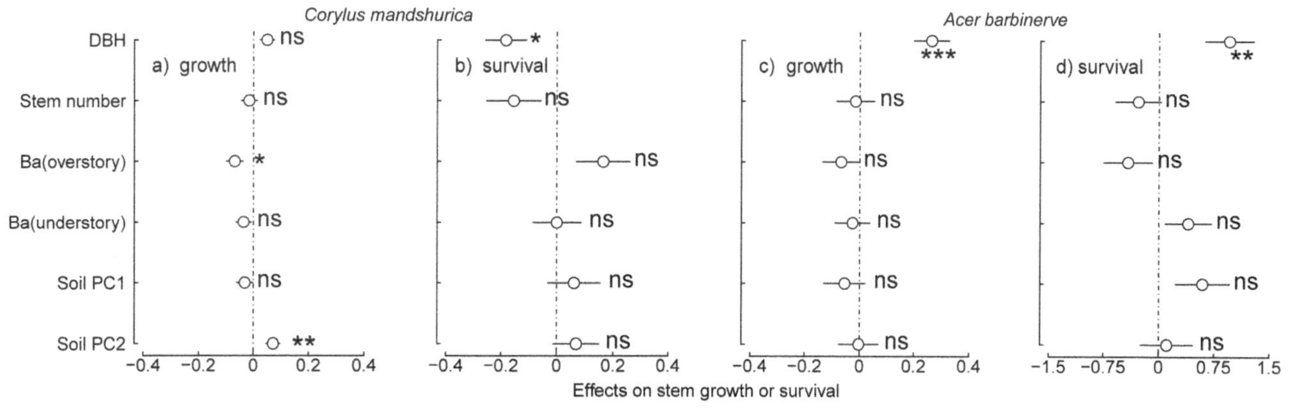

Figure 3. Effects of dbh, stem number in a shrub, neighborhood and soil variables on stem growth or survival of *C. mandshurica* and *A. barbinerve* in the secondary forest plot. *Circles* show coefficient estimate for each parameter, with 2S.E. indicated by *horizontal lines*. See Figure legend 2 for Ba, asterisks and ns.

Multi-stemmed growth form was advantageous in some cases. The presence of multiple stems increased the growth of both species and survival of *C. mandshurica* in the old growth forest. Although the stems within multi-stemmed shrubs have more competition, shrubs with more stems have better resource acquisition, which may benefit stem performance. In multi-stemmed plants, the storage of carbohydrates is generally shared among stems [36], and this may be an effective strategy to buffer plants against stresses and improve survival [37] or growth. However, the advantage of multiple stems was not found in the secondary forest. While this study was unreplicated for stand age, and we can draw no firm conclusions. We hypothesize that there might be stronger competition from local neighbors in the old growth relative to the secondary forest. This conclusion is supported by the interaction between stem number and basal area of overstory neighbors in the survival of *C. mandshurica*. This interaction suggests that the effect of multiple stems is more important in areas of the forest with larger trees, that are probably also darker and more resource-limited. Also, random small-scale gap formation is more common in the old growth forest, and more patchily distributed light levels may be more favorable to multi-stemmed individuals [38].

There were considerable differences between the two species in the effects of overstory neighborhood basal area in the old growth forest. In fact, denser overstory neighborhoods significantly reduced stem growth and survival of *C. mandshurica* in the old growth forest but had little effect on *A. barbinerve*. Also, *A. barbinerve* performed better than *C. mandshurica* in the old growth forest (average of growth rate: 0.280 ± 0.004 $cm^2 \cdot yr^{-1}$ vs. 0.198 ± 0.002 $cm^2 \cdot yr^{-1}$; average probability of survival: 0.803 ± 0.001 vs. 0.721 ± 0.001). Since higher basal area of neighbors indicates the presence of larger trees that may produce more shading, these results suggest that *A. barbinerve* may be better adapted to the understory environment in the old growth forest, while *C. mandshurica* is probably more sensitive to suppression with a smaller tolerance for shading. Consistent with previous studies [39–41], *C. mandshurica*, the less shade-tolerant species, also had higher mortality rates in these two forests.

While our data only provides limited information regarding the differences between forests, it was clear that shrub dynamics were different in the two forests. For example, the performance of the two species was clearly influenced by basal area of overstory or understory neighbors in the old growth forest, but not the secondary forest. The old growth forest in our study is characterized by overstory trees with larger diameter and greater basal area relative to the secondary forest (Table 1), which is consistent with previous research [42]. Because of crowding and shading from neighboring large trees, there is less available space in the understory of the old growth than the secondary forest. This may explain why the density of the local neighborhood was more important, and why both our study species had lower percent of surviving stems in the old growth forest relative to the secondary forest. The content of organic matter, total phosphorus and total potassium were significantly higher in the old growth forest, suggesting that soil nutrients are more available in old growth than developing phase. This observation is consistent with previous research in temperate forests [20]. This difference in soil characteristics among forests may explain why both our study species grew more in the old forest and their growth was positively correlated with soil variables in the old growth but not in the secondary forest.

Overall, we found that multi-stemmed growth form was an advantageous trait in the understory for *C. mandshurica* and partly for *A. barbinerve* in the old growth forest. Also, overstory neighbors influenced stem growth and survival for *C. mandshurica* but not for *A. barbinerve* in the old growth forest, which may suggest differences in shade tolerance among these two species. Differences in shade-tolerance among species commonly found in the understory have been reported before [40], but our study improves on previous studies in that we compared their dynamics in two forests with different structure. In our study, the old growth forest is characterized by overstory trees with larger diameter and greater basal area, and also more available soil nutrients relative to the secondary forest. According to the succession theory, the diversity of the secondary forest is not as stable as the old growth forest [43]. There may be more stochastic processes affecting dynamics of the two multi-stemmed understory shrub species in the secondary forest. Hence, we found that the effects of neighborhood and soil variables on growth and survival were not as strong in the secondary forest as in the old forest.

Generally, shrub species have different growth strategies than tall tree species. Many shrub species have more stems through vegetative reproduction, while overstory tree species are able to grow up into the canopy layer and sprout less frequently. Some authors have suggested that multi-stemmed growth form can be beneficial to understory species [10], while others have suggested it can be detrimental to stem dynamics [11]. Here we show that the effects of multiple stems differ for different species. Furthermore, even within the same species the benefits might be facultative, differing among forests and neighborhood compositions.

Supporting Information

File S1 Supporting tables and figure. Table S1. Soil variable loadings for the two PCAs. Table S2. AIC, △AIC, BIC and △BIC values of stem growth and survival models of the two species for understory neighbors across different scales. The most likely models are shown in bold (red). Table S3. AIC, △AIC, BIC and △BIC values of stem growth and survival models of the two species for overstory neighbors across different scales. The most likely models are shown in bold (red). Figure S1. Variograms illustrating the spatial autocorrelation in residuals of the GLMMs for stem growth and survival of the two species in the old growth and secondary forest plots.

Acknowledgments

The Smithsonian Tropical Research Institute supported Tania Brenes-Arguedas through most of this collaboration, and provided the facilities for Xuejiao Bai to work on the manuscript with Tania Brenes-Arguedas for one month. We thank Drs. Liza S. Comita and Simon A. Queenborough for helpful suggestions on data analysis. We thank Drs. Ben Bond-Lamberty, Erin Kurten and three anonymous reviewers for critical comments for the manuscript. We also thank Buhang Li and Zhaochen Zhang for their field work and data collection. The publication of this manuscript has been approved by all co-authors.

Author Contributions

Conceived and designed the experiments: XB ZH. Performed the experiments: XB JY XW FL ZY SS DX. Analyzed the data: XB TBA DX. Contributed reagents/materials/analysis tools: ZH JY XW FL ZY DX. Wrote the paper: XB TBA.

References

1. Beckage B, Clark JS, Clinton BD, Haines BL (2000) A long-term study of tree seedling recruitment in southern Appalachian forests: the effects of canopy gaps and shrub understories. Canadian Journal of Forest Research 30: 1617–1631.
2. O'Brien MJ, O'Hara KL, Erbilgin N, Wood DL (2007) Overstory and shrub effects on natural regeneration processes in native Pinus radiata stands. Forest Ecology and Management 240: 178–185.
3. Chen YL, Cui XY, Zhu N, Guan JY (1998) Status and role of the bush layer and main shrub species in nutrient cyclings of Tilia amurensis Pinus koraiensis forest. Journal of Northeast Forestry University 26: 7–13.
4. Guan JY, Chen YL, Zhu N, Zhuo LH, Liu Y (1999) Status and role of the shrub layer and its main shrub species in nutrient cycling of Mongolian oak forest. Bulletin of Botanical Research 19: 100–110.
5. Nilsson M-C, Wardle DA (2005) Understory vegetation as a forest ecosystem driver: evidence from the northern Swedish boreal forest. Frontiers in Ecology and the Environment 3: 421–428.
6. Kanno H, Hara M, Hirabuki Y, Takehara A, Seiwa K (2001) Population dynamics of four understorey shrub species during a 7-yr period in a primary beech forest. Journal of Vegetation Science 12: 391–400.
7. Kroons Hd, Hutchings MJ (1995) Morphological plasticity in clonal plants: the foraging concept reconsidered. Journal of Ecology 83: 143–152.
8. Fujiki D, Kikuzawa K (2006) Stem turnover strategy of multiple-stemmed woody plants. Ecological Research 21: 380–386.
9. Wilson BF (1995) Shrub stems: form and function. In: Garter BL, editor. Plant stems: physiology and functional morphology. New York: Academic Press. pp. 91–102.
10. Tanentzap AJ, Mountford EP, Cooke AS, Coomes DA (2012) The more stems the merrier: advantages of multi-stemmed architecture for the demography of understorey trees in a temperate broadleaf woodland. Journal of Ecology 100: 171–183.
11. Matsushita M, Tomaru N, Hoshino D, Nishimura N, Yamamoto S-I (2010) Factors affecting the production, growth, and survival of sprouting stems in the multi-stemmed understory shrub Lindera triloba. Botany 88: 174–184.
12. Weiner J (1990) Asymmetric competition in plant populations. Trends in Ecology & Evolution 5: 360–364.
13. Wright SJ (2002) Plant diversity in tropical forests: a review of mechanisms of species coexistence. Oecologia 130: 1–14.
14. Gracia M, Montané F, Piqué J, Retana J (2007) Overstory structure and topographic gradients determining diversity and abundance of understory shrub species in temperate forests in central Pyrenees (NE Spain). Forest Ecology and Management 242: 391–397.
15. McKenzie D, Halpern CB, Nelson CR (2000) Overstory influences on herb and shrub communities in mature forests of western Washington, U.S.A. Canadian Journal of Forest Research 30: 1655–1666.
16. Kanno H, Seiwa K (2004) Sexual vs. vegetative reproduction in relation to forest dynamics in the understorey shrub, Hydrangea paniculata (Saxifragaceae). Plant Ecology 170: 43–53.
17. Bellingham PJ, Sparrow AD (2009) Multi-stemmed trees in montane rain forests: their frequency and demography in relation to elevation, soil nutrients and disturbance. Journal of Ecology 97: 472–483.
18. Clarke PJ, Knox KJE, Wills KE, Campbell M (2005) Landscape patterns of woody plant response to crown fire: disturbance and productivity influence sprouting ability. Journal of Ecology 93: 544–555.
19. Finegan B (1984) Forest succession. Nature 312: 109–114.
20. Wirth C (2009) Old-growth forest: function, fate and value: a synthesis. In: Wirth C, Gleixner G, Heimann M, editors. Old-growth forests: Function, fate and value. New York, Berlin, Heidelberg: Springer.
21. Armesto JJ, Carmona M, Celis-Diez JL, Gaxiola A, Gutiérrez AC, et al. (2007) Old-growth temperate rain forests of South America: Conservation, plant-animal interactions, and baseline biogeochemical processes. In: Wirth C, Heimann M, Gleixner G, editors. Old Growth Forests: Function, Fate and Value. Berlin, Germany: Springer Verlag. pp. 367–390.
22. Shao GF, Schall P, Weishampel JF (1994) Dynamic simulations of mixed broadleaved-Pinus koraiensis forests in the Changbaishan Biosphere Reserve of China. Forest Ecology and Management 70: 169–181.
23. Stone R (2006) Ecology - A threatened nature reserve breaks down Asian borders. Science 313: 1379–1380.
24. Yang H, Li D, Wang B, Han J (1985) Distribution patterns of dominant tree species on northern slope of Changbai Mountain. Research of Forest Ecosystem 5: 1–14.
25. Yang X, Xu M (2003) Biodiversity conservation in Changbai Mountain Biosphere Reserve, northeastern China: status, problem, and strategy. Biodiversity and Conservation 12: 883–903.
26. Xu HC (2001) Natural forests of pinus koraiensis in China. Beijing: China Forestry Publishing House.
27. Bai XJ, Li BH, Zhang J, Wang LW, Yuan ZQ, et al. (2010) Species composition, structure and spatial distribution of shrub species in broad-leaved Korean pine (Pinus koraiensis) mixed forest. Chinese Journal of Applied Ecology 21: 1899–1906.
28. Condit R, Ashton PS, Manokaran N, LaFrankie JV, Hubbell SP, et al. (1999) Dynamics of the forest communities at Pasoh and Barro Colorado: comparing two 50-ha plots. Philosophical Transactions of the Royal Society of London Series B: Biological Sciences 354: 1739–1748.
29. Young SS, Carpenter C, Zhi-Jun W (1992) A Study of the Structure and Composition of an Old Growth and Secondary Broad-Leaved Forest in the Ailao Mountains of Yunnan, China. Mountain Research and Development 12: 269–284.
30. Ribeiro PJ, Diggle PJ (2011) geoR: Analysis of geostatistical data. 1.6-34 ed.
31. Baayen RH (2011) languageR: Data sets and functions with "Analyzing Linguistic Data: A practical introduction to statistics". 1.2 ed.
32. Bates D, Maechler M, Bolker B (2010) lme4: Linear mixed-effects models using S4 classes. R Package Version: 0.999375-37 ed.
33. Grothendieck G (2010) sqldf: Perform SQL Selects on R Data Frames. 0.3-5 ed.
34. Nock CA, Baker PJ, Wanek W, Leis A, Grabner M, et al. (2011) Long-term increases in intrinsic water-use efficiency do not lead to increased stem growth in a tropical monsoon forest in western Thailand. Global Change Biology 17: 1049–1063.
35. Anning AK, McCarthy BC (2013) Competition, size and age affect tree growth response to fuel reduction treatments in mixed-oak forests of Ohio. Forest Ecology and Management 307: 74–83.
36. Del Tredici P (2001) Sprouting in temperate trees: A morphological and ecological review. The Botanical Review 67: 121–140.
37. Kobe RK (1997) Carbohydrate allocation to storage as a basis of interspecific variation in sapling survivorship and growth. Oikos 80: 226–233.
38. Koop H (1987) Vegetative reproduction of trees in some European natural forests. Plant Ecology 72: 103–110.
39. Harcombe PA, Marks PL (1983) Five years of tree death in a Fagus-Magnolia forest, southeast Texas (USA). Oecologia 57: 49–54.
40. Yoshida T, Kamitani T (1997) The stand dynamics of a mixed coppice forest of shade-tolerant and intermediate species. Forest Ecology and Management 95: 35–43.
41. Whitney GG (1984) Fifty years of change in the arboreal vegetation of heart's content, an old-growth hemlock-white pine-northern hardwood stand. Ecology 65: 403–408.
42. Ziegler SS (2000) A comparison of structural characteristics between old growth and postfire second-growth hemlock-hardwood forests in Adirondack Park, New York, U. S. A. Global Ecology & Biogeography 9: 373–389.
43. Hao ZQ, Tao DL, Zhao SD (1994) Diversity of higher plants in broad-leaved Korean pine and secondary birch forests on northern slope of Changbai Mountain. Chinese Journal of Applied Ecology 5: 16–23.

2

Land Use and Land Cover Change Dynamics across the Brazilian Amazon: Insights from Extensive Time-Series Analysis of Remote Sensing Data

João M. B. Carreiras[1,2*], Joshua Jones[3], Richard M. Lucas[3,4], Cristina Gabriel[1]

1 Tropical Research Institute (IICT), Lisboa, Portugal, **2** National Centre for Earth Observation (NCEO), Centre for Terrestrial Carbon Dynamics (CTCD), University of Sheffield, Sheffield, South Yorkshire, United Kingdom, **3** Department of Geography and Earth Sciences, Aberystwyth University, Aberystwyth, Ceredigion, United Kingdom, **4** Centre for Ecosystem Science, School of Biological, Earth and Environmental Sciences, University of New South Wales, Sydney, New South Wales, Australia

Abstract

Throughout the Amazon region, the age of forests regenerating on previously deforested land is determined, in part, by the periods of active land use prior to abandonment and the frequency of reclearance of regrowth, both of which can be quantified by comparing time-series of Landsat sensor data. Using these time-series of near annual data from 1973–2011 for an area north of Manaus (in Amazonas state), from 1984–2010 for south of Santarém (Pará state) and 1984–2011 near Machadinho d'Oeste (Rondônia state), the changes in the area of primary forest, non-forest and secondary forest were documented from which the age of regenerating forests, periods of active land use and the frequency of forest reclearance were derived. At Manaus, and at the end of the time-series, over 50% of regenerating forests were older than 16 years, whilst at Santarém and Machadinho d'Oeste, 57% and 41% of forests respectively were aged 6–15 years, with the remainder being mostly younger forests. These differences were attributed to the time since deforestation commenced but also the greater frequencies of reclearance of forests at the latter two sites with short periods of use in the intervening periods. The majority of clearance for agriculture was also found outside of protected areas. The study suggested that a) the history of clearance and land use should be taken into account when protecting deforested land for the purpose of restoring both tree species diversity and biomass through natural regeneration and b) a greater proportion of the forested landscape should be placed under protection, including areas of regrowth.

Editor: Bruno Hérault, Cirad, France

Funding: This study was supported by the "REGROWTH-BR - Remote sensing of regenerating tropical forests in Brazil: mapping and retrieving biophysical parameters" project (Ref. PTDC/AGR-CFL/114908/2009), funded by the Foundation for Science and Technology (FCT, http://www.fct.pt/index.phtml.en), Portugal. J.M.B. Carreiras' work was partially funded by NERC National Centre for Earth Observation (Ref: R8/H12/82). The funders had no role in study design, data collection and analysis, decision to publish, or preparation of the manuscript.

Competing Interests: The authors have declared that no competing interests exist.

* Email: j.carreiras@sheffield.ac.uk

Introduction

Humans are increasingly changing the state and dynamics of the Earth system, affecting processes within and between the biosphere, hydrosphere and atmosphere. Climate change is a recognized consequence of such changes and would be occurring at a faster rate if several ecosystem services at the global scale were absent [1]. For example, less than half of the total amount of carbon dioxide (CO_2) released to the atmosphere each year remains there because of removal by terrestrial and ocean carbon (C) sinks [2]. Losses of biodiversity from terrestrial ecosystems have also been substantial.

The role of terrestrial ecosystems in the global C cycle had been widely recognized in the literature (e.g., [3]), with particular recognition given to the pan-tropical belt [4,5]. In the past two decades, two main international conventions have sought to establish mechanisms aimed at stabilizing greenhouse gases (GHG) concentrations in the atmosphere (the United Nations Framework Convention on Climate Change; UNFCCC) and protecting biological diversity (Convention on Biological Diversity; CBD). Specifically, the alarmingly high rate of tropical land use and land cover change (LULCC) [6] and resulting biodiversity loss with further severe consequences for ecosystem function and structure [7] has driven the UNFCCC to establish several investment mechanisms and market based C transactions. These are related to the enhancement of forest C stocks and the decrease of deforestation and forest degradation, while promoting sustainable development in developing countries (UNFCCC Non-Annex I countries). Such mechanisms include the UNFCCC Clean Development Mechanism (CDM) initiative and the post-Kyoto Protocol Reduced Emissions from Deforestation and Degradation (REDD+) program (e.g., [8]). These efforts are proving successful in many regions leading to greater conservation of the intact forests of the Neotropical, Afrotropical, Australasia and Indo-Malay biogeographical regions [9] and the carbon and biodiversity they contain. However, many of the deforested and degraded lands of the tropics are also capable of supporting forests and hence, in addition, there is potential to restore some of the ecosystem values (e.g., carbon amounts and biodiversity) lost through previous disturbances [10,11]. This capacity depends in part, however, on the history of forest clearance and land use that has occurred and hence a clearer understanding of the potential of

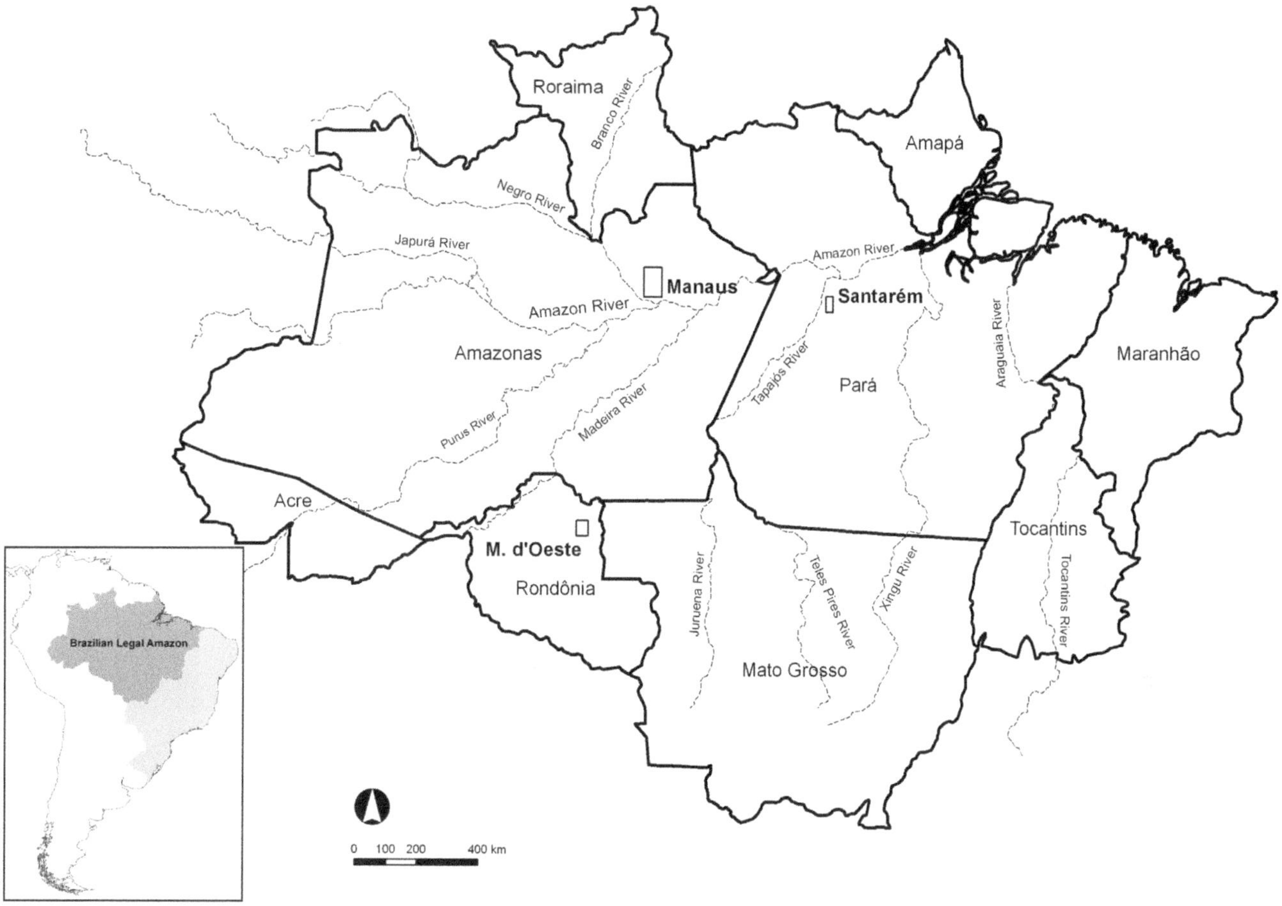

Figure 1. Location of the three study sites. Location of the three study sites: Manaus, Santarém and Machadinho d'Oeste (M. d'Oeste), and the insertion (bottom left) of the Brazilian Legal Amazon (dark grey) in Brazil (light grey) and South American continent. Also shown are the major rivers included in the Amazon basin.

deforested areas to recover ecosystems is needed. This study addresses these issues by focusing on sites in the Brazilian Legal Amazon (BLA) that have experienced deforestation and variable histories of subsequent land use.

The BLA covers an area of approximately 5,000,000 km^2 and consists primarily of primary tropical forest (rain and seasonal forest) [12]. Brazil's National Institute for Space Research (Instituto Nacional de Pesquisas Espaciais, INPE) has, since 1988, undertaken annual mapping of deforestation within the BLA using remote sensing data [13], which has been highly variable across the region. For example, an average deforestation rate of 21,050 km^2 yr^{-1} was reported between 1977 and 1988, which decreased subsequently to 11,030 km^2 yr^{-1} up to 1991; from 1991 to 2004 the deforestation rate increased to 27,772 km^2 yr^{-1} (reaching a record high of 29,059 km^2 yr^{-1} from 1994 to 1995) and then decreased to a record low of 4,571 km^2 yr^{-1} in 2012 [13]. Currently, much of the deforested area is under agricultural use (following disturbance mainly with slash-and-burn practices) but in many areas, the land has been abandoned and frequently supports forests at different stages of regeneration [14].

Several estimates of the deforested land occupied with regeneration have been generated, mainly from interpretation of remote sensing data, but also using transition modeling. Data from the National Oceanic and Atmospheric Administration (NOAA) Advanced Very High Resolution Radiometer (AVHRR) from 1988–1991 were used by Stone et al., 1994 [15] to generate a land cover map of South America, and subsequently Schroeder and Winjum, 1995 [16] used this dataset to estimate the extent of forest regeneration in the BLA at $\sim 151 \times 10^3$ km^2. Fearnside, 1996 [17], using a matrix of annual transition probabilities, estimated that approximately 48% ($\sim 195 \times 10^3$ km^2) of the deforested landscape in 1990 supported forest regeneration. Lucas et al., 2000 [18] using NOAA AVHRRR data estimated that $\sim 158 \times 10^3$ km^2 supported some type of forest regeneration in the period 1991–1994. Cardille and Foley, 2003 [19] used a matrix of annual transition probabilities to estimate that 36% ($\sim 91 \times 10^3$ km^2) of the area deforested between 1980 and 1995 in the entire Amazon river drainage basin was in some stage of secondary succession forest. Carreiras et al., 2006 [20] exploited a time-series of 12 monthly composite images of the year 2000, derived from the *Satellite Pour l'Observation de la Terre* (SPOT-4) VEGETATION sensor and estimated that, in 2000, $\sim 140 \times 10^3$ km^2 of land was occupied by regenerating forests. Neeff et al., 2006 [21] used large-area land cover maps derived from remote sensing datasets to generate estimates of regrowth extent within the BLA and concluded that regrowth increased from $\sim 29 \times 10^3$ km^2 in 1978 to $\sim 161 \times 10^3$ km^2 in 2002. Whilst providing information on the extent of regenerating forests at certain points in time, the dynamics of regenerating forests and the land on which they occupy has only been attempted using remote sensing by a few studies (e.g., [22]). However, such information is needed across large areas as regenerating forests may represent one of the prime

Table 1. Landsat Multi-spectral Scanner (MSS), Thematic Mapper (TM) and Enhanced TM (ETM+) data available for the three Amazonian sites.

Manaus (Path 231, Row 62)		Santarém (Path 227, Row 62)		Machadinho d'Oeste (Path 231, Row 67)	
Date[1]	Sensor	Date[1]	Sensor	Date[1]	Sensor
19730707	MSS	19840824	TM	19840617	TM
19770731	MSS	19850726	TM	19860810	TM
19780822	MSS	19860729	TM	19870712	TM
19790703	MSS	19870716	TM	19890717	TM
19830709	MSS	19880803	TM	19900618	TM
19850604	TM	19890822	TM	19910925	TM
19880815	TM	19900809	TM	19940816	TM
19890802	TM	19910711	TM	19950803	TM
19910808	TM	19931020	TM	19960704	TM
19920607	TM	19951010	TM	19970723	TM
19941019	TM	19960825	TM	19980624	TM
19950920	TM	19970727	TM	19990729	TM
19960720	TM	19980815	TM	20010803	TM
19990713	TM	19990903	TM	20030724	TM
20010827	ETM+	20000905	TM	20050713	TM
20020830	ETM+	20010916	ETM+	20060716	TM
20030809	TM	20030829	TM	20070703	TM
20060716	TM	20050701	TM	20080806	TM
20070804	TM	20060805	TM	20090809	TM
20080806	TM	20070621	TM	20100625	TM
20090910	TM	20081130	TM	20110612	TM
20100727	TM	20090712	TM		
20110831	TM	20100629	TM		

[1]Date format is yyyymmdd.

mechanisms by which biomass, biological diversity and ecosystem services can be restored [23–25].

The aim of this study was to establish the extent to which the age class distribution of regenerating forests was determined by the history of deforestation and subsequent use and management of the land, thereby giving insight into the potential for forest restoration and conservation into the future. This was achieved by comparing Landsat sensor data classified into the broad amalgamated categories of mature forest, non-forest (i.e., agriculture, including stock pasture) and secondary forest over decadal periods and time-separated by a maximum of four years. Three study areas were chosen to perform the analysis: Manaus (Amazonas state), Santarém (Pará state) and Machadinho d'Oeste (Rondônia state). Several measures were used to compare the three areas since the inception of widespread agricultural practices in the region (1970s for Manaus and Santarém and 1980s for Machadinho d'Oeste) up to the present: Deforestation (conversion from mature forest to non-forest or from secondary forest to non-

Table 2. Error matrix obtained from the accuracy assessment of the 2007 land cover map of the Manaus site (MF – mature forest, NF – non-forest, SF – secondary forest).

	Classified				
Reference	**MF**	**NF**	**SF**	**Total**	**Omission error**
MF	186	4	10	200	0.070
NF	8	185	7	200	0.075
SF	11	2	187	200	0.065
Total	205	191	204	600	
Commission error	0.093	0.031	0.083		

The overall accuracy was 0.930.

Table 3. Error matrix obtained from the accuracy assessment of the 2010 land cover map of the Santarém site (MF – mature forest, NF – non-forest, SF – secondary forest).

	Classified				
Reference	**MF**	**NF**	**SF**	**Total**	**Omission error**
MF	196	0	4	200	0.020
NF	0	198	2	200	0.010
SF	23	11	166	200	0.170
Total	219	209	172	600	
Commission error	0.105	0.053	0.035		

The overall accuracy was 0.93.

forest) and regrowth rates (conversion from non-forest to regeneration), the age of secondary forest, the period of active land use prior to abandonment to regeneration, and the frequency of clearance. The study builds on that of Prates-Clarke et al., 2009 [38] in that the time-series was extended from 2003 to 2011 for Manaus and Santarém and a new time-series was generated for Machadinho d'Oeste.

Study Areas

For all three sites (Manaus, Santarém and Machadinho d'Oeste; Figure 1), the natural vegetation prior to disturbance was primary (undisturbed) *terra firme* (i.e., non-flooded) forest. In each case, the canopy heights averaged between 25 and 35 m, with emergent trees exceeding 50 m at the more productive sites (e.g., Manaus). Species diversity is also high, often exceeding 225 species ha^{-1} [26], with an estimated mean above ground biomass of approximately 270 Mg ha^{-1} [27].

The first site, north of Manaus, the capital of Amazonas state, occupied an area of 5,042 km^2 (2°33′11″ S, 60°5′7″ W, Figure 1) and included the Instituto Nacional de Pesquisas da Amazônia (INPA) and the Smithsonian Institution (SI) Biological Dynamics of Forest Fragments Project (BDFFP) research sites (established in 1979) [28], the Adolfo Ducke and Walter Egler forest reserves, and other state and federal environmental protection areas. Significant deforestation began in the area following construction in the early 1970s of the BR-174 highway connecting Manaus with Boa Vista in Roraima. Most deforestation activity occurred either side of the highway, with this fuelled primarily by agricultural expansion. Recognition of the opportunity to study the impacts of clearance and forest fragmentation on biodiversity as well as ecosystem services spurred the preservation of fragments of various dimensions (typically less than 2 km^2) prior to felling of the surrounding tropical forest. Many of these fragments were, however, reconnected because of rapid regrowth of forests following abandonment of many clearances from the mid-1980s onwards. Up until the present, a substantial amount of research has been conducted, with these informing future management strategies across the Amazon [28].

The second site (3°10′5″ S, 54°55′42″ W) covered an area of 1,118 km^2 and was located approximately 80 km to the south of Santarém, the second main city in the Pará state (Figure 1). This study area is within the Tapajós National Forest, between the Tapajós River and the BR-163 highway connecting Santarém to the state capital of Mato Grosso, Cuiabá. This conservation unit was created in 1974 and has been recognized as a model for sustainable forest management, including logging activities, at levels ranging from individuals to communities (e.g., [29,30]). Compared to Manaus, forests are less dense with a mixed arrangement of broadleaves and palm trees such as Açaí (*Euterpe oleracea*) and Babaçu (*Orbignya phalerata*) being commonplace [31].

The third site (9°32′56″ S and 62°6′27″ W) occupied an area of 1,780 km^2 and was located mostly within the Machadinho d'Oeste municipality, Rondônia state. This municipality originated from the former Machadinho settlement project, deployed by the Brazilian federal government (through the *Instituto Nacional de Colonização e Reforma Agrária*, INCRA) in 1982 as part of the POLONORDESTE Program [32]. As such, the majority of its inhabitants are dependent on agriculture for subsistence [33]. Open rain forest is the dominant vegetation type in the municipality [32]. According to data from the Brazilian Ministry

Table 4. Error matrix obtained from the accuracy assessment of the 2010 land cover map of the Machadinho d'Oeste site (MF – mature forest, NF – non-forest, SF – secondary forest).

	Classified				
Reference	**MF**	**NF**	**SF**	**Total**	**Omission error**
MF	200	0	0	200	0.000
NF	0	197	3	200	0.015
SF	19	37	144	200	0.280
Total	219	234	147	600	
Commission error	0.087	0.158	0.020		

The overall accuracy was 0.90.

Table 5. Area (ha) and relative incidence (RI, %) of perennial (PC) and tree (TC) crops with the mapped area of mature forest (MF), non-forest (NF) and secondary forest (SF).

	Manaus (2007)		Machadinho d'Oeste (2010)			
	TC		PC		TC	
Classified	Area (ha)	RI (%)	Area (ha)	RI (%)	Area (ha)	RI (%)
MF	139	0.0	31	0.1	0	0.0
NF	398	1.0	540	0.9	14	0.0
SF	4,237	4.4	1,323	2.1	3	0.0
Total	4,774		1,894		17	

Estimates were based on the land cover map closest to the date that was used to delineate crop types using very high resolution (VHR) imagery.

of the Environment (MMA), this site includes several conservation units, mainly extractive reserves, which were established in and after 1995.

The area of the Manaus site was approximately three to four times the size of the other two study areas for several reasons. Deforestation in Manaus started in the early 1970s and land conversion from primary forest to agriculture/pasture was much more scattered. By comparison, deforestation patterns were more concentrated at Santarém (either side of and in proximity to a highway) and Machadinho d'Oeste (a planned settlement). A larger area was also selected to encompass older deforestation within the BDFFP study area and more recent clearing towards the city of Manaus, thereby allowing a wide range of LULCC processes to be captured.

At all three sites, much of the history of deforestation and land use has been captured by optical (primarily Landsat) sensors during the dry season, when cloud cover is minimal. In Manaus, a moderately strong dry season occurs between June and October [28], in Santarém between May and October [34], and in Machadinho d'Oeste between April and November [32]. Rainfall is generally higher at Manaus (between 1,900 and 2,500 mm annually) and lower at Santarém and Machadinho d'Oeste with an annual average of 2000 mm. The mean annual temperatures at the sites range from 25°C to 26°C [32,34,35]. The distribution of forests types is determined in part by the topography and geology. The topography is moderately flat (up to 160 m elevation) north of Manaus and is divided by a large number of waterlines. Soils are nutrient-poor (i.e., Oxisols), with these being typical of many across the Amazon basin [36]. The area south of Santarém is also moderately flat, ranging from approximately 50 to 240 m, and consists of nutrient poor soils (i.e., Oxisols and Ultisols) [34,37]. The terrain is moderately undulated at Machadinho d'Oeste, ranging from approximately 90 to 370 m, and the dominant soil types are Alfisols, Oxisols, Ultisols, and Alluvial [32].

Data

For Manaus, Landsat Multi-spectral Scanner (MSS), Thematic Mapper (TM) and Enhanced Thematic Mapper Plus (ETM+) data were acquired between 1973 and 2011 (path 231, row 62; Table 1). The Landsat MSS data acquired between 1973 and 1983 were only available as hard-copy prints [38]. The remaining Landsat TM and ETM+ scenes (1985 onwards) were provided in digital format from both INPE and the United States Geological Service (USGS).

In the previous study of Prates-Clark et al., 2009 [38], scenes acquired between 1985 and 2003 were used to study forest regeneration in this area. In this study, the time-series was extended to 2011 using six additional annual scenes, which were relatively cloud free (<20%), or cloud was distributed in only certain areas of the image. In total, the time-series extended over an approximate 40-year period from 1973 (when the deforested area was minimal) to 2011. For Santarém, Prates-Clark et al., 2009 [38] again used Landsat TM and ETM+ imagery acquired between 1984 and 2003 (path 227, row 62), with most scenes unaffected by substantive cloud cover. The time-series was extended using Landsat TM data acquired for each year from 2005 to 2010. Coverage for 2011 was actively sought but no data with minimal cloud coverage were available. Whilst most changes at Santarém were associated with deforestation for agriculture and pasture, extensive wildfires damaged the surrounding forests in 1992 and 1997. The last fire episode was associated with the 1997–1998 El Niño Southern Oscillation (ENSO) event. The time-series classification for Machadinho d'Oeste had not been undertaken previously and so 21 new time-series digital images acquired between 1984 and 2011 were obtained from the USGS. Gaps in the time-series ranged from between one (70%) to approximately three years (5%). Cloud cover was minimal on all but two dates, although the vast majority of the deforested area was visible within these.

Methods

4.1. Image pre-processing

Landsat images acquired by the USGS were orthorectified and calibrated to units of spectral radiance (W m^{-2} sr^{-1} μm^{-1}) and then calibrated to top of atmosphere (TOA) reflectance using calibration factors and equations provided by Chander et al., 2009 [39]. Images acquired by INPE were processed to TOA reflectance and geometrically corrected using Environment for Visualizing Images (ENVI) software (Exelis Visual Information Solutions, Boulder, CO, USA), a third order polynomial and a nearest neighbor transformation [38]. Each Landsat image was then subsetted to encompass the main area of deforestation and the intersection of all images in the time series.

4.2. Image classification

As with Prates-Clark et al., 2009 [38], each digital image within the time-series for each site was classified into mature forest (MF), non-forest (NF), and secondary forest (SF), with the second category including crops (herbaceous and woody) and pastures. Areas of open water were also mapped and a common mask was applied to all dates in the time-series. Several classification approaches were utilized with the intention of providing the best

A

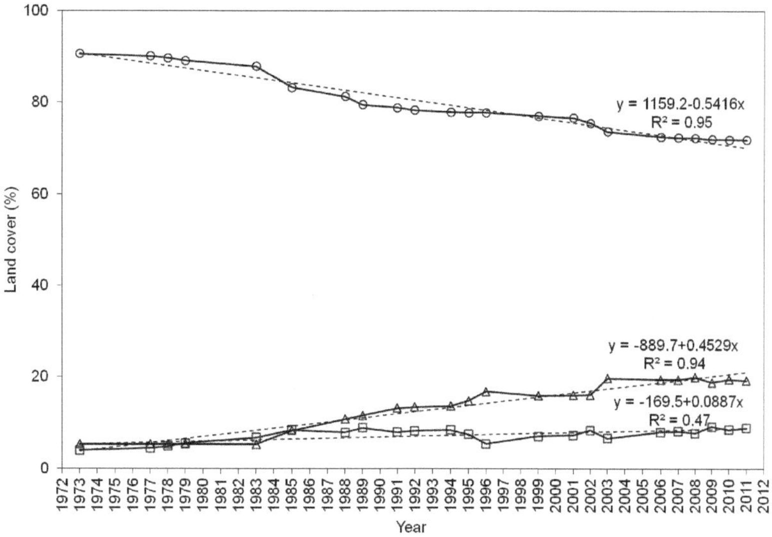
Land cover (%)
Year
y = 1159.2-0.5416x
R² = 0.95
y = -889.7+0.4529x
R² = 0.94
y = -169.5+0.0887x
R² = 0.47

B

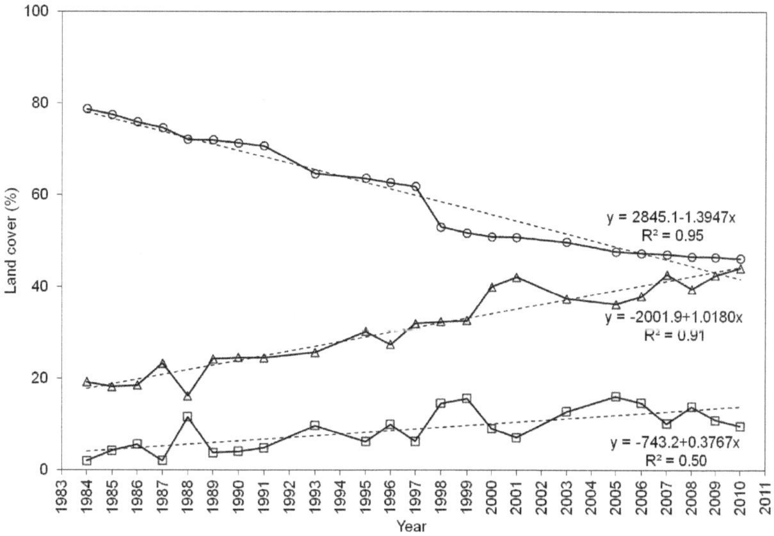
Land cover (%)
Year
y = 2845.1-1.3947x
R² = 0.95
y = -2001.9+1.0180x
R² = 0.91
y = -743.2+0.3767x
R² = 0.50

C

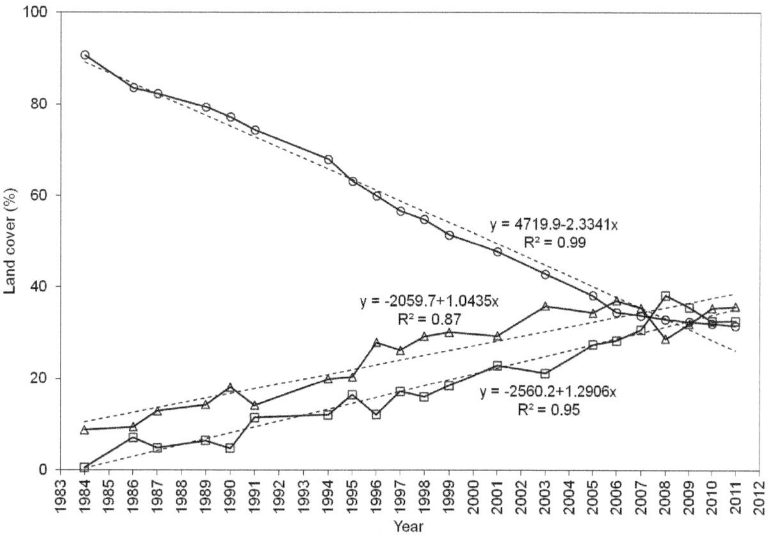
Land cover (%)
Year
y = 4719.9-2.3341x
R² = 0.99
y = -2059.7+1.0435x
R² = 0.87
y = -2560.2+1.2906x
R² = 0.95

Figure 2. Land cover change (%) across the three selected sites. Land cover change (%) across the three selected sites; A) Manaus (1973–2011), B) Santarém (1984–2010), and C) Machadinho d'Oeste (1984–2011). Each dashed line represent the linear fit of the proportion of each land cover class as a function of the year; also showing are the corresponding equations and coefficient of determination (R^2).(○ - mature forest, □ - non-forest, Δ- secondary forest).

classification of the desired categories. At Manaus, and for digital images acquired before 2006, supervised classifications based on the minimum distance or maximum likelihood algorithms were applied (e.g., [40]), with regions of interest representing the main cover types defined around a number of target areas described on the basis of field observations or through reference to very high resolution data. For the Landsat TM images acquired between 2006 and 2011, an object oriented classification (e.g., [41]) was followed and involved a decision-rule classification applied within the eCognition software (Trimble Geospatial Imaging, Munich, Germany). The rule base method used data from the available Landsat sensor bands as well as data layers derived from spectral indices, namely the Normalized Difference Vegetation Index (NDVI) and Normalized Difference Water Index (NDWI). The classification of MF was refined with a cloud-free mask of the MF area obtained through classification of the most-cloud-free and recent image in the time-series. Images acquired on the 31st August, 2011, 29th June, 2010 and 12th June 2011 were used to generate the MF mask, on the assumption that the land cover type had remained as MF for the whole of the time-series. If MF was covered by cloud in earlier images of the time-series, this was subsequently classified as MF. At Manaus, SF were older compared to the other sites and the structure of the upper canopy of the regenerating forests led to a spectral response in the near and shortwave infrared channels that was similar to MF; hence, reference to images earlier in the time-series was necessary [18,22,42]. At Santarém, the Landsat data acquired between 1984 and 2003 were classified using a fuzzy logic approach that was applied to the original Landsat bands as well as derived fractional images (shade/moisture, vegetation and soil). Each pixel of the Landsat data acquired between 2005 and 2010 was classified into one of the three main land cover classes (described above) using the random forests machine learning classification algorithm (e.g., [43]) and training data. This algorithm has a number of options (e.g., [43]) that were tested for each date to obtain the best possible classifier. As with Santarém, all Landsat TM dates for Machadinho d'Oeste were classified using the random forests algorithm. In all cases, the objective was to obtain the best classification of NF, MF and SF for each year of observation.

4.3. Post-classification

The comparison of the time-series of classified images (as MF, NF, or SF) identified several cases where some pixels classified as SF or NF in a given date were classified as MF in the following date. As this sequence is not plausible, in-house Interactive Data Language (IDL) code (Exelis Visual Information Solutions, Boulder, CO, USA) was written to identify these cases by comparing the classification over two consecutive dates. Where SF or NF areas were classified as such in one year and as MF in the next, the MF class was reallocated to SF. These areas were not reclassified as NF because it was assumed that spectral confusion between NF and MF was unlikely. This procedure was similar to that undertaken by Roberts et al., 2002 [44], who also considered "disallowed transitions" between cover types.

4.4. Deforestation and regrowth rates

Annual rates of change from MF (or SF) to NF (deforestation) and from NF to SF (regeneration) were calculated for consecutive dates in the time-series using equations (1) and (2) [45]

$$r = \frac{1}{t_2 - t_1} \ln \frac{A_2}{A_1} \tag{1}$$

$$R = \frac{A_1 - A_2}{t_2 - t_1} \tag{2}$$

where r and R are the annual rate of change expressed in percentage per year and hectares per year respectively; A_1 and A_2 are the MF (or SF, in the case of deforestation occurring in regeneration areas) or NF cover areas (hectares) at time t_1 and t_2 respectively (time period). These rates were used to characterize the temporal evolution of change across the three sites, to detect events (primarily clearance) and describe general trends.

4.5. Age of secondary forest, period of active land use, and frequency of clearance

Algorithms were written and implemented in IDL to compare the classifications of MF, NF, and SF between dates. Subsequently, datasets relating to the history and dynamics of land use, namely the age of secondary forest (ASF), the period of active land use (PALU) prior to abandonment, and the frequency of clearance (FC) were generated for each site and for each year in which an image had been acquired. The ASF was estimated by summing the time (in years) that each pixel was occupied by SF since the last clearance event. However, when SF occurs in the first date of the time series, and it is not subsequently cleared, the ASF can only be considered as a minimum age as the exact date of land abandonment is not known. The PALU was defined as the difference (in years) between the time of initial forest clearance and the onset of regeneration. However, where reclearance of regenerating forest had occurred, the PALU was calculated by summing the period since the last reclearance event until the forest cover had re-established. The FC was estimated by summing the frequency of transitions from MF (or SF) to NF. For both the Manaus and Santarém study areas, fire scars were evident within some clearances by the lower near and shortwave infrared reflectance compared to the original vegetated surface [46]. However, many burned areas acquired a vegetated cover quite rapidly [47], and so only a partial fire history could be retained.

To better spatially represent and discuss the main results, classes of ASF, PALU, and FC were generated and used to assess their temporal evolution. Each metric was classified into three classes as to identify a lower, middle and upper interval range. However, these intervals do have some degree of subjectivity. ASF classes were defined as initial (≤5 years), intermediate (6–15 years) and advanced (≥16 years), adapted from Lucas et al., 2000 [18]. PALU classes were defined as short (≤2 years), medium (3–4 years) and long (≥5 years), with these representing various different crop or pasture cycles. FC classes were defined as low (1 time), medium (2 times) and high (≥3 times), representing several land use patterns in terms of the frequency of the deforestation -> agriculture/pasture -> abandonment -> re-growth temporal sequence.

Table 6. Land cover proportion (%) of mature forest (MF), non-forest (NF) and secondary forest (SF) in each year of the time-series at Manaus, Santarém and Machadinho d'Oeste.

Manaus				**Santarém**				**Machadinho d'Oeste**			
Year	**MF**	**NF**	**SF**	**Year**	**MF**	**NF**	**SF**	**Year**	**MF**	**NF**	**SF**
1973	91	4	5	1984	79	2	19	1984	91	1	9
1977	90	4	5	1985	78	4	18	1986	84	7	9
1978	90	5	5	1986	76	6	19	1987	82	5	13
1979	89	6	5	1987	75	2	23	1989	79	6	14
1983	88	7	5	1988	72	12	16	1990	77	5	18
1985	83	8	8	1989	72	4	24	1991	74	11	14
1988	81	8	11	1990	71	4	25	1994	68	12	20
1989	79	9	12	1991	71	5	24	1995	63	16	20
1991	79	8	13	1993	65	10	26	1996	60	12	28
1992	78	8	13	1995	64	6	30	1997	57	17	26
1994	78	8	14	1996	63	10	27	1998	55	16	29
1995	78	7	15	1997	62	6	32	1999	51	18	30
1996	78	5	17	1998	53	15	32	2001	48	23	29
1999	77	7	16	1999	52	16	33	2003	43	21	36
2001	77	7	16	2000	51	9	40	2005	38	27	34
2002	76	8	16	2001	51	7	42	2006	35	28	37
2003	74	7	20	2003	50	13	37	2007	34	31	35
2006	73	8	19	2005	48	16	36	2008	33	38	29
2007	72	8	19	2006	47	15	38	2009	32	36	32
2008	72	8	20	2007	47	10	43	2010	32	33	35
2009	72	9	19	2008	47	14	40	2011	32	33	36
2010	72	9	19	2009	46	11	43				
2011	72	9	19	2010	46	10	44				

Table 7. Summary of deforestation and regrowth rates between consecutive dates in the Manaus time-series (MF – mature forest; SF – secondary forest).

	MF annual rate of deforestation		SF annual rate of deforestation		Annual rate of deforestation		Annual rate of regrowth	
	(% yr^{-1})	(ha yr^{-1})	(% yr^{-1})	(ha yr^{-1})	(% yr^{-1})	(ha yr^{-1})	(% yr^{-1})	(ha yr^{-1})
Minimum	0.1	277	0.0	0	0.1	691	0.0	0
Maximum	2.9	13,451	27.7	8,453	3.8	17,936	133.9	24,738
Mean	0.7	3,123	6.4	3,231	1.3	6,355	34.1	6,017
Median	0.4	1,591	5.0	2,725	0.9	4,513	24.8	4,793
Inter-quartile range	0.4	2,172	5.1	3,006	1.3	5,881	31.9	6,156
Standard deviation	0.8	3,726	6.5	2,699	1.1	4,905	33.7	5,756

Table 8. Summary of deforestation and regrowth rates between consecutive dates in the Santarém time-series (MF – mature forest; SF – secondary forest).

	MF annual rate of deforestation		SF annual rate of deforestation		Annual rate of deforestation		Annual rate of regrowth	
	(% yr^{-1})	(ha yr^{-1})	(% yr^{-1})	(ha yr^{-1})	(% yr^{-1})	(ha yr^{-1})	(% yr^{-1})	(ha yr^{-1})
Minimum	0.4	235	0.5	194	0.5	448	10.5	832
Maximum	15.6	11,046	82.3	11,352	17.7	16,952	261.5	12,257
Mean	2.1	1,598	16.7	3,189	4.9	4,787	78.5	4,675
Median	1.2	963	10.7	2,769	4.0	4,236	71.3	3,361
Inter-quartile range	1.3	1,085	16.6	2,350	3.0	2,831	79.6	4,031
Standard deviation	3.2	2,292	18.7	2,849	4.4	4,188	61.5	3,332

Table 9. Summary of deforestation and regrowth rates between consecutive dates in the Machadinho d'Oeste time-series (MF – mature forest; SF – secondary forest).

	MF annual rate of deforestation		SF annual rate of deforestation		Annual rate of deforestation		Annual rate of regrowth	
	(% yr^{-1})	(ha yr^{-1})	(% yr^{-1})	(ha yr^{-1})	(% yr^{-1})	(ha yr^{-1})	(% yr^{-1})	(ha yr^{-1})
Minimum	1.2	748	0.0	17	1.5	2,386	3.7	103
Maximum	9.9	9,223	88.1	16,711	17.7	18,202	84.2	17,389
Mean	3.9	4,142	19.5	5,289	7.2	9,286	25.8	6,804
Median	3.5	4,171	14.3	4,273	6.0	7,436	19.1	6,172
Inter-quartile range	3.4	4,057	13.5	7,114	6.7	9,571	18.0	4,609
Standard deviation	2.3	2,394	19.5	4,467	4.4	5,248	21.5	4,141

4.6. Accuracy assessment and area calibration of the time-series of classified images

The most recent very high resolution (VHR) imagery was used to carry out the accuracy assessment of the time-series classifications at each site. As VHR data were not available to perform the accuracy assessment for all the classified images in the time-series, the accuracy in the classification of the image that was closest in time was assumed to be similar for the remaining images. To quantify the accuracy of the 2010 classification for Santarém, reference was made to 5 m SPOT-5 panchromatic data from 2009 and 2011 whilst the 2007 and 2010 classifications for Manaus and Machadinho d'Oeste respectively were validated using GeoEye imagery available on Google Earth.

At each site, 200 random points located at the center of larger polygons (greater than 6 ha) were generated for each class. The points on the classification were compared with the VHR imagery from the same date. At Santarém, it was assumed that if the same land cover was present in both the 2011 and 2009 image, the land cover in 2010 would be the same. The accuracy assessment was reported as a standard error matrix, including the overall accuracy and omission and commission errors (e.g., [48]). From the same VHR imagery and past field studies of the study sites (e.g., [32,38]), woody agriculture crops had established on some of the deforested areas. At Manaus, a large area of tree crops (TC, mainly oil palm) was progressively planted in the north west of the site from the 1983 to 1989, whilst at Machadinho d'Oeste, perennial crops (PC, mainly coffee plantations) were more common. To assess how well the areas of MF, NF and SF were discriminated using the classification, these areas were delineated manually from the VHR imagery and used in the generation of the error matrices.

The most straightforward way of estimating total areas for each class was to count the number of pixels in the MF, NF and SF classes with respect to the ground truth data. This is called "naïve estimation" in the remote sensing literature [49]. As a matter of fact, areal estimates obtained this way are biased (e.g., [50]). Some post-classification methods have been used to improve these estimates, namely the so-called calibration techniques that are divided into classical and inverse methods. The information contained in the error matrix was used to correct for misclassification bias [50]. Walsh and Burk, 1993 [50] carried out a simulation study and concluded that the inverse method consistently performed better than the classical method and lead to unbiased estimates for total areas. The idea behind this calibration technique was to use the misclassification probabilities among classes to revise the proportions given by pixel counting (p_i) [51]. According to Tenenbein, 1972 [51] and Walsh and Burk, 1993 [50], the revised (calibrated) proportion of the total area occupied by class i (π_i) can be estimated with equation (3)

$$\pi_i = \sum_{j=1}^{k} \left[(f_j + n_{.j})/N \right] \left(n_{ij} / n_{.j} \right) \qquad (3)$$

where i represents the observed class, j the predicted class, k the number of classes, n the number of observations in the validation sample, N the combined number of observations in the validation and satellite datasets, f_j the (N-n) in class j, n_{ij} the number of observations in class i that were classified as j, and $n_{.j}$ the sum of all training observations predicted as class j. This method was applied to calibrate the area estimates obtained from each date of classified images using the error matrices that were produced for the three sites.

Table 10. An assessment of the dynamics of land use at Manaus, Santarém and Machadinho d'Oeste (figures in brackets represent the number of years).

		Sites		
Indicators	**Impact**	**Manaus**	**Santarém**	**Machadinho d'Oeste**
a) Clearance of MF only with little or no deforestation over SF	High pressure for new land	1973–1985 (12)	Pre-1984	1984–1987 (3)
b) Area of MF cleared>SF cleared.	Pressure for new land with some contained re-use of the existing deforested area (through reclearance)	1985–1989 (4)	1986–1987 (1)	1987–1990 (3)
		2002–2003 (1)	1991–1993 (2)	1991–1996 (5)
			1997–1998 (1)	1997–1998 (1)
			2006–2007 (1)	2001–2003 (2)
			2008–2009 (1)	
c) Area of MF cleared<SF cleared	Greater re-use of existing deforested land but still requirement for more land	1989–2002 (13)	1984–1986 (2)	1990–1991 (1)
		2003–2011 (11)	1987–1991 (4)	1996–1997 (1)
			1993–1997 (4)	1998–2001 (3)
			1998–2006 (8)	2003–2011 (8)
			2007–2008 (1)	
			2009–2010 (1)	
d) Area of MF and SF cleared>area of SF regenerating		1973–1985 (12)	1984–1986 (2)	1984–1986 (2)
		1988–1989 (1)	1987–1988 (1)	1987–1989 (2)
		1991–1994 (3)	1989–1993 (4)	1990–1995 (5)
		1996–2002 (6)	1995–1996 (1)	1996–1997 (1)
		2003–2007 (4)	1997–1999 (2)	1998–2001 (3)
		2008–2009 (1)	2001–2005 (4)	2003–2008 (5)
		2010–2011 (1)	2007–2008 (1)	2010–2011 (1)
e) Reclearance of SF but no clearance of MF	Contained re-use of area already deforested	No years	No years	No years
f) Area of SF regeneration>MF and SF deforestation	Net abandonment of land to regenerating forests	1985–1988 (3)	1986–1987 (1)	1986–1987 (1)
		1989–1991 (2)	1988–1989 (1)	1989–1990 (1)
		1994–1996 (2)	1993–1995 (2)	1995–1996 (1)
		2002–2003 (1)	1996–1997 (1)	1997–1998 (1)
		2007–2008 (1)	1999–2001 (2)	2001–2003 (2)
		2009–2010 (1)	2005–2007 (2)	2008–2010 (2)
			2008–2010 (2)	

Results

5.1. Accuracy assessment of land cover classification

The error matrices resulting from the accuracy assessment over Manaus, Santarém, and Machadinho d'Oeste are presented in Table 2, 3 and 4 respectively. For all three sites, the overall accuracy was high (above 0.90), although some relatively high omission and commission errors were identified, especially in Santarém and Machadinho d'Oeste. For Manaus, the omission and commission errors were always below 10%. For Santarém, a higher omission error in the SF class was detected (17.0%), mainly due to misclassification as MF (11.5%), which reflects also in the high commission error in the MF class (10.5%). At Machadinho d'Oeste, a major source of error also comes from a high omission error in the SF class (28%), with this mainly being a consequence of misclassification as NF (18.5%) and to a lesser extent as MF (9.5%); a high commission error was also observed in the NF class (15.8%), with this being due to misclassification with SF.

In Manaus, TC (4,774 ha) was mostly incorrectly classified as SF (89%), illustrating the similarity in their spectral signatures, but accounted only for 4.4% of the SF area in 2007 (Table 5). However, most of the TC classified as SF in Manaus was associated with the oil palm plantation situated in the north west of the study site. The majority of TC in Machadinho d'Oeste was correctly classified as NF, although the overall area was very small (17 ha). Conversely, 70% of the PC (mainly coffee plantations) at Machadinho d'Oeste was misclassified as SF, although this only accounted for 2.1% the entire SF area in 2010.

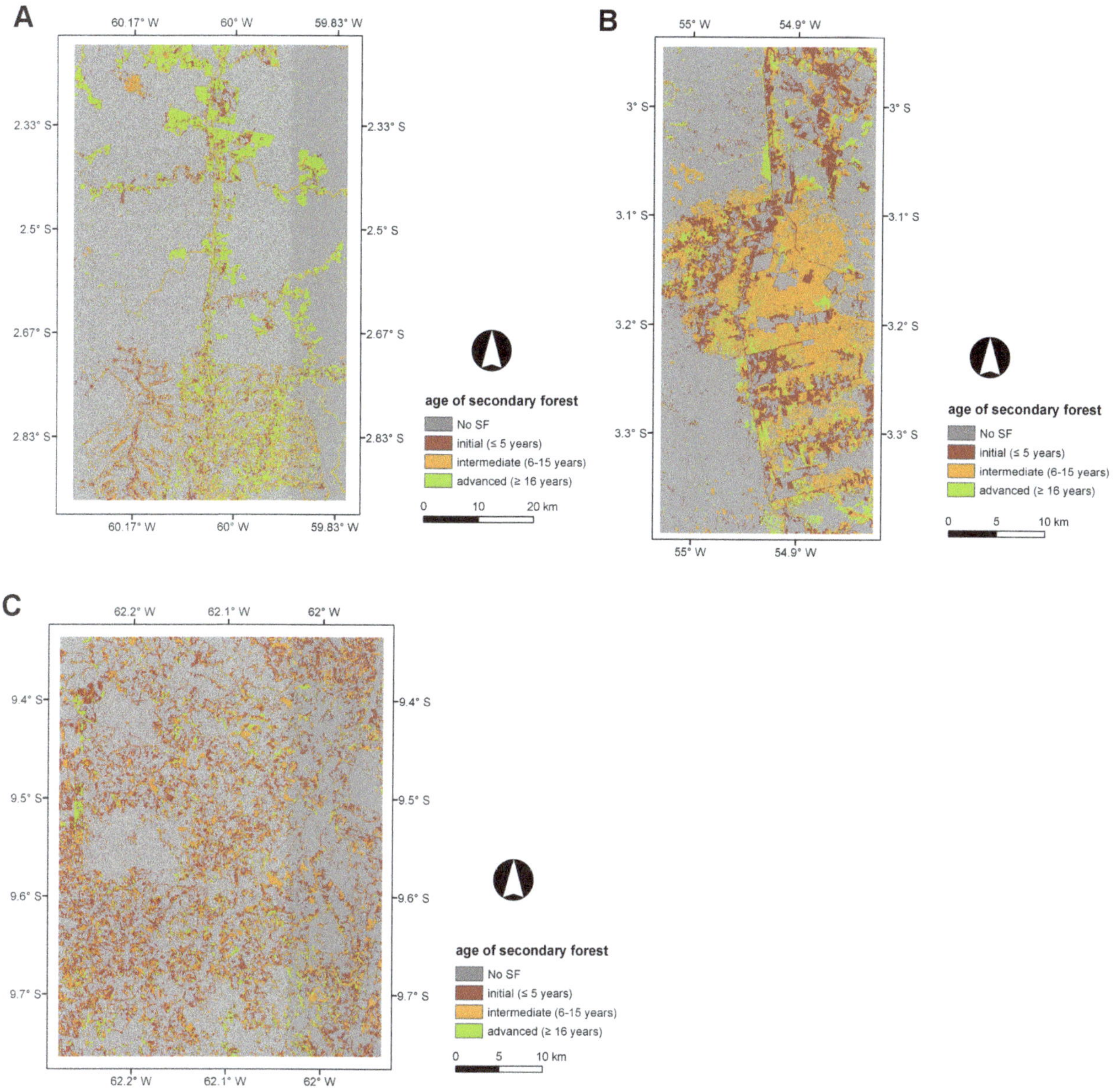

Figure 3. Maps displaying the age of secondary forest (ASF, years) across the three selected sites. Maps displaying the age of secondary forest (ASF) across the three selected sites for areas undergoing secondary forest (SF) in the last year of the corresponding time-series; A) Manaus (2011), B) Santarém (2010), and C) Machadinho d'Oeste (2011).

5.2. Land cover change

The evolution of land cover change in Manaus, Santarém and Machadinho d'Oeste is depicted in Figure 2A, 2B and 2C respectively and in Table 6, which shows that the most rapid change occurred in Machadinho d'Oeste, followed by Santarém and Manaus. For all sites, there was a strong correspondence between the land cover percentage and time (year) for the MF class, with the rate decreasing by an average of 0.54% in Manaus to 1.39% in Santarém and 2.33% in Machadinho d'Oeste. For all sites, a corresponding increase in the area deforested occurred, with the land cover alternating between NF and SF in all cases. The change rate in the NF class increased from 0.09% (Manaus) and 0.38% (Santarém) to 1.29% (Machadinho d'Oeste) whilst the SF change rate progressively increased from Manaus (0.45%) to Santarém (1.02%) and Machadinho d'Oeste (1.04%). The relationships between the proportion (%) of MF, NF and SF and time (year) were significant at all sites ($P<0.001$).

5.3. Deforestation and regrowth rates

Tables depicting relative and absolute deforestation and regeneration rates between consecutive dates for Manaus, Santarém and Machadinho d'Oeste are presented as Supporting Information (Table S1, S2 and S3 respectively in File S1) and a summary is given in Tables 7, 8 and 9 respectively. The maximum relative annual rate of deforestation over MF was observed at Santarém, with a 15.6% yr^{-1} loss between 1997 and 1998, with this associated with the extensive wildfires in 1997, and the highest average annual rate of deforestation was in Machadinho d'Oeste (3.9% yr^{-1}). The maximum relative rates of SF clearance were in Machadinho d'Oeste (88.1% yr^{-1}, 1990–1991) and Santarém

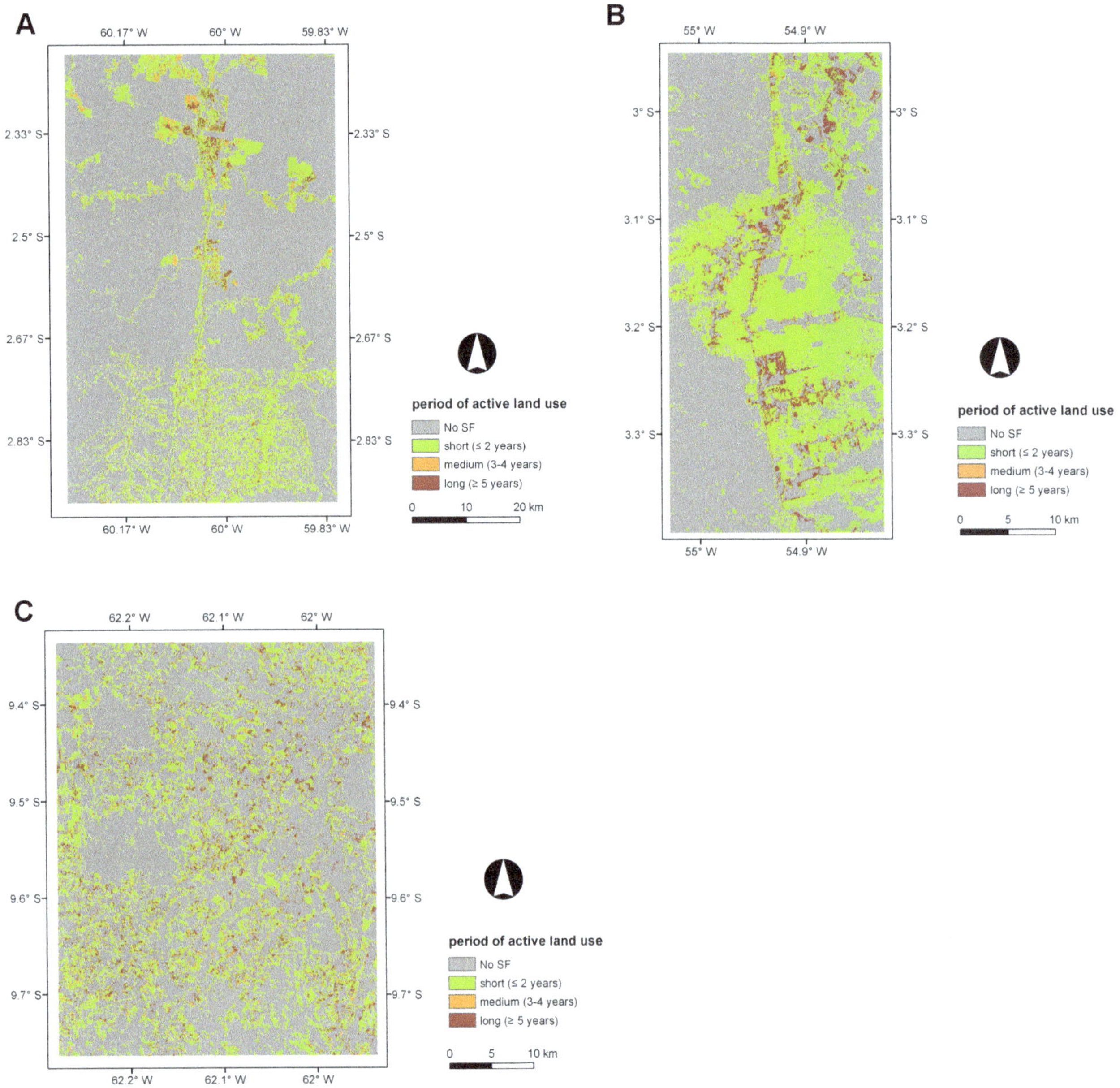

Figure 4. Maps displaying the period of active land use (PALU, years) across the three selected sites. Maps displaying the period of active land use (PALU, years) across the three selected sites for areas undergoing secondary forest (SF) in the last year of the corresponding time-series; A) Manaus (2011), B) Santarém (2010), and C) Machadinho d'Oeste (2011). An area of 667 ha (1.5%) with a PALU of zero in the Santarém site was included in the short PALU class (≤2 years) and corresponds to areas that were SF from 1984 to 2010.

(82.3% yr^{-1}, 1987–1988), with average rates being 19.5% and 16.7% for these sites respectively. The maximum and the highest average relative rates of regeneration were observed at Santarém, with 261.5% yr^{-1} (1997–1998) and 78.5% yr^{-1} respectively. The lowest relative rates of regrowth were at Machadinho d'Oeste, with these averaging 25.8% yr^{-1}. The average overall relative deforestation rates (combined MF and SF) were greatest for Machadinho d'Oeste (7.2% yr^{-1}), followed by Santarém (4.9% yr^{-1}) and Manaus (1.4% yr^{-1}). On average, deforestation rates over SF were higher than those over MF, indicating reclearance at all sites and particularly Machadinho d'Oeste (averaging 19.5% yr^{-1}) and Santarém (averaging 16.7% yr^{-1}).

Absolute areas (annual or based on periods of observation) are useful when comparing deforestation and regeneration rates between sites and the processes of change. Several key indicators are then listed in Table 10 together with the dates for each site to which these apply. For all sites, a trend from indicators a) to f) was observed with the majority being cleared initially, regenerating forests progressively establishing with these then recleared to varying degrees, and, of these, several are then abandoned to regrowth again.

5.4. Age of secondary forest (ASF), period of active land use (PALU), and frequency of clearance (FC)

Maps of the ASF, PALU, and FC for the last date of the time-series are presented in Figures 3, 4 and 5 for Manaus (2011), Santarém (2010) and Machadinho d'Oeste (2011) respectively and the proportion of areas with respect to these classes is indicated in

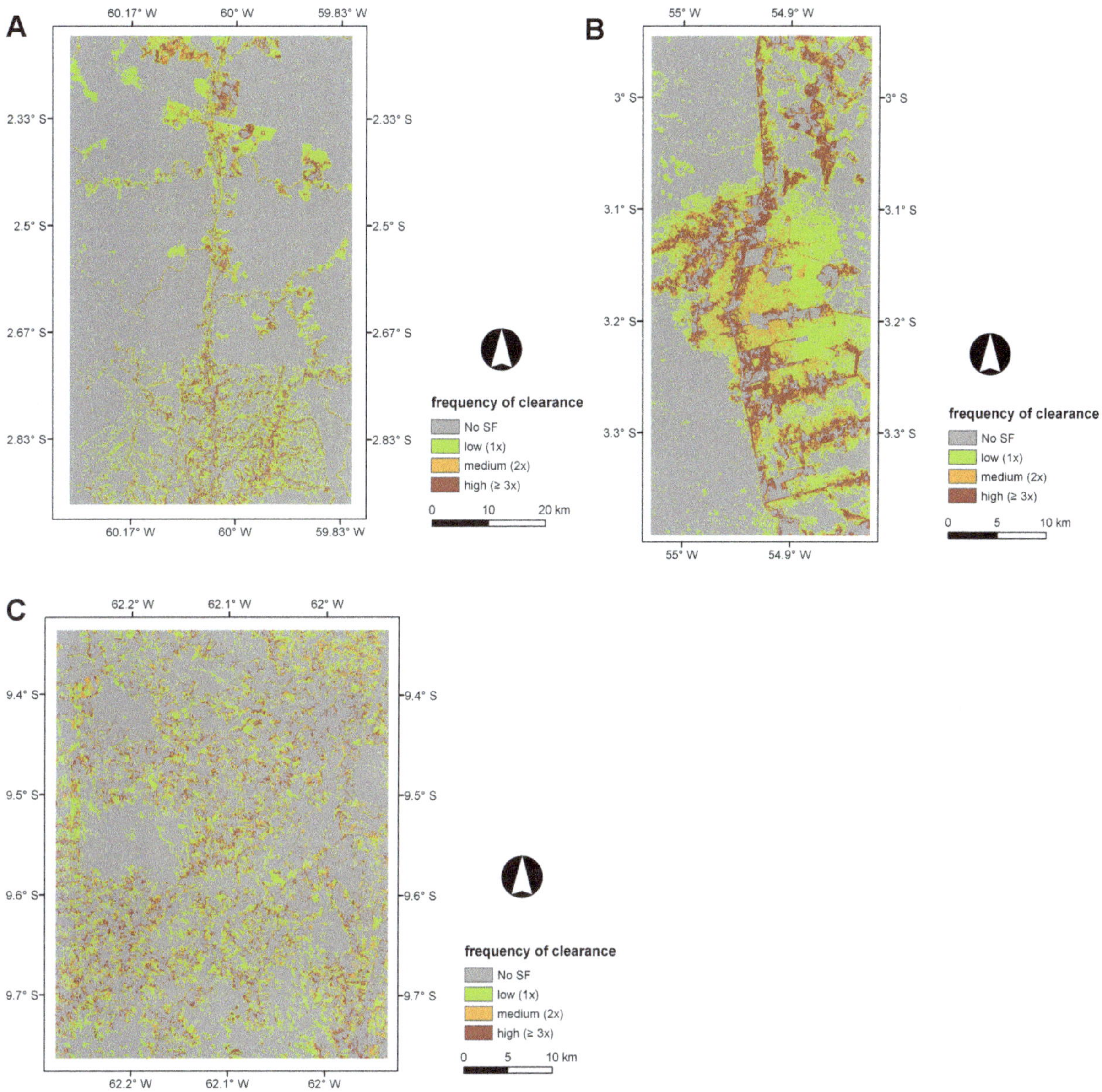

Figure 5. Maps displaying the frequency of clearance (FC) across the three selected sites. Maps displaying the frequency of clearance (FC) across the three selected sites for areas undergoing secondary forest (SF) in the last year of the corresponding time-series; A) Manaus (2011), B) Santarém (2010), and C) Machadinho d'Oeste (2011). In Manaus, Santarém and Machadinho d'Oeste an area of 103 ha (0.1%), 726 ha (1.6%), and 55 ha (0.1%), respectively, with a PALU of zero was included in the low FC class (1 time) and corresponded to areas that were non-forest (NF) in the first date of the time-series and no clearance has occurred in the secondary forest (SF) that persisted until the end of the time-series. In Santarém, the 726 ha also included areas that were already SF in 1984 that persisted until 2010.

Figure 6. At Manaus, regrowth forests were comparatively older, with 50% occupied by forests ≥16 years (Figure 6A). At Santarém, 57% of forests were aged 6–15 years, and at Machadinho d'Oeste, the majority of the area occupied by SF aged ≤5 years (46%). Hence, the three sites contain different distributions of age classes, with this attributable to differences in the time since deforestation but also the PALU and the frequency of clearance events. At all sites, the majority of land has been used actively for ≤2 years (Figure 6B), either following the initial clearance of the primary forest or subsequent reclearance events, but the deforestation had commenced at different times. At Manaus, the PALU was also ≤2 years for over 64% of the area and many of the forests had been able to regenerate undisturbed, with some approaching 40 years in 2011, and had been cleared mainly on one (65%) occasion (Figure 6C). At Santarém, 88% of the area under regrowth had been used actively for ≤2 years prior to the last abandonment and the majority (53%) cleared only once. At Machadinho d'Oeste, the PALU was typically ≤2 years (75%), and again, forests were relatively young because most of the clearance occurred only on one (57%) occasion. Hence, different typologies of SF were observed at each site.

As a result of combining the 3-class PALU and FC maps for the last date of the time-series, a proxy for land use intensity in each site was generated (Table 11). At Manaus, of the 65% of land occupied by SF in 2011 that was subjected to only one clearing event, 41% had a short PALU but a significant proportion (16%

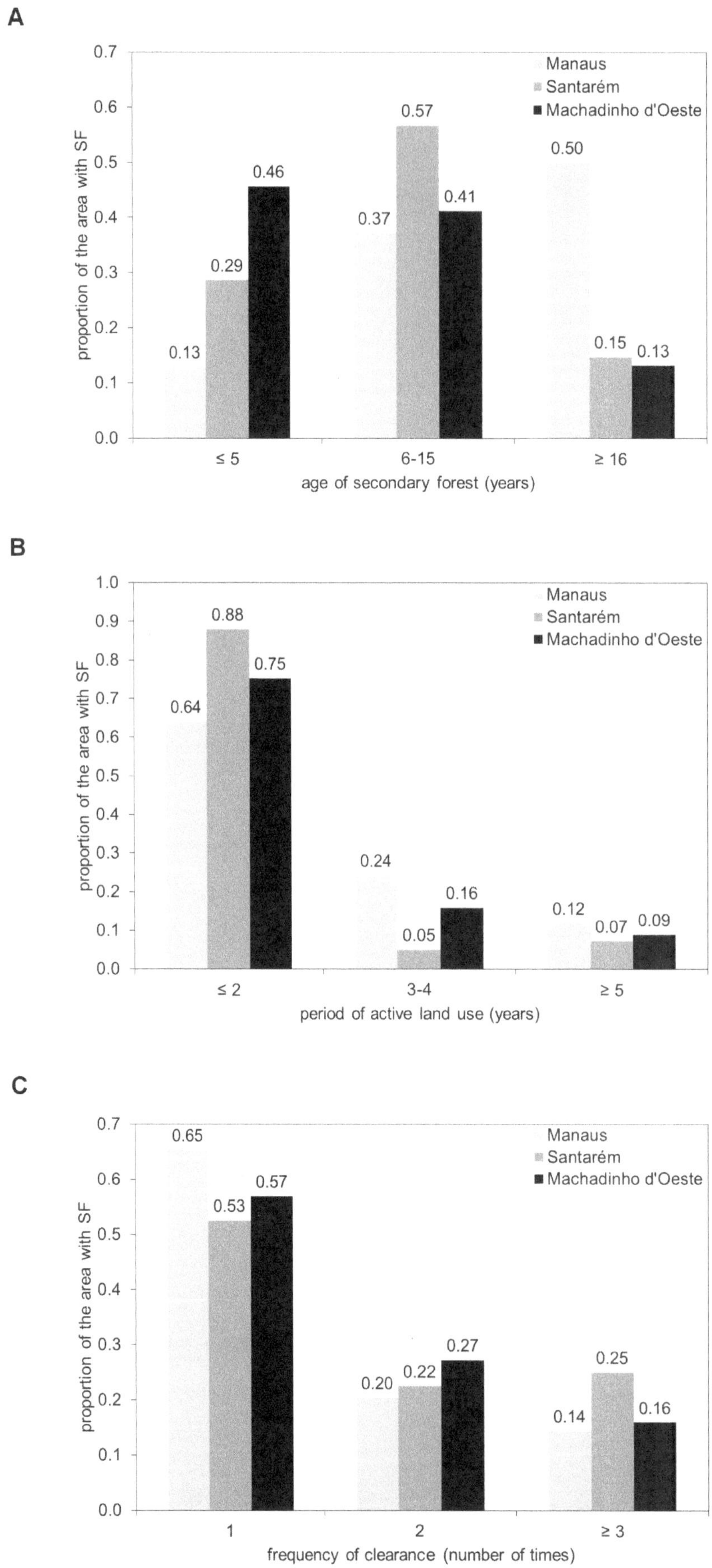
A
0.7
0.6
0.5
0.4
0.3
0.2
0.1
0.0
proportion of the area with SF
Manaus
Santarém
Machadinho d'Oeste
0.13
0.29
0.46
0.37
0.57
0.41
0.50
0.15
0.13
≤ 5
6-15
≥ 16
age of secondary forest (years)
B
1.0
0.9
0.8
0.7
0.6
0.5
0.4
0.3
0.2
0.1
0.0
proportion of the area with SF
Manaus
Santarém
Machadinho d'Oeste
0.64
0.88
0.75
0.24
0.05
0.16
0.12
0.07
0.09
≤ 2
3-4
≥ 5
period of active land use (years)
C
0.7
0.6
0.5
0.4
0.3
0.2
0.1
0.0
proportion of the area with SF
Manaus
Santarém
Machadinho d'Oeste
0.65
0.53
0.57
0.20
0.22
0.27
0.14
0.25
0.16
1
2
≥ 3
frequency of clearance (number of times)

Figure 6. Classes of age of secondary forest, period of active land use, and frequency of clearance for the three sites. Classes of A) age of secondary forest (ASF), B) period of active land use (PALU), and C) frequency of clearance (FC) for the three sites, as a proportion of the area with secondary forest (SF) in the last date of the time-series. The proportion of the area with SF in the first class of PALU (≤2 years) in the Santarém site includes a small proportion (1.5%) corresponding to areas of SF persisting in all dates of the time-series. The proportion of the area with SF in the first class of FC (1x) in Manaus, Santarém and Machadinho d'Oeste includes a small percentage (0.1%, 1.6%, 0.1%, respectively) corresponding to those cases where non-forest (NF) was observed in the first date of the time-series and no clearance has occurred (i.e., the only transition was from NF to SF).

and 9%) had a medium and long PALU respectively. Santarém is paradigmatic, as almost all land with SF in 2010 that was cleared only once was subjected to a short PALU. Assuming that a high FC and short and medium PALU will represent a higher land use intensity, then Manaus had 13% of the area undergoing SF in 2011 in these conditions, while Santarém and Machadinho d'Oeste had 20% and 15% in 2010 and 2011 respectively.

To investigate whether these typologies were dependent on the last date of the time-series, the same analysis was performed for all the dates in the time-series. At Manaus, the proportion of the area with SF in the ≤5 year ASF class decreased irregularly with the onset of regeneration up to approximately 15% in the last date of the time-series (Figure 7A), with this pattern being mirrored for the 6–15 year ASF class, and the ≥16 year ASF class steadily increasing up to around 50% in 2011. At Santarém (Figure 7B), the proportion of the area with SF in the initial ASF class decreased to around 30% in 2010, again mirrored by an overall increase to about 60% in the 6–15 year ASF class. As with Manaus, the proportion of the area with SF in the advanced ASF class increased from 15 years following the onset of regeneration to cover approximately 15% in 2010. In Machadinho d'Oeste (Figure 7C), the proportion of SF in the initial ASF class systematically decreased from the onset of regeneration leveling at around 45% in 2011, and an opposite pattern was observed in the intermediate ASF class, with a systematic increase since the onset of regeneration up to approximately 40% in 2011. As in Manaus and Santarém, a steady increase in the proportion of area with regeneration in the advanced ASF class was observed, from 17 years after the onset of regeneration up to around 15% in 2011. By combining the initial and intermediate ASF classes, a more consistent but asymptotic increase in the area of SF was observed, with this compensating for the interplay between these ASF classes, which occurs because of reclearance. The general trend across all sites is a decrease in the proportion of the area with SF in the initial ASF class up to the present, with a simultaneous increase in the proportion of the intermediate and advanced ASF classes. No asymptotic pattern is visible in the relationship between the proportion of the area with SF and the number of years following the onset of regeneration for this variable in the three selected sites.

At Manaus, the proportion of areas with a short PALU was initially high (100%) but, over time, this area decreased to about 40% but then increased subsequently to 60%, where it remained constant 32 years from the onset of regeneration (Figure 8A). This decrease was associated with an increase in the area with a medium and long PALU (14–16 years after the onset of SF regeneration). At Santarém (Figure 8B), the proportion with short PALU remained high, with this suggesting clearance of SF was common practice. The areas with medium and long PALU remained relatively low (<5–10%) but increased slightly towards the end of the time-series. At Machadinho d'Oeste, the temporal pattern was similar to that observed at Manaus; after 12–14 years, the proportion of short, medium and long PALU stabilized around 75%, 15% and 10% respectively.

The FC over areas undergoing SF also varied between sites (Figure 9). At Manaus, the SF had been cleared only once until about 14 years after the onset of SF, but this proportion decreased, stabilizing at approximately 65% in 2011. This occurred because SF began to be cleared again 15 years after the onset of regeneration and again after 19 years (i.e., 3 FC events), with these areas stabilizing at approximately 20% and 15% respectively. At

Table 11. Percentage of area with SF in the last date of the time-series, per combination of PALU and FC classes.

PALU (classes)	FC (classes)			Total
Manaus	**Low (1x)**	**Medium (2x)**	**High (≥3x)**	
Short (≤2 years)	41	13	10	64
Medium (3–4 years)	16	5	3	24
Long (≥5 years)	9	2	1	12
Total	66	20	14	
Santarém	**Low (1x)**	**Medium (2x)**	**High (≥3x)**	
Short (≤2 years)	51	20	17	88
Medium (3–4 years)	1	1	3	5
Long (≥5 years)	0	2	5	7
Total	52	23	25	
Machadinho d'Oeste	**Low (1x)**	**Medium (2x)**	**High (≥3x)**	
Short (≤2 years)	48	17	11	75
Medium (3–4 years)	5	7	4	16
Long (≥5 years)	4	4	1	9
Total	57	28	16	

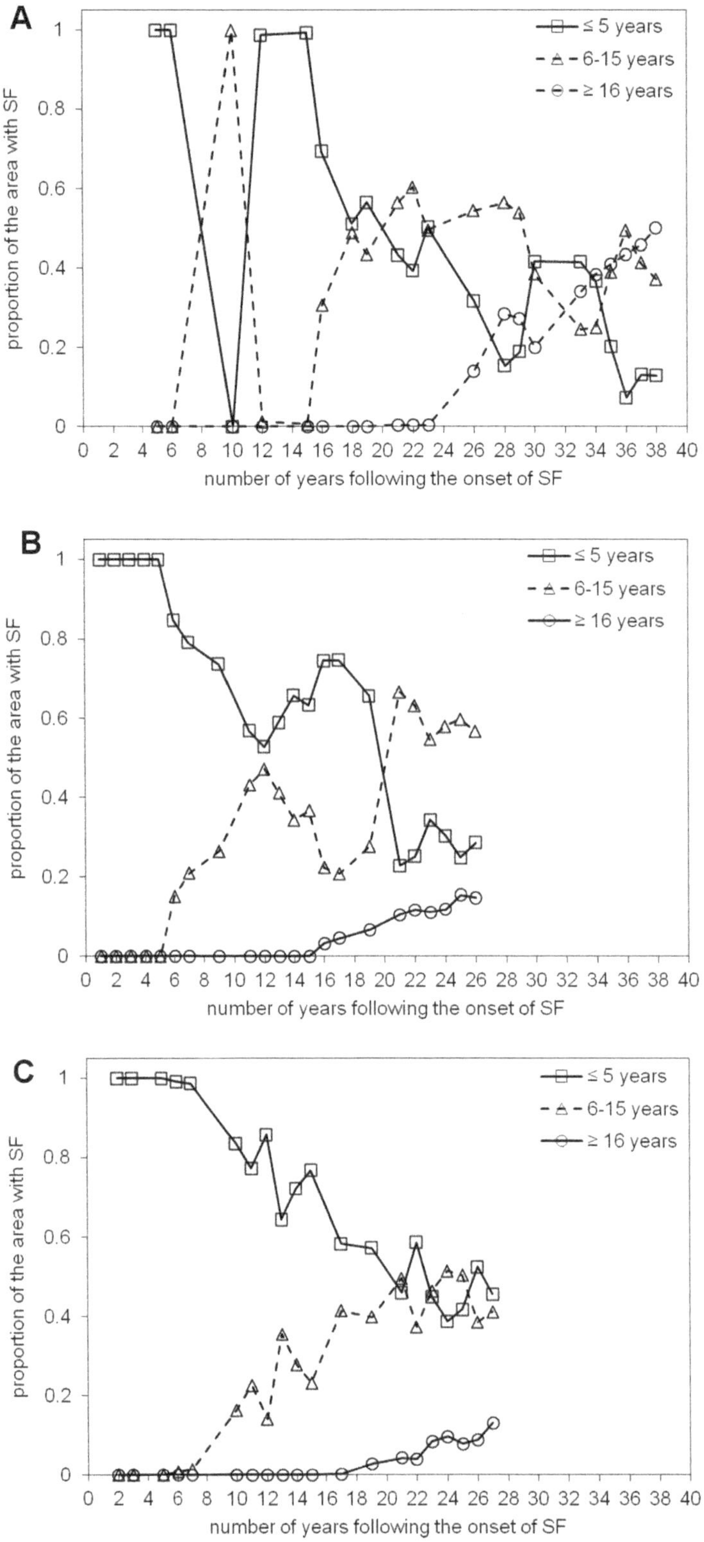

Figure 7. Proportion of the area with secondary forest (SF) as a function of the number of years following the onset of SF for the three classes of age of secondary forest. Proportion of the area with secondary forest (SF) as a function of the number of years following the onset of SF for the three classes of age of secondary forest (ASF) in A) Manaus, B) Santarém, and C) Machadinho d'Oeste: initial, ≤5 years; intermediate, 6–15 years; advanced, ≥16 years.

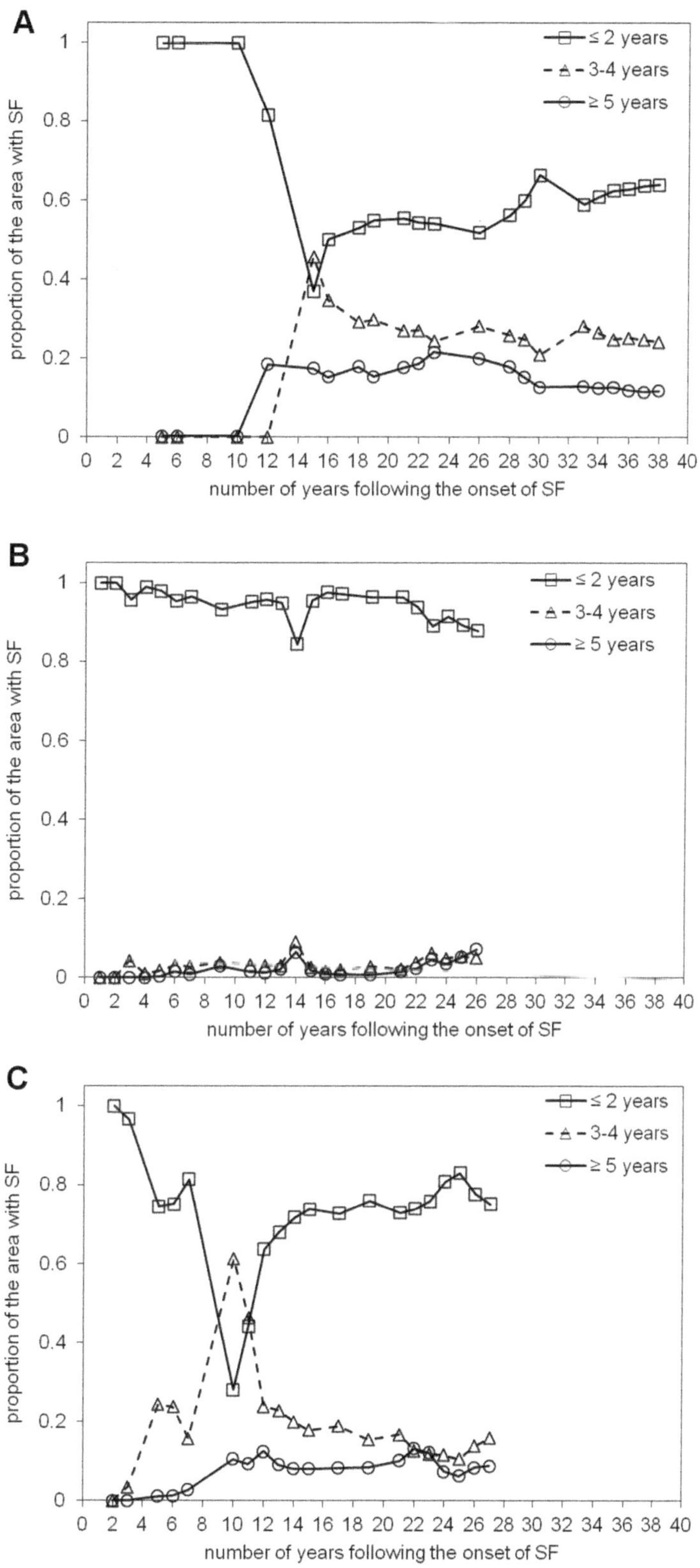

Figure 8. Proportion of the area with secondary forest (SF) as a function of the number of years following the onset of SF for the three classes of period of active land use. Proportion of the area with secondary forest (SF) as a function of the number of years following the onset of SF for the three classes of period of active land use (PALU) in A) Manaus, B) Santarém, and C) Machadinho d'Oeste: short, ≤2 years; medium, 3–4 years; long, ≥5 years.

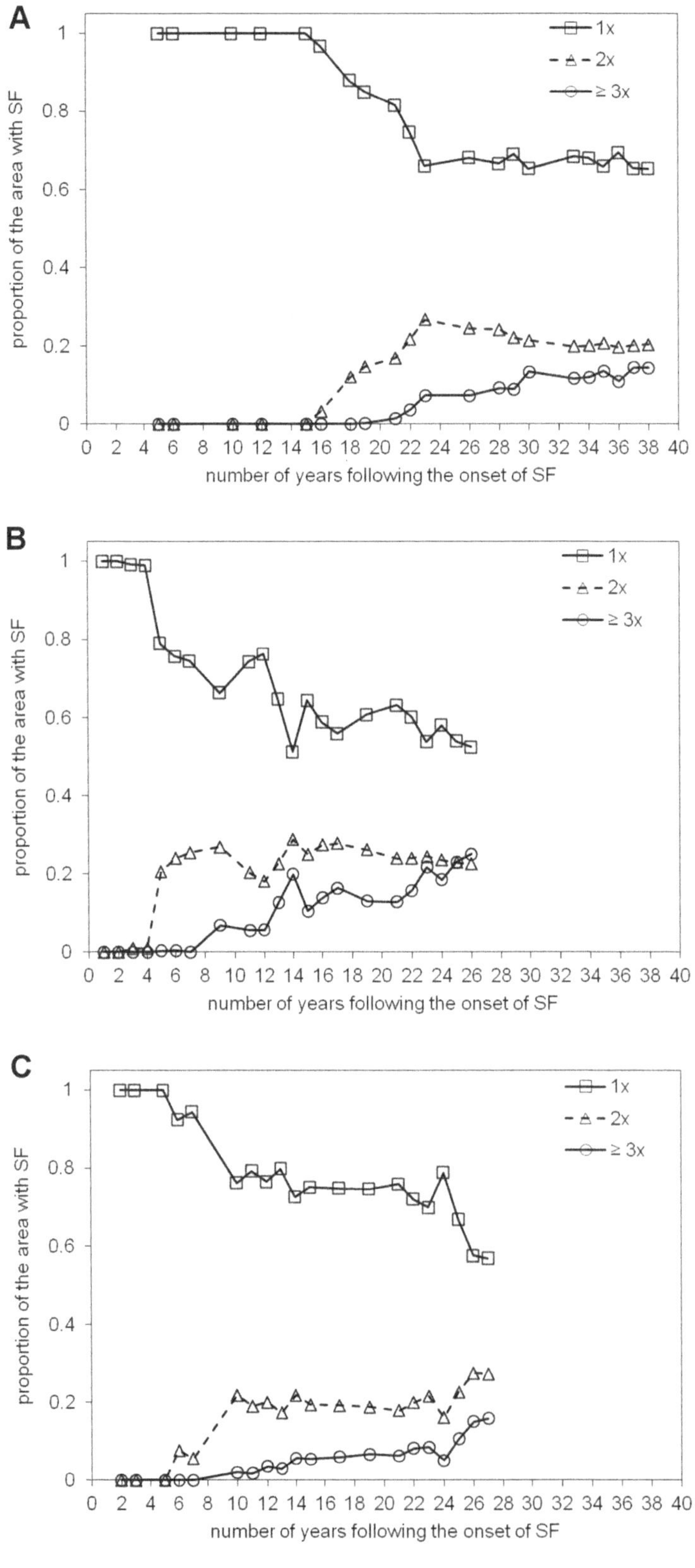

Figure 9. Proportion of the area with secondary forest (SF) as a function of the number of years following the onset of SF for the three classes of frequency of clearance. Proportion of the area with secondary forest (SF) as a function of the number of years following the onset of SF for the three classes of frequency of clearance (FC) in A) Manaus, B) Santarém, and C) Machadinho d'Oeste: low, 1 time; medium, 2 times; high, ≥3 times.

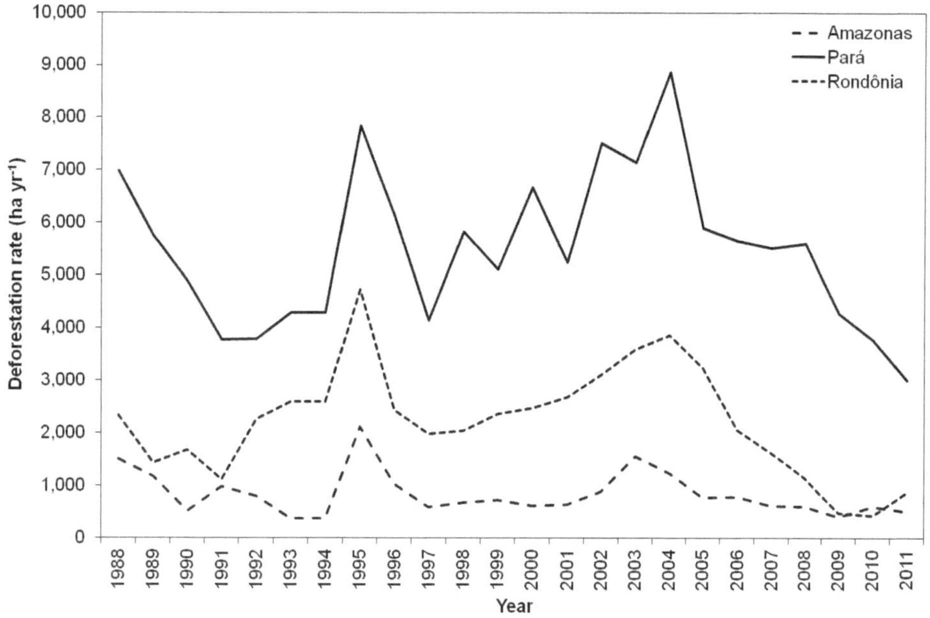

Figure 10. Deforestation rates in the Amazonas, Pará and Rondônia states according to INPE, 2013 [13]. Deforestation rates since 1988 (ha yr^{-1}) in the Amazonas, Pará and Rondônia states according to the data reported by INPE, 2013 [13].

Santarém, SF was cleared within 2–4 years from the onset of regeneration and continually thereafter. This was reflected in the increase in the proportion of SF areas cleared twice (after 4 years) and three times (after 7 years) with these both stabilizing at approximately 20%. At Machadinho d'Oeste, the proportion of areas with only a clearing event decreased to 55% 28 years after the onset of regeneration. The medium and high FC classes commenced after 4 years and 7 years respectively with the proportions of these remaining relatively constant (25% and 15%).

Discussion

6.1. Accuracy assessment of the land cover time-series

Several studies carried out in the Amazon have acknowledged some issues in discriminating SF from MF and, sometimes, NF. Lucas et al., 2000 [18] reported omission and commission errors ranging from 55–75% and 51–76%, respectively, associated with a greater confusion between NF and SF rather than between MF and SF. Carreiras et al., 2006 [20] reported an omission and commission error in the SF class of 66% and 43%, respectively, due to confusion between the SF, MF and NF classes. These authors concluded that misclassification among SF, MF, and NF was understandable, since from a spectral standpoint, SF is a transitional class between NF (i.e., agriculture/pasture) and MF. The initial stages of regeneration are spectrally similar to agriculture/pasture. Conversely, final stages of regrowth are more related to MF. The values reported in the two previous studies are much higher than those from this study, and that could be associated with the spatial resolution of the datasets that were used: ~1-km SPOT 4 VEGETATION and NOAA AVHRR vs. the 30-m spatial resolution Landsat TM and ETM+ data. The spatial arrangement of SF in the BLA, dominated by small previously abandoned areas, especially in Rondônia (e.g., [52]), creates an additional difficulty when trying to discriminate these from the surrounding land cover classes with coarse spatial resolution data. Kimes et al., 1999 [53] used 20-m spatial resolution SPOT High Resolution Visible (HRV) data between 1986 and 1994 to discriminate MF, NF and SF in Rondônia and reported an accuracy of 95% in the SF class with misclassification happening mostly with the MF class. Metzger, 2002 [54] used Landsat TM data over the Bragantina region (Pará state) in 1996 to map various land cover classes and reported omission and commission errors in SF age classes ranging from 8–11% and 7–10%, respectively, with misclassification occurring mainly from confusion with MF. Vieira et al., 2003 [55] used Landsat 7 ETM+ data to map different stages of SF in São Francisco do Pará (northeastern Pará state) and reported omission and commission errors between 17–25% and 22–40% respectively, due mainly to misclassification among SF stages. Kuplich, 2006 [56] used Landsat TM and Synthetic Aperture Radar (SAR) data to discriminate various land cover classes in Manaus (included the BDFFP) and reported omission and commission errors in the SF age classes of 10–99% and 32–80% respectively, but those errors decreased when the SF age classes were aggregated into a unique class (78% and 25%, respectively). Although errors are still high in some studies (e.g., [18,20,56]), others have been able to discriminate SF from MF and NF with a high degree of accuracy (e.g., [53–55]).

Prates-Clark et al., 2009 [38] validated the 1995 and the 2001 land cover maps over Manaus and Santarém respectively using field-based reference data. For the validation of the 1995 land cover map over Manaus, field data collected in 1995 were used [22,42] and the authors reported an overall accuracy of ~99%, with omission errors in the MF, NF and SF classes of 0.3%, 3.5% and 1.9%, respectively, and commission errors of 0.5%, 7.0% and 9.5%, respectively. The major source of error in the SF class (higher commission error) was misclassification as MF class. At Santarém, field data collected in 2002 [57] were used to validate the 2001 land cover map. The overall accuracy was 88%, the omission errors were 2.2%, 1.1% and 4.2% for the MF, NF and SF classes respectively, and the commission errors were 0.6%, 28.8% and 71.9% respectively. The particularly high commission error in the SF class was essentially the consequence of misclassification as MF. The NF class displayed a high commission error that was related to misclassification as both MF and SF. Higher errors indicated in Prates-Clark et al., 2009 [38] for the Santarém site were also depicted in our accuracy assessment of the 2010 land cover map, although of much lower magnitude.

A major limitation of the accuracy assessment carried out in this study it that only one date (classification) of each time-series per study area was validated with very high spatial resolution data. Nevertheless, a 10-fold cross-validation procedure was implemented for those dates classified with the random forests algorithm, i.e., the 21 Landsat TM images of the Machadinho d'Oeste time-series (1984–2011) and the post 2003 Santarém Landsat TM images (2005–2010). In Machadinho d'Oeste, the average overall accuracy was ~98%, the omission errors of the MF, NF and SF classes ranged from 0.0–4.4%, 0.0–2.1% and 0.7–7.6% respectively, and the commission errors were between 0.0–4.2%, 0.0–0.3% and 1.7–7.1% respectively. At Santarém, the 2005–2010 time-series of classified Landsat TM data resulted in an average overall accuracy of 95%, the omission errors of the MF, NF and SF classes ranged between 1.8–6.7%, 0.0–0.9% and 8.1–16.7%, respectively, and the commission errors were between 3.7–7.5%, 0.0–0.6% and 4.1–13.2% respectively. Both the omission and commission errors of the NF class were very low, thus indicating that the major source of error of the SF class was misclassification as MF.

6.2. Differences of land cover change patterns among sites

Extensive deforestation across the BLA started in the 1970s and was concentrated along the southern and eastern rims of the region [58]. The main deforestation drivers were conversion of forest to cropland and/or cattle ranching, which were carried out both by small farmers and large landholders [59]. According to the data provided by INPE, 2013 [13], the states where the three sites are located displayed different deforestation rates (Figure 10). Deforestation rates were always higher in Pará, followed by Rondônia and Amazonas. However, since 2003–2005, deforestation rates have been decreasing, with Rondônia experiencing a rapid decrease leading to values closer to those observed in the Amazonas state. Several factors contributed to this decline, namely the improvement of market-driven environmental governance [60]. Pará experienced the highest number of settled families by agrarian reform in the nine states composing the BLA (~31,000 settled families per year in the period 2003–2006), and the number has been increasing since the 1960s [61]. Although the vast majority of deforestation in the BLA can be tracked to large landholders occupying the land for cattle ranching, the implementation of planned settlements is not negligible at all, especially in Rondônia, which is known by its small farmers' radial, fishbone and watershed deforestation patterns (e.g., [58]). For example, up to the mid-1990s, the number of settled families in Rondônia was only second to Pará, with 1,423 settled families per year against 1,462 settled families per year in Pará [61].

The three study areas have different socio-economic drivers leading to deforestation, different land uses and subsequent regeneration. The Manaus site is mostly included in the Manaus municipality, although 24% is also included in the Rio Preto da Eva municipality. This study area basically encompasses several environmental protection areas. As such, it seems that these conservation areas have been critical at preserving most of the forest in the region. The conservation units (CUs) limits accessed from the Brazilian Ministry of the Environment (MMA, available at http://mapas.mma.gov.br/i3geo/datadownload.htm) were used to estimate the ratio of relative incidence of deforestation (NF and SF) outside and inside the CUs in the last date of the time-series for each site. The three sites had a significant proportion of the area included in the CUs: 43% in Manaus, 37% in Santarém and 23% in Machadinho d'Oeste. In Manaus, the relative incidence of deforestation outside the CUs was twice that inside CUs and in Santarém and Machadinho d'Oeste that value increased to five times. In fact, several studies have concluded that protected areas in the Amazon region are indeed effective at reducing deforestation rates (e.g., [60,62,63]). Figure 2A illustrates the lower amount of land cover change in Manaus, with approximately 20% of MF lost between 1973 and 2011, when compared with Santarém (more than 30% of MF lost between 1984 and 2010) and Machadinho d'Oeste (~40% of MF lost between 1984 and 2011). Deforestation in the Manaus study area started in the 1970s, with the construction of the BR-174 highway connecting Manaus and Boa Vista. Early deforestation is traceable along this highway, but many areas were abandoned as a result of poor soil fertility; indeed, many small farmers moved to Roraima, which was considered to have younger fertile soils [64]. The area of forest in 2011 was approximately 69% and 81% of the territory of the Manaus and Rio Preto da Eva municipalities respectively [13]. Figure 2A shows that the proportion of MF in the Manaus site decreased from approximately 90% in 1973 to 71% in 2011, which is similar to the value reported by INPE, 2013 [13] for the Manaus municipality.

The Santarém site is located south of the city of Santarém, in the Belterra municipality, along the BR-163 highway, and including part of the Tapajós National Forest. This highway links Santarém and Cuiabá in the south. According to Scatena et al., 1996 [65] and Brondizio and Moran, 2012 [66], this region has a tradition of agriculture and agroforestry since the early 1900s, which has further escalated as a consequence of the construction of the BR-163 in the 1970s. In fact, of the three sites Santarém had the lowest proportion of MF at the beginning of the time-series (~80%) and the highest proportion of SF (~20%), thus suggesting that deforestation was commonplace in the area in the early 1980s. According to INPE, 2013 [13], approximately 68% of the forest area in the Belterra municipality was remaining in 2011, whereas we report a proportion of MF in 2010 of 46%. A vast majority of the Belterra municipality is included in the Tapajós National Forest, mainly the part that is west of the BR-163 highway. So, it is not surprising that the proportion of remaining forest in this municipality is higher than the proportion of remaining forest in the study area that includes the fraction east of the BR-163 highway that is not a protected area.

The Machadinho d'Oeste site (Rondônia state) is included mostly in the Machadinho d'Oeste municipality, although 20% overlaps the Vale do Anari municipality. Of the three sites, this was the one with the highest degree of land cover change, with a reduction of ~60% in the MF cover from 1984 (91%) to 2011 (32%). Deforestation in this region started in the 1980s with the implementation of settlements by INCRA, of which Machadinho d'Oeste and Vale do Anari are just two examples [33]. According to Batistella and Moran, 2005 [33], the proportion of MF in 1988 and 1998 in Machadinho d'Oeste was around 80% and 66% respectively, while our estimates were 82% in 1987 (1988 was not mapped) and 55% in 1998. INPE, 2013 [13] reports an area of MF in the Machadinho d'Oeste municipality in 2011 of approximately 61%, and our study estimated the percentage of MF in the Machadinho d'Oeste study area as 32%. Google Earth data from 2013 shows that the areas of the Machadinho d'Oeste municipality to the east and north of the study area have much less deforestation and so this is a possible explanation for the higher proportion of MF detected by Batistella and Moran, 2005 [33] in 1998 and INPE, 2013 [13] up to 2011. Also, Batistella and Moran,

2005 [33] identified the land cover types occurring in the deforested areas of Machadinho d'Oeste. In 1988, approximately 2% was covered by SF, the value increasing to approximately 13% in 1998; our study mapped SF in 1987 and 1998 and the proportion was ~13% and ~29% respectively. Again, when considering the entire Machadinho d'Oeste municipality, there are extensive areas to the north and east of the study area that remained intact, with these covered by MF. Nevertheless, there is a coincident temporal trend that is the increase of SF from 1988 to 1998.

6.3. Land abandonment and the emergence of secondary forest

Land abandonment in the BLA is a consequence of a range of factors that have changed across the last decades since the inception of large scale deforestation in the region. Namely these are; reduced crop productivity as a consequence of poor soil fertility; lack of financial incentives, migratory patterns; non-traditional land uses and market fluctuations [67]. At Manaus, in 2011, land abandonment resulted in SF that was comprised mainly of areas with advanced ASF (50% ≥16 years), although large areas of intermediate ASF (37% between 6–15 years) were also present (Figure 3A and Figure 6). This suggests that the conservation areas were not only effective at preserving MF but also SF areas. As mentioned before, deforested areas in this study area were abandoned mainly because of poor soil fertility, with settlers moving to the nearby state of Roraima [64]. It is clear from Figure 3A that the higher proportion of advanced ASF in the study area is in the northern half of the site. These areas undergoing regeneration in 2011 had mostly short PALU (64%) as a consequence of intensification of land use and requirement for new land, and 65% was deforested only once. In comparison, the Santarém site, which included parts of the Tapajós National Forest, had regrowth dominated by intermediate ASF in 2010. Those areas experienced essentially short PALU (88% ≤2 years) and low FC (52% deforested once). According to Brondizio and Moran, 2012 [66], land abandonment in the region, and associated regeneration, was high until 1999. At this point, large-scale soybean cultivation started, mainly because of the decay of primary and secondary roads, lack of social services, and limited access to water. On the other hand, land abandonment in Machadinho d'Oeste resulted in SF that was mainly in the initial (46%) and intermediate (41%) ASF classes, with these areas experiencing mostly (75%) short PALU and low FC (57%). This was a planned settlement and most of the deforested area is under agriculture and or pasture use, suggesting that a vast majority of the area undergoing SF could be indeed forest fallow that might be under cattle ranching use or being subjected to subsistence agriculture in a crop/fallow cycle [66].

One important aspect worth mentioning is the large proportion of SF in the Santarém site that resulted from the 1997 wildfires. This area of regeneration was not the consequence of land abandonment but a natural process following fire-induced disturbance. This wildfire was just one among many others that occurred in the BLA as a consequence of the 1997–1998 *El Niño* Southern Oscillation (ENSO) event (e.g., [68]). In fact, the proportion of NF in 1997 (prior to the wildfire) was around 6% and in 1998 had increased to approximately 15%, mainly as a consequence of that large wildfire, perfectly visible in Figure 3B as the orange irregular zone of intermediate ASF in the centre of the study area. As a consequence of the natural regeneration process, the SF proportion increased to approximately 44% in 2000 (~32% in 1997). Therefore, approximately 20% of the area of SF in 2010 was estimated as resulting from natural regeneration following that wildfire event.

6.4. Implications of prior land use for biomass accumulation and biodiversity restoration in secondary forests

Regenerating forests in the three selected sites have been classified in terms of ASF, PALU prior to abandonment and FC. Prior LULCC practices have a fundamental influence on the vegetation regenerating following land abandonment in the Amazon [10,22,69–75]. Vegetation structure, species composition and dominance are just some of the parameters of the regenerating vegetation that have been studied to identify the impact of prior land use practices (e.g., [76]). As such, these have an impact on the capability of these forests to accumulate biomass (and to act as a C sink at a greater or lesser extent) and to restore biodiversity (e.g., [77]).

Uhl et al., 1988 [71] studied areas undergoing secondary succession following pasture abandonment in Pará (Paragominas municipality) and identified major patterns related to prior land use intensity; light use sites, characterized by a lower PALU (0–4 years) and with no reclearance or just one slash-and-burn episode, had high species richness dominated by pioneer species (e.g., *Cecropia* sp. and *Solanum* sp.), and a ~13 m closed canopy with an AGB accumulation rate of 10 Mg ha^{-1} yr^{-1} (8-year old sites); comparatively, moderate use sites, having higher PALU (6–12 years) and with 1 to 5 clearing episodes, had lower species richness dominated by a 7–8 m *Vismia* sp. partially developed canopy and an AGB accumulation rate of 5 Mg ha^{-1} yr^{-1}; heavy use sites, subjected to large-scale mechanized operations, had poor species richness (*Solanum* sp. and *Cecropia* sp.), mainly composed by scattered trees and an AGB accumulation rate of only 0.6 Mg ha^{-1} yr^{-1}. Mesquita et al., 2001 [75] studied regeneration pathways in a region north of Manaus that is included in our Manaus site (BDFFP) and identified major differences related to prior land use. Basically, areas that were deforested and abandoned immediately were dominated by *Cecropia* spp. trees, whereas those that were abandoned after some years of pasture use were dominated by *Vismia* spp. trees. Lucas et al., 2002 [22], also identified two distinct regeneration pathways in a region north of Manaus and in the same way as Mesquita et al. [75]; the authors identified *Vismia* spp. dominated regeneration associated to sites with more intensive land use and those dominated by *Cecropia* spp. to less intensive land use. According to Chazdon et al., 2007 [74], in terms of AGB accumulation, tropical secondary forests are characterized by a rapid growth rates in the first years after land (agriculture or pasture) abandonment. Furthermore, Zarin et al., 2001 [78] showed that there is a strong correlation between the AGB of secondary forests and the number of years following abandonment (i.e., ASF), soil texture and climate data in the Amazon region. On this pretext, the secondary forests in Manaus have the potential to accumulate more AGB, followed by Santarém and Machadinho d'Oeste. In 2011 (the last date of the time-series), Machadinho d'Oeste had approximately 45% of the area of SF with less than five years of age, and around 40% with less than three years of age. Machadinho d'Oeste was initially a settlement project implemented by the Brazilian government and today most of its people are dependent on subsistence agriculture [32]. Therefore, it is possible that some of the area mapped as SF was indeed forest fallow, part of a crop/fallow system of subsistence agriculture, which have been acknowledged by several studies (e.g., [79,80]).

Concluding Remarks

An accurate assessment of large-scale land use and land cover change dynamics over remote areas (e.g., Amazon) can only be carried out with an adequate monitoring system. Remote sensing data, in this case annual or quasi-annual time-series of high spatial resolution optical data (Landsat program), has proven its ability to accurately discriminate MF, SF and NF over three sites in the BLA experiencing several decades of deforestation. Lower deforestation rates and a greater proportion of intermediate and advanced ASF classes were characteristic of the Manaus site. On the other hand, Machadinho d'Oeste had the highest deforestation rates and lower regrowth rates, with most of its regeneration occurring in areas abandoned over the past 5 years. LULCC at Santarém displayed an intermediate behavior, with intermediate deforestation rates and the highest regrowth rates, which lead to regeneration dominated by areas of intermediate ASF. Conservation units were effective at reducing deforestation at all three sites. The temporal evolution of the spatial arrangement of the various parcels of land identified as one of the three classes could provide new insights about the fragmentation patterns in the region (Ewers et al. 2013 [81]).

Although several generalizations about the type and composition of secondary forest occurring in each of the three sites can be made, a correct assessment can only be made after an *in situ* assessment. Therefore, the research presented here will provide the opportunity to assess the influence of previous LULCC dynamics on the biomass accumulation and biodiversity restoration and to further investigate hypothesis related to differences among the three sites. A field campaign is foreseen for 2014 to collect forest inventory data that will be used to study the influence of prior LULCC on the capability of regrowth areas to accumulate AGB and to restore biodiversity. In the end, a framework related to the influence of prior LULCC on the biomass accumulation and biodiversity will be built and could be used to inform land management policies in the region.

Acknowledgments

The United States Geological Survey (USGS) and Brazil's National Institute for Space Research (INPE) are particularly thanked for providing free access to the Landsat sensor data used in this study. Nathan Thomas (Department of Geography and Earth Sciences, Aberystwyth University, Aberystwyth, United Kingdom) is particularly acknowledged for reviewing earlier versions of the manuscript.

Author Contributions

Conceived and designed the experiments: JC RL. Performed the experiments: JC JJ RL CG. Analyzed the data: JC JJ RL. Contributed reagents/materials/analysis tools: JC JJ RL CG. Wrote the paper: JC JJ RL.

References

1. Falkowski P, Scholes RJ, Boyle E, Canadell J, Canfield D, et al. (2000) The global carbon cycle: A test of our knowledge of earth as a system. Science 290: 291–296.
2. Raupach M (2011) Pinning down the land carbon sink. Nature Climate Change 1: 148–149.
3. Heimann M, Reichstein M (2008) Terrestrial ecosystem carbon dynamics and climate feedbacks. Nature 451: 289–292.
4. Pan Y, Birdsey R, Fang J, Houghton R, Kauppi P, et al. (2011) A Large and Persistent Carbon Sink in the World's Forests. Science 333: 988–993.
5. Gibson L, Lee T, Koh L, Brook B, Gardner T, et al. (2011) Primary forests are irreplaceable for sustaining tropical biodiversity. Nature 478: 378–381.
6. Achard F, Eva H, Stibig H, Mayaux P, Gallego J, et al. (2002) Determination of deforestation rates of the world's humid tropical forests. Science 297: 999–1002.
7. Hooper D, Adair E, Cardinale B, Byrnes J, Hungate B, et al. (2012) A global synthesis reveals biodiversity loss as a major driver of ecosystem change. Nature 486: 105–108.
8. Lederer M (2011) From CDM to REDD+ – What do we know for setting up effective and legitimate carbon governance? Ecological Economics 70: 1900–1907.
9. Olson D, Dinerstein E, Wikramanayake E, Burgess N, Powell G, et al. (2001) Terrestrial ecoregions of the worlds: A new map of life on Earth. Bioscience 51: 933–938.
10. Chazdon RL (2003) Tropical forest recovery: legacies of human impact and natural disturbances. Perspectives in Plant Ecology Evolution and Systematics 6: 51–71.
11. Cardinale B, Duffy J, Gonzalez A, Hooper D, Perrings C, et al. (2012) Biodiversity loss and its impact on humanity. Nature 486: 59–67.
12. Goulding M, Barthem R, Ferreira EJG (2003) The Smithsonian atlas of the Amazon. Washington, D.C.: Smithsonian Books. 253 p.
13. INPE (2013) Projeto PRODES: Monitoramento da Floresta Amazônica Brasileira por Satélite. Instituto Nacional de Pesquisas Espaciais. Available: http://www.obt.inpe.br/prodes/index.php. Accessed 11 March 2014.
14. Davidson EA, de Araujo AC, Artaxo P, Balch JK, Brown IF, et al. (2012) The Amazon basin in transition. Nature 481: 321–328.
15. Stone TA, Schlesinger P, Houghton RA, Woodwell GM (1994) A map of the vegetation of South America based on satellite imagery. Photogrammetric Engineering and Remote Sensing 60: 541–551.
16. Schroeder PE, Winjum JK (1995) Assessing Brazil carbon budget: II. Biotic fluxes and net carbon balance. Forest Ecology and Management 75: 87–99.
17. Fearnside P (1996) Amazonian deforestation and global warming: Carbon stocks in vegetation replacing Brazil's Amazon forest. Forest Ecology and Management 80: 21–34.
18. Lucas RM, Honzak M, Curran PJ, Foody GM, Milne R, et al. (2000) Mapping the regional extent of tropical forest regeneration stages in the Brazilian Legal Amazon using NOAA AVHRR data. International Journal of Remote Sensing 21: 2855–2881.
19. Cardille J, Foley J (2003) Agricultural land-use change in Brazilian Amazonia between 1980 and 1995: Evidence from integrated satellite and census data. Remote Sensing of Environment 87: 551–562.
20. Carreiras JMB, Pereira JMC, Campagnolo ML, Shimabukuro YE (2006) Assessing the extent of agriculture/pasture and secondary succession forest in the Brazilian Legal Amazon using SPOT VEGETATION data. Remote Sensing of Environment 101: 283–298.
21. Neeff T, Lucas RM, dos Santos JR, Brondizio ES, Freitas CC (2006) Area and age of secondary forests in Brazilian Amazonia 1978–2002: An empirical estimate. Ecosystems 9: 609–623.
22. Lucas RM, Honzak M, Amaral ID, Curran PJ, Foody GM (2002) Forest regeneration on abandoned clearances in central Amazonia. International Journal of Remote Sensing 23: 965–988.
23. Fearnside PM (2000) Global warming and tropical land-use change: Greenhouse gas emissions from biomass burning, decomposition and soils in forest conversion, shifting cultivation and secondary vegetation. Climatic Change 46: 115–158.
24. Houghton RA (2005) Aboveground forest biomass and the global carbon balance. Global Change Biology 11: 945–958.
25. Silva JMN, Carreiras JMB, Rosa I, Pereira JMC (2011) Greenhouse gas emissions from shifting cultivation in the tropics, including uncertainty and sensitivity analysis. Journal of Geophysical Research-Atmospheres 116. doi:10.1029/2011JD016056.
26. Laurance S, Laurance W, Andrade A, Fearnside P, Harms K, et al. (2010) Influence of soils and topography on Amazonian tree diversity: a landscape-scale study. Journal of Vegetation Science 21: 96–106.
27. Saatchi S, Houghton R, Alvala R, Soares J, Yu Y (2007) Distribution of aboveground live biomass in the Amazon basin. Global Change Biology 13: 816–837.
28. Laurance WF, Camargo JLC, Luizao RCC, Laurance SG, Pimm SL, et al. (2011) The fate of Amazonian forest fragments: A 32-year investigation. Biological Conservation 144: 56–67.
29. van Gardingen PR, Valle D, Thompson I (2006) Evaluation of yield regulation options for primary forest in Tapajos National Forest, Brazil. Forest Ecology and Management 231: 184–195.
30. Bacha CJC, Rodriguez LCE (2007) Profitability and social impacts of reduced impact logging in the Tapajos National Forest, Brazil - A case study. Ecological Economics 63: 70–77.

31. Espírito-Santo FDB, Shimabukuro YE, Aragão LEOC, Machado ELM (2005) Analysis of the floristic and phytosociologic composition of Tapajós national forest with geographic support of satellite images. Acta Amazonica 35: 155–173.
32. Miranda EE (2013) Sustentabilidade Agrícola na Amazonia - 23 anos de monitoramento da agricultura em Machadinho d'Oeste (RO). Embrapa Monitoramento por Satélite. Available: http://www.machadinho.cnpm.embrapa.br/. Accessed 11 March 2014.
33. Batistella M, Moran EF (2005) Dimensões humanas do uso e cobertura das terras na Amazônia: uma contribuição do LBA. Acta Amazonica 35: 239–247.
34. Silver WL, Neff J, McGroddy M, Veldkamp E, Keller M, et al. (2000) Effects of soil texture on belowground carbon and nutrient storage in a lowland Amazonian forest ecosystem. Ecosystems 3: 193–209.
35. Bierregaard RO (2001) Lessons from Amazonia: the ecology and conservation of a fragmented forest. New Haven, USA: Yale University Press. 478 p.
36. Laurance WF, Fearnside PM, Laurance SG, Delamonica P, Lovejoy TE, et al. (1999) Relationship between soils and Amazon forest biomass: A landscape-scale study. Forest Ecology and Management 118: 127–138.
37. Keller M, Varner R, Dias JD, Silva H, Crill P, et al. (2005) Soil-atmosphere exchange of nitrous oxide, nitric oxide, methane, and carbon dioxide in logged and undisturbed forest in the Tapajos National Forest, Brazil. Earth Interactions 9: 1–28.
38. Prates-Clark C, Lucas R, dos Santos J (2009) Implications of land-use history for forest regeneration in the Brazilian Amazon. Canadian Journal of Remote Sensing 35: 534–553.
39. Chander G, Markham B, Helder D (2009) Summary of current radiometric calibration coefficients for Landsat MSS, TM, ETM+, and EO-1 ALI sensors. Remote Sensing of Environment 113: 893–903.
40. Lillesand TM, Kiefer RW (2000) Remote sensing and image interpretation. New York, USA: John Wiley & Sons. 724 p.
41. Navulur K (2007) Multispectral image analysis using the object-oriented paradigm. Boca Raton, USA: CRC Press/Taylor & Francis. 165 p.
42. Lucas R, Xiao X, Hagen S, Frolking S (2002) Evaluating TERRA-1 MODIS data for discrimination of tropical secondary forest regeneration stages in the Brazilian Legal Amazon. Geophysical Research Letters 29: 42-1-42-4.
43. Hastie T, Tibshirani R, Friedman JH (2009) The elements of statistical learning: data mining, inference, and prediction. New York (USA): Springer. 745 p.
44. Roberts D, Numata I, Holmes K, Batista G, Krug T, et al. (2002) Large area mapping of land-cover change in Rondonia using multitemporal spectral mixture analysis and decision tree classifiers. Journal of Geophysical Research-Atmospheres 107: LBA 40-1-LBA 40-18.
45. Puyravaud J (2003) Standardizing the calculation of the annual rate of deforestation. Forest Ecology and Management 177: 593–596.
46. Pereira MC, Setzer AW (1993) Spectral characteristics of fire scars in Landsat-5 TM images of Amazonia. International Journal of Remote Sensing 14: 2061–2078.
47. Cochrane M, Souza C (1998) Linear mixture model classification of burned forests in the Eastern Amazon. International Journal of Remote Sensing 19: 3433–3440.
48. Foody GM (2009) Sample size determination for image classification accuracy assessment and comparison. International Journal of Remote Sensing 30: 5273–5291.
49. Gallego FJ (2004) Remote sensing and land cover area estimation. International Journal of Remote Sensing 25: 3019–3047.
50. Walsh TA, Burk TE (1993) Calibration of satellite classifications of land area. Remote Sensing of Environment 46: 281–290.
51. Tenenbein A (1972) A double sampling scheme for estimating from misclassified multinomial data with applications to sampling inspection. Technometrics 14: 187–202.
52. Ferraz SFD, Vettorazzi CA, Theobald DM, Ballester MVR (2005) Landscape dynamics of Amazonian deforestation between 1984 and 2002 in central Rondonia, Brazil: assessment and future scenarios. Forest Ecology and Management 204: 67–83.
53. Kimes DS, Nelson RF, Salas WA, Skole DL (1999) Mapping secondary tropical forest and forest age from SPOT HRV data. International Journal of Remote Sensing 20: 3625–3640.
54. Metzger JP (2002) Landscape dynamics and equilibrium in areas of slash-and-burn agriculture with short and long fallow period (Bragantina region, NE Brazilian Amazon). Landscape Ecology 17: 419–431.
55. Vieira ICG, de Almeida AS, Davidson EA, Stone TA, de Carvalho CJR, et al. (2003) Classifying successional forests using Landsat spectral properties and ecological characteristics in eastern Amazonia. Remote Sensing of Environment 87: 470–481.
56. Kuplich TM (2006) Classifying regenerating forest stages in Amazonia using remotely sensed images and a neural network. Forest Ecology and Management 234: 1–9.
57. Prates-Clark CC (2004) Remote sensing of tropical regenerating forests in the Brazilian Amazon. PhD dissertation. Aberystwyth, Wales, UK: University of Wales.
58. Fearnside PM (2005) Deforestation in Brazilian Amazonia: History, rates, and consequences. Conservation Biology 19: 680–688.
59. Kirby KR, Laurance WF, Albernaz AK, Schroth G, Fearnside PM, et al. (2006) The future of deforestation in the Brazilian Amazon. Futures 38: 432–453.
60. Nepstad DC, McGrath DG, Soares B (2011) Systemic Conservation, REDD, and the Future of the Amazon Basin. Conservation Biology 25: 1113–1116.
61. Pacheco P (2009) Agrarian Reform in the Brazilian Amazon: Its Implications for Land Distribution and Deforestation. World Development 37: 1337–1347.
62. Rosa IMD, Purves D, Souza C Jr, Ewers RM (2013) Predictive Modelling of Contagious Deforestation in the Brazilian Amazon. PLoS ONE 8(10): e77231. doi:10.1371/journal.pone.0077231.
63. Nolte C, Agrawal A, Silvius KM, Soares-Filho BS (2013) Governance regime and location influence avoided deforestation success of protected areas in the Brazilian Amazon. Proceedings of the National Academy of Sciences of the United States of America 110: 4956–4961.
64. Fearnside PM, Graca P (2006) BR-319: Brazil's manaus-porto velho highway and the potential impact of linking the arc of deforestation to central Amazonia. Environmental Management 38: 705–716.
65. Scatena FN, Walker RT, Homma AKO, deConto AJ, Ferreira CAP, et al. (1996) Cropping and fallowing sequences of small farms in the "terra firme" landscape of the Brazilian Amazon: A case study from Santarem, Para. Ecological Economics 18: 29–40.
66. Brondizio ES, Moran EF (2012) Level-dependent deforestation trajectories in the Brazilian Amazon from 1970 to 2001. Population and Environment 34: 69–85.
67. Perz SG, Skole DL (2003) Social determinants of secondary forests in the Brazilian Amazon. Social Science Research 32: 25–60.
68. Nepstad DC, Verissimo A, Alencar A, Nobre C, Lima E, et al. (1999) Large-scale impoverishment of Amazonian forests by logging and fire. Nature 398: 505–508.
69. Uhl C (1987) Factors controlling succession following slash-and-burn agriculture in Amazonia. Journal of Ecology 75: 377–407.
70. Buschbacher R, Uhl C, Serrao EAS (1988) Abandoned pastures in eastern Amazonia. II. Nutrient stocks in the soil and vegetation. Journal of Ecology 76: 682–699.
71. Uhl C, Buschbacher R, Serrao EAS (1988) Abandoned pastures in eastern Amazonia. I. Patterns of plant succession. Journal of Ecology 76: 663–681.
72. Saldarriaga JG, West DC, Tharp ML, Uhl C (1988) Long-term chronosequence of forest succession in the upper Rio Negro of Colombia and Venezuela. Journal of Ecology 76: 938–958.
73. Uhl C, Jordan CF (1984) Succession and nutrient dynamics following forest cutting and burning in Amazonia. Ecology 65: 1476–1490.
74. Chazdon RL, Letcher SG, van Breugel M, Martinez-Ramos M, Bongers F, et al. (2007) Rates of change in tree communities of secondary Neotropical forests following major disturbances. Philosophical Transactions of the Royal Society B-Biological Sciences 362: 273–289.
75. Mesquita R, Ickes K, Ganade G, Williamson G (2001) Alternative successional pathways in the Amazon Basin. Journal of Ecology 89: 528–537.
76. Uhl C, Clark K, Clark H, Murphy P (1981) Early plant succession after cutting and burning in the upper Rio Negro region of the Amazon basin. Journal of Ecology 69: 631–649.
77. Chazdon RL (2008) Beyond deforestation: Restoring forests and ecosystem services on degraded lands. Science 320: 1458–1460.
78. Zarin DJ, Ducey MJ, Tucker JM, Salas WA (2001) Potential biomass accumulation in Amazonian regrowth forests. Ecosystems 4: 658–668.
79. Brown S, Lugo A (1990) Tropical Secondary forests. Journal of Tropical Ecology 6: 1–32.
80. Lu DS, Moran E, Batistella M (2003) Linear mixture model applied to Amazonian vegetation classification. Remote Sensing of Environment 87: 456–469.
81. Ewers RM, Didham RK, Pearse WD, Lefebvre V, Rosa IMD, et al. (2013) Using landscape history to predict biodiversity patterns in fragmented landscapes. Ecology Letters 16: 1221–1233.

3

Inter-Annual Variation in Characteristics of Endozoochory by Wild Japanese Macaques

Yamato Tsuji*

Primate Research Institute, Kyoto University, Inuyama, Aichi, Japan

Abstract

Endozoochory is important to the dynamics and regeneration of forest ecosystems. Despite the universality of inter-annual variation in fruit production, few studies have addressed the sign (seed predation versus seed dispersal) and strength (frequency and quantity) of fruit-frugivore interaction and the effectiveness of endozoochory in response to the long-term temporal context. In this study I evaluated the characteristics of endozoochorous dispersal by wild Japanese macaques *Macaca fuscata* inhabiting deciduous forest in northern Japan for five different years. I collected 378 fecal samples from the macaques in fall (September to November) and quantified the proportion of feces containing seeds, number of seeds per fecal sample, ratio of intact seeds, and seed diversity. The proportion of feces containing seeds of any species (five-year mean: 85.9%, range: 78–97%) did not show significant inter-annual variation, while species-level proportions did. The intact ratio of seeds (mean: 83%, range: 61–98%) varied significantly both between years and between months, and this varied among dominant plant species. The number of seeds per fecal sample (mean: 78, range: 32–102) varied monthly but did not between years, and the seed diversity (mean: 0.66, range: 0.57–0.81) did not show significant inter-annual variation, both of which were attributed to longer duration of macaques' gastro-intestinal passage time of seeds exceed their feeding bouts. This study demonstrated that frequency and success of seed dispersal over seed predation of macaque endozoochory showed inter-annual variation, indicating low specificity across the seed–macaque network. The temporal variability in the quality of seed dispersal may provide evidence of high resilience in response to fluctuating environmental conditions in the temperate forests.

Editor: Cédric Sueur, Institut Pluridisciplinaire Hubert Curien, France

Funding: This study was supported by a Grant-in-Aid from the Japan Society for the Promotion of Science (No. 23780160) and the Corporative Research Fund of the Primate Research Institute, Kyoto University. The funders had no role in study design, data collection and analysis, decision to publish, or preparation of the manuscript.

Competing Interests: The author has declared that no competing interests exist.

* Email: ytsuji1002@gmail.com

Introduction

The context-dependence of species interactions is a key area of ecological studies [1–2]. At finer level interactions between a pair of species, temporal variations in the biotic and abiotic conditions can lead to the context-dependence of the strength and sign (i.e., its position along the mutualism-antagonism continuum) of the interactions [2–3]. To address the ecological and/or evolutionary meaning of the interactions, it makes sense to document these context-specific phenomena rather than to consider the average outcome over time [3].

Fruits, including fleshy fruits and nuts, are consumed by a wide assemblage of frugivores, especially mammals and birds [4–6], which are the principal seed dispersers of fleshy-fruited plant species in forest ecosystems [7]. In general, fruit production is not stable over time, but fluctuates not only seasonally but also inter-annually [8–12]. Large temporal changes in fruiting often cause corresponding long-term variation in the diet of frugivorous animals [13–16]. This in turn can affect the quantity and quality of seed dispersal, and hence the overall dispersal effectiveness over time [17]. However, despite worldwide research efforts on fruit–frugivore interactions [18–19], there have been few studies which addressed the variation in the sign (in terms of seed predation versus seed dispersal) and the strength (frequency and quantity) of fruit–frugivore interactions at the local level [17] but see [2]. To accurately evaluate the fruit–frugivore interaction and the effectiveness of endozoochory, studies covering several years with differing food environment conditions are necessary.

In this study I evaluated the relative importance of inter-annual variations in the sign and strength of fruit–frugivore interactions between fleshy-fruited plants and wild Japanese macaques, *Macaca fuscata* (Cercopithecidae, order Primates). Previous studies of this species have revealed that they disperse seeds of various plant species, especially in fall (September to November) through defecation [20–22], and they show inter-annual variation in the degree of fruit-eating [14,23]. In addition, it has been known that fruiting of many plant species on which the macaques feed (berries) show inter-annual variation [24–25]. I, therefore, expected that the macaques would modulate the quantitative and qualitative components of seed dispersal effectiveness via their feeding behavior at different times.

Specifically, I quantified the following four aspects of endozoochory: 1) the proportion of feces containing seeds, 2) the number of seeds per fecal sample, 3) the ratio of intact seeds within a single fecal sample, and 4) the seed species diversity within each single fecal sample in fall for five years (between 2000 and 2008). Based on the inter-annual variation in fruiting of many dietary plants in northern Japan [24–25], I predicted that the intact ratio, diversity, frequency of seed occurrence, and number of dispersed seeds of each fruit–macaque pair would not be stable over time and that there would be a low specificity across the fleshy-fruited plants–macaque network in the temperate forest.

Results

During the study period I collected 388 macaque fecal samples in total, from which I picked up 31,166 seeds from 15 different species (Table 1). I summarized characteristics of each species: life form, fruiting months, size of seeds and fruits, and handling technique by the macaques in Table 2.

Seed appearance ratios (*AR*s) for all species combined ranged from 78% (in 2008) to 97% (in 2005), and showed no significant variations among months and among years. An effect of interaction between year and month was not detected (Table 3). At the finer (i.e., species) level, however, I obtained significant inter-annual variations in the *AR*s; for example, in October, the *AR* of *Viburnum dilatatum* (lower in 2000), *Vitis flexuosa* (higher in 2005 and 2008), *Pourthiaea villosa* (higher in 2000 and 2007), *Malus tschonoskii* (higher in 2000 and 2008), and *Swida macrophylla* (lower in 2004 and 2005) showed significant inter-annual variations in the *AR*s (Fig. 1).

The mean number of seeds per single fecal sample (*SN*) ranged from 32 (in 2000) to 102 (in 2004) (Table 1). Mean (± SD) *SN* differed monthly (Table 3), and was the greatest in November (153±190), middle in October (50±81), and the smallest in September (38±50) (Fig. 2). Effects of the year, the dominant plant species (*DPS*), and all interactions (year × month, year × *DPS*, month × *DPS*, year × month × *DPS*) on the *SN* were not significant (Table 3).

The seed intact ratio (*IR*) ranged from 61% (in 2000) to 98% (in 2005) (Table 1). Effects of year, month, *DPS*, and, and all interactions on the *IR* were also significant (Table 3); that is, the effect of the *DPS* on the *IR* varied inter-annually and monthly. For example, the *IR*s of seeds composed mainly of *Swida macrophylla* and *Malus tschonoskii* were lower in 2004 and in 2007, respectively, while the *IR*s of seeds composed mainly of *Viburnum dilatatum* and *Rosa multiflora* were stable throughout the study period. Further, the degree of inter-annual variation in the *IR* was larger in September: *Berchemia racemosa*, *Cornus kousa*, and *Swida macrophylla* were typical examples (Fig. 3).

Seed diversity (H') ranged from 0.57 (in 2004) to 0.81 (in 2000). The effects of year, month, *DPS*, and all interactions on the H' were not significant (Table 3).

Discussion

Among the four aspects of endozoochory I examined, the intact ratio of seeds (*IR*) changed significantly among years and among major plant species, and the degree of them varied among dominant plants. Further, the proportion of feces containing seeds (*AR*) at the species level also changed among years. These points imply that the quality and quantity of the endozoochory performed by wild Japanese macaques are not species-specific, but vary with the temporal (i.e., monthly and inter-annually) context, as pointed out for other animals [2,17].

Table 1. Summary of the data obtained from the faecal samples collected during five study years.

Year/Month	Number of faecal samples	Total number of seeds	*AR* (%)	*SN* (mean±SD)	*IR* (%±SD)	Seed diversity (H'±SD)
2000	66	2079	87.9	31.6±58.6	60.6±32.8	0.81±0.30
October	66	2079	87.9	31.6±58.6	60.6±32.8	0.81±0.30
2004	72	7099	83.3	101.6±189.7	87.1±21.3	0.57±0.30
September	3	10	100.0	3.3±3.2	42.9±51.5	0.00±0.00
October	24	2250	87.5	93.9±164.8	90.3±20.1	0.35±0.36
November	45	4839	80.0	108.6±207.0	88.9±14.9	0.47±0.39
2005	57	3720	96.5	65.3±109.0	97.6±9.1	0.66±0.33
September	11	446	81.8	32.6±51.0	90.2±20.2	0.42±0.39
October	31	1302	100.0	42.0±40.6	98.7±4.7	0.47±0.41
November	15	1972	100.0	131.5±191.4	99.9±0.2	0.31±0.36
2007	77	7426	83.1	96.5±132.5	86.4±20.2	0.63±0.32
September	29	1322	89.7	45.6±55.3	86.2±24.7	0.49±0.33
October	25	2424	100.0	97.0±75.3	80.0±17.7	0.75±0.34
November	23	3680	56.5	160.2±209.0	99.3±1.6	0.47±0.31
2008	116	10842	78.4	94.2±123.8	84.4±24.4	0.65±0.33
September	36	1234	50.0	68.6±46.2	78.4±27.6	0.44±0.24
October	41	1359	82.9	33.3±39.2	76.9±27.3	0.78±0.36
November	39	8249	100.0	213.6±145.2	93.8±16.7	0.58±0.33
5- Year Total	388	31166				
Mean			85.9	77.8±122.7	83.2±21.6	0.66±0.32

Seed diversity has been calculated at the Shannon Winer index (H').
AR: appearance ration of seeds, *SN*: Number of seeds per faecal sample, *IR*: intact ratio of seeds.

Table 2. Seed found within fecal samples of Japanese macaques on Kinkazn, northern Japan (fall in 2000, 2004, 2005, 2007, and 2008).

Species	Family	Life from[a]	Fruiting[b]	SHT[c]	Seed size[d] mm	Seed size[d] mm³	# Seeds per fruit[d]	Fruit size[e] mm	Fruit size[e] mm³	Relative volume of seeds (%)[f]
Berchemia racemosa	Rhamnaceae	V	-Sep	S	3.7	26.5	1.0	5.0	65.4	40.5
Cornus kousa	Cornaceae	H	Sep–Oct	D	4.5	47.7	1.0	15.7	2026.3	2.4
Schisandra nigra	Schisandraceae	V	Sep–Oct	S	4.4	44.6	2.0	–	810.5	11.0
Tubocapsicum anomalum	Solanaceae	H	Sep–Nov	S	1.3	1.2	117.8	–	–	–
Swida macrophylla	Cornaceae	HT	Sep–Nov	C	3.4	20.6	2.0	–	97.4	42.3
Malus tschonoskii	Rosaceae	HT	Sep–Nov	S	3.5	22.4	7.5	–	4717.1	3.6
Pourthiaea villosa	Rosaceae	S	Sep–Nov	S	2.7	10.3	3.3	–	325.9	10.4
Zanthoxylum piperitum	Rutaceae	S	Sep–Nov	C	2.8	11.5	1.0	4.3	41.6	27.6
Amphicarpaea brachteata	Leguminosae	H	Sep–Nov	C	1.6	2.1	2.8	–	–	–
Vitis flexuosa	Vitaceae	V	Sep–Nov	S	3.5	22.4	3.3	9.0	381.7	19.4
Sorbus japonica	Rosaceae	H	Sep–Nov	S	3.1	15.6	2.0	9.0	63.9	48.8
Rosa multiflora	Rosaceae	S	Oct–Nov	S	2.2	5.6	7.0	7.5	220.9	17.7
Vibrnum dilatatum	Caprifoliaceae	S	Oct–Nov	S	3.3	18.8	1.0	6.0	113.1	16.6
Perilla frutescens	Labiatae	H	Oct–Nov	C	1.7	2.6	4.0	–	–	–
Viscum album	Loranthaceae	HP	Nov–	S	4.0	33.5	1.0	–	–	–

Life form, fruiting months, sizes of both seeds and fruits, and seed handling technique by the macaques are also shown.
[a]H: herbaceous plants, HT: high tree, HP: hemi-parasite, V: vine, S: shrub, obtained from [22].
[b]Tsuji, unpublished data.
[c]Seed handling technique (C: crunching, D: discarded, S: swallowing) obtained from [22].
[d]obtained from [22].
[e]obtained from [15,42–43].
[f]calculated by (total seed volume/fruit volume × 100).

Table 3. Factors affecting seed apparent ratio (AR), seed number per fecal sample (SN), intact ratio of seeds (IR), and seed diversity (H') revealed by the GLM analyses using month, year, dominant plant species (DPS), and their interactions as independent variables.

Variables	Main effects						Interactions							
	Month		Year		*DPS*		Month × Year		Month × *DPS*		Year × *DPS*		Month × Year × *DPS*	
Apparent ratio (*AR*)	$z=-1.263$	$p=0.207$	$z=1.393$	$p=0.371$	–	–	$z=0.930$	$p=0.353$	–	–	–	–	–	–
Number of seeds (*SN*)	$z=1.978$	$p=0.048$*	$z=0.899$	$p=0.369$	$z=1.203$	$p=0.229$	$z=-0.778$	$p=0.436$	$z=-1.299$	$p=0.194$	$z=-0.621$	$p=0.535$	$z=0.649$	$p=0.516$
Intact ratio (*IR*)	$z=2.365$	$p=0.018$*	$z=2.531$	$p=0.011$*	$z=2.073$	$p=0.038$*	$z=-2.362$	$p=0.018$*	$z=-2.006$	$p=0.045$*	$z=-2.072$	$p=0.038$*	$z=2.005$	$p=0.045$*
Seed diversity (*H'*)	$t=0.840$	$p=0.402$	$t=0.209$	$p=0.835$	$t=0.367$	$p=0.714$	$t=1.539$	$p=0.134$	$t=0.663$	$p=0.512$	$t=0.491$	$p=0.627$	$t=-0.725$	$p=0.474$

DPS: dominant plant species (see methods in text), *: $p<0.05$.

The cause for the inter-annual variation in the species-level *AR* was elucidated. The most likely factor affecting this is inter-annual variation in fruit production, though I do not have data supporting this. In general, Japanese macaques feed on various kinds of fleshy-fruited plants, probably due to their opportunistic feeding nature [26–27]. Furthermore, the macaques flexibly shift fruit species consumption in response to fruit availability [14,23]. Many fleshy fruited plants in my study site are known to show inter-annual variation in fruiting [24–25]; in the years when some fleshy fruits are scarce, they relied on alternative food sources such as unripe fruits or seeds, or the subset of plants bearing fruits during that year *sensu* [9]. Such temporal context-dependence would lead to a lack of specificity in the plant–macaque network, which could enhance its resilience to fluctuating environmental conditions in the temperate region. From this viewpoint, fleshy-fruited plants in the temperate forest are likely to lessen the risk of a completely inappropriate seed treatment by the macaques, and is likely to increase the probability of successful dispersal whatever conditions prevail, as observed in Mediterranean ecosystems [2].

IR also changed among years. This was similar to the case of *Cercopithecus* monkeys (guenons) inhabiting tropical Africa, which altered their seed handling technique (swallowing or masticating) according to fruit availability, and *IR* decreased in the year when fruits were scarce and the guenons performed mastication of the seeds within fruits [9]. In the temperate region, seasonality in fruiting is quite clear [28], and the macaques can find some fruits in the fruiting season (fall). Thus, it is not likely that inter-annual changes in the fruit-handling technique by the macaques occur. Rather, inter-annual changes in the *IR* of seeds would be attributed to the type of available fruits at any time. It is known that the size of seeds eaten by macaques has a negative effect on the *IR* [22]; the *IR* of seeds composed mainly of large-sized species is lower than that of small-sized seeds. Thus, in years when only fruit species with relatively large seeds (>3 mm, for example) were available, *IR* would become lower. This tendency might be strong in September in which number of potentially consumed fruits is smaller (11 species over 8 year study) than October and November (20 species for each month, [22]). My result supported this idea.

Unlike the above two characteristics, the number of seeds per fecal sample (*SN*) and seed diversity (*H'*) per fecal sample was consistent throughout the study period (though the former variable changed monthly). It is known that the gastrointestinal passage time of captive Japanese macaques is 37–54 h [29–30], which is far longer than the feeding bout length in a fruit patch (ca. 5–25 min) [31–32], and single fecal samples contain seeds of multiple plant species ingested across several feeding bouts. Also, the number of seeds within a single fruit (<10 seeds, except for *Tubocapsicum anomalum* having >100 seeds per fruit), does not vary among 15 species investigated here (Table 2). Therefore, it is unlikely that feeding on specific fruits affects the density of seeds within single feces/at defecation sites, and consequent competition among seedlings over resources.

Mutualisms are typically characterized by costs and benefits being highly dependent on the community context to which they belong [3]. The Japanese macaques have broad diets, and are not closely dependent on the limited number of fleshy-fruited plants in any season [26–27]. Thus, the ecological variability and conditional outcomes of plant–macaque interactions should be treated as less-extreme fitness costs in less-obligate interactions *sensu* [3]. For several plants, specific dispersal patterns performed by the macaques in extreme years can have a more significant

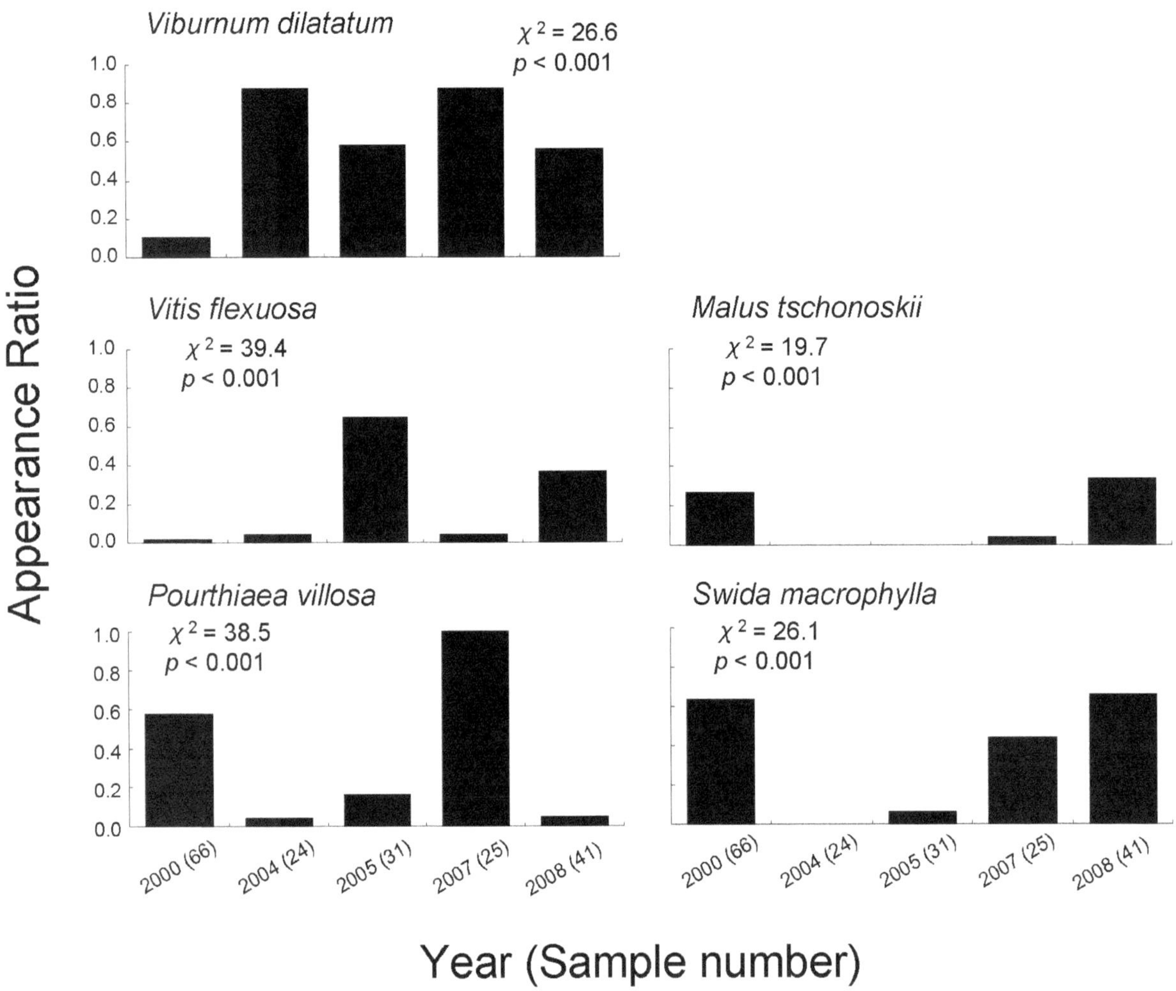

Figure 1. Inter-annual change in the seed appearance ratio of five major plant species from fecal samples collected in October. Samples were collected between 2000 and 2008. Figures in parentheses represent the number of analyzed samples in a given year. Results of statistical analyses (chi-square tests) are also shown.

implication than the patterns of typical years. Therefore, evaluating the performance of the dispersers over a longer time scale is important. This possibility can be tested by estimating seed rain and evaluating germination and growth across multiple years when the fruiting and related feeding behavior of dispersers differ. From this viewpoint, long-term monitoring of fruiting of major diet fruits should be conducted.

This study revealed that variations in the frequency and success of seed dispersal over seed predation are affected by the temporal context, while abundance and diversity of seeds are not. My conclusion supports the generally accepted idea of the context-dependence of species interactions [1,3]. The finding that the context-dependence of endozoochorous dispersal does occur over time seems to be a highly generalized process at other sites in the forest ecosystems. Studies of primates and other frugivorous mammals in temperate regions to date have shown that home range size and its utilization (e.g., [23,33–35]) changes inter-annually. By combining the results of those studies and the present one, researchers can draw conclusions about patterns of seed dispersal, such as the distance from mother trees and the degree of seed concentration in different years (e.g., [36]).

Materials and Methods

Ethics statement

I did not need to specific permission for conducting behavioral observation of the macaques and fecal sample collection on Kinkazan Island because this island is considered as public property. My field study did not involve endangered or protected animal/plant species. The research methodology complied with protocols approved by the guidelines (Guide for the Care and Use of Laboratory Primates, Second Edition) of Primate Research Institute, Kyoto University, Japan and the legal requirements of Japan.

Study site and subjects

Kinkazan Island (38.3 N, 141.6 E) is located 700 m from the Oshika Peninsula, northern Japan. The total area of the island is ca. 9.6 km^2, and the highest peak is 450 m a.s.l. The monthly

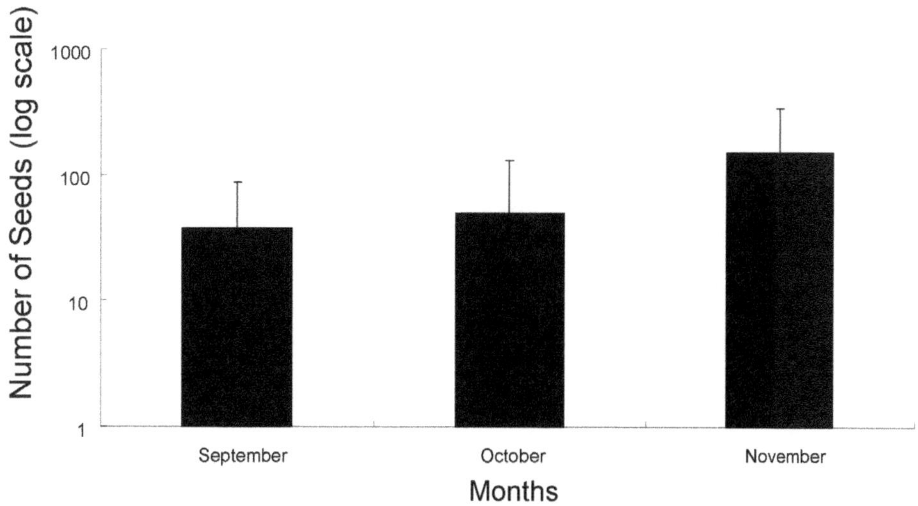

Figure 2. Monthly change in mean (±SD) number of seeds per fecal sample of the Japanese macaques on Kinkazan Island, northern Japan. Samples were collected between 2000 and 2008.

mean air temperature on the island ranges from 2.5°C in February to 22.3°C in August. Based on the climate conditions, the year was divided into four seasons: spring (March–May), summer (June–August), fall (September–November), and winter (December–February) [22]. Deciduous forests of *Fagus crenata* (Fagaceae) dominate the higher elevations (>150 m), whereas a mixture of deciduous forests of *Carpinus* spp. (both *C. tschonoskii* and *C. laxiflora*, Betulaceae) and coniferous forests of *Abies firma* (Pinaceae) cover the lower elevations (<150 m) on the island. Approximately 200–250 Japanese macaques belonging to six troops (A, B_1, B_2, C_1, C_2, and D) inhabit the island [33].

Fecal sample collection and analysis

I collected fresh fecal samples from adult females in the fall in 2000, 2004, 2005, 2007, and 2008 during the behavioral observation of troop A (16 surveys in total, 2000: October, 2004–2008: September, October, and November), details of the behavioral observation are shown in my previous studies [37–38]. I was in contact with the macaques (ca. 10 m in distance) during my observation. Each fecal sample was thoroughly mixed with water and rinsed through 0.5 mm sieves under fresh water. I then picked up all seeds from the fecal samples, identified the seeds to the species level, and counted the seeds of each species. Seed identification was based on my reference collection that had been prepared previously [22,39].

I evaluated the seed characteristics based on previous studies [22,40]. I first calculated the number of dispersal events, which was defined as the number of fecal samples containing seeds, for each year for each plant species. I also parsed data by month because fruit vary in abundance during specific weeks and three months window seemed broad. Since I did not collect fecal samples in equal quantities across the five years (range: 57–116), I calculated seed appearance ratios (*AR* hereafter), obtained by dividing the dispersal events by the number of fecal samples examined for each year for each target species. I calculated the *AR* both for all species combined and for each species separately. For the latter analyses I used fecal samples collected in October to eliminate the difference in collecting effort among fall months in each year. For each fecal sample from which seeds appeared, I counted the number of seeds per fecal sample (*SN*). I further calculated the seed intact ratio (*IR*), that is, the ratio of seeds with no apparent physical damage after gut passage, as a qualitative index of the efficacy of seed dispersal. The *IR* was obtained by dividing the total number of intact seeds by the total number of seeds in a given sample. Finally, as an index of seed diversity, I calculated the Shannon-Wiener Index (H') for each fecal sample containing seeds as: $H' = -\Sigma p_i \ln p_i$, where, p_i is the proportion of seeds from species i relative to the total number of seeds in a given fecal sample.

The data I used in this study is given in Table S1.

Statistical analysis

To evaluate the variation in the *AR*, I fitted a generalized linear model (GLM) in which the response variable was binary (feces containing seeds or not, independently of plant species; binomial error family) and the fixed effect was year and month. I performed chi-square tests for independence to examine inter-annual changes in the *AR* in October for each fleshy-fruited species separately. To evaluate the differences in the *NS*, I ran another GLM with count data (negative binomial error family). Only feces containing seeds were included in this model. The response variable was the total number of seeds found in each fecal sample, and the explanatory variables were year, month, and "dominant plant species" (*DPS* hereafter), defined as the most abundant seed species in a fecal sample [2]. In this analysis, I also considered the effect of interaction between year, month, and the *DPS*. In order to test the differences in the *IR*, I fitted another GLM with the negative binomial error family. The response variable was the number of intact seeds in each fecal sample (overall number of seeds were treated as an offset term), and the explanatory factors were the same as the second GLM model. Finally, to analyze the seed diversity in feces, I again ran GLM (Gamma error family) with seed diversity as a response variable and the fixed effects were the same as those in the second and third GLM models. I performed these GLMs using the "MASS" and "multcomp" packages in the statistical software R.15.2. [41]. The significance levels of these analyses were set at 5%.

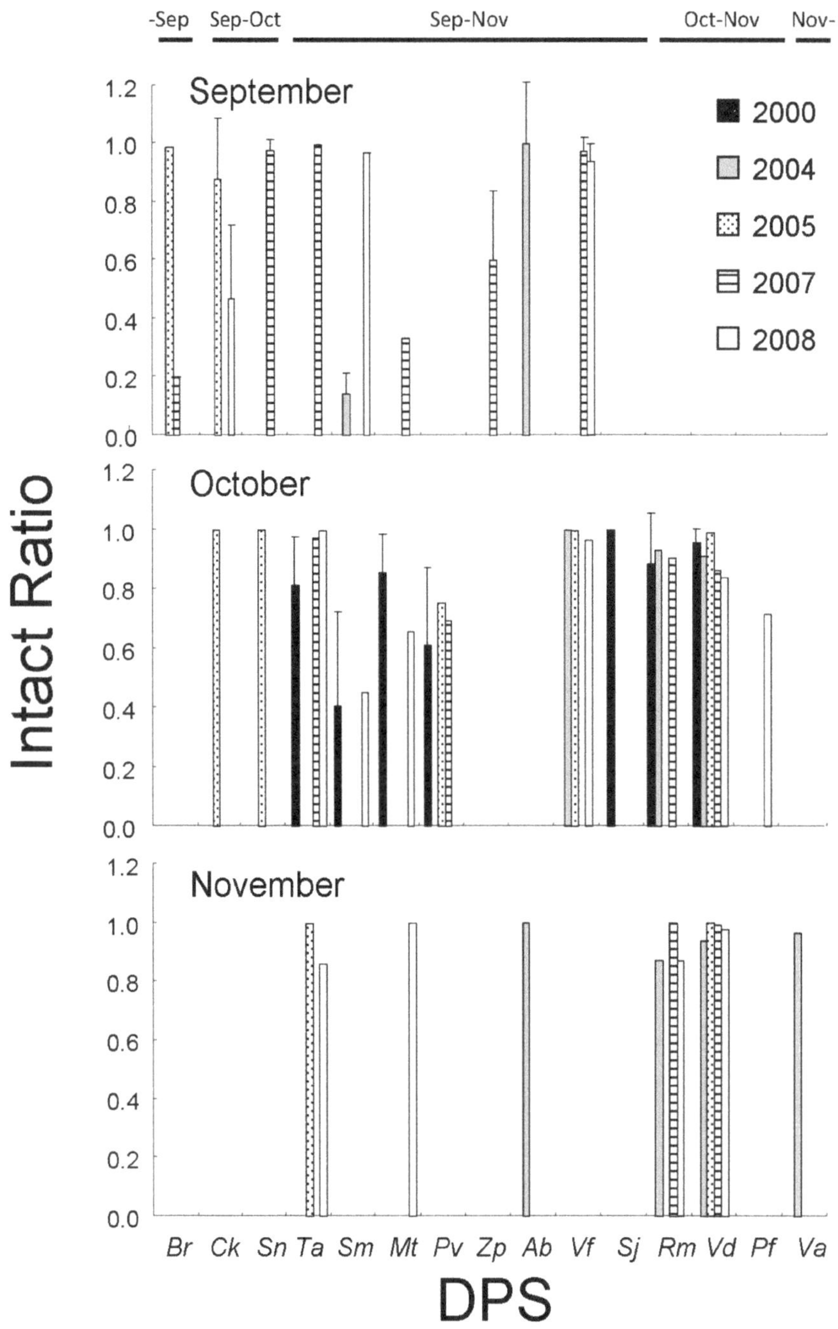

Figure 3. Inter-annual change in mean (±SD) intact ratio of seeds of the Japanese macaques for the dominant plant species in September (top), October (middle), and November (bottom). The intact ratio is obtained by dividing total number of intact seeds by total number of seeds in a given sample. Fruiting months of each species are shown on the top of the figure. Date from 2000 is only from October. *Br = Berchemia racemosa, Ck = Cornus kousa, Sn = Schisandra nigra, Ta = Tubocapsicum anomalum, Sm = Swida macrophylla, Mt = Malus tschonoskii, Pv = Pourthiaea villosa, Zp = Zanthoxylum piperitum, Ab = Amphicarpaea brachteata, Vf = Vitis flexuosa, Sj = Sorbus japonica, Rm = Rosa multiflora, Vd = Viburnum dilatatum, Pf = Perilla frutescens, Va = Viscum album.*

Supporting Information

Table S1 A. Number of all seeds detected from fecal analyses collected during five study years. Seed diversity has been calculated at the Shannon Winer index (*H'*). **B.** Number of intact seeds detected from fecal samples collected during five study years. Intact ratio (*IR*) is also shown.

Acknowledgments

I thank the members of the Kinkazan Deer Research Group, especially Drs. S Takatsuki and M Minami, and N Ohnishi for their cooperation with fecal sample collection; K Sato, H Ebihara, T Ochiai, N Noguchi, T Tatewaki, and T Harasaki for their assistance with fecal sample analyses; MA Huffman for his constructive comments on an earlier version of this manuscript; and Dr. K Izawa and his colleagues for their support during fieldwork on Kinkazan Island.

Author Contributions

Conceived and designed the experiments: YT. Performed the experiments: YT. Analyzed the data: YT. Contributed reagents/materials/analysis tools: YT. Wrote the paper: YT.

References

1. Holland JN, DeAngelis DL (2009) Consumer-resource theory predicts dynamic transitions between outcomes of interspecific interactions. Ecol Lett 12: 1357–1366.
2. Perea R, Delibes M, Polko M, Suarez-Esteban A, Fedriani JM (2013) Context-dependent fruit–frugivore interactions: partner identities and spatio-temporal variations. Oikos 122: 943–951.
3. Bronstein JL (1994) Conditional outcomes of mutualistic interactions. Trend Ecol Evol 9: 214–217.
4. Wilson MF (1993) Mammals as seed-dispersal mutualists in North America. Oikos 67: 159–176.
5. Corlett RT (1998) Frugivory and seed dispersal by vertebrates in the oriental (Indomalayan) region. Biol Rev 73: 413–448.
6. Hickey JR, Flynn RW, Buskirk SW, Gerow KG, Willson MF (1999) An evaluation of a mammalian predator, *Martes americana*, as a disperser of seeds. Oikos 87: 499–508.
7. Traveset A, Robertson AW, Rodoriguez-Perez J (2007) A review on the role of endozoochory in seed germination. In: Dennis AJ, Schupp EW, Green RJ, Westcott DA, editors. Seed dispersal: theory and its application in a changing world. Cambridge: CABI. 78–103.
8. Kelly D (1994) The evolutionary ecology of mast seeding. Trends Ecol Evol 9: 465–470.
9. Kaplin BA, Moermond TC (1998) Variation in seed handling by two species of forest monkeys in Rwanda. Am J Primatol 45: 83–101.
10. Shibata M, Tanaka H, Nakashizuka T (1998) Causes and consequences of mast seed production of four co-occurring *Carpinus* species in Japan. Ecology 79: 54–64.
11. Suzuki W, Osumi K, Masaki T (2005) Mast seeding and its spatial scale in *Fagus crenata* in northern Japan. For Ecol Manag 205: 105–116.
12. Sun IF, Chen YY, Hubbell SP, Right SJ, Noor NSMD (2007) Seed predation during general flowering events of varying magnitude in a Malaysian rain forest. J Ecol 95: 818–827.
13. Mattson DJ, Blanchard BM, Knight RR (1992) Yellowstone grizzly bear mortality, human habituation, and white bark-pine seed crops. J Wildl Manage 56: 432–442.
14. Tsuji Y, Fujita S, Sugiura H, Saito C, Takatsuki S (2006) Long-term variation in fruiting and the food habits of wild Japanese macaques on Kinkazan Island, northern Japan. Am J Primatol 68: 1068–1080.
15. Takahashi K, Shiota T, Tamatani H, Koyama M, Washitani I (2008) Seasonal variation in fleshy fruit use and seed dispersal by the Japanese black bear (*Ursus thibetanus japonicus*). Ecol Res 23: 471–478.
16. Koike S (2010) Long-term trends in food habits of Asiatic black bears in the Misaka Mountains on the Pacific coast of central Japan. Mamm Biol 75: 12–28.
17. Schupp EW, Jordano P, Gomez M (2010) Seed dispersal effectiveness revisited: a conceptual review. New Phytol 188: 333–353.
18. Leavey DJ, Silva WR, Galetti M (2002) Seed dispersal and frugivory: ecology, evolution and conservation. Cambridge: CABI.
19. Dennis AJ, Schupp EW, Green RJ, Westcott DA (2007) Seed dispersal: theory and its application in a changing world. Cambridge: CABI.
20. Otani T, Shibata E (2001) Seed dispersal and predation by Yakushima macaques, *Macaca fuscata yakui*, in a warm temperate forest of Yakushima Island, southern Japan. Ecol Res 15: 133–144.
21. Terakawa M, Isagi Y, Matsui K, Yumoto T (2009) Microsatellite analysis of the maternal origin of *Myrica rubra* seeds in the feces of Japanese macaques. Ecol Res 24: 663–670.
22. Tsuji Y, Sato K, Sato Y (2011) The role of Japanese macaques (*Macaca fuscata*) as endozoochorous seed dispersers on Kinkazan Island, northern Japan. Mamm Biol 76: 525–533.
23. Hill DA, Agetsuma N (1995) Supra-annual variation in the influence of *Myrica rubra* fruit on the behavior of a troop of Japanese macaques in Yakushima. Am J Primatol 35: 241–250.
24. Komiyama A, Wada K, Kuga H (1991) Annual fluctuation of seed-fall amounts in the basin of Yokoyu River, Shiga Heights-*Fagus*, *Quercus*, and *Cornus* species in the utilization area of Japanese monkeys-. Bull Exp For Gifu Univ 56: 165–174.
25. Takita M, Mizui N, Tarazawa K, Umeki K (1998) Annual fluctuation of seed production in thirty-five deciduous broad-leaved species. Bull Hokkaido For Res Inst 35: 31–41.
26. Hill DA (1997) Seasonal variation in the feeding behavior and diet of Japanese macaques (*Macaca fuscata yakui*) in lowland forest of Yakushima. Am J Primatol 43: 305–322.
27. Go M (2010) Seasonal changes in food resource distribution and feeding sites selected by Japanese macaques on Koshima Islet, Japan. Primates 51: 149–158.
28. Hanya G, Tsuji Y, Grueter CC (2013) Fruiting and flushing phenology in Asian tropical and temperate forests: implications for primate ecology. Primates 54: 101–110.
29. Tsuji Y, Morimoto M, Matsubayashi K (2010) Effects of the physical characteristics of seeds on gastrointestinal passage time in captive Japanese macaques. J Zool 280: 171–176.
30. Sawada A, Sakaguchi E, Hanya G (2011) Digesta passage time, digestibility, and total gut fill in captive Japanese macaques (*Macaca fuscata*): effects food type and food intake level. Int J Primatol 32: 390–405.
31. Maruhashi T, Saito C, Agetsuma N (1998) Home range structure and inter-group competition for land of Japanese macaques in evergreen and deciduous forests. Primates 39: 291–301.
32. Terakawa M, Matsui K, Hamada T, Noma N, Yumoto T (2008) Reduced seed dispersal effectiveness in the large-seeded tree *Myrica rubra* in the absence of the Japanese macaque on Tanegashima Island, Japan. Jap J Cons Ecol 13: 161–167.
33. Izawa K (2009) Research on wild Japanese macaques. Tokyo: Dobutsusha.
34. Arimoto I, Goto Y, Nagai C, Furubayashi K (2011) Autumn food habits and home-range elevations of Japanese black bears in relation to hard mast production in the beech family in Toyama prefecture. Mamm Study 36: 199–208.
35. Kozakai C, Yamazaki K, Nemoto Y, Nakajima A, Koike S, et al. (2011) Effect of mast production on home range use of Japanese black bears. J Wildl Manag 75: 867–875.
36. Koike S, Masaki T, Nemoto Y, Kozakai C, Yamazaki K, et al. (2011) Estimate of the seed shadow created by the Asiatic black bear *Ursus thibetanus* and its characteristics as a seed disperser in Japanese cool-temperate forest. Oikos 120: 280–290.
37. Tsuji Y, Takatsuki S (2009) Effects of yearly change in nut fruiting on autumn home range use of Japanese macaques on Kinkazan Island, northern Japan. Int J Primatol 30: 169–181.
38. Tsuji Y, Takatsuki S (2012) Inter-annual variation in nut abundance is related to agonistic interactions of foraging female Japanese macaques (*Macaca fuscata*). Int J Primatol 33: 489–512.
39. Nakayama S, Inokuchi M, Minamitani T (2000) Seeds of wild plants in Japan. Sendai: Tohoku University Press.
40. Kunz BK, Linsenmair KE (2008) The role of the olive baboon (*Papio anubis*, Cercopithecidae) as seed disperser in a savanna-forest mosaic of West Africa. J Trop Ecol 24: 235–246.
41. R Development Core Team (2012) R: a language and environment for statistical computing. Version 2.15.0. Vienna, Austria: R Foundation for Statistical Computing.
42. Nakanishi H (1996) Fruit color and fruit size of bird-disseminated plants in Japan. Vegetatio 123: 207–218.
43. Masaki T, Takahashi K, Sawa A, Kado T, Naoe S, et al. (2012) Fleshy fruit characteristics in a temperate deciduous forest of Japan: how unique are they? J Plant Res 125: 103–114.

4

Microclimate Impacts Survival and Prevalence of *Phytophthora ramorum* in *Umbellularia californica*, a Key Reservoir Host of Sudden Oak Death in Northern California Forests

Matthew V. DiLeo¤, Richard M. Bostock, David M. Rizzo*

Department of Plant Pathology, University of California Davis, Davis, California, United States of America

Abstract

Phytophthora ramorum, an invasive pathogen and the causal agent of Sudden Oak Death, has become established in mixed-evergreen and redwood forests in coastal northern California. While oak and tanoak mortality is the most visible indication of *P. ramorum*'s presence, epidemics are largely driven by the presence of bay laurel (*Umbellularia californica*), a reservoir host that supports both prolific sporulation in the winter wet season and survival during the summer dry season. In order to better understand how over-summer survival of the pathogen contributes to variability in the severity of annual epidemics, we monitored the viability of *P. ramorum* leaf infections over three years along with coincident microclimate. The proportion of symptomatic bay laurel leaves that contained viable infections decreased during the first summer dry season and remained low for the following two years, likely due to the absence of conducive wet season weather during the study period. Over-summer survival of *P. ramorum* was positively correlated with high percent canopy cover, less negative bay leaf water potential and few days exceeding 30°C but was not significantly different between mixed-evergreen and redwood forest ecosystems. Decreased summer survival of *P. ramorum* in exposed locations and during unusually hot summers likely contributes to the observed spatiotemporal heterogeneity of *P. ramorum* epidemics.

Editor: Boris A. Vinatzer, Virginia Tech, United States of America

Funding: This research was supported by the USDA Forest Service Pacific Southwest Research Station, the Gordon and Betty Moore Foundation, and the Department of Homeland Security. The funders had no role in study design, data collection and analysis, decision to publish, or preparation of the manuscript.

Competing Interests: The authors have declared that no competing interests exist.

* Email: dmrizzo@ucdavis.edu

¤ Current address: Keygene Incorporated, Rockville, Maryland, United States of America

Introduction

Sudden Oak Death, caused by the oomycete *Phytophthora ramorum*, has killed potentially millions of native oak (*Quercus* spp.) and tanoak (*Notholithocarpus densiflorus*) trees in Oregon and northern California since the mid-1990s [1,2]. This invasive pathogen has become established primarily in two forest types: mixed-evergreen forests, dominated by coast live oak (*Quercus agrifolia*), and redwood forests, dominated by redwood (*Sequoia sempervirens*) and tanoak. These two forests types occur in a fine-scale mosaic that is superimposed on the diverse topography of this region. *P. ramorum*-infested forest sites are most commonly found within 30 km of the Pacific Coast and San Francisco Bay, where mild winter wet seasons and plentiful fog provide a conducive environment for sporulation and infection of aerial host tissues [1–3].

The most dramatic symptoms of *P. ramorum* infection include bole cankers on oaks and both bole cankers and shoot dieback on tanoaks, leading to mortality and stem failure of mature trees [4]. Relatively few spores are produced from infected tanoaks, however, and most California oak species appear to provide an epidemiological dead end for this pathogen [2,5]. Instead, the establishment and reproduction of *P. ramorum* in California forests is primarily driven by California bay laurel (*Umbellularia californica*), an evergreen reservoir host that is common in both mixed-evergreen and redwood forests [5,6]. *P. ramorum* causes a minor leaf blight on bay laurel, often infecting a large proportion of the leaves in the lower, visible canopy, yet does not appear to cause significant injury to the host [7]. Infected bay laurel leaves are capable of producing many thousands of splash-dispersed sporangia and zoospores during the winter wet season and appear to be the primary vehicle by which the pathogen survives the summer dry season [5,8,9].

The viability of *P. ramorum* in infected bay laurel leaves has been found to decrease progressively during the summer dry season in California forests [3,10]. One previous study [3] also documented a more pronounced decrease in pathogen viability over the summer in bay laurel leaves from a mixed-evergreen forest relative to a nearby redwood forest. While these two sites did not differ in temperature or humidity, the water potential of infected trees was found to be lower in the mixed-evergreen forest [3]. In the present study, the survival of *P. ramorum* within infected bay laurel leaves was measured over three summers coincident with the collection of ecophysiological and microclimate data in order to identify environmental conditions that inhibit over-summer survival of the pathogen. An improved understanding of the over-summer survival will help define the ecological limits of *P. ramorum* and inform Sudden Oak Death risk management.

Materials and Methods

Research site locations and characteristics

Sites were located in the Sonoma and Mayacmas mountain ranges within Sonoma County, CA. This region is subject to a Mediterranean climate, consisting of cool, wet winters and hot, dry and intermittently foggy summers. Vegetation occurs as a patchwork of annual grasslands, mixed-evergreen forests and redwood forests. 12 sites were selected from a larger set of existing long-term study sites outfitted with weather-monitoring data loggers [11]. These 12 sites were chosen to represent the diverse topographic, climatic and ecological environments in which bay trees are found while allowing sufficient access to trails and roads as required for water potential measurements. Three to nine bay laurel trees with ramorum blight symptoms were tagged at each site and followed through the duration of the study. A total of 25 bay laurel trees were tagged in each of mixed-evergreen and redwood forest. An initial set of 18 trees was tagged in the spring of 2005 and the remaining 32 trees were tagged in the spring of 2006. Mixed-evergreen forests were located within Fairfield Osborn Preserve (FOP, 122° 35′41″W, 38° 20′37″N), Sugarloaf Ridge State Park (SRSP, 122° 31′41″W, 38° 26′30″N) and Annadel State Park (ASP, 122° 62′36″W, 38° 42′92″N). Redwood forests were located in Jack London State Historic Park (JLSHP, 122° 33′16″W, 38° 21′2″N) and the Sonoma Development Center (SDC, 122° 32′10″W, 38° 20′52″N). All field work was completed under permits issued by California State Parks (SRSP, ASP, JLSHP, SDC), Sonoma State University (FOP), and California Department of Food and Agriculture. Field measurements were made in July, August, September and October during 2005, 2006 and 2007.

Microclimate and physiological measurements

Data loggers were used to record the temperature and humidity of the air at each site hourly [Onset Corporation, Bourne, MA, USA; Campbell Scientific, Logan, UT, USA] [R. Meentemeyer, UNC-Charlotte, unpublished]. All sensors were placed approximately 1 m above the ground and were shielded from the sun by the forest canopy or a protective solar shield to prevent inflated temperature measurements. Most tagged trees were located within a few meters of their corresponding data logger and all were within 200 m. Canopy coverage above each tree was manually estimated by eye. Aspect and slope were measured at the base of each tree and were used to calculate insolation, an estimation of the amount of solar radiation that an area of land receives based on the angle of the land relative to the position of the sun [12]. Elevation data were obtained from topographic maps [R. Meentemeyer, UNC-Charlotte, personal communication]. Vapor pressure deficit was calculated from temperature and relative humidity data. Average vapor pressure deficit, average temperature and the number of days where temperature exceeded 30°C were all calculated for each 30 day period preceding each time point.

Leaf water potential of individual bay laurel leaves was measured with a pump-up pressure chamber [PMS Instrument Company, Albany, OR, USA]. All measurements were obtained between 11 AM and 1 PM during a three-day window surrounding each time point in order to minimize the influence of short term weather variation. This restriction limited the number and location of potential study sites; while representative of local environments, all sites were relatively easily accessed by vehicle and foot. Leaf water potential measurement procedures were adapted from McCutchan and Shackel [13]. Variation in leaf water potential within a given canopy at a given time point was generally 0.01 MPa or less. Minimum leaf water potential, occurring in August or September, was recorded for each tree each year.

Pathogen isolation and quantitative image analysis

P. ramorum commonly causes tip and edge necrosis on bay laurel leaves that consumes 5–10% leaf area. Ten symptomatic leaves were arbitrarily selected for removal from each tree at each time point. Symptomatic leaf images were captured with a scanner and analyzed with APS Assess© [American Phytopathological Society, St. Paul, MN]. The total leaf area and percent necrotic area was recorded. Subsequently, half of each necrotic lesion (including adjacent chlorotic and symptomless leaf tissue) was cut from each leaf and plated on the selective medium PARP (cornmeal agar with pimaricin, ampicillin, rifampicin, and pentachloronitrobenzene) [14]. *In vitro* hydration treatments did not increase isolation frequency, as may occur when chlamydospores are present [9]. Excised lesion margins were sampled from all parts of each symptomatic leaf, including necrotic spots in the interior of the leaf when present. Leaves were not surface sterilized. Following a 3–4 week incubation at 18°C, *P. ramorum* mycelia were identified by colony morphology and the presence of large chlamydospores [1]. It is assumed that the inability to isolate *P. ramorum* from symptomatic leaves is solely due to the death of the pathogen, therefore pathogen survival was recorded where *P. ramorum* was successfully isolated from symptomatic leaf material. This is supported by previous work, which found isolation frequency to be well correlated with pathogen viability, as measured with RT-PCR [15]. Although other *Phytophthora* species are known from bay laurel in California forests [24], no other *Phytophthora* species were isolated from bay laurel leaves in this study.

Data analysis

Statistical analysis was performed in three stages. First, systemic differences between mixed-evergreen and redwood forest sites were tested. Leaf water potential, percent canopy cover, elevation and insolation were compared between bay laurel trees located in mixed-evergreen and redwood forest sites using the Wilcoxon rank sum test [JMP IN 5.1 (Cary, NC)]. Variation in leaf area, lesion area, percent lesion area, average temperature, average vapor pressure deficit and number of days exceeding 30°C were compared between trees located in mixed-evergreen and redwood forest sites and over the summer using repeated measures ANOVA.

Associations between these ecophysiological variables and *P. ramorum* survival were assessed using a linear mixed effects model [16]. *P. ramorum* survival, the percentage of symptomatic leaves from which *P. ramorum* was successfully isolated at each time point, was arcsine square root transformed. Year was included as a blocking factor, while month, number of days exceeding 30°C, leaf water potential, percent canopy coverage and lesion area were included with main effects. Redundant measured variables were omitted to avoid biasing the final model. Thus, elevation was omitted as it is primarily associated with differences in temperature. Temperature was included only as the number of days that exceeded 30°C, as this has been shown to be directly detrimental to *P. ramorum* [17]. Vapor pressure deficit was omitted as it is strongly associated with leaf water potential, a more direct measurement of leaf water status. Percent canopy coverage was included instead of insolation as the former was judged to be a better proxy for the amount of full sun that each canopy was exposed to. Finally, lesion area was included while percent leaf area was omitted as the ability of the pathogen to survive the summer may be influenced by the mass of mycelium or number of propagules within the leaf. Random effects were included for both site and tree nested within site. Correlation was allowed among the measurements for each tree within each year to account for possible dependence, assuming compound symmetry. Interaction effects were not examined due to the complexity of the model. Model assessment was performed using graphical analysis of residuals and a Shapiro-Wilks test for normality. Statistical significance was declared at the 0.05 level. All analyses were implemented using the MIXED procedure in SASR 9.1 for Windows (Cary, NC).

Finally, each variable found to have a significant correlation with *P. ramorum* survival was further analyzed by linear regression. This analysis was done on averaged data for each site. The average percent *P. ramorum* survival calculated for each site in October was compared to the average minimum leaf water potential, the average percent canopy coverage and the average number of days exceeding 30°C at each site. r^2 values were calculated for each regression using Microsoft Office Excel 2003 (Redmond, WA).

All data sets used in this study are available from http://datadryad.org/resource/.

Results

P. ramorum survival

The percentage of symptomatic bay laurel leaves that contained viable *P. ramorum* infections fell from 87% in July 2005 to 25% in September 2005 and remained below 33% for all time points in 2006 and 2007 (Figure 1). Over three summers, *P. ramorum* survived within an average of 24% of symptomatic bay laurel leaves from mixed-evergreen forest sites and 26% of leaves from redwood forest sites. Average end of summer (October) *P. ramorum* survival was 16% in mixed-evergreen forest sites and 21% in redwood forest sites, a difference that is not statistically significant.

Associations with forest type

Descriptive statistics were calculated for all trees and for trees partitioned by forest type (Table 1). Although the survival of *P. ramorum* did not differ significantly between mixed-evergreen and redwood forest sites, other variables did. Redwood forest sites had significantly less negative bay laurel leaf water potential ($P<.0001$), greater canopy cover ($P = 0.0027$) and lower elevation ($P<.0001$). Mixed-evergreen and redwood forest sites did not differ significantly by average temperature ($P = 0.1262$) or the number of days exceeding 30°C ($P = 0.3925$), although vapor pressure deficit was greater in mixed-evergreen forest sites ($P = 0.0355$). While bay laurel trees in redwood-forest sites had larger leaves ($P = 0.0008$) and larger lesions ($P = 0.0004$), the percent of the leaf consumed by necrosis did not vary between forest types ($P = 0.4415$). Lesion area and percent lesion area both decreased during the summer ($P = 0.0002$ and $P<0.0001$, respectively), while leaf area did not ($P = 0.8612$).

Mixed model ANOVA

Five variables were found to have significant associations with *P. ramorum* survival and were retained in the final model: year ($P = .0074$), month ($P = .0153$), leaf water potential ($P<.0001$), number of days exceeding 30°C ($P<.0001$) and percent canopy coverage ($P<.0001$) (Table 2). Lesion area ($P = .9741$) was not found to be significant. *P. ramorum* survival was positively associated with canopy cover and leaf water potential and negatively associated with the number of days over 30°C.

Post-hoc Analysis

Linear regression of *P. ramorum* survival by leaf water potential ($r^2 = 0.21$) and canopy cover ($r^2 = 0.08$) both showed positive correlations. Linear regression of survival by the number of days exceeding 30°C showed a negative correlation ($r^2 = 0.35$). Overall, survival of *P. ramorum* in symptomatic bay laurel leaves was

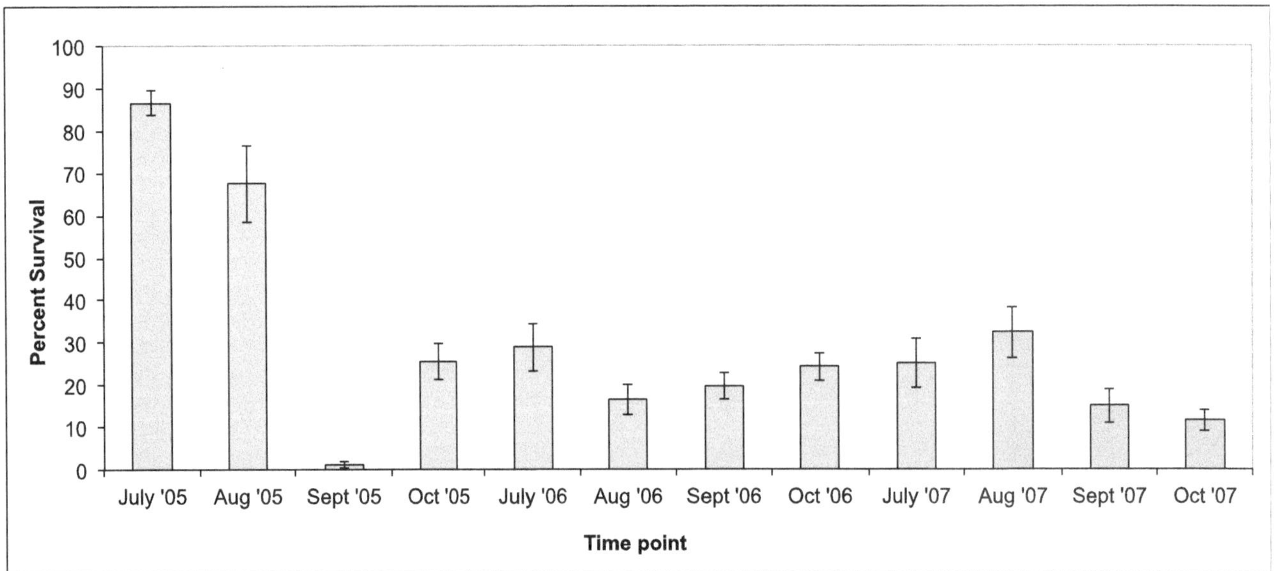

Figure 1. Average percent survival of *P. ramorum* in symptomatic bay laurel leaves across three years. The percent of symptomatic bay laurel leaves from which the pathogen could be recovered declined during the summer of 2005, as has been documented in previous studies, but remained low for the following two years. Specific weather events occurring in the study region may account for the observed pattern (± standard error).

Table 1. Comparison of environmental variables between mixed-evergreen and redwood forest sites across three years.

Site Variables	Mixed-evergreen sites			Redwood sites			*p* value
	Range	Mean	S.E.	Range	Mean	S.E.	
Daily mean temperature (°C)	12.2–27.8	17.8	0.18	12.4–21.4	16.9	0.13	0.1262
Daily mean VPD (kpa)	0.4–2.1	0.8	0.02	0.3–1.2	0.7	0.01	0.0355
Number of days >30°C	0–17	4.3	0.27	0–13	3.3	0.22	0.3925
Elevation (m)	185–610	472	25	206–328	251	7	<.0001
Insolation (kWh/(m^{2}d))	0.5–1.2	1.1	0.03	0.5–1.2	1.0	0.04	0.1307
Percent canopy cover	20–75	37	3	25–60	50	2	0.0027
Leaf water potential (MPa)	−1.4– −2.6	−2.1	0.1	−0.8– −2.2	−1.4	0.1	<.0001
Leaf area (cm^2)	7.2–30.8	14.2	0.38	11.6–32.9	20.2	0.32	0.0008
Lesion area (cm^2)	0.2–1.9	0.8	0.03	0.2–2.9	1.3	0.04	0.0004
Percent lesion area	1.0–20.5	6.0	0.31	1.8–15.3	6.6	0.21	0.4415

p values were generated from Wilcoxon rank sum tests and repeated measures ANOVA, as described in the Materials and Methods. S.E. indicates standard error.

associated with less negative water potential, greater canopy cover and fewer days with maximum hourly temperature over 30°C during the study period.

Discussion

The percentage of symptomatic bay laurel leaves that contained viable *P. ramorum* infections decreased markedly during the summer of 2005 as previous work has suggested; falling from almost 90% to below 30% [3]. Although it was expected that this pattern would have been repeated in 2006 and 2007, survival instead remained below 33% for the remainder of this study. Such year-to-year variation in disease prevalence has been documented in previous studies of *P. ramorum* in these ecosystems. For example, prolonged spring rainfall, experienced in the study region in 2005, was linked to prolific sporulation from, and new infections of, bay laurel leaves well into June [3,5]. In contrast, the much drier spring of 2004 had dramatically lower levels of sporulation and infection of bay laurel leaves [3,5]. Likewise, the relatively dry spring in 2007 may partially explain the poor survival of *P. ramorum* observed in this year. In this context, the poor survival observed during the relatively wet spring of 2006 is surprising, though corroborated by an unusually early ceasing of *P. ramorum* sporulation in April (unpublished data). Whatever factors beyond rainfall limited survival and sporulation in the spring of 2006 was likely compounded by the intense heat wave that this region experienced during June of the same year. While in both 2005 and 2007, most sites had not been exposed to temperatures exceeding 30°C at all, in 2006 each site had experienced between 3 and 11 such days before the first survival time point was taken, possibly masking an earlier peak in infection viability in 2006.

The survival of *P. ramorum* within bay laurel leaves was negatively correlated with the number of days exceeding 30°C and positively correlated with percent canopy cover and leaf water potential. As these three variables inevitably interact, the statistical model cannot definitively distinguish between variables that are responsible for the death of the pathogen and those merely associated with it. However, the known biology of this system strongly implicates heat, exacerbated by solar heating, in the progressive death of *P. ramorum* within bay laurel leaves. It is very unlikely that water potential directly impacts the survival of *P. ramorum* in bay leaves as this and other *Phytophthora* species have been found to not only survive, but grow well at water potentials well below those recorded in this study [18,19]. Alternatively, *in vitro* studies have found that exposure of *P. ramorum* to temperatures over 30°C decreases survival quantitatively over several hours [17]. Temperatures over 40°C, which occasionally were recorded in this study, have been shown to kill *P. ramorum* cultures in minutes to hours [17,20]. Another study found that *P. ramorum* could be recovered specifically from chlamydospores and infected leaf disks over a week at 20 or 30°C, but that recovery quickly declined at 35°C and did not occur at 40°C [21]. Furthermore, it is important to consider that temperatures recorded in this study were obtained from sensors that were shielded from the sun. Sun exposure can quickly raise leaf temperatures up to 20°C above ambient in still air when little transpiration occurs, which is likely common during Mediterranean summers [22,23]. It is interesting to note that for a few trees with canopies that were evenly split between high sun and deep shade environments, *P. ramorum* was consistently recovered at much higher proportions from leaves that were immersed in shade. Trees that were fully exposed to sun rarely produced successful isolations, and consequently, a few of these trees were

Table 2. Mixed model analysis of variance.

Effect	*F*	*P>F*
Year	5.16	0.0074
Month	3.53	0.0153
Leaf water potential	314.90	<.0001
Canopy cover	69.64	<.0001
Number of days above 30°C	15.76	<.0001
Lesion area	0.00	0.09741

Ecophysiological and site variables were associated with over-summer survival of *P. ramorum* in symptomatic bay laurel leaves.

not included in this final analysis due to a lack of symptomatic leaves at later time points (unpublished data).

Previous studies have confirmed that *P. ramorum* performs best in mild, wet and shaded environments, much like other foliar Phytophthoras [14]. The current study illustrates the critical role that microclimate plays in the survival of individual *P. ramorum* leaf infections through the summer dry season. While rainfall was not recorded during this study due to its absence in the summer dry season, it is clear that wet season rainfall also plays an important role in determining the prevalence of the pathogen during the summer months. Increased wet season rainfall has been associated with both increased inoculum production and pathogen survival in these forests and appears at least partially responsible for the greater inoculum production and infection observed in redwood-tanoak compared to mixed-evergreen forest sites [3]. Inoculum production and infection in mixed-evergreen sites appear particularly sensitive to wet season rainfall, as extended spring rains have been associated with levels of pathogen activity similar to that commonly seen in redwood-tanoak sites [5,6]. Moreover, these extended spring rains may be required for *P. ramorum* epidemics to produce exponential growth, as commonly occurs in redwood-tanoak sites [3].

Just as a particularly conducive wet season is capable of supporting large amounts of sporulation and new infections, this study demonstrates that exposure to hot, sunny dry-season conditions appears capable of significantly reducing local *P. ramorum* populations, though not necessarily removing the potential for future epidemics [24,25]. Such reductions in local *P. ramorum* populations likely contributes to delays in wet season epidemics and may ultimately limit pathogen prevalence at the beginning of the following dry season [3]. The implications of these seasonal dynamics on the long term spread and establishment of *P. ramorum* in changing California forest ecosystems remain to be seen.

Acknowledgments

We thank Ross Meentemeyer and Brian Anacker for access to microclimate data, Ken Shackel and Tedmund Swiecki for discussions of plant water status and assistance with the pressure bomb and Jerome Braun for help with statistical analyses. Finally, we thank Elizabeth Fichtner for assistance with much of the preliminary work this project was based on and Jenny Davidson for access to pre-published results.

Author Contributions

Conceived and designed the experiments: MD DR RB. Performed the experiments: MD. Analyzed the data: MD. Contributed reagents/materials/analysis tools: DR RB. Wrote the paper: MD DR RB.

References

1. Rizzo DM, Garbelotto M, Davidson JM, Slaughter GW, Koike ST (2002) *Phytophthora ramorum* as the cause of extensive mortality of Quercus spp. and *Lithocarpus densiflorus* in California. Plant Disease 86: 205–214.
2. Rizzo DM, Garbelotto M, Hansen EA (2005) *Phytophthora ramorum*: Integrative research and management of an emerging pathogen in California and Oregon forests. Annual Review of Phytopathology 43: 309–335.
3. Davidson JM, Patterson HA, Wickland AC, Fichtner EJ, Rizzo DM (2011) Forest type influences transmission of *Phytophthora ramorum* in California oak woodlands. Phytopathology 101: 492–501.
4. McPherson BA, Mori SR, Wood DL, Storer AJ, Svihra P, et al. (2005) Sudden oak death in California: Disease progression in oaks and tanoaks. Forest Ecology and Management 213: 71–89.
5. Davidson JM, Patterson HA, Rizzo DM (2008) Sources of inoculum for *Phytophthora ramorum* in a redwood forest. Phytopathology 98: 860–866.
6. Davidson JM, Wickland AC, Patterson HA, Falk KR, Rizzo DM (2005) Transmission of *Phytophthora ramorum* in mixed-evergreen forest in California. Phytopathology 95: 587–596.
7. DiLeo MV, Bostock RM, Rizzo DM (2009) *Phytophthora ramorum* does not cause physiologically significant systemic injury to California bay laurel, its primary reservoir host. Phytopathology 99: 1307–1311.
8. Fichtner EJ, Lynch SC, Rizzo DM (2009) Survival, dispersal, and potential soil-mediated suppression of *Phytophthora ramorum* in a California redwood-tanoak forest. Phytopathology 99: 608–619.
9. Fichtner EJ, Lynch SC, Rizzo DM (2007) Detection, distribution, sporulation, and survival of *Phytophthora ramorum* in a California redwood-tanoak forest soil. Phytopathology 97: 1366–1375.
10. Hüberli D, Hayden KJ, Calver M, Garbelotto M (2012) Intraspecific variation in host susceptibility and climatic factors mediate epidemics of sudden oak death in western US forests. Plant Pathology 61: 579–592.
11. Condeso TE, Meentemeyer RK (2007) Effects of landscape heterogeneity on the emerging forest disease sudden oak death. Journal of Ecology 95: 364–375.
12. Buffo J, Fritschen LJ, Murphy JL (1972) Direct solar radiation on various slopes from 0 to 60 degrees north latitude. Res Pap PNW-RP-142.
13. McCutchan H, Shackel KA (1992) Stem-water potential as a sensitive indicator of water stress in prune trees (*Prunus domestica* L. cv. French). Journal of the American Society for Horticultural Science 117: 607–611.
14. Erwin DC, Ribeiro OK (1996) *Phytophthora* diseases worldwide: American Phytopathological Society Press, St. Paul, MN.
15. Chimento A, Cacciola SA, Garbelotto M (2011) Detection of mRNA by reverse transcription PCR as an indicator of viability in *Phytophthora ramorum*. Forest Pathology 42: 14–21.
16. Littell RC, Milliken GA, Stroup WW, Wolfinger RD (1996) SAS system for mixed models. Cary, NC, USA: SAS institute Inc.
17. Browning M, Englander L, Tooley PW, Berner D (2008) Survival of *Phytophthora ramorum* hyphae after exposure to temperature extremes and various humidities. Mycologia 100: 236–245.
18. DiLeo MV, Fichtner EJ, Bostock R, Rizzo DM (2006) Effects of water potential on the growth and survival of *Phytophthora ramorum*. Phytopathology. Quebec City, Quebec, Canada.
19. Duniway JM (1983) Role of physical factors in the development of *Phytophthora* diseases. In: Erwin DC, Bartnicki-Garcia S, Tsao PH, editors. Phytophthora its biology, taxonomy, ecology, and pathology. St. Paul: The American Phytopathological Society. 175–187.
20. Englander L, Browning M, Tooley PW (2006) Growth and sporulation of *Phytophthora ramorum* in vitro in response to temperature and light. Mycologia 98: 365–373.
21. Tooley PW, Browning M, Berner D (2008) Recovery of *Phytophthora ramorum* following exposure to temperature extremes. Plant Disease 92: 431–437.

22. Ansari AQ, Loomis WE (1959) Leaf temperatures. Botanical Society of America. 713–717.
23. Fuchs M (1990) Infrared measurement of canopy temperature and detection of plant water-stress. Theoretical and Applied Climatology 42: 253–261.
24. Beh MM, Metz MR, Frangioso KM, Rizzo DM (2012) The key host for an invasive forest pathogen also facilitates the pathogen's survival of wildfire in California forests. New Phytologist 196: 1145–1154.
25. Konzanitas M, Garbelotto M (2012) Long-term monitoring of disease progression and the population genetics of *Phytophthora ramorum* within the SFPUC watershed in San Mateo County. Fifth Sudden Oak Death Science Symposium. Petaluma, California.

5

A Novel Moisture Adjusted Vegetation Index (MAVI) to Reduce Background Reflectance and Topographical Effects on LAI Retrieval

Gaolong Zhu[1,2], Weimin Ju[2,3]*, J. M. Chen[2,4], Yibo Liu[5]

1 Department of Geography, Minjiang University, Fuzhou, China, **2** International Institute for Earth System Sciences, Nanjing University, Nanjing, China, **3** Institute for Climate and Global Change Research, Nanjing University, Nanjing, China, **4** Department of Geography, University of Toronto, Toronto, Ontario, Canada, **5** Jiangsu Key Laboratory of Agricultural Meteorology, College of Applied Meteorology, Nanjing University of Information Science and Technology, Nanjing, China

Abstract

A new moisture adjusted vegetation index (MAVI) is proposed using the red, near infrared, and shortwave infrared (SWIR) reflectance in band-ratio form in this paper. The effectiveness of MAVI in retrieving leaf area index (LAI) is investigated using Landsat-5 data and field LAI measurements in two forest and two grassland areas. The ability of MAVI to retrieve forest LAI under different background conditions is further evaluated using canopy reflectance of Jack Pine and Black Spruce forests simulated by the 4-Scale model. Compared with several commonly used two-band vegetation index, such as normalized difference vegetation index, soil adjusted vegetation index, modified soil adjusted vegetation index, optimized soil adjusted vegetation index, MAVI is a better predictor of LAI, on average, which can explain 70% of variations of LAI in the four study areas. Similar to other SWIR-related three-band vegetation index, such as modified normalized difference vegetation index (MNDVI) and reduced simple ratio (RSR), MAVI is able to reduce the background reflectance effects on forest canopy LAI retrieval. MAVI is more suitable for retrieving LAI than RSR and MNDVI, because it avoids the difficulty in properly determining the maximum and minimum SWIR values required in RSR and MNDVI, which improves the robustness of MAVI in retrieving LAI of different land cover types. Moreover, MAVI is expressed as ratios between different spectral bands, greatly reducing the noise caused by topographical variations, which makes it more suitable for applications in mountainous area.

Editor: Lalit Kumar, University of New England, Australia

Funding: This work was supported by the National Natural Science Foundation of China (Grant No. 41271354), the National Basic Research Program of China (Grant No. 2010CB950702), and the Program for New Century Excellent Talents in Fujian Province University. The funders had no role in study design, data collection and analysis, decision to publish, or preparation of the manuscript.

Competing Interests: The authors have declared that no competing interests exist.

* Email: juweimin@nju.edu.cn

Introduction

In recent decades, numerous spectral vegetation indices (VIs) derived from remotely sensed data have been developed to monitor the Earth's vegetation cover and retrieve vegetation parameters such as leaf area index (LAI), fractional vegetation cover, biomass, and photosynthetic activity [1–4]. These VIs are often algebraic combinations of spectral reflectance in the red and near infrared (NIR) wavebands, for example, the most commonly used simple ratio (SR) [5] and normalized difference vegetation index (NDVI) [6]. They have been proved to be better than a single spectral band alone for estimating biophysical parameters of vegetation. However, the effectiveness of their applications is limited to different degrees by the effect of perturbing factors such as atmospheric conditions, topography, illumination and viewing geometry, sensor calibration, and soil background [7–8]. In forest ecosystems, the background often refers to all materials below the tree canopy (overstorey), including understorey (grass, shrub), moss, litter, and soil [9].

Varieties of VIs have been developed to reduce part of noise caused by these perturbing factors. Based on the so-called soil line concept [10], several soil adjusted VIs have been proposed to correct the perturbation of soil background, such as PVI [11], SAVI [12], WDVI [2], TSAVI [13], MSAVI [14], OSAVI [8], and GESAVI [15] (see Table 1). These two-band SAVI family indices appear to be less sensitive to soil brightness changes, but are more applicable to retrieving biophysical parameters of vegetation with relatively homogeneous canopies such as grasslands and croplands [7–8]. In order to enhance the sensitivity to vegetation change and minimize the noise caused by other factors, some three-band vegetation indices were developed by incorporating the blue or shortwave infrared (SWIR) bands. For instance, the atmospherically resistant vegetation index (ARVI) was designed to minimize atmospheric noise [16], and enhanced vegetation index (EVI) can reduce both the effects of atmospheric condition and soil background [17]. The modified normalized difference vegetation index (MNDVI) [18] and reduced simple ratio (RSR) [19] which combine the reflectance in the red, NIR, and SWIR bands are able to reduce the background effects. Although these indices might have some advantages for specific purposes, they are not based on the band ratio form and much of the noise may be retained or even enhanced [20–21]. It has been recognized that taking ratios between different spectral bands has the advantage of reducing unwanted noise caused by topography

Table 1. Formulas of several commonly used vegetation indices.

Vegetation index	Formula	Reference
Simple ratio (SR)	$SR = NIR/R$	[5]
Normalized difference vegetation index (NDVI)	$NDVI = (NIR-R)/(NIR+R)$	[6]
Perpendicular vegetation index (PVI)	$PVI = (NIR-aR-b)/(a^2+1)^{1/2}$	[11]
Soil adjusted vegetation index (SAVI)	$SAVI = (NIR-R)/(NIR+R+L)(1+L)$	[12]
Weighted difference vegetation index (WDVI)	$WDVI = NIR-aR$	[2]
Transformed soil adjusted vegetation index (TSAVI)	$TSAVI = a(NIR-aR-b)/(R+a(NIR-b)+X(1+a^2))$	[13]
Modified soil adjusted vegetation index (MSAVI)	$MSAVI = (2NIR+1-((2NIR+1)^2-8(NIR-R))^{1/2})/2$	[14]
Optimized soil adjusted vegetation index (OSAVI)	$OSAVI = (NIR-R)/(NIR+R+Y)$	[8]
Generalized soil adjusted vegetation index (GESAVI)	$GESAVI = (NIR-aR-b)/(R+Z)$	[15]
Atmospherically resistant vegetation index (ARVI)	$ARVI = (NIR-RB)/(NIR+RB)$, $RB = R-\gamma(B-R)$	[16]
Modified normalized difference vegetation index (MNDVI)	$MNDVI = NDVI\times(SWIR_{max}-SWIR)/(SWIR_{max}-SWIR_{min})$	[18]
Enhanced vegetation index (EVI)	$EVI = 2.5\times((NIR-R)/(NIR+6R-7.5B+1))$	[17]
Reduced simple ratio (RSR)	$RSR = SR\times(SWIR_{max}-SWIR)/(SWIR_{max}-SWIR_{min})$	[19]
Moisture adjusted vegetation index (MAVI)	$MAVI = (NIR-R)/(NIR+R+SWIR)$	This paper

Note: *B*, *R*, *NIR*, and *SWIR* are the surface reflectance in the blue, red, near infrared, and shortwave infrared bands, respectively. $SWIR_{max}$ and $SWIR_{min}$ are the maximum and minimum surface reflectance in the SWIR band, respectively. *SWIRmax* and *SWIRmin* are defined as the 1% minimum and maximum cutoff points in the histogram of the SWIR band reflectance here. *a* and *b* are the slope and intercept of the soil line, respectively. *L*, *X*, *Y*, and *Z* are soil background adjusted factors. γ is an atmospheric self-correcting factor which depends on aerosol types.

because the topographic effects often make the reflectances in the different bands change in similar proportions in the same direction. Therefore, efforts should be made to develop three-band index in band-ratio form.

In this study, we develop a new three-band VI in band-ratio form, namely a moisture adjusted vegetation index (MAVI) (see Table 1). The sensitivity of MAVI to LAI is evaluated using LAI measured in two forest and two grassland areas and Landsat-5 data. Furthermore, the background reflectance effects of MAVI on forest canopy LAI retrieval are also investigated using the canopy reflectance of Jack Pine and Black Spruce forests with different background conditions simulated by the 4-Scale model [22]. Finally, we investigate how MAVI responds to topographical variations relative to other VIs and whether this ratio principle helps reduce topographical influence on VIs.

Materials and Methods

Ethics Statement

Field LAI measurements were performed in Tiantongshan Mountain forest, Maoershan Mountain forest, Hulunbeier grassland, and Xinlinhaote grassland (described in Section 2.2 and Section 2.3). The ground measurement permits were issued by Tiantongshan Forest Ecosystem Observation and Research Station in Zhejiang Province, Maoershan Forest Ecosystem Observation and Research Station in Heilongjiang Province, Hulunbeier Grassland Ecosystem Observation and Research Station of the Ministry of Agriculture, and Inner Mongolia Grassland Ecosystem Research Station the Chinese Academy of Sciences, respectively.

2.1 Design of MAVI

The SWIR band within the 1.5–1.75 μm range provides valuable complementary information relative to visible and NIR, regarding the geometrical structure of the canopy and on the optical properties of the underlying soil [23]. SWIR reflectance is strongly related to the canopy-equivalent water thickness, which provides a possibility of inferring canopy closure from remotely sensed data [24]. Panigrahy and Parihar proved that classification accuracy of crops was significantly improved by incorporating SWIR reflectance [25]. This band has been used to develop vegetation indices together with red and NIR bands in many previous studies [26–27]. Nemani et al. and Brown et al. used SWIR reflectance to modify NDVI and SR, resulting in a decreased sensitivity to background noise while improving their correlations with LAI [18–19]. Lymburner et al. indicated that the specific leaf area vegetation index ($SLAVI = NIR/(R+SWIR)$) had a strong positive correlation with specific leaf area index [28].

Because SWIR reflectance is sensitive to water content in the canopy as the canopy liquid water absorption is linearly related to LAI [29], a new vegetation index incorporating SWIR reflectance for improving LAI retrieval, namely the moisture adjusted vegetation index (MAVI) is developed empirically as:

$$MAVI = (NIR - R)/(NIR + R + SWIR) \tag{1}$$

which can be rewritten as

$$MAVI = NDVI/(1 + SWIR/(NIR + R)) \tag{2}$$

In general, an increase in the vegetation amount (LAI) causes a decrease in the SWIR and red reflectance and an increase in the NIR reflectance and NDVI. Because the absolute amount of the change in the red reflectance is usually less than that in the NIR and SWIR reflectance, the value of $SWIR/(NIR+R)$ in Equation (2) decreases as the LAI increases. Consequently, the numerator and denominator in Equation (2) show an inverse relationship when the vegetation amount changes. This indicates MAVI potentially has higher sensitivity to LAI than NDVI. MAVI is expressed as ratios between the reflectance of the three bands, which gives it the potential to minimize the influences of external

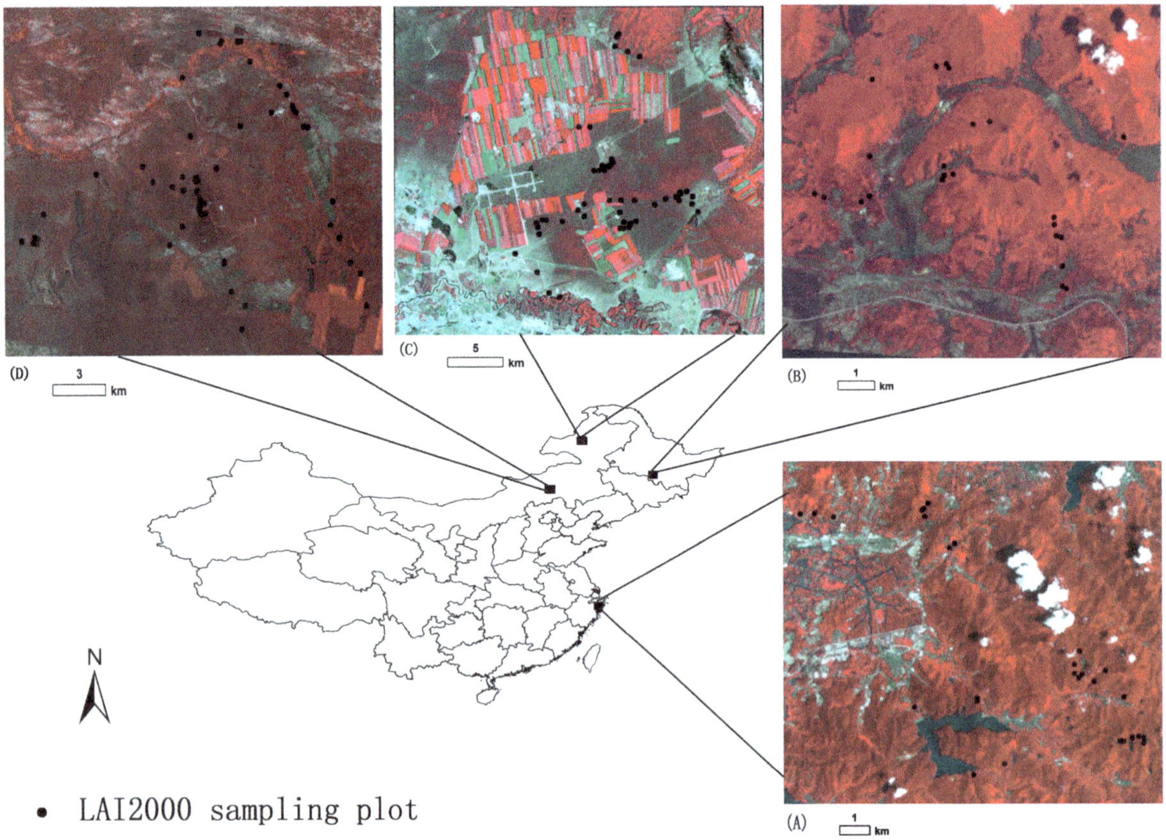

Figure 1. Areas in which LAI was measured. The effectiveness of the newly developed moisture adjusted vegetation index (MAVI) is tested using LAI measured in two grassland and two forest study areas in China: (A) Tiantongshan Mountain forest, (B) Maoershan Mountain forest, (C) Hulunbeier grassland, (D) Xinlinhaote grassland. The images composed of reflectance in bands 4, 3 and 2 from Landsat-5 TM are also shown.

perturbing factors, such as the changes of illumination and observation geometry, complex topography, and instrument calibration defects.

2.2 Study Areas

The effectiveness of MAVI is tested using LAI measured in two grassland and two forest areas (Figure 1). Table 2 summarizes the major characteristics of these four areas. Tiantongshan Mountain lies in Zhejiang Province of eastern China. The study area is located in the subtropical monsoon climate zone with a mean annual temperature of 16.2°C and mean annual precipitation of 1374.7 mm [30]. The elevations are mostly below 500 m. The forests consist of evergreen and deciduous broadleaf species and coniferous species. Maoershan Mountain is located in Heilongjiang Province of northern China. It covers an area of 266.2 km^2. The mean annual temperature and precipitation in this area are 2.8°C and 723.8 mm, respectively [31]. The elevation in this hilly highland varies from 250 m to 817 m with a mean elevation of 428 m. It is covered by a regenerative forest with various types, including broadleaf, needle-leaf, and mixed forest. The forests are approximately 50 years old.

The two grassland areas are near Hulunbeier City and Xilinhaote City of the Inner Mongolian Plateau. The elevation in Hulunbeier grassland varies from 650 m to 700 m. The mean annual temperature varies from −2.0°C to −1.0°C, and annual precipitation ranges from 350 mm to 400 mm [32]. It is a representative meadow steppe ecosystem with the height of the grass ranging from 0.3 m to 0.5 m. The average elevation in the Xilinhaote grassland is approximately 1100 m. The Xilinhaote grassland belongs to the semiarid grassland climate region with a mean annual temperature of 2.0°C and mean annual precipitation of about 350 mm, mainly occurring from June to August [33]. This site is a typical steppe ecosystem in northern China.

2.3 LAI Measurement

Field LAI measurements at Maoershan and Tiantongshan were carried out from 12 to 20 July and from 19 to 25 September, 2009, respectively. Each area had 23 plots in needle-leaf forests, broadleaf forests, and mixed forests which represent major vegetation species in the study areas. These plots were located on relatively flat terrains and composed of relatively homogeneous vegetation species. Two LAI-2000 (Li-Cor, Nebraska, USA) units were used to measure LAI. One unit was set up at an open location close to the observation plots to acquire *A* readings (above-canopy) at 30 s intervals. The other one was used to record *B* readings below the canopy at each sampling plot. A 270° view cap was used on each of LAI-2000 units to avoid the influence of the operator on the sensor. At each sampling plot, two 50 m parallel transects separated by 25 m were laid in the centre of the plot. Along each transect, measurements were made just above the ground level at 6 equally spaced spots using LAI-2000. At each spot, two *B* readings were made at random locations within a 1 m diameter circle. For each sampling spot, LAI was calculated using the *A* and *B* readings acquired simultaneously. Because of the

Table 2. The information on the study areas and the Landsat-5 TM images used in this study.

Study areas	Latitude range	Longitude range	Vegetation type	Field dates	No. of Plots	LAI range	TM scene date	Path/row
Tiantongshan	29.7755°–29.8553°N	121.6956°–121.8103°E	forest	19–25 Sep., 2009	23	3.41–6.66	18 Aug., 2009	118/39
Maoershan	45.2659°–45.3232°N	127.4957°–127.6047°E	forest	12–20 Jul., 2009	23	1.61–6.55	24 Jun., 2009	117/28
Hulunbeier	49.2744°–49.4772°N	119.9148°–120.1401°E	grassland	21–26 Jun., 2010	52	0.46–4.06	21 Jun., 2010	123/26
Xilinhaote	43.5025°–43.631°N	116.5456°–116.7898°E	grassland	29 Jun.–4 Jul., 2010	51	0.65–4.7	21 Jun., 2010	123/30

relatively severe multiple scattering effect of blue wavelength on the data of the fifth ring of LAI-2000, only the data of rings1–4 was used to calculate LAI. The LAI value for a plot was computed as the mean of 24 LAI measurements (2 LAI/spot×6 spot/transect×2 transect). LAI measurements were conducted when the sky was overcast or in hours just after sunset. The values of measured LAI in Maoershan and Tiantongshan range from 1.61 to 6.55 and from 3.41 to 6.66, respectively.

The LAI in Hulunbeier and Xilinhaote grasslands was measured from 21 to 26 June and from 29 June to 4 July, 2010, respectively. In the study areas of Xilinhaote and Hulunbeier, 51 and 52 plots were set up, respectively. Two 40 m parallel transects were placed at a distance of 25 m in each plot. Five sampling spots were spaced in 10 m intervals along each transect. A narrow groove about 0.05 m deep was created at each sampling spot to allow the upper surface of the LAI-2000 sensor head to be placed at the same level as the ground surface to measure LAI of the short grass. At each plot, the mean LAI value was obtained from 12 LAI-2000 readings arranged in a sequence: starting with one *A* reading (above-grass reading), followed by 10 *B* readings (below-grass reading in each groove), and ended with another *A* reading. Once again the 270° view cap was used to avoid operator interference with the sensor. All measurements were made near sunrise, sunset, or when overcast. The LAI of Hulunbeier and Xilinhaote grasslands is shown to vary from 0.46 to 4.06 and from 0.65 to 4.7, respectively. A GPS device was used to survey the geographical position of the centre of each plot in these four study areas.

2.4 Remotely Sensed Data Processing

Four Landsat-5 Thematic Mapper (TM) scenes covering these four study areas were downloaded from United States Geological Survey (USGS) (Table 2). These images were registered to within half a pixel with ground control points recorded in the field campaign. Radiometric correction was made using the gain and offset parameters of each band included in the Landsat-5 TM header files. Surface reflectance was obtained after atmospheric correction using the 6S code [34], with inputs of a continental air-mass, mid-latitude summer, a uniform target, and 40 km atmospheric visibility. Surface reflectance images were projected to the UTM/WGS84 coordinate system with a 30 m spatial resolution. The LAI sampling plots were positioned on the surface reflectance images using latitude/longitude coordinates measured by GPS. Because the plots were approximately 50 m×50 m and the pixel size of TM image was 30 m, the average surface reflectance in red, NIR, and SWIR bands was extracted from a 3 by 3 pixels window centered on each plot. The average surface reflectance in bands 3, 4 and 5 from Landsat-5 TM of each plot was used to calculate VIs.

2.5 VI-LAI Modeling

The semi-empirical exponential function based on the modified Beer's law is used to fit VI-LAI relationships:

$$VI = VI_\infty - (VI_\infty - VI_g)\exp(-K_{VI}LAI) \tag{3}$$

where VI_∞ is the asymptotic value of a specific VI when LAI reaches infinity (in fact, this limiting value is always reached when LAI approaches 8); VI_g is the VI value of the bare soil ($LAI = 0$). The difference between VI_∞ and VI_g controls the dynamic range of the VI. K_{VI} is an extinction coefficient determining the sensitivity of VI to LAI.

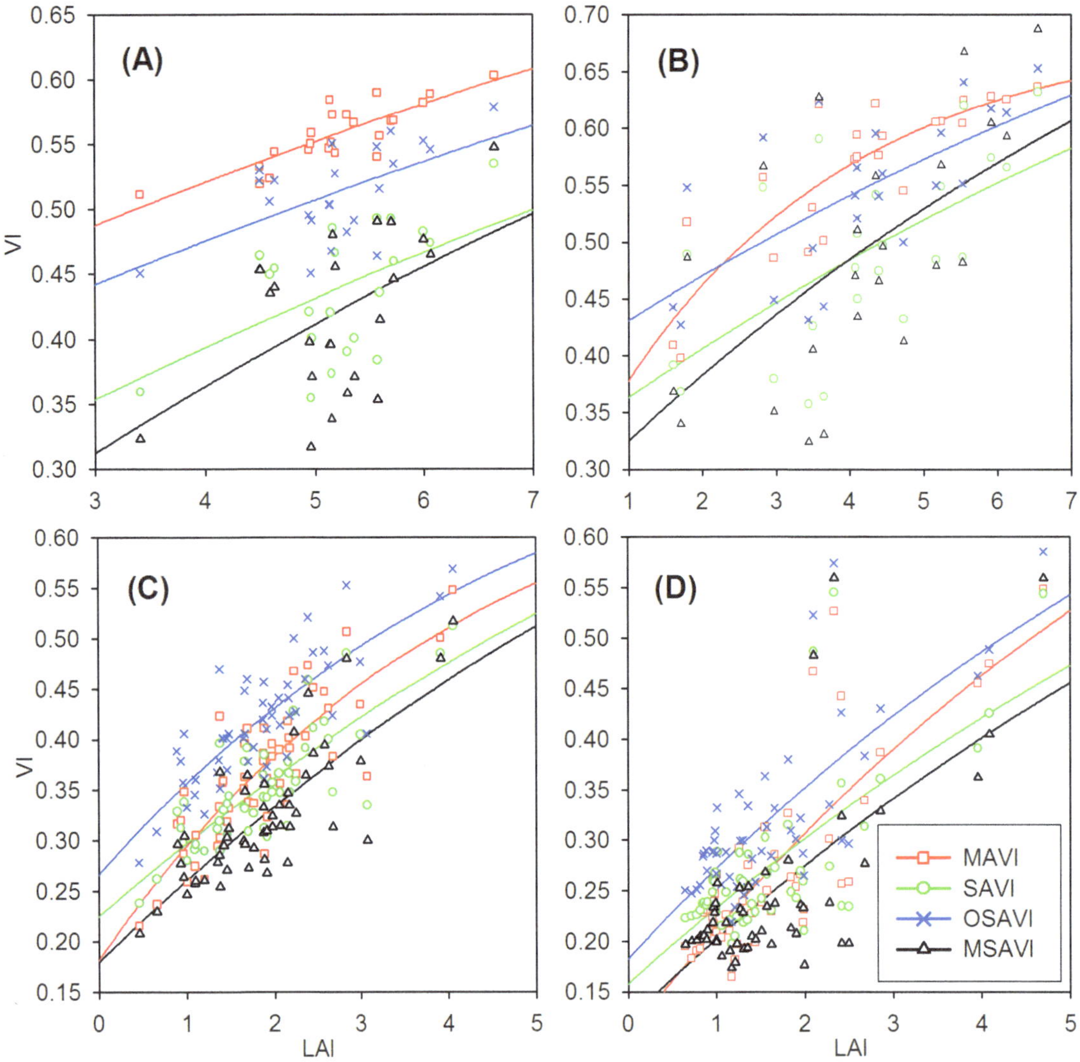

Figure 2. The best fitted relationships between LAI and vegetation indices. The MAVI and three soil adjusted vegetation indices (SAVI, OSAVI, and MSAVI) are compared in the four study areas: (A) Tiantongshan, (B) Maoershan, (C) Hulunbeier, and (D) Xilinhaote. The statistics of the best fitted VI-LAI relationships are listed in Table 3. MAVI produces a higher R^2, smaller normalized RMSE of retrieved LAI compared with the three soil adjusted vegetation index in both forest and grassland areas.

The Marquardt-Levenberg algorithm [35] is used to determine the optimal fitting parameters of each selected VI (NDVI, SAVI, OSAVI, MSAVI, RSR, MNDVI, and MAVI, see Table 1) in the four study areas in terms of the largest adjusted coefficient of determination (R^2) of each VI-LAI relationship. For each VI, these fitted parameters depend on illumination and viewing geometries, soil optical properties, and leaf inclination [13], [36]. The performance of MAVI is assessed through comparing its ability to predict LAI with the abilities of other VIs to predict LAI. The soil adjusted factors of these VIs (Table 1) are set as the values recommended by their authors ($L = 0.5$, $X = 0.08$, $Y = 0.16$, and $Z = 0.35$).

2.6 Assessing Sensitivity of VIs to Forest Background Reflectance

Forest background reflectance has an important influence on the accuracy of canopy LAI estimation in the case of low to intermediate canopy cover [9]. The canopy reflectance of Jack Pine and Black Spruce forests with different background conditions simulated by Brown et al. [19] is used here to investigate the sensitivity of VIs to background disturbances. They employed the 4-Scale model [22] to simulate the canopy reflectance of Jack Pine and Black Spruce forests with different types of backgrounds. The in situ spectroradiometric measurements of the backgrounds, including lichen, sphagnum moss, and forest soils, taken in the two sites are used in the simulations of canopy reflectance. A synthetic background of 50% of water and 50% of moss spectra, representing an extreme case, is also used to simulate the canopy reflectance of Black Spruce forest. In the simulations, the canopy LAI of Jack Pine and Black Spruce forests is allowed to change from 0.5 to 6 at an interval of 0.5 to analyze the change in the sensitivity of VIs to background reflectance with canopy density. A complete description of these modeled data can be found in Brown et al. [19].

The sensitivity of VIs to background reflectance (T_{VI}) can be characterized as [37]:

$$T_{VI} = (\sigma_{VI}/\sigma) \times 100\% \quad (4)$$

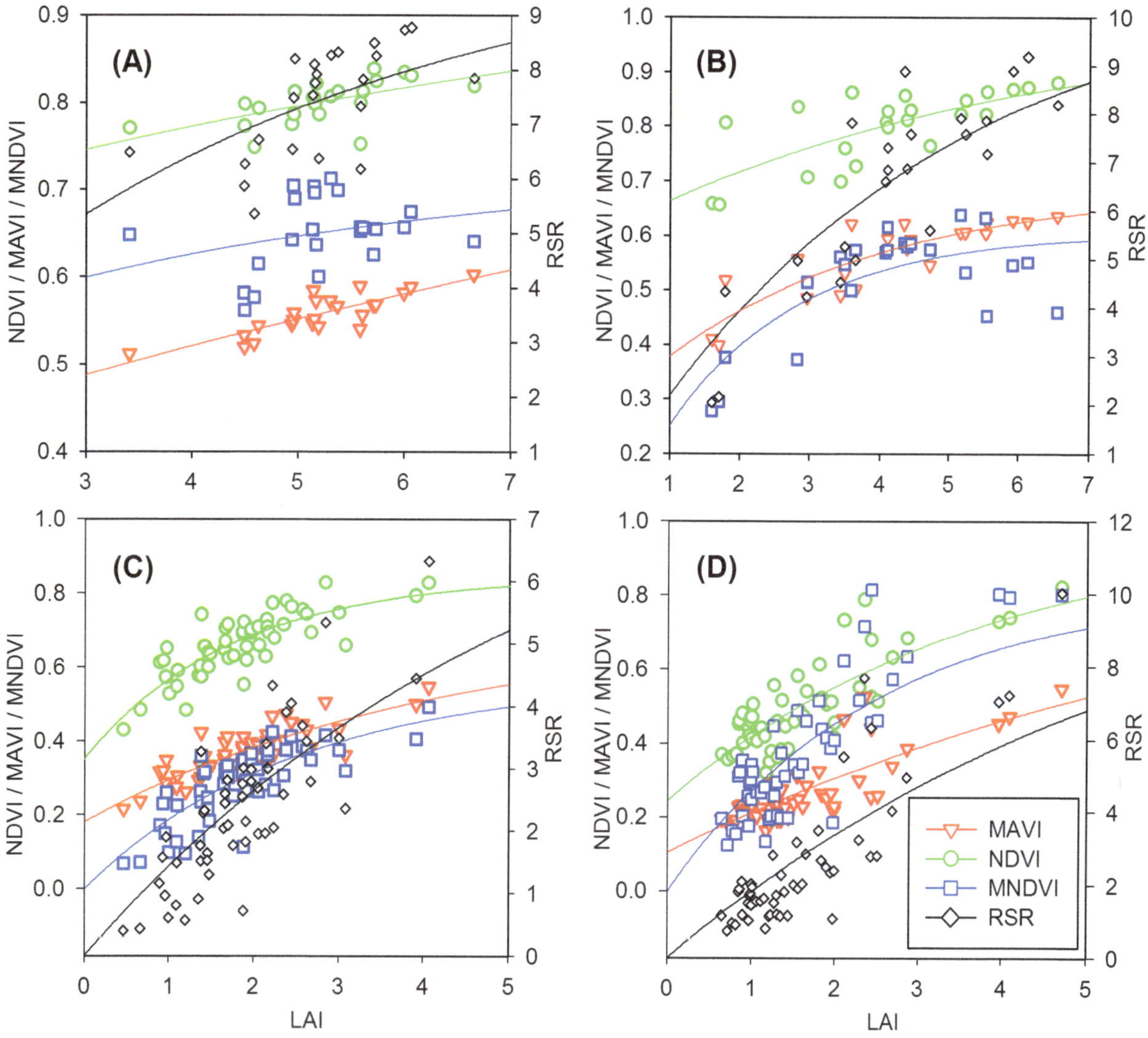

Figure 3. The best fitted relationships between LAI and vegetation indices. The MAVI and three selected vegetation indices (NDVI, MNDVI, and RSR) are compared in the four study areas: (A) Tiantongshan, (B) Maoershan, (C) Hulunbeier, and (D) Xilinhaote. MAVI produces the largest R^2 and the smallest normalized RMSE of estimated LAI in Tiantongshan and Hulunbeier. The performance of MAVI based on R^2 and RMSE is only slightly second to RSR in Maoershan and Xilinhaote. These results prove that MAVI has stable correlations with LAI under different cover types through incorporating the SWIR reflectance in band-ratio form.

where σ_{VI} refers to the standard deviation of VI values corresponding to a given LAI value and σ is the standard deviation of the VI over the whole range of measured LAI. A smaller T_{VI} value indicates higher efficiency and low sensitivity to background reflectance for a VI.

2.7 Assessing Topographical Effects on VIs

The study area in Maoershan Mountain is selected to assess the topographical effects on VIs. It is a low elevation highland with a mean slope of 14.2°. As the altitude increases, the slope and forest age increase slowly due to relatively frequent deforestation at flat terrains and lower altitudes. In assessing topographical effects on VIs, pixels with slopes from 3° to 27° are binned into 5-degree slope intervals. The pixels with slopes more than 27° are excluded since they are too few in the study area. The five classes are tagged by the median value of the slope range of each class hereafter (i.e., 5°, 10°, 15°, 20°, 25°). In each class, the mean values at every aspect angle of the selected VI are calculated. The study area is covered by high density forests with almost closed canopies, so the effects of soil background on VIs are negligible. Because the study area is relatively small, the atmospheric conditions of each pixel can be assumed identical, and as the study area is composed of relatively homogeneous vegetation species, the variations of a VI within each slope class are assumed a result of topography.

The coefficient of variation (CV) is used to evaluate the effects on VIs resulting from topography variation. It is defined as,

$$CV = \sigma_{\text{slope}} / Mean \times 100\% \tag{5}$$

where σ_{slope} and $Mean$ are the standard deviation and the mean value of VI corresponding to a given slope class, respectively. The CV value represents the noise caused by topographic variations. A smaller CV value indicates a smaller topographic effect.

As the second criterion, the topographic effects on VIs can also be expressed in the following way:

Table 3. The best fitted VI-LAI relationships and their statistics.

Study areas	VI	$VI = VI_\infty - (VI_\infty - VI_g)\exp(-K_{VI}LAI)$	STDV	$(VI_{max} - VI_{min})$	R^2	RMSE	Normalized RMSE
Tiantongshan	NDVI	$y = 1-0.354\exp(-0.11x)$	0.025	0.090	0.3703	0.0205	0.2278
	SAVI	$y = 1-0.7837\exp(-0.064x)$	0.048	0.180	0.2600	0.0431	0.2389
	OSAVI	$y = 1-0.672\exp(-0.0619x)$	0.035	0.128	0.3291	0.0305	0.2381
	MSAVI	$y = 1-0.8693\exp(-0.078x)$	0.061	0.231	0.2571	0.0549	0.2380
	RSR	$y = 10.5-10.5\exp(-0.237x)$	0.990	3.424	0.3648	0.8274	0.2417
	MNDVI	$y = 0.7299-0.2589\exp(-0.2271x)$	0.042	0.151	0.0879	0.0425	0.2806
	MAVI	$y = 1-0.6268\exp(-0.067x)$	0.024	0.091	**0.6975**	0.0137	**0.1501**
Maoershan	NDVI	$y = 1-0.4002\exp(-0.1724x)$	0.068	0.224	0.5338	0.0487	0.2171
	SAVI	$y = 1-0.6835\exp(-0.0703x)$	0.082	0.275	0.3902	0.0674	0.2454
	OSAVI	$y = 1-0.6115\exp(-0.0715x)$	0.070	0.225	0.4409	0.0546	0.2423
	MSAVI	$y = 1-0.7391\exp(-0.0901x)$	0.108	0.363	0.3802	0.0893	0.2459
	RSR	$y = 11-11\exp(-0.222x)$	2.022	7.143	**0.7580**	1.0435	**0.1461**
	MNDVI	$y = 0.6052-0.6052\exp(-0.5351x)$	0.101	0.360	0.6281	0.0644	0.1789
	MAVI	$y = 0.6904-0.4266\exp(-0.3107x)$	0.068	0.238	**0.7417**	0.0361	**0.1517**
Hulunbeier	NDVI	$y = 0.8482-0.4956\exp(-0.5686x)$	0.086	0.398	0.6667	0.0505	0.1269
	SAVI	$y = 1-0.7754\exp(-0.0978x)$	0.056	0.274	0.6496	0.0340	0.1239
	OSAVI	$y = 0.776-0.5105\exp(-0.196x)$	0.060	0.291	0.6622	0.0357	0.1228
	MSAVI	$y = 1-0.8206\exp(-0.1041x)$	0.063	0.309	0.6382	0.0389	0.1258
	RSR	$y = 9-9\exp(-0.1739x)$	1.234	5.927	0.6386	0.7569	0.1277
	MNDVI	$y = 0.5713-0.5713\exp(-0.3985x)$	0.100	0.424	0.6449	0.0607	0.1430
	MAVI	$y = 0.7205-0.5394\exp(-0.2354x)$	0.069	0.332	**0.6924**	0.0391	**0.1176**
Xilinhaote	NDVI	$y = 1-0.7552\exp(-0.2631x)$	0.120	0.503	**0.6647**	0.0712	**0.1415**
	SAVI	$y = 1-0.8425\exp(-0.0936x)$	0.078	0.348	0.5071	0.0562	0.1615
	OSAVI	$y = 1-0.8181\exp(-0.1163x)$	0.083	0.366	0.5923	0.0541	0.1479
	MSAVI	$y = 1-0.8784\exp(-0.0957x)$	0.086	0.386	0.4741	0.0640	0.1658
	RSR	$y = 12.5-12.5\exp(-0.1579x)$	1.995	9.297	**0.6669**	1.1749	**0.1264**
	MNDVI	$y = 0.825-0.825\exp(-0.4007x)$	0.186	0.691	**0.6858**	0.1065	0.1541
	MAVI	$y = 1-0.8952\exp(-0.1275x)$	0.092	0.384	**0.6627**	0.0544	**0.1418**

Note: STDV is the standard deviation of each VI corresponding to the whole range of measured LAI. $(VI_{max}-VI_{min})$ is the difference between the maximum and minimum value of each VI corresponding to the whole range of measured LAI. The last column gives the normalized RMSE (i.e., $RMSE/(VI_{max}-VI_{min})$) for each VI. Values in bold font indicate high performance.

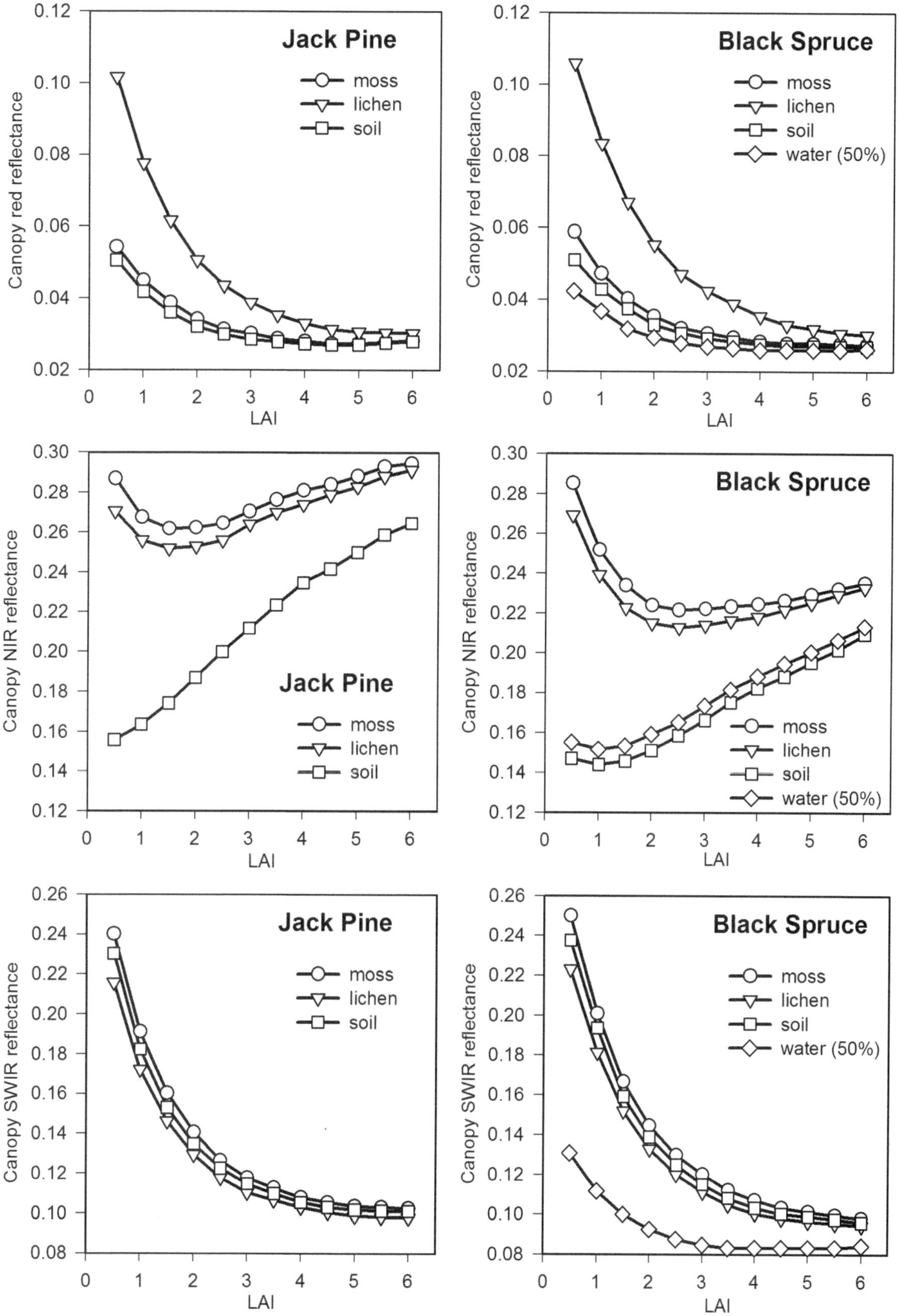

Figure 4. Canopy reflectance of Jack Pine and Black Spruce forests against LAI for different backgrounds. The canopy reflectance of Jack Pine and Black Spruce forests is simulated using the 4-Scale model against LAI for different backgrounds (moss, lichen, and forest soil). A synthetic background consisting of 50% of water and 50% of moss is also included for the Black Spruce forest [19]. The sensitivity of MAVI to background reflectance disturbances is investigated using these modeled results.

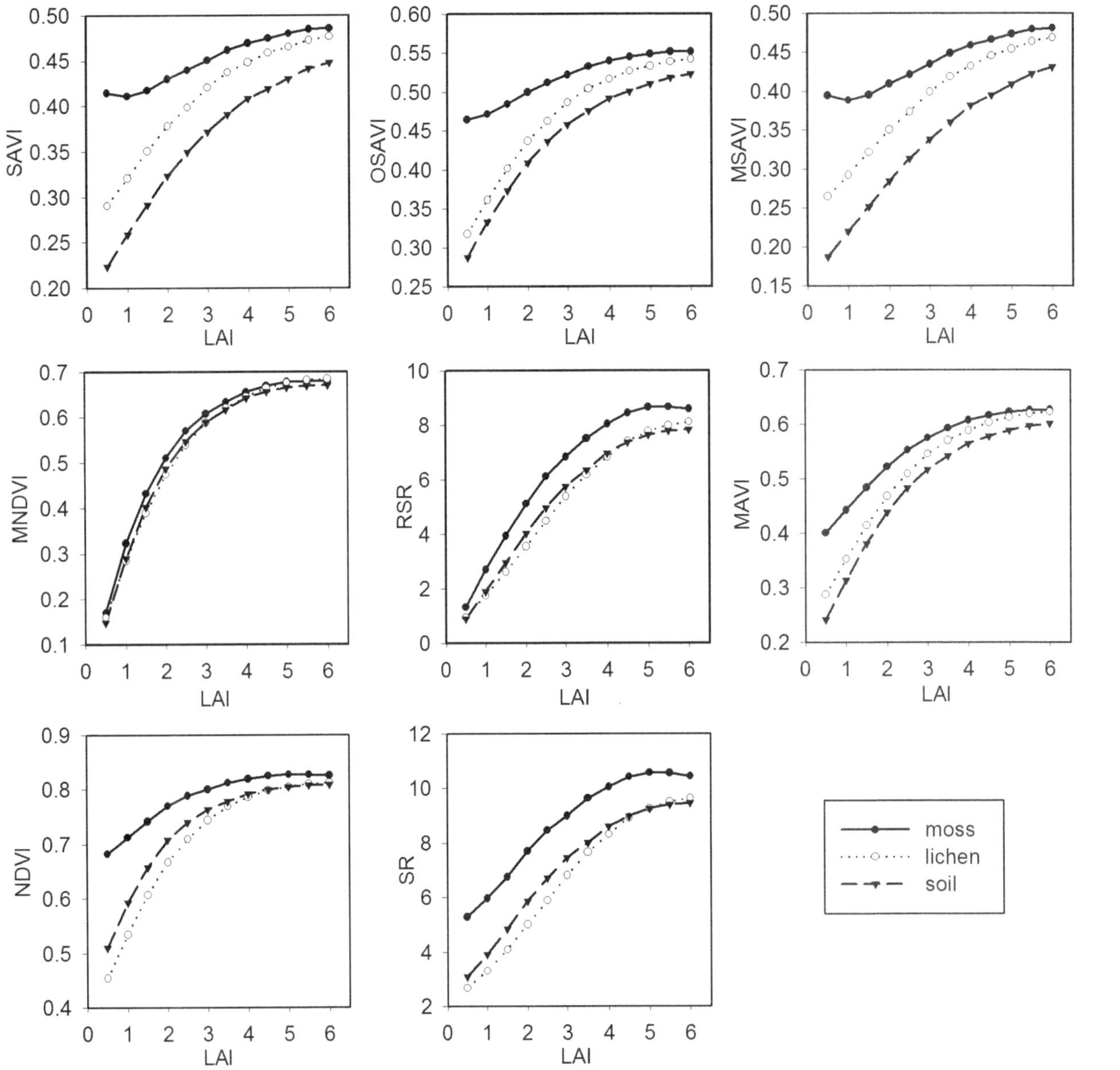

Figure 5. Background reflectance effects on vegetation indices at different LAI values in Jack Pine forest. The effects of different backgrounds (moss, lichen, and forest soil) on the selected vegetation indices (SAVI, OSAVI, MSAVI, MNDVI, RSR, MAVI, NDVI, and SR) are simulated using the 4-Scale model for the different LAI levels in Jack Pine forest. The forest background strongly affects the values of SAVI, OSAVI, MSAVI, and NDVI as the LAI values are less than 2. MAVI and SR can reduce the effects of forest backgrounds at low LAI values. RSR and MNDVI show the smallest background reflectance effects among these vegetation indices.

$$VI = m \times \cos i + n \quad (6)$$

$$\cos i = \cos\theta_s \cos\theta + \sin\theta_s \sin\theta \cos(\varphi_s - A) \quad (7)$$

where i is the incidence angle defined as the angle between the direction of the sun and the local surface normal, which represents one of the most important perturbations in remote sensing over mountainous terrain [38]. m and n represent the slope and intercept of the linear regression line between VI and cos$i$; θ_s and φ_s are the solar zenith angle and the solar azimuth angle, respectively; θ and A are the slope and aspect of an inclined pixel, respectively. These are derived from the 30 m DEM of the study area. The coefficient of determination (R^2) of the regression represents the sensitivity of VI to topographic variations. The R^2 value increases as the topographic effect increases.

Results and Discussions

3.1 Relationships between VIs and LAI

Seven VIs including NDVI, SAVI, OSAVI, MSAVI, RSR, MNDVI, and MAVI, full descriptions found in Table 1, are selected and calculated from the TM surface reflectance image to investigate their sensitivity to LAI. Figure 2 and Figure 3 show the best fitted relationships between the selected VIs and LAI in the four study areas. All VIs are almost linearly correlated with LAI in

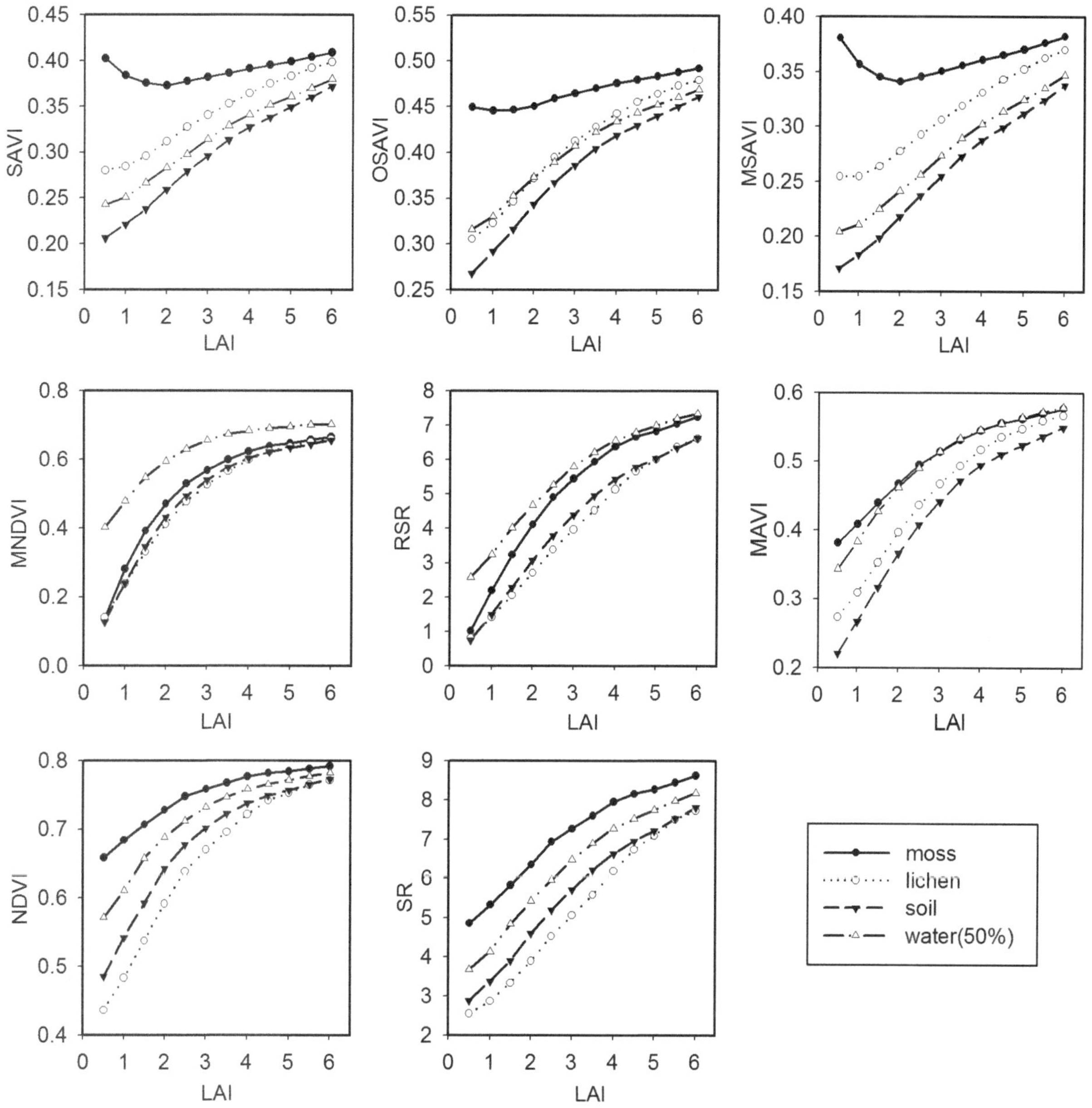

Figure 6. Background reflectance effects on vegetation indices at different LAI values in Black Spruce forest. The effects of different backgrounds (moss, lichen, forest soil, and the mixed background of water and moss) on the selected vegetation indices (SAVI, OSAVI, MSAVI, MNDVI, RSR, MAVI, NDVI, and SR) are simulated using the 4-Scale model for the different LAI levels in Black Spruce forest. The results are similar to those of Figure 5, but RSR and MNDVI do not perform much better than other vegetation indices in reducing the effect of the mixed background of water and moss.

the Tiantongshan, in which vegetation density is relatively high and in a smaller range compared with other areas. RSR, SAVI, OSAVI, and MSAVI also show approximately linear correlations with LAI in the remaining three study areas. NDVI, MNDVI and MAVI increase rapidly with increasing LAI at low LAI values, and increase slowly with increasing LAI at high LAI values.

Table 3 shows the best fitted relationships between the selected VIs and LAI in the four study areas. In order to facilitate the comparison, the RMSE for each VI is normalized by the difference between the maximum and minimum value of each VI corresponding to the whole range of measured LAI. The fitted parameters in the VI-LAI relationships vary considerably with VIs and cover types. The dynamic ranges (VI_∞-VI_g) of these selected VIs are quite different in the four study areas. This is particularly true in Xilinhaote where the dynamic range is relatively larger. Among these two-band VIs, MSAVI has the largest dynamic range values of 0.8693, 0.7391, 0.8206, and 0.8784 in Tiantongshan, Maoershan, Hulunbeier, and Xinlinhaote, respectively, while NDVI has the smallest dynamic range values of 0.354, 0.4002, 0.4956, and 0.7552 in these areas. These results are in agreement with the previous findings of Wu et al. [36]. NDVI presents R^2 values of 0.3703, 0.5338, 0.6667, and 0.6647 in the four study areas, which are larger than those of the three soil adjusted VIs (SAVI, OSAVI, and MSAVI), and has smaller normalized RMSE of 0.2278, 0.2171, and 0.1415 in Tiantongshan, Maoershan, and Xinlinhaote, respectively, indicating this

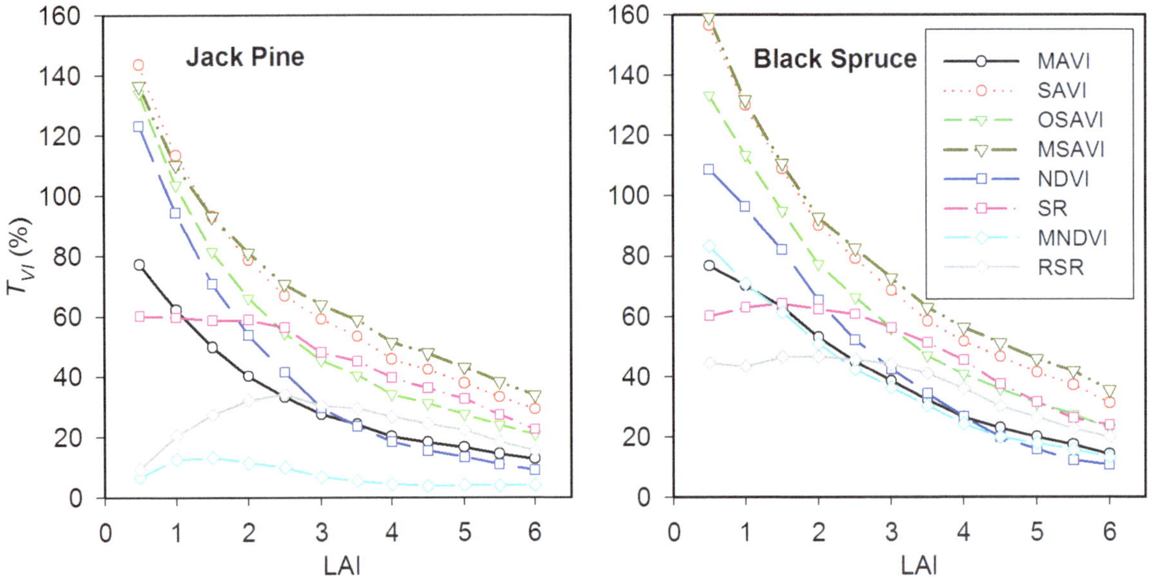

Figure 7. Sensitivity of different vegetation indices to forest background reflectance. The background reflectance strongly affects the values of SAVI, OSAVI, MSAVI, and NDVI in both Jack Pine and Black Spruce forests as the LAI values are less than 2, leading to larger T_{VI} values of them compared with other vegetation indices. RSR and MNDVI present small T_{VI} values at low LAI values, indicating that the background reflectance effects on them are smaller than other vegetation indices. MAVI has relatively small T_{VI} values over the entire LAI ranges in both Jack Pine and Black Spruce forests. The results demonstrate that MAVI that combines the red, NIR and SWIR reflectance can reduce the effects of background reflectance on forest canopy LAI retrieval.

ratio-based simple vegetation index has some advantages for retrieval of LAI over more complex VIs. OSAVI shows better efficiency than SAVI and MSAVI in all study areas. The three soil adjusted VIs all present stronger correlations with LAI in grassland areas than seen in forest areas, suggesting they are more applicable for retrieving LAI for vegetations with relatively homogeneous canopies such as grasslands and croplands than for vegetations with distinct canopy structure such as forests.

The three-band VIs (MNDVI, RSR, and MAVI) all perform significantly better than the two-band VIs in estimating LAI in Maoershan and Xinlinhaote. However, MNDVI and RSR are poorer predictors of LAI compared with two-band VIs in Tiantongshan and Hulunbeier, which may be due to their non-ratio forms and the difficulty in properly determining the values of $SWIR_{max}$ and $SWIR_{min}$ in them. Instead, MAVI produces the largest R^2 of 0.6975 and 0.6924 and the smallest normalized RMSE of 0.1501 and 0.1176 in Tiantongshan and Hulunbeier, respectively, indicating the superior performance of MAVI over other VIs in retrieving LAI. These results prove that MAVI has stable correlations with LAI under different cover types due to incorporating the SWIR reflectance in band-ratio form.

3.2 Sensitivity of VIs to Background Reflectance in Forest Canopy LAI Retrieval

Figure 4 shows the canopy reflectance in the red, NIR, and SWIR bands simulated by the 4-Scale model for Jack Pine and Black Spruce forests with different background types. The red and SWIR reflectance decreases with increasing LAI for both Jack Pine and Black Spruce forests over the entire range of LAI. The canopy NIR reflectance shows different curves, depending on background types. At LAI values less than 2.0, the canopy NIR reflectance decreases with increasing LAI over the moss and lichen backgrounds due to the increase in the probability of observing shadowed background, while the canopy NIR reflectance increases slowly with increasing LAI at LAI values above 2. For non-vegetative backgrounds, the canopy NIR reflectance increases with increasing LAI over the entire range of LAI. The effects of different backgrounds on the canopy reflectance in the red, NIR and SWIR bands decrease with increasing LAI. The canopy SWIR reflectance presents an inverse curvilinear trend with increasing LAI and reaches an asymptote at an LAI of approximately 5, while the canopy red reflectance reaches an asymptote at an LAI of approximately 3. The similarity of SWIR reflectance across different backgrounds (lichen, moss and soil) compared to the range shown in the red and NIR reflectance, and the large sensitivity of SWIR to LAI are the main reasons for the better performance of RSR compared with SR in estimating canopy LAI [19]. This also demonstrates that the canopy SWIR reflectance can be used to quantify the vegetation amount as well as canopy closure in most cases [23–24].

Figure 5 and Figure 6 illustrate the background reflectance effects on the calculated VIs at the different LAI levels in Jack Pine and Black Spruce forests. In general, the effects of different background reflectance on the canopy VIs decrease with increasing LAI for all VIs in both Jack Pine and Black Spruce forests, because the probability of observing background decreases with increasing LAI. The moss background strongly affects the values of SAVI, OSAVI, and MSAVI in both Jack Pine and Black Spruce forests as the LAI values are less than 2, leading to larger T_{VI} values compared with other VIs (Figure 7). Similarly, the noise induced to NDVI by forest background reflectance shows large T_{VI} values varying from 123% to 71% in Jack Pine forest and from 109% to 82% in Black Spruce forest corresponding to LAI from 0.5 to 1.5, which is in agreement with the conclusion of Nemani et al. that NDVI alone cannot be used to estimate LAI in open forest canopies [18]. SR presents relatively small T_{VI} values changing from 60% to 59% in Jack Pine forest and from 60% to 64% in Black Spruce forest corresponding to LAI from 0.5 to 1.5, indicating that the background reflectance effects on SR are

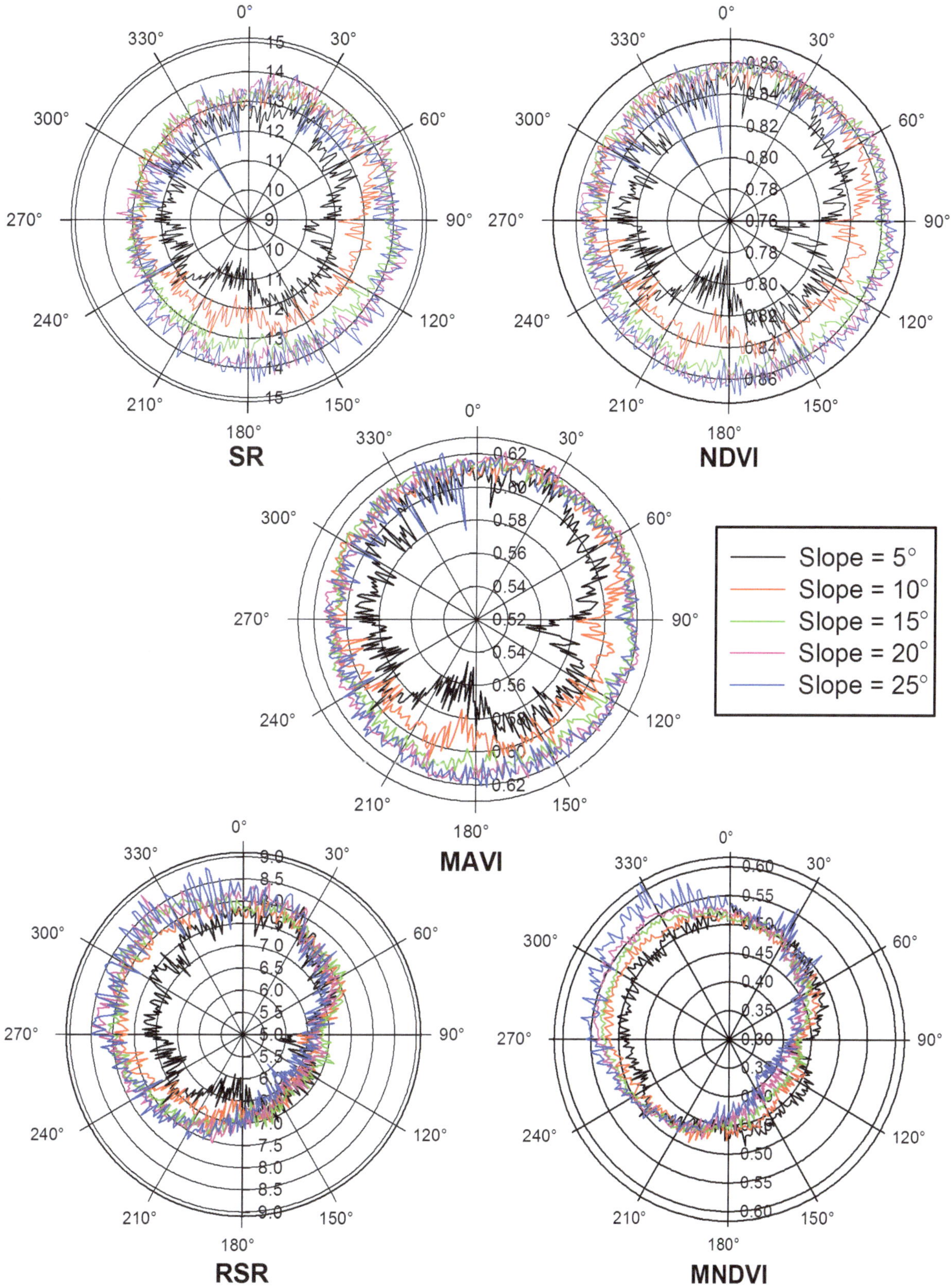

Figure 8. Variations of different vegetation indices with aspects in 5-degree slope intervals in Maoershan Mountain forest. The polar angle represents aspect, and the radius represents the mean values of each VIs at a given aspect on different slopes. Topography strongly affects RSR and MNDVI, resulting in negative biases on sun-facing slopes and positive biases on sun-backing slopes. The values of SR, NDVI, and MAVI increase as the slope increases, which is similar to the changes of forest age with slope. It can be inferred that vegetation indices expressed in band-ratio form are able to remove a large proportion of topographical noise.

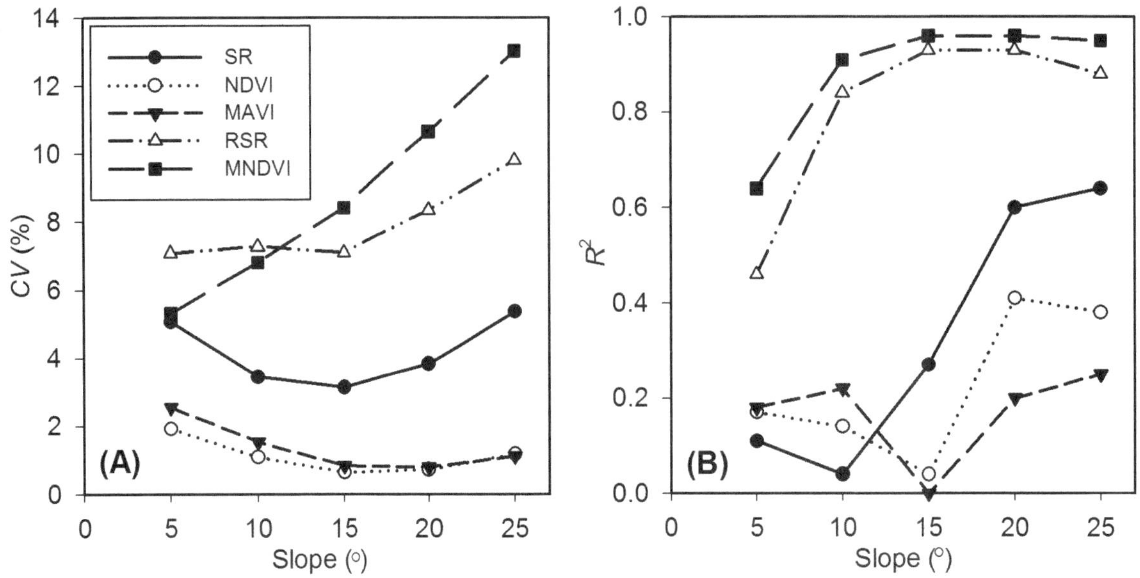

Figure 9. Effects of slope variations on different vegetation indices. Note: (A) the coefficient of variation (*CV*) of each vegetation index varies with slopes, (B) the R^2 values of linear correlations between vegetation indices and the cosine of the incidence angle vary with slopes. The *CV* values of MNDVI vary from 5.32% to 13.02% corresponding to the slopes from 5° to 25°, which shows the largest topographical noise among all the selected vegetation indices. RSR has the second largest *CV* values ranging from 7.09% to 9.81%. SR presents a medium *CV* values in the range from 3.14% to 5.37%. The *CV* values are quite small for NDVI and MAVI ranging from 0.64% to 1.92% and from 0.80% to 2.56%, respectively, implying that NDVI and MAVI can remove much of topographic noise through expressing in band-ratio form. The conclusions based on R^2 are also similar.

smaller than those on NDVI, which supports the statement by Chen and Cihlar that SR has the highest correlation with LAI in boreal forests [39]. RSR and MNDVI can largely reduce the effects of moss, lichen, and forest soil backgrounds on calculated VIs at low LAI values (Figure 5–7), mainly due to the fact that the SWIR reflectance in RSR and MNDVI can scale down the increased NIR response in open stands associated with understory vegetation or other highly reflective backgrounds [18]. However, RSR and MNDVI do not perform much better than other VIs in reducing the effect of the mixed background of water and moss since SWIR is sensitive to wet background leading to the relatively low SWIR reflectance (Figure 6). On the other hand, the efficiency of RSR and MNDVI to reduce background reflectance effects is strongly controlled by the maximum and minimum values of SWIR reflectance in them, which is difficult to determine properly in many practical applications.

MAVI has relatively small T_{VI} values varying from 77% to 13% in Jack Pine forest and from 76% to 14% in Black Spruce forest over the entire LAI ranges (Figure 7). The results demonstrate that MAVI that combines the red, NIR and SWIR reflectance can reduce the effects of backgrounds on forest canopy LAI retrieval. Its underlying physical mechanism is similar to those of RSR and MNDVI, i.e., SWIR reflectance is highly sensitive to LAI at low LAI values and has the largest values in open forest canopies, which acts as an adjustment factor in the denominator of MAVI to moderate the increased NIR in open stands (Figure 4). However, the canopy SWIR reflectance is also sensitive to the wetness of background, especially at low LAI values. Fortunately, the wetness of background affects not only the canopy SWIR reflectance but also the canopy NIR and red reflectance (Figure 4). The ratio form of MAVI makes it possible to constrain the effects of the change in the wetness of background partially and to be suitable for retrieving LAI for forests with different types of backgrounds.

3.3 Topographic Effects on VIs

Figure 8 shows the band-ratio VIs (SR, NDVI, and MAVI) and the non-band-ratio VIs (RSR and MNDVI) change with aspect at different slopes. When the slope is 5°, each VI varies greatly with aspect due to relatively large vegetation variations. When the slope is larger than 10°, the changes of SR, NDVI, and MAVI at a given slope are quite different from those of RSR and MNDVI. The values of SR, NDVI, and MAVI increase as the slope increases, which is similar to the changes of forest age with slope. However, the values of RSR and MNDVI decrease on sun-facing slopes and increase on sun-backing slopes as the slopes increase. At a given slope, the values of RSR and MNDVI of sun-facing slopes are much smaller than those of sun-backing slopes. The larger the slope, the larger the difference with regard to sun-facing and sun-backing slopes. Because forest ages are almost the same at all aspect according to the ground reference data, the values of each VI at a given slope should be independent of the aspect relative to the sun. Therefore, it can be inferred that VIs expressed in band-ratio form are able to remove a large proportion of topographical noise. Topography strongly affects RSR and MNDVI, resulting in negative biases on sun-facing slopes and positive biases on sun-backing slopes.

Figure 9 depicts quantitatively the topographic effects on the selected VIs. The *CV* values of MNDVI vary from 5.32% to 13.02% corresponding to the slopes from 5° to 25°, showing the largest topographical noise among all the selected VIs. RSR has the second largest *CV* values ranging from 7.09% to 9.81%. SR presents medium *CV* values in the range from 3.14% to 5.37%. The *CV* values are quite small for NDVI and MAVI ranging from 0.64% to 1.92% and from 0.80% to 2.56%, respectively, implying that NDVI and MAVI can remove much of the topographic noise through expressing in band-ratio form.

In general, the R^2 values of linear correlations between VI and the cosine of the incidence angle (cos*i*) increase as the slope increases (Figure 9). The R^2 values are in the range from 0.46 to

0.93 for RSR and from 0.64 to 0.96 for MNDVI in the study area, indicating that VIs not based on band ratios are influenced strongly by topographic variations. When the slope is above 20°, the largest R^2 values are 0.64 for SR and 0.41 for NDVI. The results demonstrate that even VIs based on band-ratios still include significant topographic effects on steep slopes, confirming the findings of Burgess et al. [40] and Kusaka and Sakane [41]. When the slope is 25°, the R^2 value of MAVI is 0.25, which proves that MAVI performs rather effectively in removing the effects of slopes. SR is more sensitive to topographic variations on steep slopes than NDVI and MAVI, which may be due to the relatively different topographic effects on the red and NIR bands. Therefore, the topographic effects should be removed before the applications of VIs that are not based on band ratios, even when slopes are small. The topographic effects on VIs expressed in band-ratio form can usually be ignored on small slopes, but careful topographic corrections are needed when they are used over the rugged surface.

Conclusions

In this study, we develop a new three-band moisture adjusted vegetation index (MAVI). Its performance is evaluated against commonly used two-band VIs (NDVI, SR, SAVI, MSAVI, OSAVI) and three-band VIs (MDNVI, RSR) with field measurements made in two forest and two grassland areas in China. The reflectance data simulated by the 4-Scale model is also use to investigate the background reflectance effects of MAVI on forest canopy LAI retrieval. The following conclusions can be drawn from this study:

(1) MAVI is suitable for retrieving LAI using remote sensing images. It produces a higher R^2, smaller normalized RMSE of retrieved LAI compared with two-band VIs in both forest and grassland areas. The superior performance of MAVI over two-band VIs is mainly due to its integration of the signals from the red, NIR, and SWIR bands sensitive to the greenness, chlorophyll, and water content of the vegetation.

(2) MAVI can reduce the background reflectance effects on forest canopy LAI retrieval as effectively as RSR and MNDVI. It outperformed RSR and MNDVI for retrieving LAI in the four study areas without the need for inputting the maximum and minimum SWIR values, which are notoriously difficult to determine.

(3) Topography strongly affects VIs that are not based on band ratios, such as RSR and MNDVI. Since MAVI is expressed as ratios between spectral bands, it can greatly reduce the noise caused by topographical variations, which makes it suitable for application in mountainous area.

In this study, validation shows the robustness of MAVI in retrieving LAI of forests and grasses. Because the SWIR reflectance is also affected by the wetness of soils, especially when vegetation density is low, the robustness of MAVI may need further validation using more data from other ecosystems.

Acknowledgments

The authors would like to acknowledge Maoershan Forest Ecosystem Observation and Research Station in Heilongjiang Province, Tiantongshan Forest Ecosystem Observation and Research Station in Zhejiang Province, Inner Mongolia Grassland Ecosystem Research Station the Chinese Academy of Sciences (IMGERS), and Hulunbeier Grassland Ecosystem Observation and Research Station of the Ministry of Agriculture for supporting ground measurement. We also thank United States Geological Survey (USGS) for providing Landsat-5 data in the study.

Author Contributions

Conceived and designed the experiments: JMC WJ. Performed the experiments: GZ WJ YL JMC. Analyzed the data: GZ WJ. Contributed reagents/materials/analysis tools: WJ JMC GZ. Wrote the paper: GZ WJ JMC.

References

1. Asrar G, Fuchs M, Kanemasu ET, Hatfield JL (1984) Estimating absorbed photosynthetic radiation and leaf area index from spectral reflectance in wheat. Agronomy Journal 76: 300–306.
2. Clevers JPW (1989) The applications of a weighted infrared-red vegetation index for estimating leaf area index by correcting for soil moisture. Remote Sensing of Environment 29: 25–37.
3. Myneni RB, Williams DL (1994) On the relationship between fAPAR and NDVI. Remote Sensing and Environment 49: 200–211.
4. Bannari A, Morin D, Bonn F, Huete AR (1995) A review of vegetation indices. Remote Sensing Reviews 13: 95–120.
5. Jordan CF (1969) Derivation of leaf area index from quality of light on the forest floor. Ecology 50: 663–666.
6. Rouse JW, Haas RH, Schell JA, Deering DW (1974) Monitoring vegetation systems in the great plains with ERTS. In Proceeding of Third Earth Resources Technology Satellite Symposium 1, Greenbelt, USA, NASA SP-351, 309–317.
7. LePrieur D, Verstraete MM, Pinty B (1994) Evaluation of the performance of various vegetation indices to retrieve cover from AVHRR data. Remote Sensing Reviews 10: 265–284.
8. Rondeaux G, Steven M, Baret F (1996) Optimization of soil-adjusted vegetation indices. Remote Sensing of Environment 55: 95–107.
9. Pisek J, Chen JM, Miller JR, Freemantle JR, Peltoniemi JI, et al. (2010) Mapping forest background reflectance in a boreal region using multiangle compact airborne spectrographic imager data. IEEE Transactions on Geoscience and Remote Sensing 48: 499–510.
10. Baret F, Jacquemoud S, Hanocq JF (1993) The soil line concept in remote sensing. Remote Sensing Reviews 7: 65–82.
11. Richardson AJ, Wiegand CL (1977) Distinguishing vegetation from soil background information. Photogrammetric Engineering and Remote Sensing 43: 1541–1552.
12. Huete AR (1988) A soil adjusted vegetation index (SAVI). Remote Sensing of Environment 25: 295–309.
13. Baret F, Guyot G (1991) Potentials and limits of vegetation indices for LAI and APAR assessment. Remote Sensing of Environment 35: 161–173.
14. Qi J, Chehbouni A, Huete AR, Kerr YH, Sorooshian S (1994) A modified soil adjusted vegetation index. Remote Sensing of Environment 48: 119–126.
15. Gilabert MA, González-Piqueras J, García-Haro FJ, Meliá J (2002) A generalized soil-adjusted vegetation index. Remote Sensing of Environment 82: 303–310.
16. Kaufman YJ, Tanre D (1992) Atmospherically resistant vegetation index (ARVI) for EOS-MODIS. IEEE Transactions on Geoscience and Remote Sensing 30: 261–270.
17. Liu HQ, Huete AR (1995) A feedback based modification of the NDVI to minimize canopy background and atmospheric noise. IEEE Transactions on Geoscience and Remote Sensing 33: 457–465.
18. Nemani R, Pierce L, Running S, Band L (1993) Forest ecosystem processes at the watershed scale: sensitivity to remotely sensed leaf area index estimates. International Journal of Remote Sensing 14: 2519–2534.
19. Brown LJ, Chen JM, Leblanc SG, Cihlar J (2000) A shortwave infrared correction to the simple ratio: an image and model analysis. Remote Sensing of Environment 71: 16–25.
20. Chen JM (1996) Evaluation of vegetation indices and a modified simple ratio for boreal applications. Canadian Journal of Remote Sensing 22: 229–242.
21. Matsushita B, Yang W, Chen J, Onda Y, Qiu G (2007) Sensitivity of the enhanced vegetation index (EVI) and normalized difference vegetation index (NDVI) to topographic effects: A case study in high-density cypress forest. Sensors 7: 2636–2651.
22. Chen JM, Leblanc SG (1997) A four-scale bi-directional reflectance model based on canopy architecture. IEEE Transactions on Geoscience and Remote Sensing 35: 1316–1337.
23. Baret F, Guyot G, Begue A, Maurel P, Podaire A (1988) Complementarity of middle infrared with visible and near-infrared reflectance for monitoring wheat canopies. Remote Sensing of Environment 26: 213–225.
24. Butera C (1986) A correlation and regression analysis of percent canopy closure and TM spectral response for selected forest sites in San Juan national forest, Colorado. IEEE Transactions on Geoscience and Remote Sensing 24: 122–129.
25. Panigrahya S, Parihara JS (1992) Role of middle infrared bands of Landsat thematic mapper in determining the classification accuracy of rice. International Journal of Remote Sensing 13: 2943–2949.

26. Gao BC (1996) NDWI – a normalized difference water index for remote sensing of vegetation liquid water from space. Remote Sensing of Environment 58: 257–266.
27. Gong P, Pu R, Biging G, Larrieu MR (2003) Estimation of forest leaf area index using vegetation indices derived from Hyperion hyperspectral data. IEEE Transactions on Geoscience and Remote Sensing 41: 1355–1362.
28. Lymburner L, Beggs PJ, Jacobson CR (2000) Estimation of canopy-average surface-specific leaf area using Landsat TM data. Photogrammetric Engineering and Remote Sensing 66: 183–191.
29. Vescovo L, Gianelle D (2008) Using the MIR bands in vegetation indices for the estimation of grassland biophysical parameters from satellite remote sensing in the Alps region of Trentino (Italy). Advances in Space Research 41: 1764–1772.
30. Huang JJ, Wang XH, Yan ER (2007) Leaf nutrient concentration, nutrient resorption and litter decomposition in an evergreen broad-leaved forest in Eastern China. Forest Ecology and Management 239: 150–158.
31. Ning B, Gong WF, Fan WY (2009) Evaluation of land use suitability of Maoershan forest farm based on RS and GIS. Journal of Northeast Forestry University 37: 56–58 (in Chinese with English abstract).
32. Li J, Zhang DG, Zhang HB, Li G, Yang GX, et al. (2010) A study on the variations of pivotal photosynthetic active radiation parameters of a Leymus chinensis meadowsteppe in Hulunber. Acta Prataculturae Sinica 19: 243–250 (in Chinese with English abstract).
33. Pan QM, Bai YF, Han XG, Yang JC (2004) Studies on the fate of labelled nitrogen applied to a Leymus chinensis community of typical steppe in Inner Mongolia Grassland. Acta Phytoecologica Sinica 28: 665–671 (in Chinese with English abstract).
34. Vermote E, Tanré D, Deuzé JL, Herman M, Morcette JJ (1997) Second simulation of the satellite signal in the solar spectrum: an overview. IEEE Transactions on Geoscience and Remote Sensing 35: 675–686.
35. Marquardt DW (1963) An algorithm for least squares estimation of parameters. Journal of the Society of Industrial and Applied Mathematics 11: 431–441.
36. Wu JD, Wang D, Bauer ME (2007) Assessing broadband vegetation indices and QuickBird data in estimating leaf area index of corn and potato canopies. Field Crops Research 102: 33–42.
37. Gilabert MA, González-Piqueras J, García-Haro J, Meliá J (1998) Designing a generalized soil-adjusted vegetation index (GESAVI). In E.T. Engman (Ed.), Remote sensing for agriculture, ecosystems, and hydrology, Proceedings of SPIE, Washington, USA, vol. 3499, 396–404.
38. Proy C, Tanré D, Deschamps PY (1989) Evaluation of topographic effects in remotely sensed data. Remote Sensing of Environment 30: 21–32.
39. Chen JM, Cihlar J (1996) Retrieving leaf area index of boreal conifer forests using Landsat TM images. Remote Sensing of Environment 55: 153–162.
40. Burgess DW, Lewis P, Muller J-PAL (1995) Topographic effects in AVHRR NDVI data. Remote Sensing of Environment 54: 223–232.
41. Kusaka T, Sakane M (1997) Estimation of topographic effects in NDVI data obtained from satellite images. IEEE Proc. IGARSS'97, PP. 972–974.

6

Leaf Phenological Characters of Main Tree Species in Urban Forest of Shenyang

Sheng Xu, Wenduo Xu, Wei Chen, Xingyuan He*, Yanqing Huang, Hua Wen

State Key Laboratory of Forest and Soil Ecology, Institute of Applied Ecology, Chinese Academy of Sciences, Shenyang, China

Abstract

Background: Plant leaves, as the main photosynthetic organs and the high energy converters among primary producers in terrestrial ecosystems, have attracted significant research attention. Leaf lifespan is an adaptive characteristic formed by plants to obtain the maximum carbon in the long-term adaption process. It determines important functional and structural characteristics exhibited in the environmental adaptation of plants. However, the leaf lifespan and leaf characteristics of urban forests were not studied up to now.

Methods: By using statistic, linear regression methods and correlation analysis, leaf phenological characters of main tree species in urban forest of Shenyang were observed for five years to obtain the leafing phenology (including leafing start time, end time, and duration), defoliating phenology (including defoliation start time, end time, and duration), and the leaf lifespan of the main tree species. Moreover, the relationships between temperature and leafing phenology, defoliating phenology, and leaf lifespan were analyzed.

Findings: The timing of leafing differed greatly among species. The early leafing species would have relatively early end of leafing; the longer it took to the end of leafing would have a later time of completed leafing. The timing of defoliation among different species varied significantly, the early defoliation species would have relatively longer duration of defoliation. If the mean temperature rise for 1°C in spring, the time of leafing would experience 5 days earlier in spring. If the mean temperature decline for 1°C, the time of defoliation would experience 3 days delay in autumn.

Interpretation: There is significant correlation between leaf longevity and the time of leafing and defoliation. According to correlation analysis and regression analysis, there is significant correlation between temperature and leafing and defoliation phenology. Early leafing species would have a longer life span and consequently have advantage on carbon accumulation compared with later defoliation species.

Editor: David L. Roberts, University of Kent, United Kingdom

Funding: The study was supported by National Natural Science Foundation of China (31270518, 31170573), and Key Project of NNSFC (90411019), and Key Project in the National Science and Technology Pillar Program (2012BAC05B05). The funders had no role in study design, data collection and analysis, decision to publish, or preparation of the manuscript.

Competing Interests: The authors have declared that no competing interests exist.

* Email: hexy1962@126.com

Introduction

Plant leaves, as the main photosynthetic organs and the high energy converters among primary producers in terrestrial ecosystems, have attracted significant research attention. This is due to their growth and characteristics that are able to affect basic behaviours and functions of plants directly. The leaves of the urban forest tree species, in particular, show special ecological functions and aesthetic value [1]. Previous studies on leaves mainly concentrated on their external morphology, internal structure, differentiation, pigment content, leaf position, phyllotaxis, and age which are related to photosynthetic physiology [2–6]. In recent years, research concerning the ecological functions of canopy and leaf has been of increasing importance. The optimal leaf lifespan model [7] and the comprehensive ecological mechanism simulation model of leaf lifespan at large-scale were proposed to show those climate-induced vegetation changes [8–9]. The structural and ecological characteristics of plant leaves, such as: leafing time, defoliating time, and leaf lifespan determined by both (belonging to leaf phenological parameters) can all reflect the temporal and spatial resource utilisation strategies of a plant [10]. Therefore, research on plant leaf morphology at different scales, and its ecology concerning leaf lifespan and related leaf characteristics are of ecological and evolutionary significance [11].

Leaf lifespan is an adaptive characteristic formed by plants to obtain the maximum carbon in the long-term adaption process. It determines important functional and structural characteristics exhibited in the environmental adaptation of plants [12]. The importance of leaf lifespan prolongation to plants' ecological adaptation can be analysed from the perspective of carbon accumulation. Zhu *et al.* [13] studied the leaf phenology of woody plants in a deciduous oak forest in Nanjing, China. They found that the plants adopted three ways to achieve leaf lifespan prolongation, including advancing leafing, delaying defoliating, and a combination thereof. Different species took different ways and generated various leafing and defoliating phonologies. The resource utilising strategies of plants were thereby reflected. For urban forests, due to the heterogeneity of site conditions and the inter-configuration complexity of species, there are different ways for different tree species in the carbon acquisition. Typical urban forests contain three layers, including: an arbour layer, a shrub layer, and a herb layer. Before canopy closure of arbour plants, shrubs and herbaceous plants fully absorb light resources for photosynthesis and complete their life histories [14]. Leaf lifespan prolongation is one of the major mechanisms of the successful invasion of alien species. McDowell [15] found that the leaf lifespan of *Rubus* invasive species were significantly longer than the native congeners. Many successful invasive species showed earlier leafing times and later defoliating times than native species. Furthermore, the leaves of some species were positive in spring and then developed into shade leaves after canopy closure. In addition, some herbivore-infringed plant species often delayed their defoliating time to increase leaf lifespan for more carbon acquisition [16]. Sun *et al.* [17] compiled statistics about the leaf groups of *Quercus liaotungensis* in the Donglingshan Mountains in China and compared the individuals of *Q. liaotungensis*. They pointed out that those individuals leafing early, defoliated later. Wang *et al.* [18] suggested that, possibly due to different light environments, the lifespans of small trees and shrubs exceed those of arbour species in the evergreen broad-leaved forest in Tiantong, Zhejiang, China. Although the undergrowth species in temperate broad-leaved forests leaf earliest and have the longest lifespan, the leafing time and leaf lifespans of different undergrowth species varies greatly. Some undergrowth species even display later leafing times and shorter lifespans than those of canopy species [13]. In 2004, Wright *et al.* [19] analysed the leaves of 2,548 plants species of various vegetation types distributed in arctic tundra, tropical rain forest, grassland, and desert: they covered 175 sampling points world-wide. For the first time, a universal relationship between plant leaf characteristic factors was explained on a global scale. However, the leaf lifespan and leaf characteristics of urban forests were not studied up to now. In recent years, we have carried out many studies related to urban forest in Shenyang [20–24], the largest forest urban in the northeast of China. At present, approximately 19 million trees have been planted in the city [25].

In this study, the trees in the urban forest in Shenyang Arboretum, Chinese Academy of Sciences, were observed for five years to obtain the leafing phenology (including leafing start time, end time, and duration), defoliating phenology (including defoliation start time, end time, and duration), and the leaf lifespan of the main tree species. Moreover, the relationships between temperature and leaf phenology, and leaf lifespan were analysed. The results obtained provided a scientific basis for Shenyang's urban forest construction and tree species selection.

Materials and Methods

Basic condition of the research area

The research was developed in the Shenyang Arboretum, The Chinese Academy of Sciences (41° 46′ N, 123° 26′ E), which is a densely populated commercial and cultural centre of Shenyang. It covers an area of about 5 ha and shows a warm temperate sub-humid monsoon continental climate, with four distinct seasons over the same period. Its annual average temperature is 7.4°C. The average temperatures in January and July are −12.6°C and 27.5°C respectively. The maximum temperature is 38.3°C and the minimum temperature is −30.5°C. This area shows urban forest microclimate characteristics [20]. Shenyang belongs to the North China flora area. However, it also lies in the intersection zone of Changbai flora and Mongolia flora. Thus the plant species are relatively rich. In Shenyang Arboretum, arbour and shrub species counts exceed 540 and at present, Shenyang Arboretum has developed into a typical urban natural forest with multiple layers and generations [14,26]. Its forest community contains three obvious layers (arbour layer, shrub layer, and herb layer) or two layers (arbour layer and shrub layer). The arbour layer includes two sub-layers. The first sub-layer is about 12 to 22 m high. The tree species are comprise: *Phellodondron amurense*, *Robinia pseudoacacia*, *Quercus liaotungensis*, *Quercus mongolica*, *Fraxinus mandshurica*, accounting for 30% of the total trees in the arbour layer. The second sub-layer is 4 to 12 m high. The main species include: *Celtis bungeana*, *Prunus padus*, accounting for approximately 70% of the total trees in the arbour layer. The canopy density of the arbour layer is between 0.4 and 0.8. The trees having a diameter at breast height (DBH) of 4 to 32 cm are most frequent, making up 87.1% of the total, while those exceeding a DBH of 32 cm are rare, only accounting for 12.9% of the total. The main tree species in shrub layer include: *Lonicera maackii*, *Syringa velutina*, *Grewia parviflora*, *Lonicera praefloren*, *Evonymus alata*. The herb layer is mainly comprised: *Chelidonium dauricum*, *Viola yedoensis*, *Taraxacum ohwianum*, *Draba nemorosa*, *Polygonatum sibiricum*. The soil in the research area is a brown earth and is an alluvial meadow soil developed from the Hun River's deposits. The terrain is flat and free of the influence of underground water. Years of farming and 50 years of afforestation endow the forest with deep humus and fertile soil. During this experiment (2005–2009), the pH value of the soil of 0 to 40 cm deep is 6.3. The quick-acting nitrogen, quick-acting phosphorus, quick-acting potassium, and organic matter contents are: 147 mg/kg, 44.9 mg/kg, 294 mg/kg, and 2.54 mg/kg respectively.

Research methods

In reference to "Chinese Phenological Observation Method" [27] and the growth characteristics of tree leaves, leaf phonologies were recorded by the same trained observers using: visual inspection, telescopes, and other special tools. Standard samples were selected and observed once a day on their south side using a randomised branch sampling method [28]. Leaf phenology observation lasted for five years from 2005 to 2009. Thirty-eight common tree species bearing Shengyang forest zonality were chosen for phenological analysis (Table 1).

Data processing

Firstly, leaf phenology was defined and related parameters were determined. Next, the leaf counts obtained at different times were standardised by referring to the leaf numbers in leafing and defoliating phases as the maximum values [13]. The leafing start time was set as the time that the average leaf count of a tree species reaches 10% of its maximum. Leafing end time is regarded as the time in which the average leaf count of a tree species reaches 90% of its maximum. Leafing duration referred to the time between leafing start and end time. Similarly, the start time, end time, and duration of defoliation were also set with 10% and 90% as critical values. Leaf lifespan is determined as the time between 50% leafing and 50% defoliating. The leafing time and defoliating time are defined as the time points with 50% leafing and 50% defoliating respectively. Finally, the means and standard deviations of the leafing phenology, defoliating phenology, and leaf lifespan in the five years were calculated (Table 1).

Meteorological data were collected from China's Meteorological Data Sharing Service System. The meteorological data in corresponding phenological phases were selected for this tree phenology study. Shenyang experiences spring, summer, autumn,

Table 1. Leaf phonological characters of main tree species in urban forest of Shenyang (2005–2009 year).

Species	Beginning time of leaf emergence[a]	Ending time of leaf emergence	Duration of leaf emergence (d)	Beginning time of leaf abscission	Ending time of leaf abscission	Duration of leaf abscission (d)	Timing of leaf emergence	Timing of leaf abscission	Leaf life span (d)
Ginkgo biloba	43±9	68±8	21±7	209±4	225±2	16±3	45±8	218±5	173±3
Phellodondron amurense	49±5	78±4	29±4	223±5	237±5	14±1	51±5	231±4	180±5
Fraxinus mandshurica	54±3	75±3	21±2	210±2	226±3	16±5	56±3	214±15	158±14
Fraxinus rhynchophylla	49±5	71±±5	22±4	233±7	253±6	20±8	52±6	240±5	188±3
Celtis bungeana	55±10	77±10	22±4	253±3	261±2	8±2	57±3	256±2	199±15
Juglans mandshurica	46±4	71±4	25±6	220±10	232±12	12±3	48±7	226±11	17±14
Quercus liaotungensis	52±4	72±9	20±1	254±4	266±4	12±1	53±9	260±6	207±1
Ailanthus altissima	58±5	79±5	21±1	246±7	255±7	9±2	60±4	250±7	190±9
Robinia pseudoacacia	41±8	59±8	18±5	251±6	256±6	5±2	42±7	254±4	212±11
Prunus padus	21±11	51±10	30±7	217±1	228±2	11±1	23±11	224±1	201±10
Ulmus pumila	30±8	51±9	21±6	253±3	262±5	9±4	34±6	257±3	223±9
Syringa amurensis	37±8	66±6	29±6	232±3	249±5	17±7	40±5	238±4	198±7
Xanthoceras sorbifolia	47±1	74±4	27±3	239±4	251±3	12±5	50±1	245±4	195±4
Sorbus alnifolia	48±4	73±6	25±6	221±2	240±3	19±10	50±4	232±5	182±8
Tilia mandshurica	50±11	72±3	22±4	219±4	234±4	15±3	52±5	229±4	177±2
Syringa wolfi	47±10	59±5	12±3	250±5	260±2	10±2	38±7	255±2	217±4
Cotoneaster multiflorus	39±9	61±9	22±5	233±6	252±10	19±5	40±9	239±5	199±7
C. melanocarpus	32±9	60±9	28±3	232±3	246±6	14±4	35±8	239±5	204±3
Lonicera maackii	33±4	51±6	18±4	227±1	239±2	12±2	35±4	232±3	197±7
L. praeflorens	23±10	44±10	21±3	208±4	223±1	15±5	15±5	214±6	199±9
Weigela florida	52±5	72±2	20±3	229±13	250±10	21±9	54±4	250±10	196±10
Evonymus alata	51±5	66±4	15±2	237±4	251±3	14±2	53±5	244±4	191±7
Syringa velutina	33±7	54±6	21±2	250±5	263±5	13±7	35±7	256±2	221±9
Viburnum sargenti	38±9	59±9	21±2	243±2	256±3	13±3	40±10	251±2	211±9
Grewia parviflora	22±8	47±7	25±10	230±11	246±9	16±3	24±7	233±6	209±7
Quercus mongolica	51±6	78±9	27±9	241±4	251±2	18±4	54±7	253±5	199±7
Cornus alba	46±6	66±8	20±4	240±5	253±8	13±3	49±5	249±8	200±8
Alnus tinctoria	48±9	68±7	20±8	218±11	235±5	17±3	49±8	227±8	178±8
Magnolia parviflora	47±8	69±9	22±2	203±6	217±7	14±5	50±9	209±4	159±6
Salix matsudana	35±9	55±11	20±1	236±11	254±6	18±4	37±9	247±6	210±3
Spiraca trilodata	46±5	68±7	22±4	241±4	255±5	14±4	48±5	249±3	201±3
Acer ginnala	49±6	71±4	22±5	203±3	212±3	9±1	51±8	206±3	155±8
A. pseudo-sieboldianum	44±6	71±5	27±3	254±3	264±4	10±3	46±3	258±2	212±4
A. triforum	51±9	73±7	22±2	206±3	221±6	15±2	53±8	215±6	162±4
Koelreuteria paniculata	43±9	62±8	19±3	202±4	211±3	9±2	45±9	206±2	161±3
Crataegus pinnatifida	32±8	55±7	23±2	223±3	237±4	14±4	34±7	230±3	196±3
Populus alba	53±3	73±5	20±4	206±2	227±2	21±3	55±3	214±1	159±1
Berberis amurensis	38±8	62±11	24±6	232±4	252±5	20±5	42±8	245±5	203±8

[a]The timing for each leaf phenology is indicated by the days since March 1 every year from 2005 to 2009.

and winter each year: these correspond to: March to May, June to August, September to November, and December to January respectively. Coldness index (CI) was calculated by the Kira heat index formula: CI = (5 - *t*), where, *t* is the average monthly temperature over the year; CI is negative [29–30]. In this study, *t* is transformed into a ten-day average temperature (°C ten-day) [31–32]. The data were all statistically analysed and processed using MS-Excel and DPS (data process system).

Results

Leafing phenology

Statistics on the phenology observation data of the main tree species in Shengyang's urban forest [33] suggest that 73% of tree species begin to leaf at about the 15th day after sprouting (in the first ten days to the second ten days in April). However, different tree species differ greatly in leafing start time (Table 1). *G. parviflora*, *P. padus*, and *L. praeflorens* are the earliest leafing ones. They begin to leaf in the second ten days of March, which is 31 to 37 days earlier than the latest leafing ones, including *A. altissima*, *C. bungeana*, and *F.mandshurica*. Generally, the late leafing tree species leaf after blossoming, while the early leafing ones leaf before blossoming, or blossom and leaf simultaneously. Seventy-one percent of the tree species end their leafing in the first ten days to the second ten days of May. However, different tree species varies greatly in leafing end time. For example, to the second ten days in April, *G. parviflora*, *P. padus*, and *L. praeflorens* have finished their leafing, while many tree species had not yet begun to leaf. The latest leafing species, including *A. altissima*, *C. bungeana*, *P. amurense*, and *Q. mongolica* end their leafing processes by the last ten days in May, which is 26 to 35 days later than the earliest ones. The leafing process of the main tree species in this urban forest is concentrated. The leafing duration of 81% of the tree species is 18 to 25 days, and the leafing duration is 22 days on average.

Analysis of the leafing phenology of the main tree species in Shenyang's urban forest (Table 2, Figure 1) shows that leafing start time is positively correlated with leafing end time ($p<0.01$). That is to say, the earlier the leafing start time, the earlier the leafing end time; the longer the leafing duration, the later the leafing end time. However, leafing start time is independent of leafing duration. Thus it is proved to some extent that the early leafing tree species do not reduce their leafing speed and complete the leafing process earlier than those late leafing ones.

Defoliating phenology

The main tree species in Shenyang's urban forest, accounting for about 25% of all tree species, begin defoliating in late September. These tree species are mainly sub-tropical or temperate plants, such as *M. parviflora*, *K. panicullata*, *A. ginnala*, *P. alba*, *G. biloba*. The latest defoliating time is observed in the first ten days of November. Related species are mainly native temperate species, including *Q. liaotungensis*, *U. pumila*, *A. pseudo-sieboldianum*, *C. bungeana*. Due to species differences, early defoliating time is about 44 to 52 days earlier than for late species (Table 1). However, the defoliating start time of most tree species is in October, accounting for about 58% of the total. In general, the defoliating end time of tree species in Shenyang's urban forest are concentrated: 87% of tree species end their defoliation between the second ten days of October and the first ten days of November. The defoliating durations of the tree species show little differences. The defoliating duration of about 52% of tree species is between 12 and 16 days. The average defoliating duration is 14 days.

Analysis of the defoliating phenology of tree species (Table 2, Figure 2) implies that the defoliating start time shows a significant positive correlation with defoliating end time and defoliating time ($p<0.01$). The earlier the defoliating start time, the earlier the defoliating end time. However, defoliating start time has nothing to do with defoliating duration. It is suggested that, to some extent, early defoliating tree species quickly reach a defoliating peak and defoliate quickly: the difference in defoliating end time is thus decreased.

The relationship between leaf lifespan and leaf phenology

The main species in Shenyang's urban forest show an average leaf lifespan of 192 days. The leaf lifespans of 14 species exceed 200 days (Table 1), accounting for 37% of the total. *U. pumila* shows the longest leaf lifespan (223 days), while *A. ginnala* has the shortest at 155 days: a difference of 68 days. Analysis of the relationship between leaf lifespan and leaf phenology reveals that (Table 2), leaf lifespan is negatively correlated with leafing start time ($p<0.05$), leafing end time ($p<0.01$), and leafing time ($p<0.01$) at either a significant or very significant level. Moreover, leaf lifespan is positively correlated with defoliating start time ($p<0.01$), defoliating end time ($p<0.01$), and defoliating time ($p<0.01$) at a very significant level. That is to say, the earlier the leafing time and the later the defoliating time, the longer the leaf lifespan. However, leaf lifespan does not show significant correlation to leafing duration or defoliating duration. Therefore, in correlation analysis, we do not find significant correlation between the leafing time and defoliating time for these tree species (Table 2).

The relationship between air temperature and leaf phenology

Plant leaf phenology is closely correlated with temperature. However, since plant phenological phases are strictly correlated, sequenced, and synchronized, plant leafing and defoliating occurrence has different temperature requirement, respectively [34–35]. For example, plants need to experience a series of events before leafing start, such as: sap flux, bud swelling, germination. As shown in Table 3, the leafing of the main species in Shenyang's urban forest is associated with the temperature of the last ten days of March and the first ten days of April, but almost no significant correlation with temperature in winter (November to February) is found. That is to say, the leafing of the tree species in this study is mainly influenced by the temperature of the spring before leafing instead of the winter temperature. However, leafing start time is positively related to the absolute value of CI (Table 3), which is closely to the distribution of vegetation. The larger the absolute value of CI, it means the later the leafing start time with higher energy and temperature requirements. According to the relationship of average ten-day temperature and leafing phenology, the linear regression equation of the leafing phenology of the main tree species in Shenyang's urban forest is: $y = 5.827+0.493x$ ($R = 0.901$; $p<0.01$). It can thus be found that, with an increase in ten-day temperature of 1°C, leafing start time is advanced by 5 days. Nevertheless, the temperature requirements of the tree species are inconsistent due to their differing biological characteristics. The earliest leafing species, including *G. parviflora*, *P. padus*, *L. praeflorens*. leaf at an average ten-day temperature of 0.3°C in mid-March. While the latest leafing ones, such as *A. altissima*, *C. bungeana*, *F. mandshurica*. leaf at an average ten-day temperature of 11.2°C in late April. Up to mid-May, by when the ten-day temperature has reached 17.1°C, the tree species in Shenyang's urban forest have ended their leafing before warmer weather arrives.

The tree species here start to defoliate in late September. When the average ten-day temperature stays at 16.02°C, some tree

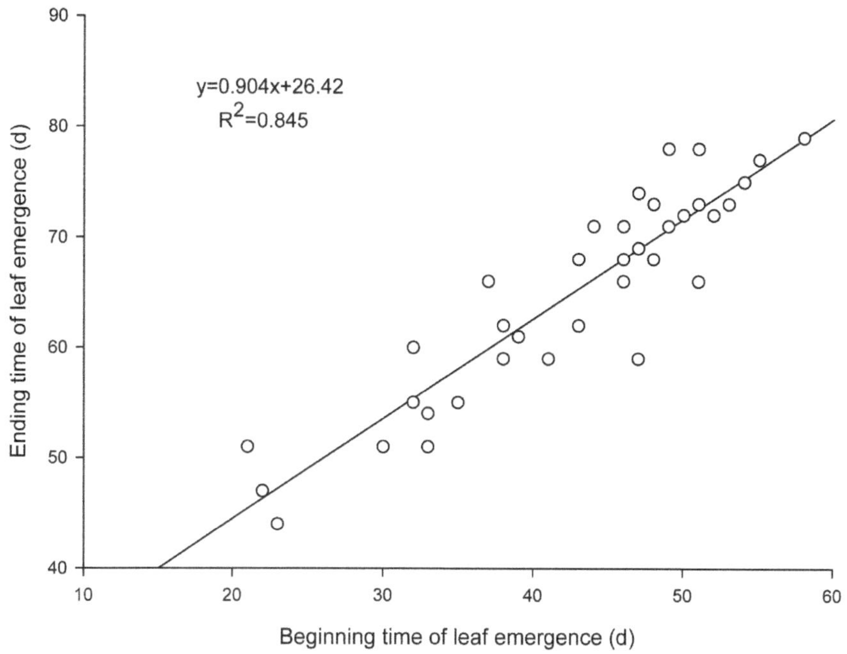

Figure 1. The relationships between beginning and ending time of leaf emergence.

species with high temperature requirements begin their defoliation, such as *K. panicullata*, *P. alba*, *G. biloba*. These species are the earliest to defoliate, accounting for about 12.36% of all defoliating tree species. Most of the tree species begin defoliation in October when the average ten-day temperature is between 7.64°C to 13.56°C. These species account for approximately 48% of the deciduous tree species. When the average ten-day temperature drops to 5.32°C in early November, 88% of the tree species enter their defoliation phase. Defoliation basically terminates at an average ten-day temperature of −2.16°C in mid-November (Figure 3). In accordance with the relationship of the defoliating phenology to average ten-day temperature, an equation is obtained: $y = 71.706 - 0.266x$; ($R = 0.898$; $p < 0.01$). It can be found that, with the average ten-day temperature decreasing by 1°C, the defoliation of the tree species studied is advanced by 3 days Moreover, when compared with leafing, defoliation is more temperature-sensitive.

Discussion

Leafing and defoliating phenologies experienced two eco-physiological processes of the periodical changes in plant leaf morphology. In these processes, plants capture carbon through leaf photosynthesis to provide for their growth, reproduction, and species maintenance [36]. Leaf lifespan, which is determined by leaf phenology, is a comprehensive index reflecting plant behaviour and function. It is considered as an adaptive strategy of plants developed in the long-term adaptation process aimed at obtaining the maximum photosynthetic yield and maintaining efficient nutrient utilisation [37]. The present study reveals that the leaf lifespan of the main tree species in Shenyang's urban forest differs due to the large differences in leafing and defoliating phenology. Leaf lifespan is significantly, and negatively, correlated with leafing start time, leafing end time, and leafing time, but is significantly, and positively, correlated with defoliating start time, defoliating end time, and defoliating time. However, leafing time and defoliating time did not show any obvious correlation. The results above suggest that plants can increase their leaf lifespan and carbon acquisition using two approaches: advancing leafing and delaying defoliating. Advancing leafing is of great ecological significances for the carbon acquisition of plants in a forest community and has often been researched. Seiwa [38] found that *A. mono* seedlings in a temperate deciduous broad-leaved forest have initiated their leafing two months before forest canopy closure. The carbon acquired in this time period was 61% to 79% of that accumulated in the whole year. Xu *et al.* [39] indicated that undergrowth invasive species *Berberis thunbergii* began to leaf two weeks before canopy closure, which is one month earlier than the native symbiotic species, *Vaccinium corymbosum*. In such a time period, *B. thunbergii* conducted photosynthesis using the early-spring sunlight in the forest. The carbon captured accounted for 36% of that accumulated in the whole growth cycle. Meanwhile, since the defoliating time of *B. thunbergii* was close to that of the native species, the advancing of leafing and long leaf lifespan of *B. thunbergii* increased its annual carbon accumulation and thus promoted the invasion. Kuppers [40] stated that the successful invasion and undergrowth settlement of *Ribes uvacrispa* were attributable to the strong competitiveness provided by early leafing instead of a high photosynthetic rate. This was because the early leafing of *R. uvacrispa* extended its leaf lifespan and advanced its carbon accumulation.

Plant leafing and defoliating processes are affected by genetic factors, environments, and biological factors such as insects, fungi. The defoliating phenology of the main species in an urban forest also determines the leaf lifespan. The later the defoliation, the longer the leaf lifespan. However, although leaf phenology is significantly correlated with leaf lifespan, defoliating phenology presents different carbon acquisition implications when compared with leafing phenology. Generally, leafing phenology means the unfolding of leaves, an increasing leaf area, an increasing leaf thickness, and finally an increasing net photosynthesis rate. Plant photosynthesis reaches a peak with the complete unfolding of leaves and then reduces with age [41]. At defoliation start time, carbon acquisition has dropped to practically zero. Measurement

Table 2. Correlation analysis of leaf phenological characters of main tree species in urban forest of Shenyang.

	Beginning time of leaf emergence[a]	Ending time of leaf emergence	Duration of leaf emergence (d)	Beginning time of leaf abscission	Ending time of leaf abscission	Duration of leaf abscission (d)	Timing of leaf emergence	Timing of leaf abscission
Leaf life span(d)	−0.499**	−0.502**	−0.022	0.707**	0.722**	−0.239	−0.544**	0.737**
Timing of leaf abscission	0.108	0.111	−0.086	0.817**	0.970**	−0.110	0.103	
Timing of leaf emergence	0.991**	0.984**	−0.124	−0.067	0.090	0.084		
Duration of leaf abscission	0.090	0.115	0.103	−0.174	0.043			
Ending time of leaf abscission	0.096	0.101	−0.115	0.813**				
Beginning time of leaf abscission	−0.066	−0.038	−0.039					
Duration of leaf emrgence	−0.250	0.121						
Ending time of leaf emergence	0.975**							

[a]The timing for each leaf phenology is indicated by the days since March 1 every year from 2005 to 2009.
**$p<0.01$.

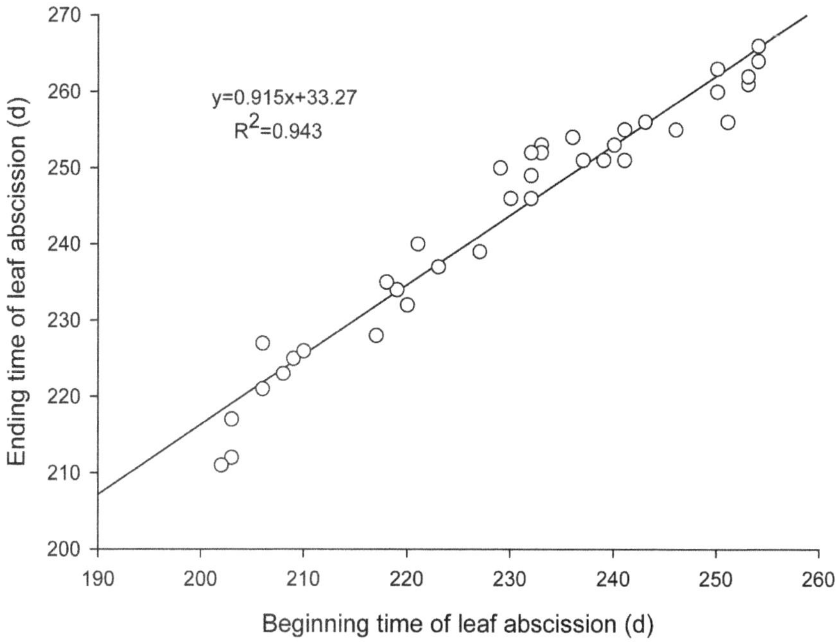

Figure 2. The relationships between beginning and ending time of leaf abscission.

Table 3. Relationships between phenology of leaf emergence and mean air temperature in ten days in urban forest of Shenyang ($n=38$).

Month[a]	Sprouting	Timing of leaf emergence
11 I	0.017	0.103
11 II	−0.096	0.101
11 III	−0.213**	−0.103
12 I	−0.012	−0.037
12 II	0.272**	0.228**
12 III	0.045	0.143**
1 I	−0.248**	−1.020
1 II	−0.134*	−0.140*
1 III	−0.232**	−0.089
2 I	−0.254**	−0.200**
2 II	−0.401**	−0.327**
2 III	−0.346**	−0.361**
3 I	−0.083	−0.111*
3 II	−0.268**	−0.288**
3 III	−0.420**	−0.430**
4 I	−0.439**	−0.552**
4 II	−0.291**	−0.453**
4 III	−0.192**	−0.373**
5 I	−0.135*	−0.265**
5 II	−0.047	−0.163**
5 III	−0.119*	−0.123*
-*CI*	0.343**	0.193**

[a]I-The first ten days of a month, II-The middle ten days of a month, III-The last ten days of a month.
**$p<0.01$,
*$p<0.05$.

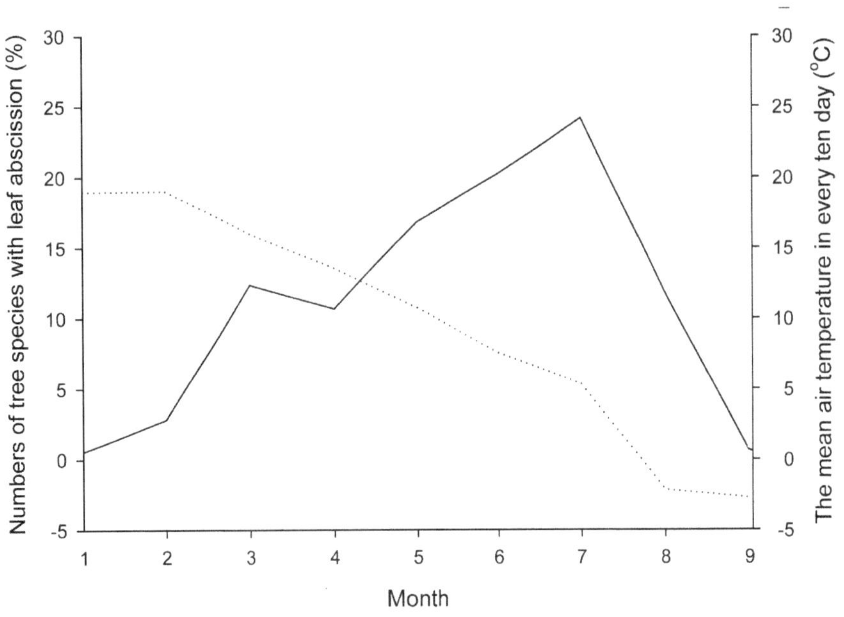

Figure 3. The relationships between leaf abscission phenology and mean air temperature in every ten days from September to November for main tree species in urban forest of Shenyang. The full line-numbers of tree species with leaf abscission (%), the dotted line-the mean air temperature in every ten day (°C).

of the photosynthetic rates of the tree species in different phenological phases implies that the photosynthetic rate of leafing phenology is higher than that during defoliation. The average photosynthetic rate is 3.75 μmol•m^{-2}•S^{-1} at the beginning of leafing, 7.93 μmol•m^{-2}•S^{-1} at the end of leafing, 2.34 μmol•m^{-2} •S^{-1} at the beginning of defoliation, and almost zero at the end of defoliation. The results fully explained why the carbon accumulated during leafing is higher than that during defoliation. Meanwhile, it is also found that some early leafing native trees also have higher photosynthetic rates. For example, the maximum instantaneous photosynthetic rate of *S. matsudana* exceeds 17.31 μmol•m^{-2}•S^{-1}, and that of *U. pumila* is 22.43 μmol•m^{-2}•S^{-1}. By contrast, some north-moving tree species of North China's flora zone leaf late and show low photosynthetic rates. For instance, the maximum instantaneous photosynthetic rates of *G. biloba*, *S. oblate*, and *W. florida* are only 9.00 μmol•m^{-2}•S^{-1}, 8.13 μmol•m^{-2}•S^{-1}, and 6.59 μmol•m^{-2}•S^{-1}, respectively. These results show that the urban tree species with different phenological phases have different abilities of carbon consequence. Therefore, leaf lifespan and leaf phenology may play an important role in tree species selection for urban forest construction and sustainable development of urban forest under global climate change.

Plant phenological phases are closely associated with temperature. The starting time of each phenological phase in the plant's growth process is significantly correlated with the temperature before the phenological phase. Generally speaking, plant phenological phases are strictly correlated, sequenced, and synchronized. The later phonological phase is based on the former [32]. Therefore, the leafing time of tree species is determined by phenological phases such as: snap flux, bud swelling, and bud cracking. The leaf-expansion phase follows the bud germination phase. Before bud germination, a bud dormancy phase need a critical low temperature stage (a chilling process) to be completed. Zhu *et al.* [42] investigated the main meteorological factors that affect the phonological phases of the tree species in natural forests in Northeast China using principal component analysis (PCA). They considered that total cumulative sunshine hours and average temperature were two dominant factors inducing tree germination. Xie *et al.* pointed out that [43] average temperature was the main meteorological factor trigging tree germination. Xu *et al.* [1] studied the leaf growth relationships for the main tree species in Shenyang's urban forest. They found that the staggering of high and low temperature periods was the main meteorological factor giving rise to tree germination, and the low temperature period before germination was of great significance as a factor therein. Moreover, the critical low temperature of the main tree species in Shenyang's urban forest is observed between late January and late February (Table 3). It is a proper low temperature stage, but not the minimum temperature in Shenyang (this occurs mid-January). In the present study, the CI in the Kira heat indices was introduced to this phenology study to explore how the temperature before a plant phenology phase influenced its germination and leafing time, and how plant phenology interacted with each other. The results showed that CI was of ecological significance for studies on spring phenology. Xu *et al.* [44] revealed that plant bud germination was negatively correlated with winter temperature. Germination was advanced with the increase of winter temperature. Besides, the reduction of the minimum winter temperature was favourable to bud dormancy breaking and thus advanced the germination phase. Hannien [45] also pointed out that the germination of the plants in spring was mainly affected by the critical low temperature in winter, and this low temperature was conductive to breaking the bud dormancy. However, Zhang [46] considered that the winter temperature was unfavourable to bud dormancy and thus postponed the blossoming and leafing time. At a plant physiology level, leaf phenology is initiated by different factors. It is generally believed that leafing is closely connected with temperature and dominated by accumulative temperature particularly, while defoliating phenology is influenced by low temperature and photo-period. It is possible that defoliation is caused by the abscisic acid (ABA) in chloroplasts generated by the interaction of low temperatures and photo-period. However, some eco-physiological mechanism problems, such as how ABA is transferred by low temperature and photo-period signals, and how ABA influences defoliation, remain to be studied.

Conclusions

In the present study, we find that there are significant correlations between leaf longevity and leaf phenology, and between temperature and leaf phenology. Early leafing species would have a longer life span and consequently have advantage on carbon accumulation compared with later defoliation species. The timing of leafing differed greatly among tree species. The early leafing species would have relatively early end of leafing. The longer it took to the end of leafing would have a later time of completed leafing. The timing of defoliation among different species varied significantly, the early defoliation species would have relatively longer duration of defoliation. We also find that if the mean temperature rise for 1°C in spring, the time of leafing would experience 5 days earlier in spring. If the mean temperature decline for 1°C, the time of defoliation would experience 3 days delay in autumn.

Author Contributions

Conceived and designed the experiments: XH SX. Performed the experiments: WX SX YH HW. Analyzed the data: WX SX. Contributed reagents/materials/analysis tools: WC YH. Contributed to the writing of the manuscript: SX WX. Contributed to the field investigation: YQH SX. Made figures by software: HW.

References

1. Xu WD, He XY, Chen W, Wen H (2006) Morphological-ecological characters and growth patterns of main tree species leaves in urban forest of Shenyang. Chin J Appl Ecol 17(11): 1999–2005.
2. Duan XH, Sun LF, Ma SR, Zu YG (2003) The studied of blade of *Adnophora potaninii* in different altitude. Bull Bot Res 23: 334–336.
3. Hang YY, Huang CH, Mu S (2004) Studies on diversity of morphological characteristics of leaves from *Dioscorea zingiberensis*. Acta Bot Yunnanica 26: 398–404.
4. Lin XQ, Zhu DF, Zhou WJ, Zhang YP (2003) Relationship between specific leaf weight and photosynthetic rate at panicle initiation stage in super hybrid rice. Chin J Rice Sci 17(3): 281–283.
5. Yang YF, Li JD (2003) Biomass allocation and growth analysis on the ramets of *Phragmites communis* populations in different habitats in the Songnen Plains of China. Chin J Appl Ecol 14: 30–34.
6. Zhang YJ, Feng YL (2004) The relationships between photosynthetic capacity and lamina mass per unit area, nitrogen content and partitioning in seedlings of two ficus species grown under different irradiance. J Plant Physiol Mol Biol 30(3): 269–276.
7. Kikuzawa K (1991) A cost-benefit analysis of leaf habit and leaf longevity of trees and their geographical pattern. Amer Nat 138: 1250–1263.
8. Gower ST, Reich PB, Son Y (1993) Canopy dynamics and aboveground production of five tree species with different leaf longevities. Tree Physiol 12: 327–345.
9. Kull O, Jarvis PG (1995) The role of nitrogen in a simple scheme to scale up photosynthesis from leaf to canopy. Plant Cell Environ 18: 1174–1182.
10. Kikuzawa K (1995) Leaf phenology as an optimal strategy for carbon gain in plants. Can J Bot 73: 158–163.
11. Reich PB, Walters MB, Ellsworth DS (1992) Leaf life-span in relation to leaf, plant, and stand characteristics among diverse ecosystems. Ecol Monogr 62: 365–392.
12. Chabot BF, Hicks DJ (1982) The ecology of leaf life spansÐ Ann Rev Ecol Sys 13: 229–259.
13. Zhu XB, Sun SC (2006) Leaf phenology of woody species in deciduous broad-leaved oak forest in Nanjing area, East China. Acta Phytoecol Sin 30: 25–32.
14. He XY, Chen W, Xu WD, Liu CF, Zhu WQ, et al. (2003) Community ecology in a typical near natural urban forest. Chin J Ecol 22(6): 162–168.
15. McDowell SCL (2002) Photosynthetic characteristics of invasive and noninvasive species of *Rubus* (*Rosaceae*). Amer J Bot 89: 1431–1438.
16. Coley PD, Balone JA (1996) Herbivory and plant defenses in tropical forests. Ann Rev Ecol Sys 27: 305–335.
17. Sun SC, Chen IZ (1998) Leaf demography of *Q. liaotungensis* in Dongling Mountain. Acta Phytoecol Sin 22: 538–544.
18. Wang XH, Zhang J, Zhang ZX (2000) Leaf longevity of evergreen broad-leaved species of Tiantong National Forest Park, Zhejiang Province. Acta Phytoecol Sin 24: 625–629.
19. Wright IJ, Reich PB, Westoby M, Ackerly DD, Baruch Z, et al. (2004) The worldwide leaf economics spectrum. Nature 428: 821–827.
20. Xu WD, He XY, Chen W, Liu CF, Sun Y (2005) Microclimate characters of urban forest in Shenyang City. Chin J Appl Ecol 16(9): 1650–1654.
21. Xu W, He X, Chen W, Wen H (2006) Morphological-ecological characters and growth patterns of main tree species leaves in urban forest of Shenyang. Chin J Appl Ecol 17(11): 1999–2005.
22. Song L, He X, Chen W, Xu W (2007) Comprehensive evaluation system of urban tree species in Shenyang. J Liaoning Tech Univ (Nat Sci Ed) 26(1): 136–139.
23. Liu C, He X, Chen W, Zhao G, Xu W (2008) Simulation of three-dimensional green biomass of urban forests in Shenyang City and the factors affecting the biomass. Chin J Appl Ecol 19(6): 1173–1178.
24. Dewen Li, Yi Shi, Xingyuan He, Guangyu Chi (2011) Seasonal variations of BVOCs emission from *Ginkgo biloba* linn in urban area. Appl Mech Mater 71–78: 2891–2894.
25. Liu C, Li X (2012) Carbon storage and sequestration by urban forests in Shenyang, China. Urban For Gree 11: 121–128.
26. Xu WD, He XY, Chen W, Zhang Y, Li HM, et al. (2004) Flora and vegetation types in the downtown area of Shenyang. Chin J Appl Ecol 14(12): 2095–2102.
27. Wan MW, Liu XZ (1979) Method of phenology observation in China. Beijing: Science Press. 136 p.
28. Maillette L (1982) Structure dynamics of silver birch. I. The fate of buds. J Appl Ecol 19: 203–218.
29. Xu WD (1983) The relation between distribution of edificator and companion in zonal vegetation and water-temperature condition in Northeast China. Chin Bull Botany 25(3): 264–274.
30. Xu WD (1985) The application of Kira's thermal index to Chinese vegetation. Chin J Ecol 4(3): 35–39.
31. Xu WD, He XY, Chen W, Hu JB, Wen H (2006) Responses of Shenyang urban tree phenology to climate warming. Chin J Appl Ecol 17(10): 1777–1781.
32. Xu WD, He XY, Chen W, Liu CF, Zhao GL, et al. (2008) Relationship between spring phenophase and global climate warming of Shenyang urban forest in the last 40 years. Chin J Ecol 27(9): 1461–1468.
33. Hu JB, Xu WD, Chen W, He XY, Wen H (2006) Phenological analysis of main tree species in Shenyang urban forest. Chin J Ecol 25(12): 1455–1459.
34. Liu J, Yang Y, Shi K, Zhang L, Zhou L, et al. (2007) Soybean phenology and its responses to climate factors from 1985 to 2005 in Jinzhou region. J Meteorol Environ 23(4): 29–32.
35. Zhao J, Zhang Y, Song F, Xu Z, Xiao L (2013) Phenological response of tropical plants to regional climate change in Xishuangbanna, south-western China. J Trop Ecol 29: 161–172.
36. Zhang L, Luo TX (2004) Advances in ecological studies on leaf lifespan and associated leaf traits. Acta Phytoecol Sin 28(6): 844–852.
37. Kikuzawa K (1995) The basis for variation in leaf longevity of plants. Vegetatio 121: 89–100.
38. Seiwa K (1998) Advantages of early germination for growth and survival of seedlings of *Acer mono* under different overstory phonologies in deciduous broad-leaved forests. J Ecol 86: 219–228.
39. Xu CY, Griffin KL, Schuster WS (2007) Leaf phenology and seasonal variation of photosynthesis of invasive *Berberis thunbergii* (Japanese barberry) and two co-occurring native understory shrubs in a northeastern United States deciduous forest. Oecologia 154: 11–21.
40. Kttppers M (1984) Carbon relations and competition between woody species in a central European hedgerow I. Photosynthetic characteristics. Oecologia 64: 332–343.
41. Kitajima K, Mulkey SS, Wright SJ (1997) Decline of photosynthetic capacity with leaf age in relation to leaf longevities for five tropical canopy tree species. Amer J Bot 85: 702–708.
42. Zhu N, Jiang H, Jin Y (1990) A phenology study on the common tree species of natural secondary forests in northeast China. Chin J Plant Ecol 14(4): 336–349.
43. Xie SC, Sheng CY, Li SC (1997) A phenologieal study on main tree species of montane humid evergreen bread-leaved forest in Ailao Mountains. Acta Ecol Sin 17(1): 51–60.
44. Xu YQ, Lu PL, Yu Q (2005) Response of tree phenology to climate change for recent 50 years in Beijing. Geogr Res 24(3): 412–420.
45. Hannien H (1995) Effects of climatic change on trees from cool and temperate regions:An ecophysiological approach to modeling of bud burst phenology. Can J Bot 73: 183–199.
46. Zhang FC (1995) Effects of global warming on plant phenological events in China. Acta Geogr Sin 50(5): 402–410.

7

Evaluating Landscape Options for Corridor Restoration between Giant Panda Reserves

Fang Wang[1,2]*, William J. McShea[1], Dajun Wang[2], Sheng Li[2], Qing Zhao[3], Hao Wang[2], Zhi Lu[2]

1 Smithsonian Conservation Biology Institute, National Zoological Park, Front Royal, Virginia, United States of America, **2** School of Life Sciences, Peking University, Beijing, China, **3** Department of Fish, Wildlife, and Conservation Biology, Colorado State University, Fort Collins, Colorado, United States of America

Abstract

The establishment of corridors can offset the negative effects of habitat fragmentation by connecting isolated habitat patches. However, the practical value of corridor planning is minimal if corridor identification is not based on reliable quantitative information about species-environment relationships. An example of this need for quantitative information is planning for giant panda conservation. Although the species has been the focus of intense conservation efforts for decades, most corridor projects remain hypothetical due to the lack of reliable quantitative researches at an appropriate spatial scale. In this paper, we evaluated a framework for giant panda forest corridor planning. We linked our field survey data with satellite imagery, and conducted species occupancy modelling to examine the habitat use of giant panda within the potential corridor area. We then conducted least-cost and circuit models to identify potential paths of dispersal across the landscape, and compared the predicted cost under current conditions and alternative conservation management options considered during corridor planning. We found that due to giant panda's association with areas of low elevation and flat terrain, human infrastructures in the same area have resulted in corridor fragmentation. We then identified areas with high potential to function as movement corridors, and our analysis of alternative conservation scenarios showed that both forest/bamboo restoration and automobile tunnel construction would significantly improve the effectiveness of corridor, while residence relocation would not significantly improve corridor effectiveness in comparison with the current condition. The framework has general value in any conservation activities that anticipate improving habitat connectivity in human modified landscapes. Specifically, our study suggested that, in this landscape, automobile tunnels are the best means to remove current barriers to giant panda movements caused by anthropogenic interferences.

Editor: Sadie Jane Ryan, SUNY College of Environmental Science and Forestry, United States of America

Funding: This research was funded by Ocean Park Conservation Foundation, Hong Kong (OPCFHK, www.opcf.org.hk) and Conservation International (CI, www.conservation.org). The funders had no role in study design, data collection and analysis, decision to publish, or preparation of the manuscript.

Competing Interests: The authors have declared that no competing interests exist.

* Email: wangf@si.edu

Introduction

Habitat loss and fragmentation are considered the primary threats to many endangered species [1]. Movement corridors can mitigate the negative effects of habitat fragmentation by connecting isolated populations [2]. Although corridor establishment and restoration are commonly proposed to compensate for the negative effects of habitat fragmentation, managers typically neglect habitat selection and migratory behavior of focal species when identifying planned location of corridors [3]. Common criticisms of corridor studies include the use of geographic metrics (e.g. slope, elevation, etc.) to measure the landscape connectivity without considering more animal-centric measures [4], or a reliance on expert opinion that makes validation and applicability to other regions problematic [5]. Validated models based on empirical data of animal habitat use should be used for predictive or planning purposes [4,5].

Giant panda (*Ailuropoda melanoleuca*) is a large mammal that requires movement corridors for long-term conservation. This species is restricted to approximately 24 isolated populations, 13 of which are considered to be at a high risk of extinction [6]. While the core populations reside within established nature reserves with intensive protection efforts, marginal populations are still losing habitat as well as population numbers [7,8]. Forest corridors have been identified and advocated for giant pandas since the 1980s, with the goal of maintaining or restoring the habitat connectivity between nature reserves [9]. However, few of the proposed corridors were found to be functional, as studies revealed that the trend of habitat fragmentation, as well as population differentiation, continued [7,10]. Giant panda corridor studies either used expert opinion to determine the environment-species relationships [11], or constructed models based on coarse resolution landscape metrics at regional scale [8,12,13], both practices which may limit the practical application of their results at specific locations.

The purpose of this study was to use a key linkage area between two known giant panda populations to evaluate landscape options for corridor establishment. This area has been identified as one of the three most important potential corridors for giant panda movement in the Qinling Mountains [9,14]. Our study aims to (1) model the habitat use of giant panda populations in a proposed corridor area; (2) predict potential giant panda movement pathways; and (3) evaluate alternative management strategies on corridor effectiveness.

Materials and Methods

Field survey was approved by the Shaanxi Forestry Department (SFD).

Study Area

Our study area is a 400 km^2 region along the Xushui River Valley (XRV) in the western Qinling Mountains, which covers three nature reserves and a key linage area between giant panda populations (Fig. 1A). Huangbaiyuan, Changqing, and eastern Niuweihe Nature Reserves are located to the east of the XRV, and harbor a population of approximate 60 giant pandas [14]. Western Niuweihe Nature Reserve, located to the west of the valley, has an estimated population of approximate 20 giant pandas [14]. Previous studies suggest that there may be dispersal movements of giant pandas between the Huangbaiyuan-Changqing population and the Niuweihe population through the XRV area [15].

The XRV is a valley corridor, where complex topography and human infrastructures (e.g., residences, agriculture practice, road, etc.) along a river valley might be barriers for the connectivity of animal populations (Fig. 1A). The proposed XRV Corridor is comprised of four dispersed communities and numerous residences along a two-lane highway (Yangxian-Taibai Highway) that bisects the valley along its north-south axis [16]. The shortest distance between the known western and eastern giant panda populations is approximate 16 km based on previous surveys (Fig. 1B) [14].

Sampling design and Data Collection

We conducted giant panda surveys using passive infrared cameras (TCB-100, Huangwu Inc., Wuhan, CHINA; Reconyx PC800, Reconyx, Wisconsin, USA). In addition, we used a sign plot method to detect evidence of giant panda activity (feces, tracks, and feeding signs) as supplementary information for the modeling process (see below). The survey season ran from April 2010 through March 2013 (35 months). We created eight 10 km^2 sampling arrays along the east-west axis of the XRV. Four arrays to the western side of the XRV were located in Huangbaiyuan and Changqing Nature Reserve, three eastern arrays were located in Niuweihe Nature Reserve, and one array was located within the XRV. Each 10 km^2 sampling array was divided into ten 1×1 km grid cells (Fig. 1). At the beginning of each sampling period, we randomly selected three grid cells in each array and placed one survey site in each selected grid cell. Field staff were allowed to select the most probable location to detect giant panda within each grid cell with the limitation of no concurrent sample locations being placed within 500 m.

Cameras were mounted on trees 40 cm above the ground and operated 24 h per day with a 20 s delay between sequential photographs. For most camera set-ups, a few drops of scent lure (Carman's Magna-Glan Lure, Sterling Fur, Ohio, USA) was applied 1–2 meters in front of the cameras to slow animal movement and to compensate for slow camera trigger times. At the end of each survey period (30–50 days, mean 38 days), the digital flash cards were collected, and the cameras were moved to the next survey locations.

A 40 m×20 m plot was surveyed for panda signs at each camera location. Field staff set up a plot centering on each camera location, and recorded all evidence of giant panda activity within the plot. Surveys were conducted before and after camera period at each location. We combined the data with camera trap data detections to generate a giant panda distribution layer, and used the layer for corridor modelling (least-cost model and circuit model).

For occupancy modelling, we reviewed previous giant panda habitat studies [17–20] and identified 13 potentially important occupancy covariates (Table 1). Geo-referenced human infrastructure data (i.e. roads, residences, and croplands) were collected using a handhold GPS unit (Garmin 60CSx, Garmin, Schaffhausen, Switzerland). We categorized residences into small (<3 households) or large (≥3 households), and roads into local (<2 lane) or provincial (≥2 lanes). We used a 30-m resolution DEM [21] to delineate elevation and slope. The SFD provided a nature reserve database with geo-referenced boundary and protected level (national, provincial, and non-nature reserves). The Taibai Forestry Company provided a forest database to categorize forest age as primary or secondary. We used 30 m Landsat data ETM+ satellite imagery (path 128, row 037, date 2008-07-06) to conduct a supervised classification of land cover across the study landscape [22], and aggregated forest composition into four classes: coniferous forest, broad-leaved forest, mixed forest and non-forest. The distribution of bamboo across the study area was surveyed by the SFD during 2007 to 2010 by conducting 413 100 m^2 vegetation plots in and outside nature reserves (Table S2, i.e. coordinates, bamboo species, and elevation). We modeled the distribution of bamboo within the study area using Maxent [23]. Maxent is a machine learning program that estimates a species' probability distribution by finding the probability distribution of maximum entropy (i.e., that is most dispersed), subject to a set of environmental constraints based on incomplete information about the species' distribution [23]. We included biophysical covariates used 300 m resolution WorldClim data [24] and four biogeographic co-variables (elevation, slope, aspect and solar radiation), and followed established protocols for their inclusion [25,26]. For detection covariates at each camera survey site, we obtained monthly mean temperature from WorldClim Data and categorized it into low (<5°C), medium (5–15°C), or high (>15°C), indicated which deployments included use of scent lure to attract animals, and categorized camera view as open or limited-view based on vegetation density.

GIS layers were standardized to the same spatial projection and resampled to a 100 m×100 m resolution, and the mean value of elevation and slope, dominate land cover type, and the distance from cell center to human infrastructure (i.e. residences, roads, and cropland) were calculated.

An independent sign transect dataset that was collected by the SFD was used to validate our occupancy model (Table S3). The SFD surveyed for giant panda sign along 47 transects in forested habitat (total distance: 54 km) during 2010 to 2012. The trails were divided into 250 m segments and segments were categorized as with (114) or without (102) giant panda sign, with the center point and attribute of each segment entered into our GIS.

Data Analysis

Our analyses included three steps. First, we constructed species occupancy modeling [27,28] to identify important occupancy covariates, and mapped habitat suitability with the resulting predictive equation. Second, we used the inverse of the habitat suitability map as a resistance surface and constructed least-cost [29] and circuit models [30] to predict the potential dispersal pathways for giant panda in this landscape. For the last step, we varied the model parameters being considered under different conservation scenarios and assessed their effectiveness.

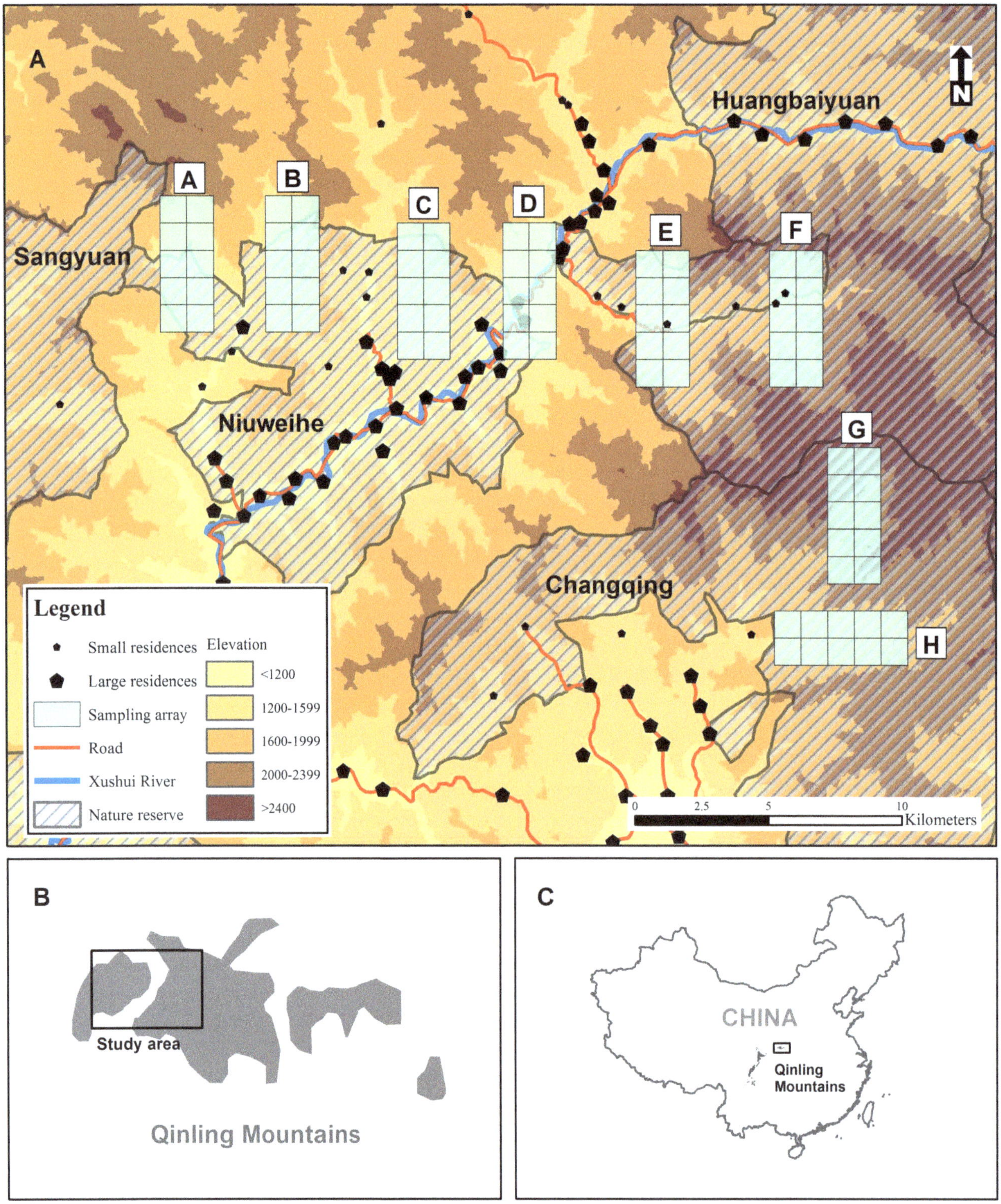

Figure 1. Sampling arrays layout in the proposed corridor area. Arrays A, B and C are located in Niuweihe Nature Reserve, arrays E, F, G, and H were located in Huangbaiyuan and Changqing Nature Reserves, and array D was located between nature reserves along the Xushui River Valley (A).The grey color in lower two figures indicates the known giant panda distribution. The study area was in the western region of Qinling Mountains (B), and located in the center of China (C).

Table 1. Covariates collected for occupancy and detection probabilities.

Name	Description
Occupancy covariates	
Elevation	Numeric (m)
Slope	Categorical (<5°, 5–20°, >20°)
Forest age	Categorical (primary, secondary)
Forest composition	Categorical (broadleaf, mixed, conifer, non-forest)
Bamboo distribution	Categorical (presence, absence)
Type of nature reserve	Categorical (not reserve, provincial, national)
Distance to small road*	Numeric (m)
Distance to large road*	Numeric (m)
Distance to any road	Numeric (m)
Distance to small residences	Numeric (m)
Distance to large residences	Numeric (m)
Distance to any residences*	Numeric (m)
Distance to cropland*	Numeric (m)
Detection covariates	
Lure	Categorical (applied, not applied)
Temperature	Categorical (<5°C, 5–15°C, >15°C)
Camera view	Categorical (open view, limited view)

*Excluded from model selection due to collinearity.

Occupancy Modelling

We divided the camera-trapping sampling period at each survey site into 5-day segments [31]. The giant panda detection history for a segment was considered as "present" if any detection was made during the 5 days, and "absent" otherwise. Prior to occupancy modelling, we examined the collinearity of variables and selected covariates that have a pairwise correlation <0.7 for later analyses. We then examined the potential non-linear relationship between occupancy probability and each occupancy covariate by constructing a series of single-variable models. For each occupancy covariate, we selected from among its original, log-transformed, and quadratic forms, and used the one with the smallest Akaike's Information Criteria or AIC [32] in later analyses. To select detection covariates we constructed a general occupancy model including all occupancy covariates and investigated all combinations of detection covariates. We selected the top observation model based on AIC values and only considered the optimal detection covariates in subsequent analyses for occupancy covariate selection [33]. All possible combinations of occupancy covariates were examined, and a model-averaging approach was applied to calculate the weight-averaged parameter values for all variables found in the top models, as indicated by delta AIC values ≤2 [32]. We then used the model-averaged parameter values to create a predictive equation for calculating occupancy probability across the entire study area. The predictive model was validated using the SFD sign transect dataset. Giant panda was considered as "present" if giant panda signs were recorded, and "absent" if no panda sign was recorded at the location or any transect within 1 km of that location. Based on this dataset, we constructed receiver operating curves (ROC) for our habitat suitability map and calculated area under ROC (AUC). All occupancy modelling was conducted using the "unmarked" package [34] in R.

Mapping Movement Pathways

We combined camera-trapping detections and sign plot records into panda presence locations. We generated the inverse of the predicted habitat suitability map as a cost surface [3] using ArcToolbox in ArcGIS 10.2 [35]. Based on this cost surface we constructed least-cost and circuit models to identify potential movement pathways. For least-cost modelling, we generated an accumulative cost surface layer and a path direction layer, and predicted least-cost-paths linking the source population and sink population using ArcToolbox in ArcGIS 10.2 [35]. In addition to the single optimal least-cost corridor, we selected five locations at each side of the river valley where giant panda were detected, and generated another 24 alternative pathways linking each possible set of points. To quantify the predicted pathways, we generated an index of movement cost per 100 meters using ArcToolbox, and calculated the mean movement cost for each pathway.

In addition to the line pathways predicted by least-cost model, we used the circuit modelling software Circuitscape 3.5 [30], and followed the method of Koen et al. [36] to model landscape connectivity for the entire XRV Corridor area.

Scenario Analysis

To compare the effectiveness of alternative restoration options, we created hypothetical landscapes and compared their movement cost to the current landscape. The scenario settings were derived from our discussion with the local forestry department and nature reserves, as well as management options presented in recent government plans [14]. The three most possible management scenarios were: (1) forest/bamboo restoration, which is widely believed to increase giant panda habitat quality due to giant panda's association with mature forest [37]; (2) extensive residence relocation, which was commonly referred to "eco-migrant", and has been conducted in numerous locations [38,39]; and (3) automobile tunnel construction, whose primary purpose in the regions is to increase transportation speeds for commercial trucks and buses [40].

For the forest/bamboo restoration scenario, we assumed with proper forest management, the young successional forest in XRV area can be restored to mature forest with a bamboo understory. For the residence relocation scenario, we assumed the 7 large residences along the Yangxian-Taibai highway were relocated. For the automobile tunnel scenario, we assumed a 1 km section of current winding road be replaced by an automobile tunnel, and the abandoned road serve as a wildlife overpass. We recalculated the movement cost surface for each scenario (current situation plus three options), and compared the mean cost of the 25 potential pathways identified via the cost analysis process using a two-way Analysis of Variance (ANOVA), followed by a paired t-test. Holm's method was applied in the paired t-test to guarantee that the family-wise type I error probability was less than 0.05.

Results

From 2010 to 2013, we surveyed 243 plot locations with a total effort of 9136 camera nights, 421 sign surveys, and 243 vegetation surveys (Table S1). Giant pandas were detected at 43 locations by camera traps and 133 locations by sign transects (Table S4).

Giant Panda Occupancy Model and Habitat Suitability Mapping

Four variables were excluded from later analyses because of their high collinearity with other covariates (Table 1). Detection probability was high when scent lure was applied or temperature was low, but was not influenced by camera view. In the selection of occupancy covariates with scent lure and temperature as detection covariates, 13 models with Δ AICs <2 were identified as top models (Table 2). All the nine occupancy covariates were included in the final model-averaged predictive equation, among which six (i.e. elevation, slope, forest age, bamboo presence, distance to road, and distance to large residences) had model weights >0.3. Partial response curves suggested that the occupancy probabilities of giant pandas were higher in areas with lower elevation, less steep slope, and the presence of bamboo and/or primary forest (Fig. 2). In addition, being close to large residences or roads reduced the occupancy probabilities of giant pandas.

Using the model-averaged predictive equation, we mapped the habitat suitability for the entire study area (Fig. 3a). The landscape had a broad fragmentation pattern: areas with predicted habitat suitability >0.5 (52% of the entire area) were mainly located further than 4 km from the XRV (55% covered by nature reserves), and areas with predicted habitat suitability <0.5 were mainly distributed near residences and roads along XRV. When comparing our prediction to the independent dataset collected by the SFD, the AUC value of 0.96 indicated strong predictive ability of our model.

Potential Corridors

The least-cost model predicted two potential movement corridors (Fig. 3b). The length of the northern corridor was 17 km, and was located directly in the XRV. The length of the southern corridor was 40 km, and followed the forest along the mountain ridge before crossing the main road in the south. The alternative pathways identified remain relatively consistent with the optimal northern corridor predicted by the least-cost model (Fig. 3b).

The circuit model predicted the habitat connectivity pattern for the entire XRV Corridor area. It predicted two potential corridors similar to the corridors predicted by least-cost models but provided additional information (Fig. 3b). The width of the northern corridor was approximate 2 km, and the width of the southern corridor was approximate 1 km. Both corridors traverse a wide extent of low probability habitat along the XRV. The circuit model also predicted a third potential pathway using the Taibai Nature Reserve to the north of the proposed XRV Corridor with a length of approximately 35 km.

Effectiveness of Alternative managements

The movement cost for optimal and alternative pathways were significantly different among scenarios (two-way-ANOVA, $F_{50, 1} = 23.29$, $p<0.001$). The dispersal cost under forest/bamboo restoration ($\bar{X} = 6453 \pm 634$ SE) and the automobile tunnel construction scenario ($\bar{X} = 5843 \pm 629$ SE) were significantly lower than dispersal cost in the current corridor ($\bar{X} = 8613 \pm 829$ SE; paired t-test: forest/bamboo restoration: $t = 10.34$, $p_{\text{holm}}<0.001$; automobile tunnel construction: $t = 13.31$, $p_{\text{holm}}<0.001$), while dispersal cost under the residence relocation scenario was not significantly lower than the current corridor condition ($\bar{X} = 8509 \pm 827$, $t = 0.44$, $p_{\text{holm}} = 1$). The automobile tunnel scenario resulted in lower dispersal cost than the forest/bamboo restoration scenario ($t = 3.41$, $p_{\text{holm}}<0.001$).

Discussion

We present a framework for using ground-based surveys to complement landscape attributes to examine large mammal habitat use across human-modified landscapes, and predict giant panda movement corridors. We based our corridor planning on predictions derived from this habitat model, a procedure widely used for conservation or land use planning [41]. Our study revealed that both natural and anthropogenic factors affect the distribution of giant pandas in the western Qinling Mountains. In particular, human infrastructures such as large residences and roads have strong negative effects and likely fragment giant panda habitat. Such results are consistent with previous studies. A negative relationship between giant panda distribution and the presence of roads and small villages was observed in adjacent areas in the Qinling Mountains [17]. Further south in the Qionglai Mountains, Li et al. [11] mapped movement corridors in Wolong Nature Reserve, and showed that a provincial highway had narrowed the movement pathway of giant pandas. Human land-use and habitat fragmentation have led to the drastic reduction of the most southern giant panda populations in the Xiaoxianling Mountains according to the genetic structure of the subpopulations [7]. Our studies did not include information on genetic structure, but our ground surveys across the XRV Corridor area found no evidence of movement between the established giant panda reserves.

Giant panda habitat selection and distribution

Studies of other large mammals across diverse ecosystems have described a broad spectrum of negative effects of roads and other human infrastructures on wildlife habitat connectivity, both direct and indirect [42]. Major roads and urbanized areas are acting as barriers for carnivore species in Poland [43], and the attributes of roads (i.e. traffic volume and road type) and housing development were correlated with the number of deer-vehicle collisions in Virginia, United States [44]. Ungulate-vehicle accidents accounted for approximately 60% of the total police-reported traffic accidents in Sweden during the 1990s, and fences along roadsides limited animal movement and access to important resources [45]. As human infrastructure is so important in regulating animal

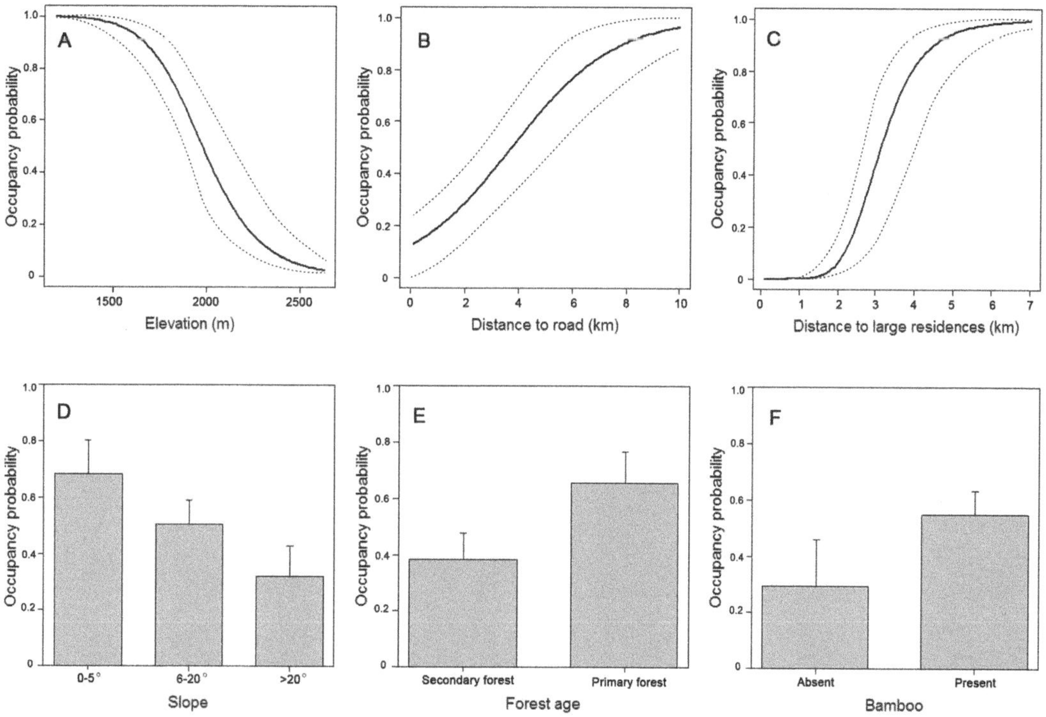

Figure 2. Partial correlation between giant panda occupancy probability and each environmental variable. The model-averaged weights is 1.00 for elevation (A), 0.46 to distance to road (B), 1.00 for distance to large residences (C), 0.30 for slope (D), 1.00 for forest age (E), and 0.45 for bamboo distribution (F).

movement, it should be considered in conservation planning for most large mammals.

In contrast to large residences and roads, giant pandas were detected near small residences. This finding opens the possibility of limits on household size being a viable management option to improve the corridor area. Household limits, and not outright housing bans, is contrary to studies suggesting intensive control of small residences within linkage areas [17]. Giant pandas and humans have co-existed within the XRV Corridor and the surrounding areas for centuries, as traditional agricultural activities only weakly infiltrated giant panda habitat [15]. Only within the past 50 years has intensive forest harvest and road construction, in combination with the government policy to aggregate families into larger residences along the river valley, altered the landscape such

Table 2. Top models for predicting the occupancy probability of giant panda.

Occupancy covariates	number of parameters	AIC	ΔAIC	AIC weight	Cumulative weight
Ele, Ftp, Dtl	5	567.16	0	0.14	0.14
Ele, Ftp, Bam, Dtl	6	567.27	0.11	0.13	0.27
Ele, Slo, Ftp, Bam, Dtl	7	567.97	0.81	0.09	0.36
Ele, Slo, Ftp, Dtl	6	568.22	1.07	0.08	0.44
Ele, Ftp, Dtr, Dtl	6	568.33	1.18	0.08	0.52
Ele, Ftp, Dtl, Dts	6	568.67	1.51	0.07	0.58
Ele, Ftp, Bam, Dtr, Dtl	7	568.72	1.56	0.06	0.65
Ele, Ftp, Bam, Ntr, Dtl	7	568.79	1.63	0.06	0.71
Ele, Slo, Ftp, Bam, Dtr, Dtl	8	568.79	1.64	0.06	0.77
Ele, Age, Ftp, Dtl	6	568.80	1.64	0.06	0.83
Ele, Slo, Ftp, Dtr, Dtl	7	568.82	1.66	0.06	0.89
Ele, Ftp, Ntr, Dtl	6	568.95	1.799	0.06	0.95
Ele, Ftp, Bam, Dtl, Dts	7	569.03	1.88	0.05	1.00

Every model contains the same detection covariates (i.e. temperature and scent lure application).

*Ele: elevation; Ftp: forest composition; Dtl: distance to large residence; Bam: bamboo distribution; Slo: slope; Dtr: distance to road; Ntr: type of nature reserve; Age: forest age; Dts: distance to small residence.

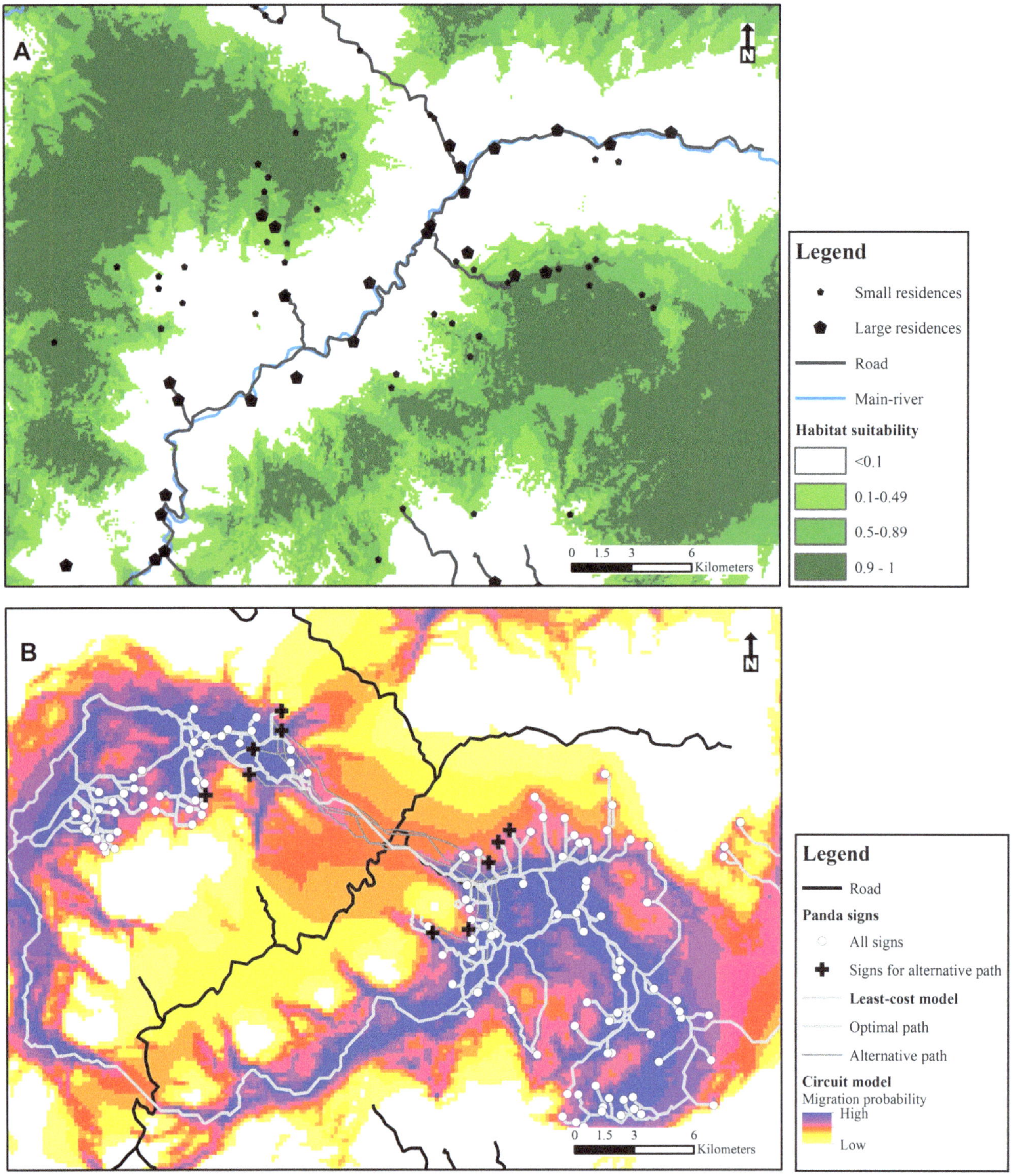

Figure 3. Maps for habitat connectivity and potential corridors. Habitat connectivity (A) predicted from occupancy model, and potential corridors (B) predicted by least-cost and circuit model.

that it might limit giant panda movements across the valley [15,16].

In determining the distribution of giant pandas across the XRV Corridor, we found that giant pandas used low elevation sites, in contrast with other studies where giant pandas selected higher-elevation areas [18,46]. Avoidance of lower elevations would make a valley corridor, such as the XRV area, less permeable for giant pandas. The differences between our study and previous efforts might be that our low-elevation areas were not all modified by human infrastructure (i.e. large residences and roads). The paradigm of giant pandas being restricted to high elevations should be reexamined, with more comprehensive investigation on anthropogenic interferences to movement across valley areas.

Interesting to us was the lack of gradation in habitat occupancy, as most habitat was either highly suitable or unsuitable for giant pandas. This result may be attributed to giant panda's avoidance to large residences along the XRV (Fig. 2C). Such infrastructure avoidance created a distinctive barrier between nature reserves along the XRV, and probably limits movement between reserve populations. In order to improve the conservation status of giant pandas, the Chinese government has focused on nature reserve creation, and established more than 50 nature reserves in the past 40 years [14]. While populations have significantly increased inside their reserves, much of the landscapes between nature reserves lack habitat protection or land-use restrictions. For the present reserve system to be viable, giant pandas must be able to move among reserves across the matrix of anthropogenic features [7,47]. In the Qinling Mountains, future conservation activities should reinforce the protection efforts at lower elevations and improve habitat linkages among core habitat areas [17]. Within the XRV, partial response curves indicate there are barriers to movement, and habitat connectivity can only be restored by removal of those barriers.

Potential use of proposed corridor

Although we found no evidence of a movement corridor across the XRV Corridor area, this area still has potential to provide for movement based on the results of both our least-cost and circuit models. We produced a cost-surface map for giant panda movement, and identified potential movement corridors connecting two giant panda populations. All the potential corridors went through the same 2 km wide section of the XRV. The section has no large residences or cropland, but abundant bamboo distribution, and thus is the best candidate for future corridor establishment efforts.

There is one caveat to our analysis. Our modelling process was limited by our reliance on detection records to determine giant panda habitat use. Using movement data from radio-collared fishers (*Martes pennanti*) to test model predictions, LaPoint et al. [48] found that least-cost and circuit models were more accurate if movement data were incorporated in the model creation. Although some telemetry studies are currently being conducted with giant pandas [15,49], these are confined to animals within reserves and do not provide movement information across the agricultural landscape between reserves. Our modelling process would be improved with telemetry data, but remains valid due to our ability to detect the animals within the agricultural matrix. In this case, field teams deploying camera-traps in association with the sign transects generated sufficient data for the modelling process, and model validation indicated a strong predictive ability.

Conservation implications

To reveal potential options for land management to improve the XRV Corridor, we used a scenario approach to explore the future consequences of conservation decisions [50]. Instead of proposing a single-best solution, our scenario approach acknowledges the uncertainty inherent in the XRV Corridor by comparing the effectiveness of alternative future management options [51]. By contrasting alternative future states into a decision-making framework, the scenario approach allows scientists to work together with decision-makers, land managers and the public to explore the long-term consequences for conservation alternates [52]. We strongly fell this is the best approach to engaging the Chinese government in discussion of the XRV Corridor area.

One of the restoration activities examined was reforestation and bamboo planting, which is considered the priority management action for giant panda habitat restoration as well as corridor establishment in several regions including the Qinling Mountains [12,13,37]. According to our models, such management actions will increase available habitat since mature forest and bamboo understory are key factors to giant panda habitat use. Such results are consistent with study from larger scale, that pandas are associated with old-growth forest [37]. However, mature forest restoration is a long-term process whose success can be slowed by multiple factors [53], and habitat restoration does not remove the impermeable barriers created by roads and large residences. Therefore, forest restoration should continue, but does not meet the urgent current need of giant panda conservation, given that the XRV Corridor area is experiencing growth of human infrastructures that will supersede any forest improvements.

The Natural Forest Conservation Program (NFCP), a new forest policy has been adopted in China, advocates residence relocation as a major means of forest ecosystem restoration [54]. Though such initiatives have created better living conditions for some communities [55], its ecological consequences remained unpredictable [56,57]. According to our scenario models, residence relocation could not meet the need to restore the habitat connectivity for giant panda in the XRV Corridor.

Our study suggests that the construction of an automobile tunnel may be the most effective approach in XRV corridor establishment, as this activity will simultaneously reduce the barriers of transportation and human residences. In this region, an automobile tunnel would divert automobile and truck transportation under the primary movement corridor. As a result, the original winding mountain highway will be abandoned and has the potential to function as a wildlife corridor. Among the 13 road structures (e.g. automobile tunnel, bridge, etc.) examined, automobile tunnels were the most effective at reducing vehicle-large ungulates collisions in the United States [58]. In the Qinling Mountains, automobile tunnel construction requires extensive landscape modification, and was found to displace goral (*Naemorhedus griseus*), serow (*Capricornis milneedwardsii*), and leopard (*Panthera pardus*) at a Qinling Mountain site [40]. However, five years after the automobile tunnel completion, giant pandas sign (footprints, dung, etc.) were found along the abandoned mountain highway, as the traffic had disappeared and the forest structure had partially recovered [59]. The lack of suitable habitat around the current highway indicates that any automobile tunnel construction would not reduce current panda habitat, but only benefit long-term conservation goals.

The automobile tunnel construction costs are usually high [58], but tunnels are increasingly used by the national and provincial governments to facilitate transportation across the Qinling Mountains [60]. In the past 10 years, more than 20 automobile tunnels were completed in Qinling Mountains, among which 5 were located in nature reserves, and 12 were located within 5 km of a nature reserves [60]. The newly planned Xian-Chengdu high-speed railroad calls for several additional tunnels in the region. While highways and high-speed rails construction usually create barriers for mammal movement, automobile tunnels within the landscape between giant panda reserves may provide opportunities to facilitate habitat connectivity by functioning as wildlife overpasses. The conservation benefits of these automobile tunnels have not been explored for giant pandas, but our models indicate they may be the best means to remove current barriers to giant panda movements.

Supporting Information

Table S1 Camera trapping data. Camera trapping data collected during 2010 to 2013. 243 camera trap deployments have been set up with a total effort of 9136 camera nights.

Table S2 Bamboo dataset surveyed by the SFD. The distribution of bamboo across the study area surveyed by the Shaanxi Forestry Department (SFD) during 2007 to 2010.

Table S3 Sign transect dataset surveyed by the SFD. The sign transect dataset that was collected by the Shanxi Forestry Department (SFD) was used to validate our occupancy model.

Table S4 Giant panda distribution data. Giant panda distribution data that was generated by combining the sign plot records with camera trap data.

Acknowledgments

We thank the staff of Huangbaiyuan Nature Reserve, Niuweihe Nature Reserve, Changqing Nature Reserve for their assistance in the field work. Z. Lingguo and Z. Xiaofeng of the Shaanxi Forestry Department helped in logistical details and permit applications.

Author Contributions

Conceived and designed the experiments: WJM FW DW HW ZL. Performed the experiments: FW SL DW. Analyzed the data: FW WJM QZ. Wrote the paper: FW WJM.

References

1. Andrén H (1994) Effects of Habitat Fragmentation on Birds and Mammals in Landscapes with Different Proportions of Suitable Habitat: A Review. Oikos 71: 355–366.
2. Downes SJ, Handasyde KA, Elgar MA (1997) The Use of Corridors by Mammals in Fragmented Australian Eucalypt Forests. Conservation Biology 11: 718–726.
3. Chetkiewicz C-LB, Boyce MS (2009) Use of resource selection functions to identify conservation corridors. Journal of Applied Ecology 46: 1036–1047.
4. Sawyer SC, Epps CW, Brashares JS (2011) Placing linkages among fragmented habitats: do least-cost models reflect how animals use landscapes? Journal of Applied Ecology 48: 668–678.
5. Zeller K, McGarigal K, Whiteley A (2012) Estimating landscape resistance to movement: a review. Landscape Ecology 27: 777–797.
6. Loucks CJ, Lü Z, Dinerstein E, Wang H, Olson DM, et al. (2001) Giant pandas in a changing landscape. Science 294: 1465–1465.
7. Zhu L, Zhan X, Wu HUA, Zhang S, Meng TAO, et al. (2010) Conservation Implications of Drastic Reductions in the Smallest and Most Isolated Populations of Giant Pandas. Conservation Biology 24: 1299–1306.
8. Wang X, Xu W, Ouyang Z (2009) Integrating population size analysis into habitat suitability assessment: implications for giant panda conservation in the Minshan Mountains, China. Ecological research 24: 1101–1109.
9. China's Ministry of Forestry & WWF (1989) A Comprehensive Report on China's Giant Panda and Its Habitat. Chengdu.
10. Zhu L, Zhang S, Gu X, Wei F (2011) Significant genetic boundaries and spatial dynamics of giant pandas occupying fragmented habitat across southwest China. Molecular Ecology 20: 1122–1132.
11. Li H, Li D, Li T, Qiao Q, Yang J, et al. (2010) Application of least-cost path model to identify a giant panda dispersal corridor network after the Wenchuan earthquake—Case study of Wolong Nature Reserve in China. Ecological Modelling 221: 944–952.
12. Shen G, Feng C, Xie Z, Ouyang Z, Li J, et al. (2008) Proposed conservation landscape for giant pandas in the Minshan Mountains, China. Conserv Biology 22: 1144–1153.
13. Yin K, Xie Y, Wu N (2006) Corridor connecting giant panda habitats from north to south in the Min Mountains, Sichuan, China. Integrative Zoology 1: 170–178.
14. State Forestry Administration (2006) The 3rd National Survey Report on Giant Panda in China. Beijing: Science Press.
15. Pan W, Lü Z, Wang D, Wang H (2001) The opportunity for the Giant Panda to exist. Peking Univ. Press, Beijing, China.
16. Zeng Z, Li J, Yan W, Fan J, Cui F, et al. (2009) Impact of Yangxian-Taibai Highway on the Movement of Giant Panda in the Qinling Mountains and Its Habitat Protection. Journal of Northwest Forestry University 24: 88–93.
17. Feng T-T, Van Manen FT, Zhao N-X, Li M, Wei F-W (2009) Habitat Assessment for Giant Pandas in the Qinling Mountain Region of China. The Journal of Wildlife Management 73: 852–858.
18. Liu X, Toxopeus AG, Skidmore AK, Shao X, Dang G, et al. (2005) Giant panda habitat selection in foping nature reserve, China. Journal of wildlife management 69: 1623–1632.
19. Swaisgood RR, Wei F, Wildt DE, Kouba AJ, Zhang Z (2010) Giant panda conservation science: how far we have come. Biology letters 6: 143–145.
20. Liu J, Linderman M, Ouyang Z, An L, Yang J, et al. (2001) Ecological Degradation in Protected Areas: The Case of Wolong Nature Reserve for Giant Pandas. Science 292: 98–101.
21. ASTER G (2011) Validation Team, 2009. ASTER Global DEM validation: summary report.
22. Muchoney D, Borak J, Chi H, Friedl M, Gopal S, et al. (2000) Application of the MODIS global supervised classification model to vegetation and land cover mapping of Central America. International Journal of Remote Sensing 21: 1115–1138.
23. Phillips SJ, Anderson RP, Schapire RE (2006) Maximum entropy modeling of species geographic distributions. Ecological modelling 190: 231–259.
24. Hijmans R, Cameron S, Parra J, Jones P, Jarvis A (2004) The WorldClim interpolated global terrestrial climate surfaces. Version 1.3.
25. Songer M, Delion M, Biggs A, Huang Q (2012) Modeling Impacts of Climate Change on Giant Panda Habitat. International Journal of Ecology 2012.
26. Tuan Mu M-N (2012) Spatiotemporal dynamics of giant panda habitat: Implications for panda conservation under a changing environment.
27. MacKenzie DI (2006) Occupancy estimation and modeling: inferring patterns and dynamics of species occurrence: Access Online via Elsevier.
28. MacKenzie DI, Nichols JD, Lachman GB, Droege S, Andrew Royle J, et al. (2002) Estimating site occupancy rates when detection probabilities are less than one. Ecology 83: 2248–2255.
29. Adriaensen F, Chardon J, De Blust G, Swinnen E, Villalba S, et al. (2003) The application of 'least-cost'modelling as a functional landscape model. Landscape and urban planning 64: 233–247.
30. McRae B, Shah V (2009) Circuitscape user guide. University of California, Santa Barbara. Available: http://www.circuitscape.org. Accessed 2014 Jul 22.
31. Li S, McShea WJ, Wang D, Shao L, Shi X (2010) The use of infrared-triggered cameras for surveying phasianids in Sichuan Province, China. Ibis 152: 299–309.
32. Burnham KP, Anderson DR (2002) Model selection and multi-model inference: a practical information-theoretic approach: Springer.
33. Erb PL, McShea WJ, Guralnick RP (2012) Anthropogenic Influences on Macro-Level Mammal Occupancy in the Appalachian Trail Corridor. PloS one 7: e42574.
34. Fiske I, Chandler R (2011) unmarked: An R package for fitting hierarchical models of wildlife occurrence and abundance. Journal of Statistical Software 43: 1–23.
35. ESRI (2013) 10.2 Desktop help. ESRI, Redlands, CA ArcGIS 10.
36. Koen EL, Bowman J, Garroway CJ, Mills SC, Wilson PJ (2012) Landscape resistance and American marten gene flow. Landscape ecology 27: 29–43.
37. Zhang Z, Swaisgood RR, Zhang S, Nordstrom LA, Wang H, et al. (2011) Old-growth forest is what giant pandas really need. Biology Letters 7: 403–406.
38. Wang Z, Song K, Hu L (2010) China's largest scale ecological migration in the three-river headwater region. AMBIO: A Journal of the Human Environment 39: 443–446.
39. Zhi-ming B (2006) The Definition, Category and Some Other Issues about Ecological Migration. Journal of the Central University for Nationalities (Philosophy and Social Sciences Edition) 1: 004.
40. Wu X, Duan W, Wang J, Zhang Y, Lin J (2010) Effect of the No. 2 Tunnel Ventilation Shaft at Qinling Zhongnanshan Highway on Wildlife Habitat. Shaanxi Forest Science and Technology 2: 5–11.
41. McSHEA WJ (2014) What are the roles of species distribution models in conservation planning? Environmental Conservation 41: 93–96.
42. Corlatti L, Hacklander K, Frey-Roos F (2009) Ability of wildlife overpasses to provide connectivity and prevent genetic isolation. Conserv Biology 23: 548–556.
43. Huck M, Jędrzejewski W, Borowik T, Miłosz-Cielma M, Schmidt K, et al. (2010) Habitat suitability, corridors and dispersal barriers for large carnivores in Poland. Acta Theriologica 55: 177–192.
44. McShea WJ, Stewart CM, Kearns LJ, Liccioli S, Kocka D (2008) Factors affecting autumn deer-vehicle collisions in a rural Virginia county. Human-Wildlife Interactions: 76.
45. Olsson MPO, Widén P, Larkin JL (2008) Effectiveness of a highway overpass to promote landscape connectivity and movement of moose and roe deer in Sweden. Landscape and Urban Planning 85: 133–139.
46. Xuezhi W, Weihua X, Zhiyun O, Jianguo L, Yi X, et al. (2008) Application of ecological-niche factor analysis in habitat assessment of giant pandas. Acta Ecologica Sinica 28: 821–828.

47. Qi D, Hu Y, Gu X, Yang X, Yang G, et al. (2012) Quantifying landscape linkages among giant panda subpopulations in regional scale conservation. Integrative Zoology 7: 165–174.
48. LaPoint S, Gallery P, Wikelski M, Kays R (2013) Animal behavior, cost-based corridor models, and real corridors. Landscape Ecology 28: 1615–1630.
49. Liu X, Skidmore AK, Wang T, Yong Y, Prins HH (2002) Giant panda movements in foping nature reserve, China. The Journal of wildlife management: 1179–1188.
50. Peterson GD, Cumming GS, Carpenter SR (2003) Scenario Planning: a Tool for Conservation in an Uncertain World. Conservation Biology 17: 358–366.
51. Postma TJBM, Liebl F (2005) How to improve scenario analysis as a strategic management tool? Technological Forecasting and Social Change 72: 161–173.
52. Thompson JR, Wiek A, Swanson FJ, Carpenter SR, Fresco N, et al. (2012) Scenario studies as a synthetic and integrative research activity for long-term ecological research. BioScience 62: 367–376.
53. Dang H, Zhang Y, Zhang K, Jiang M, Zhang Q (2010) Age structure and regeneration of subalpine fir (*Abies fargesii*) forests across an altitudinal range in the Qinling Mountains, China. Forest Ecology and Management 259: 547–554.
54. Zhang P, Shao G, Zhao G, Le Master DC, Parker GR, et al. (2000) China's Forest Policy for the 21st Century. Science 288: 2135–2136.
55. Rogers S, Wang M (2006) Environmental resettlement and social dis/re-articulation in Inner Mongolia, China. Population and Environment 28: 41–68.
56. Yeh ET (2009) Greening western China: A critical view. Geoforum 40: 884–894.
57. Wang Q, Ruan X, Pan C-D, Xu N-Y, Luo X, et al. (2006) Need for sustainability policy – a case study of the Natural Forest Conservation Program (NFCP) in the western region of Tianshan Mountain, China. The Forestry Chronicle 82: 31–39.
58. Huijser MP, Duffield JW, Clevenger AP, Ament RJ, McGowen PT (2009) Cost-benefit analyses of mitigation measures aimed at reducing collisions with large ungulates in the United States and Canada: a decision support tool. Ecology and Society 14: 15.
59. Wu P, Liu X, Shao X, Zhu Y, Cai Q (2010) GIS application in evaluating the potential habitat of giant pandas in Guanyinshan Nature Reserve, Shaanxi Province. Procedia Environmental Sciences 2: 111–122.
60. Cao Z, Su S, Gu S (2012) A Summary of Tunnel Design for the Xi'an-Hanzhong Expressway. Civil Engineering and Urban Planning 2012.pp. 56–61.

8

Soil Biochemical Responses to Nitrogen Addition in a Bamboo Forest

Li-hua Tu[1], Gang Chen[1], Yong Peng[1], Hong-ling Hu[1], Ting-xing Hu[1]*, Jian Zhang[1], Xian-wei Li[1], Li Liu[2], Yi Tang[3]

1 College of Forestry, Sichuan Agricultural University, Ya'an, Sichuan, China, **2** Personnel Department, Sichuan Agricultural University, Ya'an, Sichuan, China, **3** College of Horticulture, Sichuan Agricultural University, Ya'an, Sichuan, China

Abstract

Many vital ecosystem processes take place in the soils and are greatly affected by the increasing active nitrogen (N) deposition observed globally. Nitrogen deposition generally affects ecosystem processes through the changes in soil biochemical properties such as soil nutrient availability, microbial properties and enzyme activities. In order to evaluate the soil biochemical responses to elevated atmospheric N deposition in bamboo forest ecosystems, a two-year field N addition experiment in a hybrid bamboo (*Bambusa pervariabilis* × *Dendrocalamopsis daii*) plantation was conducted. Four levels of N treatment were applied: (1) control (CK, without N added), (2) low-nitrogen (LN, 50 kg N ha^{-1} $year^{-1}$), (3) medium-nitrogen (MN, 150 kg N ha^{-1} $year^{-1}$), and (4) high-nitrogen (HN, 300 kg N ha^{-1} $year^{-1}$). Results indicated that N addition significantly increased the concentrations of NH_4^+, NO_3^-, microbial biomass carbon, microbial biomass N, the rates of nitrification and denitrification; significantly decreased soil pH and the concentration of available phosphorus, and had no effect on the total organic carbon and total N concentration in the 0–20 cm soil depth. Nitrogen addition significantly stimulated activities of hydrolytic enzyme that acquiring N (urease) and phosphorus (acid phosphatase) and depressed the oxidative enzymes (phenol oxidase, peroxidase and catalase) activities. Results suggest that (1) this bamboo forest ecosystem is moving towards being limited by P or co-limited by P under elevated N deposition, (2) the expected progressive increases in N deposition may have a potential important effect on forest litter decomposition due to the interaction of inorganic N and oxidative enzyme activities, in such bamboo forests under high levels of ambient N deposition.

Editor: Dafeng Hui, Tennessee State University, United States of America

Funding: This project was supported by the National Natural Science Foundation of China (No. 31300522), the Specialized Research Fund for the Doctoral Program of Higher Education of China (No. 20125103120018) and the Scientific Research Fund of Sichuan Provincial Education Department of China (No. 12ZA118). The funders had no role in study design, data collection and analysis, decision to publish, or preparation of the manuscript.

Competing Interests: The authors have declared that no competing interests exist.

* Email: hutx001@163.com

Introduction

Anthropogenic nitrogen (N) deposition, primarily from food and energy production [1], has increased three- to five-fold over the last century [2], and presently adds more than 200 Tg yr^{-1} of N to terrestrial and ocean ecosystems [1]. It is predicted that N deposition will increase considerably in the 21st century with the largest increases occurring in East and South Asia [3]. Studies have shown that the increased N deposition substantially affected a range of essential processes relevant to the carbon (C) cycle in terrestrial ecosystems [4,5].

Carbon sequestration in forests depends on the balance between C fixation through plant growth and C loss through decomposition of soil organic matter (SOM) and plant litter. The response of plant growth in forests is generally positive [6–8], since the primary productivity in most forests is N limited [5]. While, the effects of N deposition on SOM and plant litter decomposition are highly controversial [9–13] due to the variability of forest properties and environmental factors. Soil extracellular enzymes (SEE) produced by microbes and plant roots were direct executors in the decomposition of litter and soil organic matter (SOM) [14]. Soil extracellular enzymes have two functions: to decompose complex substances into simple molecules and to supply sources to SEE producers [15]. Microbes and plants can never gain sources from complicated organism and the C and nutrient cycles would end without SEE [14]. Previous studies found that the response of SEE to elevated N inputs can explain the effects of exogenous N on the decomposition of SOM and litter [16]. In these studies, the best explanation for changes in decomposition rates following N deposition or N addition is the effect of N on the activity of the lignin-degrading enzyme phenol oxidase [16,17]. In general, the activities of cellulose-degrading enzymes would be stimulated after N addition, while lignin-degrading enzymes would be inhibited [18]. The response of SEE activities to N deposition in different ecosystems varied because of the diversity in nutrient availability [19], lignin content in litter [16], and the C/N ratio of litter and soil [16].

Carbon pools in plants and soils are closely linked through the nutrient cycle in forests. Thus, N deposition generally indirectly affects ecosystem processes, such as plant growth, litter decomposition and soil respiration, through the changes in soil biochemical properties such as soil nutrient availability, microbial properties and enzyme activities. Over the last decade, a considerable number of studies have demonstrated that experimental N addition has pronounced effects on soil available nutrients, soil microbial activities and structures, and soil enzyme activities [16,17,18,20–22]. However, ecosystem responses differed among ecosystem types.

It should be noted that most studies regarding the effects of N addition on soil properties have been conducted in coniferous and in broad-leaved forests. Bamboo forests are one of the most important forest types in the world [23]. China is one of the distribution centers of bamboo, and China's bamboo forests account for 15.4% (4.84 million ha in 2005) of the total area of bamboo worldwide [23]. Bamboo forests contribute about 10% of the C stocks in the living biomass of forests in China [24]. Therefore, bamboo forests play an important role in regional and global C cycling. Furthermore, bamboo forests are mainly distributed in the southern provinces of China. At present, southern China is experiencing rates of N deposition that are well above the global average [25–27]. The largest increases of N deposition in the world are projected to occur in this region over the next few decades [3]. Therefore, understanding the effect of increasing N deposition on the soil in bamboo forests is critical for predicting how the ecosystem processes regarding C cycling and nutrient circulation in bamboo ecosystems will respond to human activities.

To evaluate the initial soil biochemical responses to elevated atmospheric N deposition, we conducted experimental N addition treatments in a hybrid bamboo (*Bambusa pervariabilis* × *Dendrocalamopsis daii*) plantation over a two year period. The aim of this study is to examine the impacts of N addition on soil acidity, nutrient availability, microbial properties and soil enzyme activities.

Materials and Methods

Site description

The simulated N deposition experiment was conducted in a *Bambusa pervariabilis* × *Dendrocalamopsis daii* (hybrid bamboo) (10 ha) stand in Liujiang, Sichuan, China (29°42′25″ N, 103°14′38″ E, altitude 600 m above sea level). This region experiences an elevation-modified humid subtropical climate. The annual mean relative humidity was 86%. The mean temperature and annual precipitation in 2007, 2008, and 2009 were 16.8, 15.8, and 18.3°C, and 1758, 1847 and 1984 mm, respectively. The background wet N deposition measured in 2008 and 2009 was 82 and 113 kg N ha^{-1} $year^{-1}$, respectively [27,28]. The total atmospheric N deposition (N deposition in throughfall and stemflow in a bamboo forest) in Liujiang was approximately 131.5 kg N ha^{-1} $year^{-1}$ in 2009, which was one of the highest in the world [27]. The site was planted on former cropland in 2000 as part of the National Project of Converting Farmland to Forests (NPCFF). The site was close to the *Pleioblastus amurus* site (about 1 km apart from each other) used for the already- reported studies on carbon sequestration [29] and soil respiration [30] under N addition experiments. At the time of our study, the plant density was 13,320 stems ha^{-1} and the mean diameter at breast height (DBH) was 6 cm. The surface soil horizon (0–10 cm) contained 13.4 mg g^{-1} C and 1.58 mg g^{-1} N. The average soil depth to bedrock was approximately 1 m, and the thickness of the surface organic layer was approximately 1 cm before the experimental treatments began. The soil at the site is classified as a Dystric Purpli-Orthic Primosol (Lithic Dystrudepts according to USDA Soil Taxonomy) derived from purple sandstone and shale. There was very little shrub or herb in the understory at the time of the experiment.

The research site is owned by Sichuan Agricultural University. The field studies did not involve endangered or protected species and no specific permits were required for the described field studies.

Microclimate and litterfall measurements

Air temperature and precipitation were measured using a Davis Weather Station (Vantage Pro, Davis Inc. USA) located adjacent to the experimental area. Litterfall was collected in ten 50×50 cm nylon mesh traps placed randomly in the experimental site but outside of the treated plots. Soil temperature was measured on each sampling date using thermometer.

Experimental treatments

Nitrogen addition experiments were initiated in January 2008. Twelve plots were established and divided into four N addition treatments: control (CK; without N added), low-N (LN; 50 kg N ha^{-1} $year^{-1}$), medium-N (MN; 150 kg N ha^{-1} $year^{-1}$), and high-N (HN; 300 kg N ha^{-1} $year^{-1}$), with three replicates each. The plots measured 3×3 m, spaced at approximately about 5 m intervals, and randomly selected to receive treatments. The addition of fertilizer (NH_4NO_3) occurred monthly in twelve equal applications beginning in January 2008. During each application, the fertilizer was weighed, dissolved in 1 L of water, and applied to each plot using a portable sprayer. The control plot received 1 L of water without fertilizer.

Soil sample collection and analysis

From January to December 2009, five subsamples of surface mineral soil (0–10 cm) were monthly taken from each plot using a 27 mm diameter soil auger. Subsamples of each plot were mixed to form a composite sample and the visible roots were removed by tweezers. The soil sample was homogenized, passed through a 2 mm sieve and stored at 4°C for analysis within one week. The soil water content was determined by dried at 105°C using a subsample (10 g) from each plot. For determining soil total organic carbon (TOC) and total nitrogen (TN), air-dried subsamples were ground to sieve through the mesh with size of 0.25 mm.

The pH value was determined by a glass electrode in aqueous extracts. Soil TOC was measured by the dichromate digestion method [31]. Soil TN was determined through acid digestion, using the Kjeldahl method [32]. Available N (ammonium and nitrate) was extracted with a 2 M KCl solution, and measured colorimetrically. Soil microbial biomass C (MBC) and soil microbial biomass N (MBN) were measured using the chloroform fumigation extraction technique [33,34,35] by a total CN analyzer (Shimadzu model TOC-$V_{cPH+TNM-1}$, Kyoto, Japan). The Bray-2 method [36] was used to measure soil available P.

Invertase (β-D-fructofuranoside fructohydrolase, EC 3.2.1.26; abbreviation: βF) activity was measured with the Frankenberger Jr and Johanson method [37]. Urease (EC 3.5.1.5) activity was measured spectrophotometrically (610 nm) according to Sinsabaugh et al. [38]. The activity of acid phosphatase (AcPh; EC 3.2.1.2) was determined following published protocols [20] using 4-methyumbelliferyl (MUB) phosphate as a substrate. Phenol oxidase (PhOx; EC 1.10.3.2) and peroxidase (POD; EC 1.11.1.7) activities were measured spectrophotometrically using *L*-3, 4-dihydroxyphenylalanine (DOPA) as the substrate [20]. Catalase (CAT; EC 1.11.1.6) activity was determined with the Cohen et al. [39] method where decomposed hydrogen peroxide is measured by its reaction to excess potassium tetraoxomanganate (VII). Enzyme activity was calculated as the μmoles of substrate converted per hour (or per minute for CAT) for each gram of dried sample.

In February, April, June, August, October and December 2009, three undisturbed soil cores (100 cm^3) (0–10 cm soil horizon) were collected using the accessories of barometric process separation (BaPS) system (UMS Inc. Germany) from each plot. The soil cores were returned to the laboratory and stored on ice in coolers until analysis. Gross nitrification and denitrification rates were measured using BaPS system according to Ingwersen et al. [40]. The unit of gross nitrification and denitrification rates is $\mu g\ N\ kg^{-1}\ h^{-1}$ (based on the dry weight of the soil samples).

Statistical analyses

All analyses were conducted using SPSS 15.0 for Windows (SPSS Inc. USA). Repeated measures ANOVA with Fisher's LSD tests were performed to examine the soil temperature, moisture, pH, TOC, TN, NH_4^+-N, NO_3^--N, MBC, MBN, and activities of βF, urease, AcPh, PhOx, POD, CAT, gross nitrification rate, and denitrification rate for the different treatments. Significant effects were determined at $\alpha = 0.05$ unless otherwise stated. Mean values in the text are given ± 1SE.

The accumulative activities of enzymes were calculated by the following equation:

$$A_{acc} = \sum_{i=1}^{n} A_i \times k \times D_i$$

A_{acc} represents the accumulative activity in one year for 1 kg soil (mol kg^{-1} yr^{-1}); A_i is the mean value of enzyme activity in the month ($\mu mol\ g^{-1}\ h^{-1}$); $i = 1, 2, 3, \ldots, n$ (i.e., $i = 1$ represents January, $i = 2$ represents February); n is the number of months in one year (12); k is the unit conversion factor ($k = 0.024$ for βF, urease, AcPh, PhOx, POD; $k = 1.44$ for CAT) to converse the unit from $\mu mol\ g^{-1}\ h^{-1}$ to mol kg^{-1} day^{-1}, and D_i is the number of days in a month.

Results

Microclimate and litterfall

There were obvious seasonal variations for temperature and precipitation at the experimental site (Figs. 1a,b). Precipitation in the study period (2009) was 1984 mm and the mean annual air temperature was 18.3°C. Mean litterfall in the *P. amarus* plantation was 473±25 g m^2 $year^{-1}$ over the study period (Fig. 1b). The litterfall mass showed a strong seasonal pattern that peaked in May. The litterfall from April to June accounted for 68% of annual total litterfall. Soil temperature at 10 cm soil depth ranged from 8.3°C in January to 20.6°C in July at the control plots (Fig. 2a). The seasonal variations in soil temperature were similar to air temperature. However, soil moisture was nearly stable, ranging from 0.29 to 0.36 cm^3 H_2O cm^{-3} soil (Fig. 2b). There was no significant difference in soil temperature or soil moisture among the plots during the study period ($P = 0.795$ and 0.643, respectively).

pH, nutrient availability and microbial properties

The results of repeated measures ANOVAs indicated that the pH and the concentrations of TOC, TN, NH_4^+, NO_3^-, MBN, MBC, and AP exhibited significant seasonal patterns ($P < 0.001$, Fig. 3). Overall, the concentrations of TN, NH_4^+, NO_3^-, MBN and MBC peaked in July, and AP peaked in March. The pH values in CK, LN, MN and HN were 4.60±0.11, 4.47±0.08,

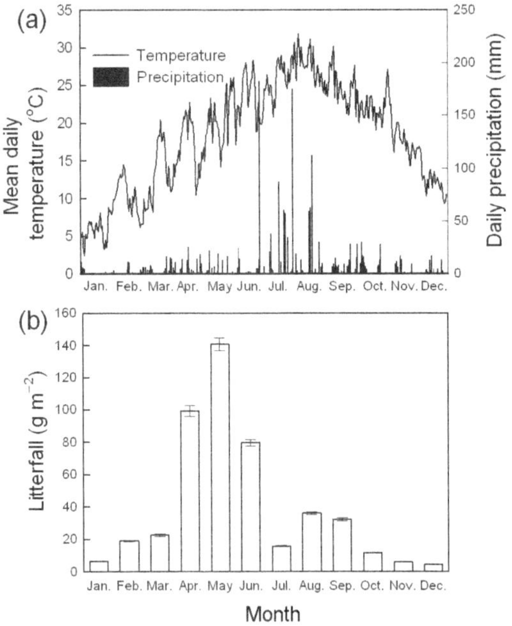

Figure 1. Seasonal variations of air temperature, precipitation and litterfall in a *Bambusa pervariabilis* × *Dendrocalamopsis daii* plantation from January to December 2009. (a) Mean daily temperature and daily precipitation; (b) monthly litterfall. Bars indicate mean ± SE, $n = 10$.

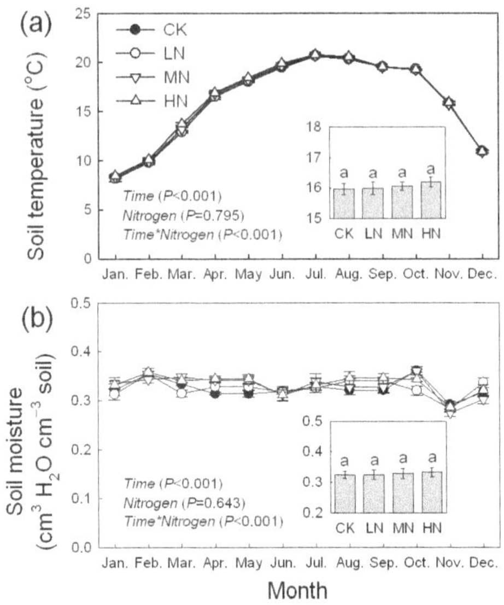

Figure 2. Seasonal variations of soil temperature and soil moisture in a *Bambusa pervariabilis* × *Dendrocalamopsis daii* plantation from January to December 2009 (mean ± SE, $n = 3$). The treatments were: CK (without N added), LN (50 kg N ha^{-1} $year^{-1}$), MN (150 kg N ha^{-1} $year^{-1}$), and HN (300 kg N ha^{-1} $year^{-1}$) with three replicate plots for each treatment. Monthly applications of NH_4NO_3 began in January 2008. (a) Soil temperature at 10 cm below surface; (b) volumetric soil moisture in the 0–20 cm soil horizon. Average values of yearly measures were exhibited in the histograms (mean ± SE, $n = 3$). Results of repeated measures ANOVAs are shown in text. Different letters indicate significant difference among N addition treatments when nitrogen effect is significant ($P < 0.05$, Fisher's least significant difference test).

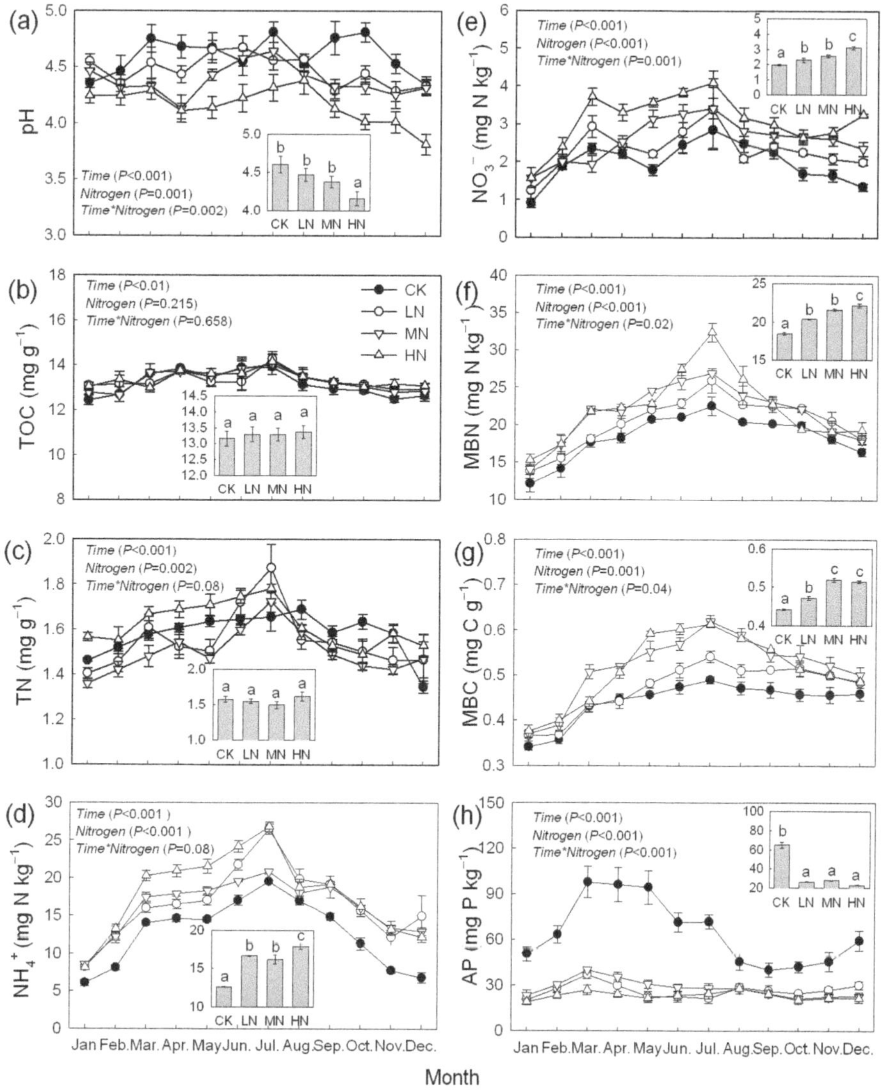

Figure 3. Seasonal variations of soil properties in a *Bambusa pervariabilis* × *Dendrocalamopsis daii* plantation from January to December 2009 (mean ± SE, *n* = 3). TOC: total organic carbon; TN: total nitrogen; MBN: microbial biomass nitrogen; MBC: microbial biomass carbon; AP: available phosphorus. Average values of yearly measures were exhibited in the histograms (mean ± SE, *n* = 3). Results of repeated measures ANOVAs are shown in text. Different letters indicate significant difference among N addition treatments when nitrogen effect is significant (*P*<0.05, Fisher's least significant difference test). CK, LN, MN and HN are as in Fig. 2.

4.38±0.08 and 4.16±0.09, respectively, and the pH in all the three treatments were significantly decreased (Fig. 3a). There was no significantly difference between treatments in TOC concentration (Fig. 3b). The addition of N did not significantly affect soil TN concentrations (Fig. 3c). The concentrations of NH_4^+, NO_3^-, MBN, and MBC increased significantly under all three N addition treatments (Figs. 3d, e, f, g). The concentration of AP decreased 57% to 64% after N added, and the differences between the control and all the N treatments were significant (Fig. 3h). Soil gross nitrification and denitrification rates exhibited significant seasonal variations (*P*<0.001), and both rates peaked in June (Fig. 4). The average gross nitrification rates for CK, LN, MN, and HN were 34.6±4.6, 46.2±4.5, 46.0±4.7, and 63.2±4.7 μg N kg^{-1} h^{-1}, respectively. The mean rates of denitrification were 57.0±5.7, 69.6±5.3, 66.7±4.8 and 79.2±5.0 μg N kg^{-1} h^{-1} in CK, LN, MN, and HN plots, respectively. The addition of N significantly increased the nitrification and denitrification rates.

Soil enzyme activities

All six enzyme activities demonstrated significant seasonal variation (P<0.01), and the addition of N changed the seasonal variation significantly (P<0.05) (Fig. 5). The peak period of activity of βF and AcPh occurred in June, and the activity of urease and CAT peaked in September. There was an obvious trough in POD activity from October to December. Results of repeated measures ANOVA tests indicated that N addition significantly increased the activity of urease and AcPh, significantly decreased the activities of PhOx and CAT significantly, and had no significant effect on βF activity. The activity of CAT decreased 31%–38% under N addition treatment, while N addition increased AcPh activity by 11%–18%.

The results of accumulative activities in 1 kg soil in one year are displayed in Table 1. Under N addition treatments, accumulative activities of *β*F and POD increased and decreased respectively, while the differences among treatments were not significant for

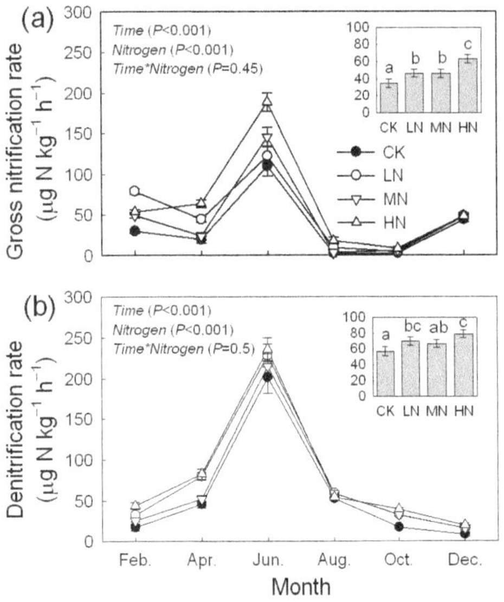

Figure 4. Seasonal variations of soil gross nitrification and denitrificantion rates in a *Bambusa pervariabilis* × *Dendrocalamopsis daii* plantation in 2009 (mean ± SE, n=3). Average values of yearly measures were exhibited in the histograms (mean ± SE, n=3). Results of repeated measures ANOVAs are shown in text. Different letters indicate significant difference among N addition treatments when nitrogen effect is significant ($P<0.05$, Fisher's least significant difference test). CK, LN, MN and HN are as in Fig. 2.

both enzyme activities. Nitrogen addition significantly increased the urease accumulative activity in the HN treatment. The accumulative activity of AcPh was stimulated, and the accumulative activities of AcPh to all N addition treatments were significantly higher than that of the control. For PhOx and CAT, the accumulative activities were significantly depressed in all the three N addition treatments.

Discussion

Nitrogen and phosphorus are two important elements for ecosystem functioning for both of them commonly limit plant growth [41]. The response of plant production to modern human influences, such as atmospheric deposition and climate change, will be mediated by changes in the availability of these elements [42]. The response of soil nutrients and organic C, especially the active portions, to N addition can reflect the effect of N input on soil C pools of plant and soil. Soil extracellular enzymes mediate the degradation, transformation and mineralization of SOM [15]. In general, hydrolytic enzymes are relevant to decomposition of SOM, and oxidative enzymes relate to the resynthesis of SOM [15]. Therefore, the variations of the two types of enzymes reflect SOM decomposition and humification in soils. In the present study, the addition of N had no significant effect on soil total N concentration, but increased the concentrations of TOC, NH_4^+, NO_3^-, MBN, MBC, and decreased AP concentration significantly. Two types of hydrolytic enzymes (urease and AcPh) were stimulated significantly, while the three oxidative enzymes were depressed significantly.

Available N in the soil mainly existed in the forms of NH_4^+ and NO_3^-, which is mainly generated from N mineralization. Urease is a type of enzyme produced by microbes or plants for accessing N. In this study, urease activity was stimulated by N addition, which suggested that the mineralization of organic N may be accelerated by N addition. It was consistent with the results in different forest ecosystems [20]. However, in a tallgrass prairie ecosystem, Ajwa et al. [43] found that N addition inhibited urease activity by approximately 15%. The rates of gross nitrification and denitrification were stimulated by N addition in this bamboo forest. The phenomenon of N addition increasing N transformation was observed in many other systems, such as temperate hardwood forests [44] and exotic annual grasslands [45]. Soil N transformation is a substrate limited process. The nitrification rate depends on the concentration of ammonia, and the nitrate concentration, C availability, as well as oxygen availability [46]. Nitrification and denitrification are main sources of N_2O emissions from soils [47]. Then, the results suggest that N_2O emission may be increased in this ecosystem. Actually, Barnard and Leadley [47] conclude in their review that the stimulation of field N_2O emissions through nitrification and denitrification processes by N additions is clearly shown by global N addition experiments conducted in agricultural and forested ecosystems. The denitrification process might be a mechanism to avoid N imbalance and reduce the negative effects from excess N in forests [48]. The atmospheric N deposition in the experimental area was extremely high compared to the rest of the world [27]. It can be anticipated that the bamboo system in this study and many other similar forest systems in this region would become N saturated systems within decades if the deposition of N continues to increase. Meanwhile, in a bamboo forest (P. amarus) adjacent to the study site, we observed very little N loss through hydrological processes, which indicates that most of the N was lost through denitrification [27]. However, the amount of N lost through denitrification may differ among forest types because of the variability of forest properties and ambient N deposition. Nitrification and denitrification play key roles in regulating soil inorganic N concentration, the production of a potent greenhouse gas N_2O, and leaching of nitrate [47]. Thus changes in both processes in response to elevated N deposition can influence N stocks and directly feed back to atmospheric and climatic change.

In contrast, the available P in soil decreased significantly after the addition of N in this study. Many previous studies suggested that in N limited systems, N addition increases microbial activity and stimulates the demand for P [18,20,43]. Almost all the N addition studies reported the upregulation of the enzymes involved in P turnover [20,21,43] and is in accordance with the results of this study. This common phenomenon was revealed by a global meta-analysis, which demonstrated that phosphatase activity was significantly greater in soils supporting symbiotic N_2-fixing plants because they could afford to invest N in P acquisition [49]. It is well established that phosphate represses the synthesis of phosphatases [50]. In this bamboo ecosystem, although the increased AcPh activity indicates an accelerated rate of transformation of organic P to inorganic P, the elevated microbial activity may have simultaneously accelerated the immobilization of inorganic P and led to the decrease of available P in the soil, and then the higher activity of AcPh. The results in this study combined with previous studies suggest that ecosystems under high levels of N deposition are moving towards being limited or co-limited by P. In fact, many ecosystems around the world are now considered to be co-limited by P [51].

Microorganisms play an important role in nutrient transformation in forest soils. Soil microbial biomass depends on total organic C content, the C stock of the litter, and microbial flora induced in the decomposition of litter and rhizospheric depositions [52]. Similar to this study, N addition increased microbial biomass in a loblolly pine (Pinus taeda) plantation [12], but several other studies have presented contradictory results [53]. There were several

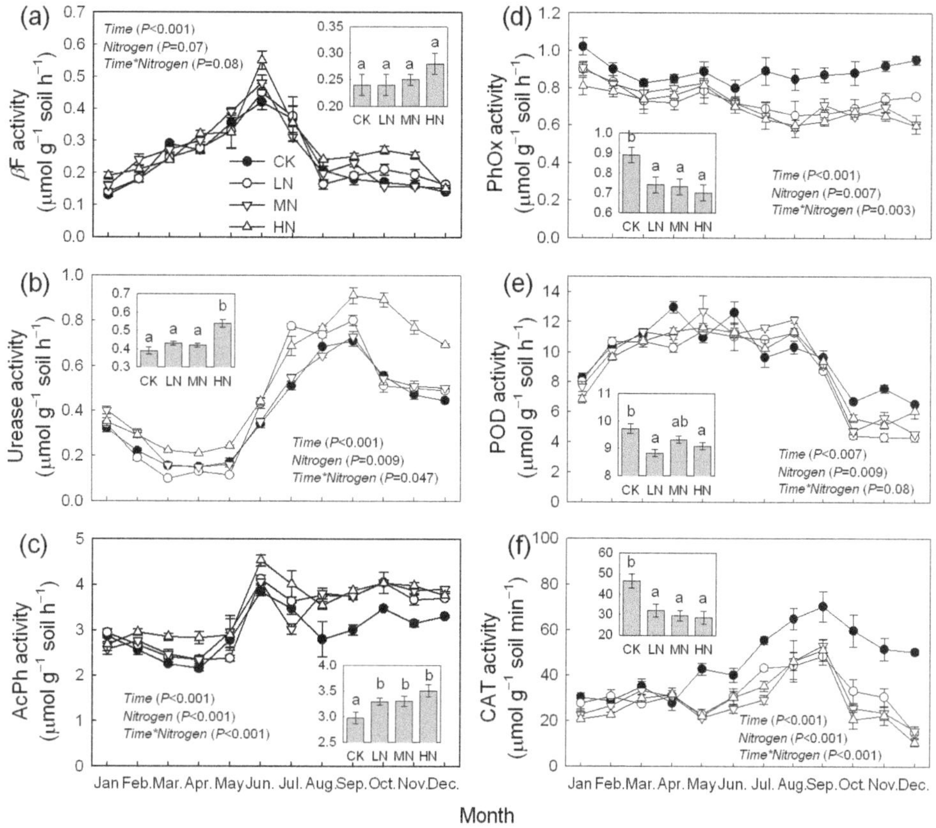

Figure 5. Seasonal variations of soil enzyme activities in a *Bambusa pervariabilis* × *Dendrocalamopsis daii* plantation from January to December 2009 (mean ± SE, *n*=3). βF: β-Fructofuranosidase; AcPh: acid phosphatase; PhOx: Phenol oxidase; POD: peroxidase; CAT: catalase. Results of repeated measures ANOVAs are shown in text. Different letters indicate significant difference among N addition treatments when nitrogen effect is significant ($P<0.05$, Fisher's least significant difference test). CK, LN, MN and HN are as in Fig. 2.

possible underlying mechanisms resulting in the phenomenon observed in this study. First, the addition of inorganic N directly provides a vital nutrient for microbial growth. Second, N addition may increase microbial biomass through stimulating the growth and metabolism of fine roots. For example, in our another N addition study, we found fine root biomass and root tissue N concentration increased significantly under N addition [29]. Plant root tissue N concentrations are generally highly correlated with root metabolism rates [54]. The increased microbial biomass in this study suggests elevated potential microbial activity and C turnover rate in this bamboo ecosystem. However, microbial biomass declined 15% on average under N addition on a global scale [55].

Soil oxidative enzymes, especially the lignin-degrading enzymes (PhOx and POD) were closely related to humification [15]. Lignin is a class of three-dimensional acrylic polymers; most of the chemical bonds in which are difficult to hydrolyzed [56]. Compared with polysaccharides and other biopolymers, lignin is highly resistant to decay [57]. There are two kinds of aerobic fungi that break down lignin: white-rot organisms (Basidiomycetes) and soft-rot organisms (Ascomycetes). Degradation carried out by the soft-rot organisms is incomplete, while the white-rot organisms are

Table 1. Results of one-way ANOVAs of accumulative soil enzyme activities in a *Bambusa pervariabilis* × *Dendrocalamopsis daii* plantation (mean ± SE, *n*=3).

Treatment	βF	urease	AcPh	PhOx	POD	CAT
	mol kg^{-1} yr^{-1}	mol kg^{-1} yr^{-1}	mol kg^{-1} yr^{-1}	mol kg^{-1} yr^{-1}	mol kg^{-1} yr^{-1}	mol kg^{-1} yr^{-1}
Control	2.10±0.16a	3.46±0.13a	26.1±0.6a	7.77±0.37b	85.0±3.0a	24453±1857b
Low-N	2.13±0.16a	3.75±0.12a	28.8±0.6b	6.47±0.35a	77.2±2.7a	16817±1648a
Medium-N	2.22±0.10a	3.65±0.13a	28.9±0.9b	6.38±0.35a	81.6±2.7a	15470±1281a
High-N	2.44±0.16a	4.74±0.15b	30.6±1.1b	6.11±0.31a	79.6±4.7a	15015±1566a

CK, LN, MN and HN denote control (without N added), low nitrogen (50 kg N ha^{-1} year^{-1}), medium nitrogen (150 kg N ha^{-1} year^{-1}), high nitrogen (300 kg N ha^{-1} year^{-1}), respectively; βF = β-Fructofuranosidase; AcPh = acid phosphatase; PhOx = Phenol oxidase; POD = peroxidase; CAT = catalase. Different letters within the same column indicate significant difference among treatments (one-way ANOVA followed by Fisher's least significant difference test, $\alpha=0.05$).

able to degrade the lignin completely [56]. The production of lignin-degrading enzymes, such as PhOx and POD, by lignin-degrading fungi decreased after inorganic N addition [17]. White-rot organisms did not synthesize lignin-degrading enzymes in the presence of low molecular N-rich compounds (for example, ammonium and amino acids) [56]. In this study, the inhibiting effect of N addition on PhOx and POD can be interpreted as the growth limitation of white-rot organisms caused by soil acidification. The optimum pH values for white-rot and soft-rot organisms are 4.0–5.0 and 6.0–7.5 [15]. Actually, a number of species of white-rot organisms with this property suggests that the lignolytic enzyme production inhibited by high N levels could be a widespread phenomenon [57]. Humus accumulated in the forest soil surface as a result of the N inhibitory effect on lignin decay [58], and the phenomenon would be more obvious when the lignin content in forest litter is higher [16].

We conducted a leaf litter decomposition experiment at the same site during this period. It demonstrated that N addition depressed the leaf litter decomposition through inhibiting the decay of lignin [13], which confirmed the inhibitory effect of N on oxidases activities in this study. Similar to this study, Sinsabaugh et al. [59] found that N deposition decreased oxidase activities significantly in a boreal forest. Analyses of laccase gene abundance and diversity indicated that decreased activity of oxidative enzymes is the result of reduced expression [60]. Freeman et al. [61] proposed that PhOx may act as an "enzymatic latch" and control C storage in many ecosystem types. For example, peatlands, which are considered enzyme-limited systems, have low PhOx activities that contribute to organic matter accumulation [61,62], while arid ecosystems appear to be substrate-limited systems where high POD and PhOx potentials are believed to limit SOM accumulation [63]. Waldrop et al. [16] conducted a simulated N deposition experiment in several temperate forest ecosystems. They found a significant relationship between changes in PhOx and changes in soil C content across three ecosystems, which indicates that the mechanism of oxidative enzyme activities control soil C storage. The results in this study suggest that more SOM may accumulate in the soil surface through depressing oxidative enzyme activities and inhibiting litter decomposition.

The inhibitory effect of N addition on oxidative enzyme activities is not consistent throughout ecosystem types. Most N addition studies have been conducted in temperate and boreal forests where soil microbial communities are dominated by Basidiomycetes [64]. Grassland soils are dominated by Glomeromycota and Ascomycota [65] and do not show decreased oxidative enzyme activities under N addition treatment [66]. Keeler et al. [18] demonstrated that the negative response of oxidative enzyme activity to N addition is not universal with a long-term N addition experiment conducted in eight different ecosystems. Furthermore, the responses of extracellular enzymes in soils and litter to N addition are not entirely in agreement, due to the differences of internal components, structures and decomposer communities. Saiya-Cork et al. [20] reported that PhOx activity increased in litter but dropped in soil under N addition.

Enzymes themselves are N-rich (protein) so the enzyme production was adjusted by soil N availability [21]. Extracellular enzymes were produced by microbes and plant roots, thus N addition may indirectly affect enzyme activities by affecting enzyme producers. For example, in our previous study conducted in a P. amurus plantation, N addition stimulated aboveground biomass, fine root biomass [29], and the growth of rhizospheric microbes. The results in this study indicate that N addition may play a positive role in plant growth of hybrid bamboo.

Overall, in this hybrid bamboo forest, N addition accelerated nitrification and denitrification rates, increased hydrolytic enzymes activities, soil N availability and microbial biomass. Nitrogen addition stimulated AcPh activity and decreased the soil P availability for the more demands of plants and soil microbes for element P under elevated N addition, suggesting this bamboo system is moving towards being limited by P or co-limited by P and N. The inhibitory effects of N addition on soil oxidative enzyme activities suggest that the expected progressive increases in N deposition may have a potential important effect on forest litter decomposition in such bamboo forests.

Author Contributions

Conceived and designed the experiments: LHT HLH TXH JZ. Performed the experiments: LHT HLH GC YP LL YT. Analyzed the data: LHT HLH TXH JZ XWL LL YT. Contributed reagents/materials/analysis tools: LHT HLH TXH. Contributed to the writing of the manuscript: LHT HLH TXH.

References

1. Galloway JN, Townsend AR, Erisman JW, Bekunda M, Cai Z, et al. (2008) Transformation of the nitrogen cycle: Recent trends, questions, and potential solutions. Science 320: 889–892.
2. Denman KL, Brasseur G, Chidthaisong A, Ciais P, Cox PM, et al. (2007) Couplings between changes in the climate system and biogeochemistry. In: Solomon S, Qin D, Manning M, Marquis M, Averyt K, et al. (eds). Climate Change 2007, The Physical Science Basis. Contribution of Working Group I to the Fourth Assessment Report of the Intergovernmental Panel on Climate Change. Cambridge University Press, Cambridge, UK. pp 499–587.
3. Dentener F, Drevet J, Lamarque JF, Bey I, Eickhout B, et al. (2006) Wild, Nitrogen and sulfur deposition on regional and global scales: A multimodel evaluation. Global Biogeochem Cy 20: doi:10.1029/2005GB002672.
4. Reay DS, Dentener F, Smith P, Grace J, Feely RA (2008) Global nitrogen deposition and carbon sinks. Nat Geosci 1: 430–437.
5. LeBauer DS, Treseder KK (2008) Nitrogen limitation of net primary productivity in terrestrial ecosystems is globally distributed. Ecology 89: 371–379.
6. Hyvönen R, Persson T, Andersson S, Olsson B, Ågren GI, et al. (2008) Impact of long-term nitrogen addition on carbon stocks in trees and soils in northern Europe. Biogeochemistry 89: 121–137.
7. Pregitzer KS, Burton AJ, Zak DR, Talhelm AF (2008) Simulated chronic nitrogen deposition increases carbon storage in Northern Temperate forests. Global Change Biol 14: 142–153.
8. De Vries W, Solberg S, Dobbertin M, Sterba H, Laubhann D, et al. (2009) The impact of nitrogen deposition on carbon sequestration by European forests and heathlands. Forest Ecol Manag 258: 1814–1823.
9. Hobbie SE (2005) Contrasting effects of substrate and fertilizer nitrogen on the early stages of litter decomposition. Ecosystems 8: 644–656.
10. Cleveland CC, Townsend AR (2006) Nutrient additions to a tropical rain forest drive substantial soil carbon dioxide losses to the atmosphere. P Natl Acad Sci USA 103: 10316–10321.
11. Allison SD, LeBauer DS, Ofrecio MR, Reyes R, Ta AM, et al. (2009) Tran, Low levels of nitrogen addition stimulate decomposition by boreal forest fungi. Soil Biol Biochem 41: 293–302.
12. Samuelson L, Mathew R, Stokes T, Feng Y, Aubrey D, et al. (2009) Soil and microbial respiration in a loblolly pine plantation in response to seven years of irrigation and fertilization. Forest Ecol Manag 258: 2431–2438.
13. Tu LH, Hu HL, Chen G, Peng Y, Xiao YL, et al. (2014) Nitrogen addition significantly affects forest litter decomposition under high levels of ambient nitrogen deposition. PLoS ONE 9: e88752.
14. Allison SD, Weintraub MN, Gartner TB, Waldrop MP (2010) Evolutionary-economic principles as regulators of soil enzyme production and ecosystem function. In: Shukla G, Varma A (eds). Soil Enzymology. New York: Springer-Verlag, pp 229–243.
15. Sinsabaugh RL (2010) Phenol oxidase, peroxidase and organic matter dynamics of soil. Soil Biol Biochem 42: 391–404.
16. Waldrop MP, Zak DR, Sinsabaugh RL, Gallo M, Lauber C (2014) Nitrogen deposition modifies soil carbon storage through changes in microbial enzymatic activity. Ecol Appl 14: 1172–1177.
17. Carreiro MM, Sinsabaugh RL, Repert DA, Parkhurst DF (2000) Microbial enzyme shifts explain litter decay responses to simulated nitrogen deposition. Ecology 81: 2359–2365.

18. Keeler BL, Hobbie SE, Kellogg LE (2009) Effects of long-term nitrogen addition on microbial enzyme activity in eight forested and grassland sites: implications for litter and soil organic matter decomposition. Ecosystems 12: 1–15.
19. Olander LP, Vitousek PM (2000) Regulation of soil phosphatase and chitinase activity by N and P availability. Biogeochemistry 49: 175–190.
20. Saiya-Cork KR, Sinsabaugh RL, Zak DR (2002) The effects of long term nitrogen deposition on extracellular enzyme activity in an *Acer saccharum* forest soil. Soil Biol Biochem 34: 1309–1315.
21. Frey SD, Knorr M, Parent JL, Simpson RT (2004) Chronic nitrogen enrichment affects the structure and function of the soil microbial community in temperate hardwood and pine forests. Forest Ecol Manag 196: 159–171.
22. Gan H, Zak DR, Hunter MD (2013) Chronic nitrogen deposition alters the structure and function of detrial food webs in a northern hardwood ecosystem. Ecol Appl 23: 1311–1321.
23. FAO (2010) Global Forest Resources Assessment 2010: Main report. Food and Agriculture Organization of the United Nations, Rome.
24. Chen X, Zhang X, Zhang Y, Booth T, He X (2009) Changes of carbon stocks in bamboo stands in China during 100 years. Forest Ecol Manag 258: 1489–1496.
25. Fang YT, Gundersen P, Vogt RD, Koba K, ChenFS, et al. (2011) Atmospheric deposition and leaching of nitrogen in Chinese forest ecosystems. J Forest Res 16: 341–350.
26. Liu X, Zhang Y, Han W, Tan A, Shen J, et al. (2013) Enhanced nitrogen deposition over China. Nature 494: 459–462.
27. Tu LH, Hu TX, Zhang J, Huang LH, Xiao YL, et al. (2013) Nitrogen distribution and cycling through water flows in a subtropical bamboo forest under high level of atmospheric deposition. PLoS ONE 8: e75862.
28. Tu LH, Hu HL, Hu TX, Zhang J, Liu L, et al. (2011) Effect of experimental nitrogen deposition on the decomposition of different litter fractions in a subtropical bamboo ecosystem. Pedosphere 21: 689–695.
29. Tu LH, Hu TX, Zhang J, Li RH, Dai HZ, et al. (2011) Short-term simulated nitrogen deposition increases carbon sequestration in a *Pleioblastus amarus* plantation. Plant Soil 340: 383–396.
30. Tu LH, Hu TX, Zhang J, Li XW, Hu HL, et al. (2013) Nitrogen addition stimulates different components of soil respiration in a subtropical bamboo ecosystem. Soil Biol Biochem 58: 255–264.
31. Kalembasa SJ, Jenkinson DSA (1973) comparative study of titrimetric and gravimetric methods for determination of organic carbon in soil, J Sci Food Agr 24: 1085–1090.
32. Grimshaw HM, Allen SE, Parkinson JA (1989) Nutrient elements. In: Allen SE (ed) Chemical Analysis of Ecological Material. Blackwell Scientific, Oxford, pp 81–159.
33. Brookes PC, Landman A, Pruden G, Jenkinson DS (1985) Chloroform fumigation and the release of soil-nitrogen—a rapid direct extraction method to measure microbial biomass nitrogen in soil. Soil Biol Biochem 17: 837–842.
34. Vance ED, Brookes PC, Jenkinson DS (1987) An extraction method for measuring soil microbial biomass C. Soil Biol Biochem 19: 703–707.
35. Wu J, Joergensen RG, Pommerening B, Chaussod R, Brookes PC (1990) Measurement of soil microbial biomass C by fumigation-extraction-an automated procedure. Soil Biol Biochem 22: 1167–1169.
36. Bray RH, Kurtz LT (1945) Determination of total, organic, and available forms of phosphorus in soils. Soil Sci 59: 39–45.
37. Frankenberger WT Jr, Johanson JB (1983) Method of measuring invertase activity in soils. Plant Soil 74: 301–311.
38. Sinsabaugh RL, Reynolds H, Long TM (2000) Rapid assay for amidohydrolase (urease) activity in environmental samples. Soil Biol Biochem 32: 2095–2097.
39. Cohen GJ, Dembiec D, Marcus J (1970) Measurement of catalase activity in tissue extracts. Anal Biochem 34: 30–38.
40. Ingwersen J, Butterbach-Bahl K, Gasche R, Papen H (1999) Barometric process separation: new method for quantifying nitrification, denitrification, and nitrous oxide sources in soils. Soil Sci Soc Am J 63: 117–128.
41. Vitousek PM, Howarth RW (1991) Nitrogen limitation on land and in the sea—how can it occur. Biogeochemistry 13: 87–115.
42. Perring MP, Hedin LO, Levin SA, McGroddy M, De Mazancourt C (2008) Increased plant growth from nitrogen addition should conserve phosphorus in terrestrial ecosystems. P Natl Acad Sci USA 105: 1971–1976.
43. Ajwa HA, Dell CJ, Rice CW (1999) Changes in enzyme activities and microbial biomass of tallgrass prairie soil as related to burning and nitrogen fertilization. Soil Biol Biochem 31: 769–777.
44. Micks P, Aber JD, Boone RD, Davidson EA (2004) Short-term soil respiration and nitrogen immobilization response to nitrogen applications in control and nitrogen-enriched temperate forests. Forest Ecol Manag 196: 57–70.
45. Sirulnik AG, Allen EB, Meixner T, Allen MF (2007) Impacts of anthropogenic N additions on nitrogen mineralization from plant litter in exotic annual grasslands. Soil Biol Biochem 39: 24–32.
46. Gumiero B, Boz B, Cornelio P, Casella S (2011) Shallow groundwater nitrogen and denitrification in a newly afforested, subirrigated riparian buffer. J Appl Ecol 48: 1135–1144.
47. Barnard R, Leadley PW (2005) Global change, nitrification, and denitrification: A review. Global Biogeochem Cy 19: doi: 10.1029/2004GB002282.
48. Xu Y, Xu Z, Cai Z, Reverchon F (2013) Review of denitrification in tropical and subtropical soils of terrestrial ecosystems. J Soils Sediments 13: 699–710.
49. Houlton BZ, Wang YP, Vitousek PM, Field CB (2008) A unifying framework for dinitrogen fixation in the terrestrial biosphere. Nature 454: 327–329.
50. Bell C, Stromberger M, Wallenstein M (2014) New insights into enzymes in the environment. Biogeochemistry 117: 1–4.
51. Elser JJ, Bracken MES, Cleland EE, Gruner DS, Harpole WS, et al. (2007) Global analysis of nitrogen and phosphorus limitation of primary producers in freshwater, marine and terrestrial ecosystems. Ecol Lett 10: 1135–1142.
52. Haynes RJ (2000) Labile organic matter as an indicator of organic matter quality in arable and pastoral soil in New Zealand. Soil Biol Biochem 32: 211–219.
53. Wallenstein MD, McNulty S, Fernandez IJ, Boggs J, Schlesinger WH (2006) Nitrogen fertilization decreases forest soil fungal and bacterial biomass in three long-term experiments. Forest Ecol Manag 222: 459–468.
54. Burton AJ, Jarvey JC, Jarvi MP, Zak DR, Pregitzer KS (2012) Chronic N deposition alters root respiration-tissue N relationship in northern hardwood forest. Global Change Biol 18: 258–266.
55. Treseder KK (2008) Nitrogen additions and microbial biomass: a meta-analysis of ecosystem studies. Ecol Let 11: 1111–1120.
56. Berg B, McClaugherty C (2008) Plant Litter: Decomposition, Humus Formation, Carbon Sequestration, second ed. Springer, New York.
57. Berg B, Matzner E (1997) Effect of N deposition on decomposition of plant litter and soil organic matter in forest systems. Environ Rev 5: 1–25.
58. Magill AH, Aber JD (1998) Long-term effects of experimental nitrogen additions on foliar litter decay and humus formation in forest ecosystems. Plant Soil 203: 301–311.
59. Sinsabaugh RL, Zak DR, Gallo M, Lauber C, Amonette R (2004) Nitrogen deposition and dissolved organic carbon production in northern temperate forests. Soil Biol. Biochem 36: 1509–1515.
60. Hassett JE, Zak DR, Blackwood CB, Pregitzer KS (2009) Are basidiomycete laccase gene abundance and composition related to reduced lignolytic activity under elevated atmospheric NO_3^- deposition in a Northern Hardwood forest? Microb Ecol 57: 728–739.
61. Freeman C, Ostle N, Kang H (2001) An enzymic 'latch' on a global carbon store. Nature 409: 149.
62. Freeman C, Ostle NJ, Fenner N, Cang H (2004) A regulatory role for phenol oxidase during decomposition in peatlands. Soil Biol Biochem 36: 1663–1667.
63. Collins SL, Sinsabaugh RL, Crenshaw C, Green LE, Porras-Alfaro A, et al. (2008) Pulse dynamics and microbial processes in aridland ecosystems. J Ecol 96: 413–420.
64. Lauber CL, Strickland MS, Bradford MA, Fierer N (2008) The influence of soil properties on the structure of bacterial and fungal communities across land-use types. Soil Biol Biochem 40: 2407–2415.
65. Porras-Alfaro A, Lipinski K, Herrera J, Natvig DO, Sinsabaugh RL (2011) Diversity and distribution of soil fungal communities in a semiarid grassland. Mycologia 103: 10–21.
66. Lyyemperumal K, Shi W (2008) Soil enzyme activities in two forage systems following application of different rates of swine lagoon effluent or ammonium nitrate. Appl Soil Ecol 38: 128–136.

9

A Generalized Approach to the Modeling of the Species-Area Relationship

Katiane Silva Conceição[1], Werner Ulrich[2], Carlos Alberto Ribeiro Diniz[3], Francisco Aparecido Rodrigues[1], Marinho Gomes de Andrade[1]*

1 Departamento de Matemática Aplicada e Estatística, Instituto de Ciências Matemáticas e de Computação, Universidade de São Paulo, São Carlos, São Paulo, Brazil, **2** Department of Animal Ecology, Nicolaus Copernicus University in Toruń, Toruń, Poland, **3** Department of Statistics, Federal University of São Carlos, São Carlos, Brazil

Abstract

This paper proposes a statistical generalized species-area model (GSAM) to represent various patterns of species-area relationship (SAR), which is one of the fundamental patterns in ecology. The approach enables the generalization of many preliminary models, as power-curve model, which is commonly used to mathematically describe the SAR. The GSAM is applied to simulated data set of species diversity in areas of different sizes and a real-world data of insects of *Hymenoptera* order has been modeled. We show that the GSAM enables the identification of the best statistical model and estimates the number of species according to the area.

Editor: Enrico Scalas, Universita' del Piemonte Orientale, Italy

Funding: K. S. Conceição would like to acknowledge FAPESP (grant # 2008/10613-1). F. A. Rodrigues would like to thank CNPq (grant # 305940/2010-4), Fapesp (grant # 2013/26416-9) and NAP eScience - PRP - USP for financial support. The authors also thank Fapesp (grant #2014/15860-8) for funding the publication of this paper. The funders had no role in study design, data collection and analysis, decision to publish, or preparation of the manuscript.

Competing Interests: The authors have declared that no competing interests exist.

* Email: marinho@icmc.usp.br

Introduction

The variation in the number of species with area, known as species-area relationship (SAR), is one of the most important ecological patterns [1]. The models of SAR enable the prediction of the number of species that coexist and share resources, as well as the impact of the extinction of species caused by habitat loss. Sampled data for a single species, or all species of a specific trophic level within a particular site have shown that the SAR has a well-defined shape, most often described by power and exponential curves [2]. The number of species in an area increases with increasing island area, but the rate of increase slows for larger islands. Many hypotheses have been proposed to explain the SAR [3,4,6]. For instance, some are based on the immigration and extinction of species [4], random sampling processes [6] or the Second Law of Thermodynamics [5].

These different hypotheses have generated many mathematical models for the description of the SAR [1–3,7–10]. The early models were based on deterministic modeling, which assumes that every set of variable states is uniquely determined by the parameters in the model. For instance, Arrhenius considered that the number of species (S) is related to area (A) through a power law form [11] (called the power-function), i.e. $S=S_0A^z$ (or $\log S=\log c+z\log A$), where S_0 represents the number of species in a unit area ($A=1$) and $0<z<1$. Due to the random nature of the sampled data, statistical modeling is more suitable for SAR description than the deterministic approach [6], therefore, many statistical models have been developed (e.g. [2]). Moreover, statistical models can be thought of as general cases of deterministic models, because the mean value of the random variable of interest yields the results of the deterministic model.

Because there exist many models to address the SAR (e.g. [3,9,12]), a natural question is how to select the best model for a given data set. To address this issue, here we integrate different models within a common framework and consider the problem of curve fitting by the transformed generalized linear model (TGLM) [13]. We propose the use of the generalized species-area model (GSAM) to describe the SAR. GSAM includes many models, such as those described in [3,9] as special cases. We also consider a model that simulates the colonization process of a region by different species and show that the GSMA has best fitted the data in comparison with traditional power-curve models. Finally, we use the data on the cumulative species richness of parasitic *Hymenoptera* from 25 nested plots in a beech forest on limestone [14]. Our results show that the GSAM can determine the best model for the data and estimate the number of species accurately.

Methods

Generalized species-area models

In species-area curves, the number of species (S) is the dependent variable and the area (A) is the explanatory variable. Some mathematical models of SAR propose that the number of species is related to the area as

$$S_i=\mu(A_i,\boldsymbol{\beta}),\quad i=1,\ldots,n, \tag{1}$$

where $\boldsymbol{\beta}$ is a p-parameter vector [1]. Function μ can be derived from laws governing the physical system that gave rise to the data. As such models are deterministic and the properties related to the random nature of variable S are neglected, the deterministic models are often inadequate due to the stochastic nature of the data [6]. The statistical modeling usually assumes that Eq. 1 can be written as

$$S_i = \mu(A_i, \boldsymbol{\beta}) + \varepsilon_i, \tag{2}$$

where $\{\varepsilon_i\}$ are independent and identically distributed random noises (*i.i.d*) — usually, $\varepsilon_i \sim N(0, \sigma^2)$. Note that the mode in Eq. 2 is a generalization of the determinic model, i.e. $E(S_i) = \mu(A_i, \boldsymbol{\beta})$. The model given by Eq. 1 can be understood as a particular case of the model $E\{\Lambda(S_i, \lambda)\} = \mu(A_i, \boldsymbol{\beta})$, where $\Lambda(\cdots, \lambda)$ is a monotonic transformation and λ is a scalar parameter defining such a transformation. For instance, in cases whose data suggested by Eq. 2 are unsatisfactory, the experimenter can assume a model with logarithmic transformation, i.e.

$$\log(S_i) = \mu(A_i, \boldsymbol{\beta}) + \varepsilon_i. \tag{3}$$

This paper proposes a new model called generalized species-area model (GSAM), which is based on the TGLM approach proposed in [13]. The GSAM works with a general parametric family of transformations from the dependent variable S to $S^{(\lambda)} = \Lambda(S; \lambda)$ and postulates that the transformed random variable $S^{(\lambda)}$ follows a continuous probability distribution belonging to the exponential family. Furthermore, the GSAM assumes that there exists some λ value such that $S^{(\lambda)}$ satisfies the usual assumptions of the generalized linear models (GLM) [15].

A suitable choice of the family of transformations enables the representation of power-curves, their recent extensions (see [11,16–19]), the models presented in [2,3,9] and the logarithmic model described in [16] as special cases of the GSAM. We have considered the Box-Cox power transformation [20], which is effective at turning skewed unimodal distributions into nearly symmetric normal-like distributions.

Let $\boldsymbol{S} = (s_1, \ldots s_n)^T$ be the vector of observations. By using

$$S^{(\lambda)} = \begin{cases} \dfrac{S^{\lambda} - 1}{\lambda} & \text{if } \lambda \neq 0 \\ \log(S) & \text{if } \lambda = 0, \end{cases} \tag{4}$$

we can obtain the transformed observations $\boldsymbol{S}^{(\lambda)} = (s_1^{(\lambda)}, \ldots s_n^{(\lambda)})^T$. The GSAM assumes that there exists some λ value such that the transformed random variables $\{S_1^{(\lambda)}, \ldots S_n^{(\lambda)}\}$ can be considered independently distributed. Each $S_i^{(\lambda)}$ follows an exponential family distribution with a probability density function of the form

$$\pi(s_i^{(\lambda)}; \theta_i, \phi) = \exp\left[\phi^{-1}\left\{s_i^{(\lambda)}\theta_i - b(\theta_i)\right\} + c(s_i^{(\lambda)}, \phi)\right], \tag{5}$$

where $b(x)$ and $c(x, \phi)$ are appropriate known functions. The dispersion parameter ϕ is assumed to be the same for all observations. The mean and variance of $S_i^{(\lambda)}$ are, respectively, $E\{S_i^{(\lambda)}\} = \mu_i = db(\theta_i)/d\theta_i$ and $Var\{S_i^{(\lambda)}\} = \phi\, V(\mu_i)$, where $V(\mu_i) = d^2 b(\theta_i)/d\theta_i^2 = d\mu_i/d\theta_i$ is the variance function. Parameter $\theta_i = \int V^{-1}(\mu_i) d\mu_i = q(\mu_i)$ is a known one-to-one function of μ_i.

The GSAM also considers a systematic component given by

$$g(\mu_i) = \eta_i = \boldsymbol{x}_i^T \boldsymbol{\beta}, \tag{6}$$

where the link function $g(\cdot)$ is a known one-to-one continuously differentiable function and $\boldsymbol{x}_i$ is a specified vector $(p \times 1)$ of the explained variables, which include the area, known functions of the area, and other environmental variables. Matrix $\boldsymbol{X}$ whose rows are vectors $\boldsymbol{x}_i^T$, $i = 1, \ldots, n$, is a specified $n \times p$ model matrix of full rank $p < n$ and $\boldsymbol{\beta} = (\beta_1, \ldots, \beta_p)^T$ is a set of unknown linear parameters to be estimated. The link function is assumed to be monotonic and differentiable.

The GSAM proposed here considers three components of structural importance: (i) the Box-Cox family of transformations (Eq. 4) in association with a more general form for the distribution of the transformed variable $S^{(\lambda)}$ (Eq. 5); (ii) a linear predictor function and (iii) a possible nonlinear link function for the regression parameters (Eq. 6). Moreover, when the variance function $V(\mu_i)$ is not constant, i.e. when the variance is correlated with mean μ, some distributions of the exponential family enable the handling of data presenting heteroscedasticity. In this context, GSAM is a generalization of the previous mathematical models that describe the SAR.

Species-area relationship models. Many models have been proposed for the description of the SAR and some can be linearized by a logarithmic transformation of the response variable (i.e. diversity of species). Models that are special cases of the GSAM have the following properties: (i) the transformation parameter in Eq. 4 is $\lambda = 0$, i.e. a log-transformation is adopted, $S^{(0)} = \log(S)$ and $\mu = E\{S^{(0)}\} = E\{\log(S)\}$ or no transformation is considered for variable S, i.e. we assume $\lambda = 1$ and $\mu = E\{S\}$; (ii) the distribution in Eq. 5 is the normal distribution; and (iii) the link function is the identity function, $g(\mu) = \mu$ and the systematic component in Eq. 6 is given by $\mu = \boldsymbol{X}\boldsymbol{\beta}$. The elements of matrix $\boldsymbol{X}$ may be area A, $\log(A)$ or additional variables, as in [12]. Very simple forms of the systematic component are given by $\mu = \beta_0 + \beta_1 \log A$ or $\mu = \beta_0 + \beta_1 A$. This special case of the GSAM can be understood as the particular cases proposed in [11,16]. A list of some models of SAR is provided.

1. Considering the stochastic nature of variable S, which represents the number of species, the power model proposed by Arrhenius [11] can be written as,

$$E\{S\} = b_0 A^{b_1}. \tag{7}$$

The logarithm of variable S yields

$$E\{\ln(S)\} = \beta_0 + \beta_1 \ln(A), \tag{8}$$

where the parameters of the power-curve in log-log space are $\beta_0 = \ln(b_0)$ and $\beta_1 = b_1$.

2. The persistence model (P1-full) proposed by Plotkin et al. [17] is given by

$$E\{S\} = b_0 A^{b_1} \exp\left\{\sum_{k=1}^{n} b_{k+1} A^k\right\}, \tag{9}$$

or, considering the logarithm of variable S,

$$E\{\ln(S)\} = \beta_0 + \beta_1 \ln(A) + \sum_{k=1}^{n} \beta_{k+1} A^k, \tag{10}$$

where $\beta_0 = \ln(b_0)$, $\beta_1 = b_1$ and $\beta_{k+1} = b_{k+1}$, $k = 1, \ldots, n$. A special case, when $n = 1$ (P1 model [17]) is given by

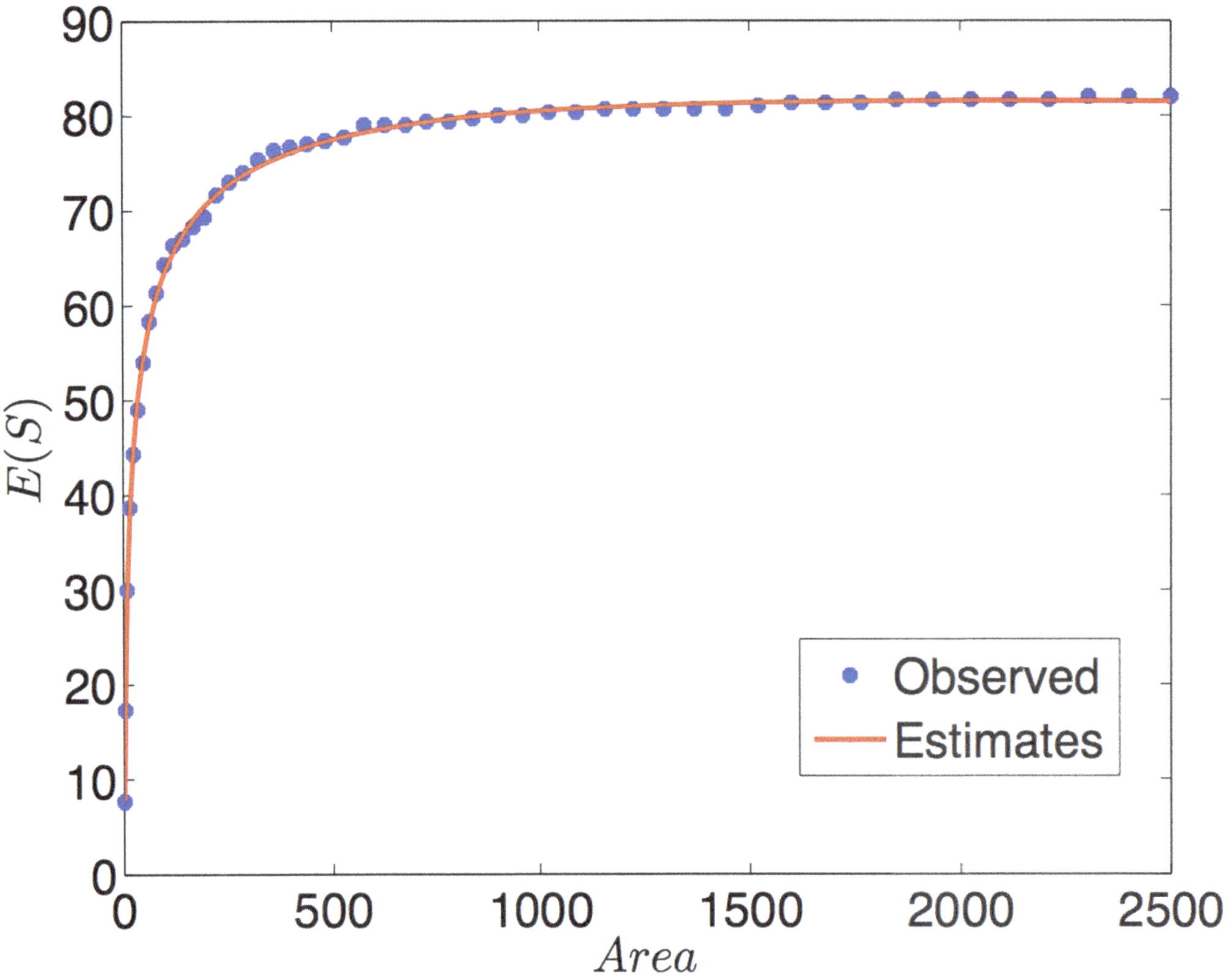

Figure 1. Data and fitted curve obtained by the GSMA model for the simulated species-area relationship.

$$E\{S\} = b_0 A^{b_1} \exp\{b_2 A\}, \tag{11}$$

or, considering the logarithm of variable S,

$$E\{\ln(S)\} = \beta_0 + \beta_1 \ln(A) + \beta_2 A \tag{12}$$

where $\beta_0 = \ln(b_0)$, $\beta_1 = b_1$ and $\beta_2 = b_2$.

3. The persistence model proposed by Ulrich and Buszko [18,19] is given by

$$E\{S\} = b_0 A^{b_1} \exp\left\{\frac{b_2}{A}\right\}, \tag{13}$$

or, considering the logarithm of variable S,

$$E\{\ln(S)\} = \beta_0 + \beta_1 \ln(A) + \frac{\beta_2}{A}, \tag{14}$$

where $\beta_0 = \ln(b_0)$, $\beta_1 = b_1$ and $\beta_2 = b_2$.

4. The polynomial power-function model proposed by Chiarucci et al. [21] is defined as

$$E\{S\} = 10^{\left(\sum_{k=0}^{n} b_k A^k\right)}, \tag{15}$$

or, considering the logarithm of variable S,

$$E\{\ln(S)\} = \sum_{k=0}^{n} \beta_k A^k, \tag{16}$$

where $\beta_k = \ln(10^{b_k})$, $k = 0,1,\ldots,n$. The quadratic power-function model proposed in [21] considers $n = 2$, i.e.

$$E\{S\} = 10^{\{b_0 + b_1 \log(A) + b_2 (\log(A))^2\}}, \tag{17}$$

or, considering the logarithm of variable S,

$$E\{\ln(S)\} = \beta_0 + \beta_1 \log(A) + \beta_2 (\log(A))^2, \tag{18}$$

where $\beta_k = \ln(10^{b_k})$, $k = 0,1,2.$.

Some species-area relationships may also be represented by linear functions, specifically:

Table 1. Some of the traditional models adjusted with Gaussian errors.

$\mu=E\{S\}$	Selection Criteria				
Models	*AIC*	*BIC*	$-2logL$	*MSE*	*MAPE*
[4]	406.59	424.09	400.59	0.608	0.305
[16]	297.87	315.34	291.87	0.069	0.086
[22]	228.63	251.93	220.63	0.016	0.050
$\mu=E\{\log(S)\}$	**Selection Criteria**				
Models	*AIC*	*BIC*	$-2logL$	*MSE*	*MAPE*
[11]	405.91	423.38	399.91	0.340	0.155
[17])	358.24	381.54	350.25	0.123	0.091
[18]	331.76	355.05	323.76	0.082	0.071
[21]	221.09	244.39	213.09	0.010	0.025

5. Linear model proposed by MacArthur and Wilson [4]

$$E\{S\}=b_0+b_1A. \tag{19}$$

6. Logarithmic function proposed by Gleason [16]

$$E\{S\}=\beta_0+\beta_1\ln(A). \tag{20}$$

7. Quadratic logarithmic function proposed by Gitay et al. [22]

$$E\{S\}=\{b_0+b_1\ln(A)\}^2, \tag{21}$$

or

$$E\{S\}=\beta_0+\beta_1\ln(A)+\beta_2(\ln(A))^2, \tag{22}$$

where $\beta_0=b_0^2$, $\beta_1=2b_0b_1$ and $\beta_2=b_1^2$

8. General power-logarithmic function proposed by Gitay et al. [22]

$$E\{S\}=\{b_0+b_1\ln(A)\}^{b_2}. \tag{23}$$

If b_2 is any real number and $|b_0|>|b_1\ln(A)|$, from Newton's generalized binomial expansion we obtain

$$\{b_0+b_1\ln(A)\}^{b_2}=(b_0)^{b_2}\sum_{k=0}^{\infty}\binom{b_2}{k}\left(\frac{b_1}{b_0}\right)^k(\ln(A))^k, \tag{24}$$

where the binomial coefficients with an arbitrary upper index can be defined as

$$\binom{b_2}{k}=\frac{b_2(b_2-1)\cdots(b_2-k+1))}{k!}=\frac{(b_2)_k}{k!}. \tag{25}$$

Therefore, the logarithmic function model can be written as

$$E\{S\}=\beta_0+\sum_{k=1}^{\infty}\beta_k(\ln(A))^k, \tag{26}$$

where $\beta_0=(b_0)^{b_2}$, $\beta_k=(b_0)^{b_2}\{(b_2)_k/k!\}(b_1/b_0)^k$.

Full-scale generalized species-area relationship model. The right side of all equations presented in the previous section always involves polynomial terms, such as $\ln(A)$, A and/or $1/A$. Here, we propose a generalization of these models by considering the right side of the full-scale model consists of three polynomials, i.e.

$$P_1(A)=\sum_{k=1}^{m}\beta_k(\ln(A))^k, \tag{27}$$

Table 2. The GSAMs fitted with different models according to the likelihood method.

Models	$g(\mu)$	Systematic component
normal	μ	$\beta_0+\beta_1\ln(A)+\beta_2(\ln A)^2+\beta_3/A$
gamma	$\frac{1}{\mu}$	$\beta_0+\beta_1\ln(A)+\beta_2(\ln A)^2+\beta_3/A+\beta_4/A^2$
I.G.	$\frac{1}{\mu^2}$	$\beta_0+\beta_1\ln(A)+\beta_2(\ln A)^2+\beta_3/A+\beta_4A^2$

Table 3. Selection criteria for the GSAMs fitted with $\hat{\lambda}$ adjusted according to the likelihood method.

Model	Parameter	Selection Criteria				
	$\hat{\lambda}$	*AIC*	*BIC*	$-2logL$	*MSE*	*MAPE*
normal	0.822(0.729; 0.922)	79.90	109.02	69.90	0.08	0.61
gamma	0.328(−0.003,0.680)	116.82	151.76	104.82	0.09	0.64
I.G.	−0.089(−0.322,0.141)	142.87	177.82	130.88	0.11	0.68

$$P_2(A)=\sum_{k=m+1}^{n}\beta_k\left(\frac{1}{A}\right)^{k-m}, \qquad (28)$$

and

$$P_3(A)=\sum_{k=m+n+1}^{q}\beta_k A^{k-(m+n)}. \qquad (29)$$

The left sides of those equations have the number of species (S) or $\ln(S)$. In order to generalize them, we have assumed that S is a random variable and considered the Box-Cox power transformation (Eq. 4).

The curves defined by the GSAM assume a linear predictor function and a nonlinear link function $g(\mu)$ for the systematic component (Eq. 6). The systematic component of the GSAM is given by

$$\begin{aligned} g(\mu) &= P_1(A)+P_2(A)+P_2(A) \\ &= \beta_0+\sum_{k=1}^{m}\beta_k(\ln(A))^k+\sum_{k=1}^{n}\beta_{m+k}\left(\frac{1}{A}\right)^k+\sum_{k=1}^{q}\beta_{m+n+k}A^k. \end{aligned} \qquad (30)$$

The GSAM has other models as special cases. For instance, the persistence function, P2-full model, is a special case of GSMA if we consider the identity link function $g(\mu)=\mu$, where $\mu=E\{\log(S)\}$, i.e.

$$\mu=\beta_0+\beta_1\ln(A)+\sum_{k=1}^{n}\beta_{k+1}\left(\frac{1}{A}\right)^k. \qquad (31)$$

The persistence model, P2-full, can be written as

$$\mu=\beta_0+\sum_{k=1}^{m}\beta_k(\ln(A))^k+\sum_{k=1}^{n}\beta_{m+k}\left(\frac{1}{A}\right)^k, \qquad (32)$$

where $\beta_0=\ln(b_0)$, $\beta_1=b_1$ and $\beta_{k+1}=b_{k+1}$, $k=1,\ldots,n$.

Although the model defined by Eq. 30 has a large number of parameters (theoretically, it can have an infinite number of parameters), in practice the fitted models have no more than six parameters. The advantage of such a model is that it enables the formulation of hypothesis testing for the choice of the parameters to be removed from those that are significant for better describing the SAR.

Model fitting. The parameters to be estimated in the GSAM are λ, $\boldsymbol{\beta}$ and ϕ (Eqs. 4 and 6). In order to obtain maximum likelihood estimates for the vector of parameters $\boldsymbol{\beta}$ and dispersion parameter ϕ, we have defined a profiled likelihood function for λ and used the same algorithm proposed in [13]. By assuming the model given by Eq. 5, the log-likelihood function for the vector of the transformed observations $\boldsymbol{S}^{(\lambda)}=(s_1^{(\lambda)},\ldots,s_n^{(\lambda)})^T$ can be written as

$$\begin{aligned} \mathcal{L}(\beta,\phi,\lambda) &= \frac{1}{\phi}\sum_{i=1}^{n}\left\{s_i^{(\lambda)}\theta_i-b(\theta_i)\right\} \\ &+\sum_{i=1}^{n}\left[c(s_i^{(\lambda)},\phi)+\log\{J(\lambda,s_i)\}\right], \end{aligned} \qquad (33)$$

where $\theta_i=\int V^{-1}(\mu_i)d\mu_i=q(\mu_i)$ and $J(\lambda,s_i)=|s_i|^{(\lambda-1)}$, $i=1,\ldots,n$ is the Jacobian of the transformation from S to $S^{(\lambda)}$.

The procedure described in [13] is used for making inferences about parameters $(\boldsymbol{\beta},\phi)$ first assuming that λ is fixed and obtains the log-likelihood equations for estimating $\boldsymbol{\beta}^{(\lambda)}$ and $\phi^{(\lambda)}$. The maximum likelihood estimates (MLE) of $\boldsymbol{\beta}$, η, μ and ϕ for a given λ are denoted by $\hat{\boldsymbol{\beta}}^{(\lambda)}$, $\hat{\eta}^{(\lambda)}=\boldsymbol{X}\hat{\boldsymbol{\beta}}^{(\lambda)}$, $\hat{\mu}^{(\lambda)}=g^{-1}(\hat{\eta}^{(\lambda)})$ and $\hat{\phi}^{(\lambda)}$, respectively. $\hat{\boldsymbol{\beta}}^{(\lambda)}$ can be calculated, without knowledge on $\phi^{(\lambda)}$, adjusting the GSAM (Eqs. 5–6) to $\boldsymbol{S}^{(\lambda)}$ by iteration.

The iteration starts with an initial set of values $\hat{\boldsymbol{\beta}}^{(\lambda)(k)}$, $k=1$, used to evaluate $\boldsymbol{W}^{(\lambda)(k)}$ and $\boldsymbol{z}^{(\lambda)(k)}$, where $\boldsymbol{W}^{(\lambda)(k)}=\text{diag}\{w_1^{(\lambda)(k)},\ldots,w_n^{(\lambda)(k)}\}$ is a diagonal matrix with weights

$$w_i^{(\lambda)(k)}=V(\mu_i^{(\lambda)(k)})^{-1}(d\mu_i^{(\lambda)(k)}/d\eta_i^{(\lambda)(k)})^2 \qquad (34)$$

Table 4. Normal GSAM model fitted by the systematic component shown in Table 2.

	β_0	β_1	β_2	β_3
Coefficients	−10.0405	14.1548	−0.9251	15.3104
(SD)	(0.4593)	(0.1713)	(0.0154)	(0.5575)

Table 5. Richness of *Hymenoptera* species in different sample areas (m^2).

Area	Species	Area	Species	Area	Species	Area	Species	Area	Species
1	55	6	133	12	246	20	274	36	379
2	89	6	157	12	204	20	310	49	409
4	167	6	147	12	200	24	311	61	454
4	116	6	146	16	258	24	311	73	473
4	148	8	226	16	260	30	344	89	521

and $z^{(\lambda)(k)} = (z_1^{(\lambda)(k)}, \ldots, z_n^{(\lambda)(k)})^T$ is a working vector whose components are given by

$$z_i^{(\lambda)(k)} = \eta_i^{(\lambda)(k)} + (s_i^{(\lambda)(k)} - \mu_i^{(\lambda)(k)}) \left(\frac{d\eta_i^{(\lambda)(k)}}{d\mu_i^{(\lambda)(k)}} \right). \quad (35)$$

The next estimate $\hat{\boldsymbol{\beta}}^{(\lambda)(k+1)}$ can be obtained by

$$\hat{\boldsymbol{\beta}}^{(\lambda)(k+1)} = (\boldsymbol{X}^T \hat{\boldsymbol{W}}^{(\lambda)(k)} \boldsymbol{X})^{-1} \boldsymbol{X}^T \hat{\boldsymbol{W}}^{(\lambda)(k)} \hat{\boldsymbol{z}}^{(\lambda)(k)}. \quad (36)$$

This new value is used to update $\boldsymbol{W}^{(\lambda)(k+1)}$ and $\boldsymbol{z}^{(\lambda)(k+1)}$ and the procedures are repeated until convergence has been achieved.

Estimating parameter $\phi^{(\lambda)}$ is more difficult than estimating $\boldsymbol{\beta}^{(\lambda)}$. In principle, $\phi^{(\lambda)}$ could also be estimated by maximum likelihood, although there may be practical difficulties associated with this task for some members of Eq. 5. Details about the technique used for finding the MLE $\hat{\phi}^{(\lambda)}$ for a fixed λ can be found in [13].

In order to obtain the MLE $\hat{\lambda}$, we replace MLE $\hat{\boldsymbol{\beta}}^{(\lambda)}$ and $\hat{\phi}^{(\lambda)}$ in (33), which results in the profile log-likelihood function $l_P(\lambda) = l(\hat{\boldsymbol{\beta}}^{(\lambda)}, \hat{\phi}^{(\lambda)}, \lambda)$. The plot of the profile likelihood function $l_P(\lambda)$ against λ for a sequence of values of λ numerically determines the MLE for λ. Once the MLE for $\hat{\lambda}$ has been obtained, it can be used to produce the unrestricted estimates $\hat{\boldsymbol{\beta}} = \hat{\boldsymbol{\beta}}^{(\hat{\lambda})}$ and $\hat{\phi} = \hat{\phi}^{(\hat{\lambda})}$.

Assuming that the estimated $\hat{\lambda}$ is known, the confidence intervals for parameters $\boldsymbol{\beta}^{(\lambda)}$ and $\phi^{(\lambda)}$ can be calculated in the usual context of the GLM and using the adjusted values $\hat{\boldsymbol{\beta}}^{(\lambda)}$ and $\hat{\phi}^{(\lambda)}$. We consider the approximate covariance matrix of $\hat{\boldsymbol{\beta}}^{(\hat{\lambda})}$ and the variance of $\hat{\phi}^{(\hat{\lambda})}$ given in [13] to make inferences about these parameters. Here, we have considered the gamma, Gaussian and inverse Gaussian distributions for the probability density function (Eq. 5)

We also performed likelihood ratio (LR) tests [23] using a statistic $w = 2\{l_P(\hat{\lambda}) - l_P(\lambda^{(0)})\}$, which has an asymptotic χ_1^2 distribution for testing $\lambda = \lambda^{(0)}$ and constructed a large sample confidence interval for λ by inverting the LR test.

Results and Discussion

Simulation of the colonization process of a region

The parameters of SAR curves are determined from the survey data. As a proof of concept we first used simulated data for 80 species placed in a 50×50 cell lattice according to a neutral model [24] without dispersal limitation, as applied in [25]. This lattice was then resampled so as to establish the shape of the SAR (Figure 1).

The adjusted models are variations of the full-scale model with six parameters, given by

$$\begin{aligned} g(\mu) = \beta_0 + \beta_1 \ln(A) + \beta_2 (\ln A)^2 \\ + \beta_3/A + \beta_4/A^2 + \beta_5 A + \beta_6 A^2. \end{aligned} \quad (37)$$

The following canonical link functions were considered: (i) $g(\mu) = \mu$ for the normal GSAM, (ii) $g(\mu) = 1/\mu$ for the gamma GSAM and (iii) $g(\mu) = 1/\mu^2$ for the inverse Gaussian GSAM.

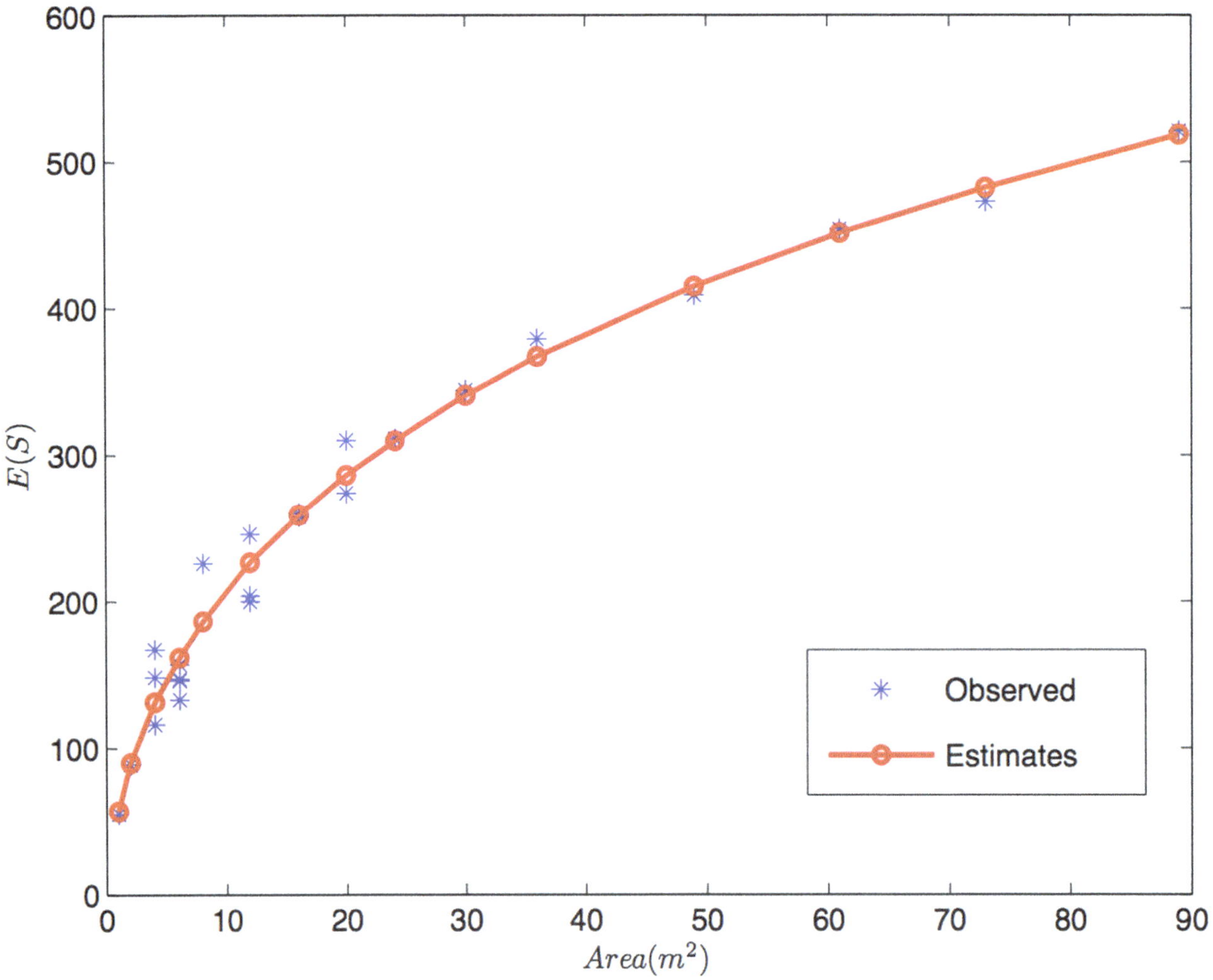

Figure 2. Data and fitted model for the real species-area relationship of *Hymenoptera* in a beech forest on limestone.

Moreover, the traditional models presented in the previous section were considered by assuming that the random variable S is normally distributed. We could estimate the mean of the transformed data $E(S^{(\lambda)})=\mu$, but to predict the expected value of the untransformed dependent variable S, when the GSAM is adjusted to the data, $E(S)$ must be estimated. The dependent variable S can be explained by subtracting μ on both sides of Eq. 4 and solving this equation for S. When $\lambda \neq 0$ we can write

$$S=(\lambda S^{(\lambda)}+1)^{1/\lambda}==(\lambda S^{(\lambda)}-\lambda\mu^{(\lambda)}+1+\lambda\mu^{(\lambda)})^{1/\lambda}= \\ =[1+\lambda\mu^{(\lambda)}]^{1/\lambda}\left\{1+\frac{\lambda}{1+\lambda\mu^{(\lambda)}}(S^{(\lambda)}-\mu^{(\lambda)})\right\}^{1/\lambda}. \quad (38)$$

The expected value of the species number S, on the original scale, can be evaluated by a first-order approximation of the binomial expansion (Eq. 38), as given in detail in [13]:

$$E(S)=(1+\lambda\mu)^{1/\lambda}\left\{1+\frac{(1-\lambda)\phi V}{2(1+\lambda\mu)^2}\right\}. \quad (39)$$

The best model can be chosen by using the AIC and BIC criteria [26], which are measurements of the relative goodness of fit of a statistical model for a given set of data. The mean square error (MSE) and mean absolute percent error (MAPE) are given, respectively, by

Table 6. Normal GSAM model fitted to SAR of *Hymenoptera*.

	β_0	β_1	β_2
Coefficients	101.9707	20.6625	−46.1669
(SD)	(9.4348)	(0.8211)	(22.3287)

$$MSE = \frac{100\%}{n\hat{\sigma}_s^2}\sum_{i=1}^{n}(s_i - \hat{s}_i)^2 \tag{40}$$

and

$$MAPE = \frac{100\%}{n}\sum_{i=1}^{n}\left|\frac{s_i - \hat{s}_i}{s_i}\right|, \tag{41}$$

where $\hat{\sigma}_s^2$ is the sample variance of S and $\hat{s}_i$ is the estimate of s_i given by Eq. 39.

Table 1 shows some of the traditional models adjusted with normal error. The selected model was the logarithmic quadratic function proposed by [21], with minimum $AIC = 221.09$, $BIC = 244.39$, $MSE = 1.02\%$ and $MAPE = 2.50\%$.

Table 2 shows the selected models adjusted by the GSAM with normal, gamma and inverse Gaussian (I.G.) distributions. Note that the adjusted models are variations of the full-scale model.

Table 3 shows the GSAM models fitted with $\hat{\lambda}$ adjusted by the profile likelihood. The selected normal GSAM has minimum $AIC = 79.90$ and $BIC = 109.02$, which are the lowest values among the adjusted models. MSE and $MAPE$ of the model are also smaller than those of the adjusted model with gamma and inverse Gaussian distributions. The adjusted value of parameter λ is $\hat{\lambda} = 0.822$ with confidence interval $(0.729; 0.922)$. Because λ is different from zero or one, there is a significant difference between the results achieved with this model or by using the traditional models given in Table 1. Therefore, for this data set, the normal GSAM is the model that has best fitted the analyzed data. The MLE estimates of the systematic component and standard-deviation (SD) of the systematic component are shown in Table 4.

Figure 1 shows the systematic observation of SAR on the original scale and the fitted curve with the adjusted GSMA models. $E(S)$ was calculated by Eq. 39 and for the adjusted GSAM we obtained $MSE = 0.08\%$ and $MAPE = 0.61\%$, respectively.

Application to real data

The GSAM model was applied to a data set that consisted of 25 observations of parasitic insects of the *Hymenoptera* order in a beech forest on limestone. *Hymenoptera* is one of the largest orders of insects that comprise sawflies, wasps, bees and ants. The total number of Hymenopteran species in Europe exceeds 20,000. The data considered here contain the summary of a long-term study of the ecology of parasitic Hymenoptera in a German beech forest, i.e. the Göttingen forest, which is approximately 120 years old and has grown over a ground limestone. The climate of the forest is typical of Central Europe and the work area covered approximately four acres. The study was conducted for 8 years (starting in 1980) in 144 square meters of forest soil.

The analysis of the SAR for Hymenoptera is essential, because the insects that belong to this order are the most important environmental agents fundamental for nutrient recycling and control of harmful species. The group is ubiquitous and it is common sense to assume that there is at least one species of parasitic insects for each species of herbivore insects [14]. Many of such species can be considered for the biological control of plague in agriculture. For instance, wasps, from *Symphyta* suborder, are plague conifers in the Northern hemisphere and several species of ants cause losses of millions of dollars for agriculture. Such insects act as special indicators and enable the inference of the diversity of arthropods of a broad spectrum of niches. *Hymenoptera* parasitoids are sensitive to environmental pollution, therefore fluctuations in their population are observed earlier than in their hosts [27]. This sensitivity makes this group an ideal candidate for studies on conservation. Therefore, the knowledge on how the number of species scales with area is fundamental for the prediction of the impact of such insect parasitic on both ecosystems and agriculture.

Table 5 shows the species richness in different sample areas (see also [14]). We modeled the data by taking into account all the models presented in previous sections. The results show that the normal GSAM with $\hat{\lambda} = 1$ was the best fitted model. No transformation of the original data was necessary:

$$E(S) = \beta_0 + \beta_1(\ln A)^2 + \beta_2/A.$$

The parameters of the fitted model are shown in Table 6. The fitted mean together with the data provided in Table 5 are shown in Figure 2. The adjusted model resulted in $AIC = 220$, $BIC = 238$, $MSE = 1.86\%$ and $MAPE = 6.69\%$, therefore, the GSMA model has proved very accurate. Our fit has provided a very good description of the increase observed in species richness and differs from the simple power-function presented in [14]. Interestingly our best fitting model includes features of the modified persistence model [18], but it has not been predicted by any recent macroecological theory, which calls for a fresh look on the patterns and constraints of spatial species distribution.

Conclusions

The generalized species-area model (GSAM) proposed here has provided a generalized model to mathematically describe the SAR. The GSMA can reduce the efforts devoted to finding the best model and can more accurately represent the effect of the area over the diversity of species than the power-curve models commonly used. This fact has been verified in simulated and real-world data.

Acknowledgments

The authors acknowledge Angela C. P. Giampedro, who provided a careful review of the text.

Author Contributions

Conceived and designed the experiments: WU KSC MGA. Performed the experiments: WU KSC MGA. Analyzed the data: KSC CARD FAR MGA. Wrote the paper: KSC WU CARD FAR MGA.

References

1. Rosenzweig ML (1995) Species diversity in space and time. Cambridge, UK.: Cambridge Univ.Press.
2. Tjørve E (2003) Shapes and functions of species-area curves: a review of possible models. J Biogeogr 30: 827–835.
3. Dengler J (2009) Which function describes the species-area relationship best? a review and empirical evaluation. J Biogeogr 36: 728–744.
4. MacArthur RH (1967) The theory of island biogeography, volume 1. Princeton University Press.
5. Würtz P, Annila A (2008) Roots of diversity relations. Journal of Biophysics 2008.
6. Connor EF, McCoy ED (1979) The statistics and biology of the species-area relationship. The American Naturalist 113: 791–833.

7. Scheiner SM (2003) Six types of species-area curves. Global Ecology & Biogeography 12: 441–447.
8. Tjørve E, Tjørve KMC (2008) The species-area relationship, self-similarity, and the true meaning of the z-value. Ecology 89: 3528–3533.
9. Tjørve E (2009) Shapes and functions of species-area curves (ii): a review of new models and parameterizations. J Biogeogr 36: 1435–1445.
10. Guilhaumon F, Mouillot D, Gimenez O (2010) mmsar: an r-package for multimodel species-area relationship inference. Ecography 33: 420–424.
11. Arrhenius O (1921) Species and area. J Ecol 9: 95–99.
12. Kallimanis AS, Mazaris AD, Tzanopoulos J, Halley JM, Pantis JD, et al. (2008) How does habitat diversity affect the species-area relationship? Global Ecol Biogeogr 17: 532–538.
13. Cordeiro GM, Andrade MG (2009) Transformed generalized linear models. J Statist Plann Inference 139: 2970–2987.
14. Ulrich W (2001) Hymenopteren in einem Kalkbuchenwald: Eine Modellgruppe zur Untersuchung von Tiergemeinschaften und ökologischen Raum-Zeit-Mustern. Schriftenr. Forschzentr. Waldökosysteme. Göttingen A 171: 249 S.
15. Dobson AJ (2010) An introduction to generalized linear models. CRC press.
16. Gleason HA (1922) On the relation between species and area. Ecology 3: 158–162.
17. Plotkin JB, Potts MD, Leslie N, Manokaran N, LaFrankie J, et al. (2000) Species-area curves, spatial aggregation, and habitat specialization in tropical forests. J Theor Biol 207: 81–99.
18. Ulrich W, Buszko J (2003) Self-similarity and the species-area relation of polish butterflies. Basic Appl Ecol 4: 263–270.
19. Ulrich W, Buszko J (2004a.) Habitat reduction and patterns of species loss. Basic Appl Ecol 5: 231–240.
20. Box GEP, Cox DR (1964) An analysis of transformation. J R Statist Soc B 26: 211–252.
21. Chiarucci A, Viciani D, Winter C, Diekmann M (2006) Effects of productivity on species-area curves in herbaceous vegetation: evidence from experimental and observational data. Oikos 115: 475–483.
22. Gitay H, Roxburgh SH, Wilson JB (1991) Species-area relations in a new zealand tussock grassland, with implications for nature reserve design and for community structure. Journal of Vegetation Science 2: 113–118.
23. Mood AM (1950) Introduction to the theory of statistics.
24. Hubbell SP (2001) The Unified Neutral Theory of Biodiversity and Biogeography. Princeton, NJ.: Princeton University Press.
25. Zillio T, Condit R (2007) The impact of neutrality, niche differentiation and species input on diversity and abundance distributions. Oikos 116: 931–940.
26. Burnham KP, Anderson DR (1998) Model selection and inference: a practical information-theoretic approach/.
27. LaSalle J, Gauld I (1991) Parasitic hymenoptera and the biodiversity crisis. Redia 74: 315–334.

10

Riparian *Ficus* Tree Communities: The Distribution and Abundance of Riparian Fig Trees in Northern Thailand

Pornwiwan Pothasin[1], Stephen G. Compton[2,3], Prasit Wangpakapattanawong[1,4]*

1 Department of Biology, Faculty of Science, Chiang Mai University, Chiang Mai, Thailand, **2** School of Biology, University of Leeds, Leeds, United Kindgom, **3** Department of Zoology and Entomology, Rhodes University, Grahamstown, South Africa, **4** World Agroforestry Center c/o Knowledge Support Center for the Greater Mekong Sub-region, Faculty of Social Sciences, Chiang Mai University, Chiang Mai, Thailand

Abstract

Fig trees (*Ficus*) are often ecologically significant keystone species because they sustain populations of the many seed-dispersing animals that feed on their fruits. They are prominent components of riparian zones where they may also contribute to bank stability as well as supporting associated animals. The diversity and distributions of riparian fig trees in deciduous and evergreen forests in Chiang Mai Province, Northern Thailand were investigated in 2010–2012. To record the diversity and abundance of riparian fig trees, we (1) calculated stem density, species richness, and diversity indices in 20×50 m randomly selected quadrats along four streams and (2) measured the distances of individual trees from four streams to determine if species exhibit distinct distribution patterns within riparian zones. A total of 1169 individuals (from *c.* 4 ha) were recorded in the quadrats, representing 33 *Ficus* species (13 monoecious and 20 dioecious) from six sub-genera and about 70% of all the species recorded from northern Thailand. All 33 species had at least some stems in close proximity to the streams, but they varied in their typical proximity, with *F. squamosa* Roxb. and *F. ischnopoda* Miq the most strictly stream-side species. The riparian forests in Northern Thailand support a rich diversity and high density of *Ficus* species and our results emphasise the importance of fig tree within the broader priorities of riparian area conservation. Plans to maintain or restore properly functioning riparian forests need to take into account their significance.

Editor: Maharaj K. Pandit, University of Delhi, India

Funding: This study was supported by the Office of the Higher Education Commission, PERDO-Climate change and Department of Biology, Faculty of Science, Chiang Mai University, Thailand. The funders had no role in study design, data collection and analysis, decision to publish, or preparation of the manuscript.

Competing Interests: The authors have declared that no competing interests exist.

* Email: prasitwang@yahoo.com

Introduction

Ficus (Moraceae) is one of the largest plant genera, with more than 750 described species distributed worldwide, mainly in tropical countries [1–3]. Fig trees occupy a diversity of habitats and display a wide range of growth forms that includes hemiepiphytes (strangling figs and banyans) and large woody climbers as well as trees and shrubs. The positions of their fruits (figs) include cauliflory (inflorescences borne on the trunk), on the surface of the soil and among the leaves, making them available to a wide range of frugivores [4]. Fig trees are often ecologically important 'Keystone' components of tropical forests, because of the large number of vertebrates that feed on their figs, more than on any other group of plants [5]. They are often abundant and are the most abundant genus in soil seed banks in tropical Asia [6,7]. Fig trees are especially prominent in the riparian zones within SE Asian forests [8,9], though only a small number of species are extreme riverine habitat specialists capable of growing on rocks in mid-stream (rheophytes). Other species are bank-side specialists that are largely restricted to riparian zones, and others are more widespread species, also found routinely in other habitats [10].

The specific identification of *Ficus* species can be difficult, but fig trees as a group are relatively easy to recognize. They are defined by their figs (syconia), unique enclosed inflorescences lined with tens to thousands of tiny, usually unisexual flowers [2,11]. *Ficus* species may be monoecious or functionally dioecious. Monoecious figs produce figs containing male flowers and female flowers with various style lengths. All trees have the same type of figs, which can produce wasps, pollen and seeds. Gynodioecious figs are functionally dioecious and produce male and female figs on separate trees [12]. Male figs contain male and short-styled female flowers [13]. The male flowers provide pollen and the female flowers nourish developing wasps, which develop in galled ovules. Male figs rarely produce seeds [14]. In contrast, female figs, which contain only long styled female flowers, exclusively produce seeds. The female flowers can be pollinated, but their long styles prevent pollinating wasps from depositing eggs into ovules [12,14]. Most fig trees are pollinated by females of one or a small number of host-specific species of fig wasp (Agaonidae) and have an obligate mutualistic relationship with them [15–22].

At least 47 *Ficus* species are native to northern Thailand [23]. Some fig trees are early successional pioneers that have been used successfully as framework species in regenerating a degraded upper watershed of evergreen and seasonal forest in Doi Suthep-Pui National Park in northern Thailand [24]. In this 'framework species' restoration method about 20 percent of planted seedlings are recommended to be fig tree species [24]. Their value for restoration stems from the birds that are attracted to feed in fig

Table 1. Geological and environmental characteristics of the study sites.

Study Site	Elevation (m asl.)	Substrate	Forest type (Maxwell & Elliot 2001)	Stream width (m)
Huay Kaew stream (HK)	472–1680	granite	deciduous dipterocarp, mixed deciduous, and evergreen forests	5.0–12.0
Mae Klang stream (MKL)	450–2000	granite	deciduous dipterocarp, mixed deciduous, and evergreen forests	7.0–20.0
Mae Sa stream (MS)	690–1490	granite	deciduous dipterocarp, mixed deciduous, and evergreen forests	7.0–18.0
Mae Ka stream (MK)	490–527	granite and limestone	deciduous dipterocarp, and mixed deciduous forests	3.5–5.0

trees, which then also disperse seeds of other tree species, thus adding to tree species diversity in rehabilitated forest areas [25]. As food resources, the all-year-round production of figs among monoecious species, often in large quantities, can potentially maintain vertebrate frugivore populations at times when other fruits are in short supply and this continuous fruiting may be a mechanism that helps maintains biodiversity in tropical rain forests [5,26–29]. Hence, the diversity and reproductive phenology of *Ficus* have special relevance for conservation and tropical forest ecosystem integrity.

Riparian zones are important interfaces (ecotones or transition zones) between streams and surrounding terrestrial habitats [30]. Natural riparian zones are among the most diverse, dynamic and complex terrestrial habitats in the world, but they are very sensitive to environmental change [30,31]. Streamside vegetation has an important function in river and stream ecosystems, as it supplies allochtonous organic matter (including leaf litter and terrestrial invertebrates), filters the nutrients and pollutants that reach the streams, helps stabilize banks, and influences water temperatures and light through shading [30,32–38]. There is often variation in plant species composition closer to streams [39], and maintenance and conservation of riparian areas therefore contributes to the diversity of the entire forest landscape [40].

Environmental disturbance to riparian forests, caused by factors such as road building, clearance and other human activities, is increasing, and has had a direct impact on native biodiversity [41]. Extensive agricultural and urban development and associated soil erosion have contributed to the decline of riparian forests in Thailand, but the significance of this decline is poorly documented. There are few estimates of how many plant species are found exclusively in riparian zones, but *Ficus* is believed to be the most diverse genus in riparian forests of Thailand [42,43]. We studied the species richness, diversity and distribution relative to stream edges of the fig trees growing in riparian zones in Northern Thailand, with the objective of improving understanding of the distribution limits and habitat requirements of this ecologically important group of species. Our objective was to answer the following questions: (1) Is there a distinct group of *Ficus* species associated with the riparian zones? And (2) Are there gradients in community metrics such as stem density or species richness as a function of distance from the streams? We hypothesized that in this forest landscape, riparian zones may provide a sufficiently different abiotic or environmental template to result in exclusively riparian tree species or community characteristics.

Methods

Study Site

Our study area was located in the Ping River basin in Northern Thailand (18° 51′ N, 98° 54′ E) (Fig. S1), which covers an area of about 35,000 km^2. It comprises an upper tributary sub-basin of the Ping River, which in turn is the largest tributary of central Thailand's Chao Phraya River. The region is mountainous with some lowlands along river tributaries, with about 60 percent of the land area having an elevation above 500 m. The climate of Northern Thailand is characterized by three distinct seasons, the rainy, the cool dry and the hot dry seasons. Mean monthly temperatures for a 25-year period (1988–2012) at the Chiang Mai meteorological station varied from 14°c in January to 36°c in April. The average annual rainfall, relative humidity, and annual temperature were 1,072.5 mm, 70%, and 25.9°c, respectively [44]. The basin is subjected to the southwest monsoon during May

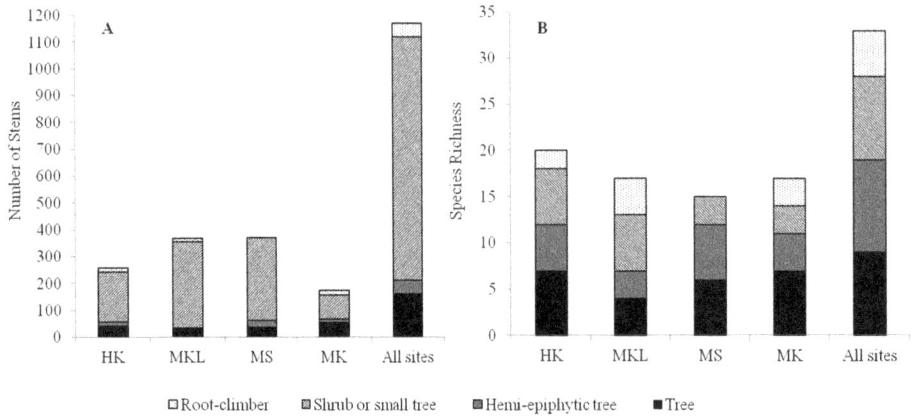

Figure 1. Stem abundance (A) and species richness (B) at four study sites: (HK) Huay Kaew stream, (MKL) Mae Klang stream, Mae Sa (MS) stream and Mae Ka stream (MK).

Table 2. Descriptive statistics for diversity indices in the study sites.

Indices	Study site	Mean* (n = 10)	S.D.	Min.	Max.
Species Richness	Huay Kaew stream	4.30[a]	1.57	2.00	8.00
	Mae Klang stream	3.60[a]	1.65	1.00	7.00
	Mae Ka stream	5.70[a]	2.67	3.00	11.00
	Mae Sa stream	5.30[a]	2.71	3.00	11.00
Abundance	Huay Kaew stream	25.80[a]	13.74	8.00	49.00
	Mae Klang stream	36.70[a]	22.87	2.00	61.00
	Mae Ka stream	17.40[a]	11.22	7.00	41.00
	Mae Sa stream	37.00[a]	18.28	19.00	74.00
Shannon	Huay Kaew stream	1.00[ab]	0.50	0.12	1.88
	Mae Klang stream	0.72[a]	0.35	0.00	1.09
	Mae Ka stream	1.42[b]	0.46	0.96	2.04
	Mae Sa stream	0.97[ab]	0.31	0.54	1.43
Simpson	Huay Kaew stream	0.55[ab]	0.25	0.05	0.85
	Mae Klang stream	0.41[a]	0.20	0.00	0.60
	Mae Ka stream	0.75[b]	0.15	0.56	0.95
	Mae Sa stream	0.51[a]	0.12	0.29	0.66
Evenness	Huay Kaew stream	0.67[ab]	0.24	0.18	0.92
	Mae Klang stream	0.54[a]	0.22	0.00	0.76
	Mae Ka stream	0.86[b]	0.10	0.71	0.98
	Mae Sa stream	0.62[a]	0.09	0.49	0.72

*Letters ([a,b]) mean that different superscripts are significantly different ($p<0.01$) within the same column.

to October and to tropical cyclonic storms from the South China Sea at the end of the rainy season between September and October. The average annual rainfall and runoff from the basin are around 1174 mm and 6815 million m^3, respectively [45]. The Ping River basin comprises tertiary continental basin-fill sediments underlain by older Paleozoic gneissic granites, Paleozoic sediments and volcanics and Mesozoic granitic rocks [46].

Stream sections of orders two to four were monitored along four perennial streams (each at least 1000 m long) during October 2010–December 2012. The stream sites were located in hill-evergreen, mixed-deciduous and deciduous-dipterocarp forests [47–49] (Table 1). They mostly have granite and gneiss as the dominant rock types and all contained large boulders, were bedrock-lined and carried only small amounts of silt and sand. The streams flow all year, with rapid temporary increases in flows in the rainy season (June–October). Stream MKL is distinct. It is larger than the other streams, has the highest elevation and it is partly fed by a small natural hot spring. This stream also has a

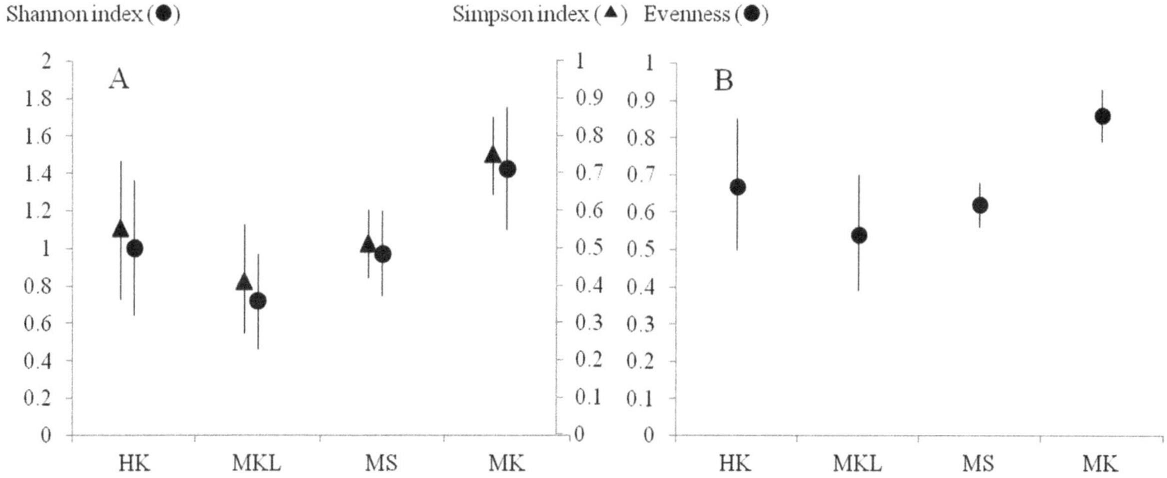

Figure 2. Estimates of diversity for riparian fig tree species at four study sites: (HK) Huay Kaew stream, (MKL) Mae Klang stream, MaeS (MS) stream and Mae Ka stream (MK). (A), mean diversity (±95% CI) for each site according to the Shannon's index (circles) and Simpson index (triangles). (B), evenness at each study site is based on the Shannon's index (circles); vertical lines indicate the 95% CI.

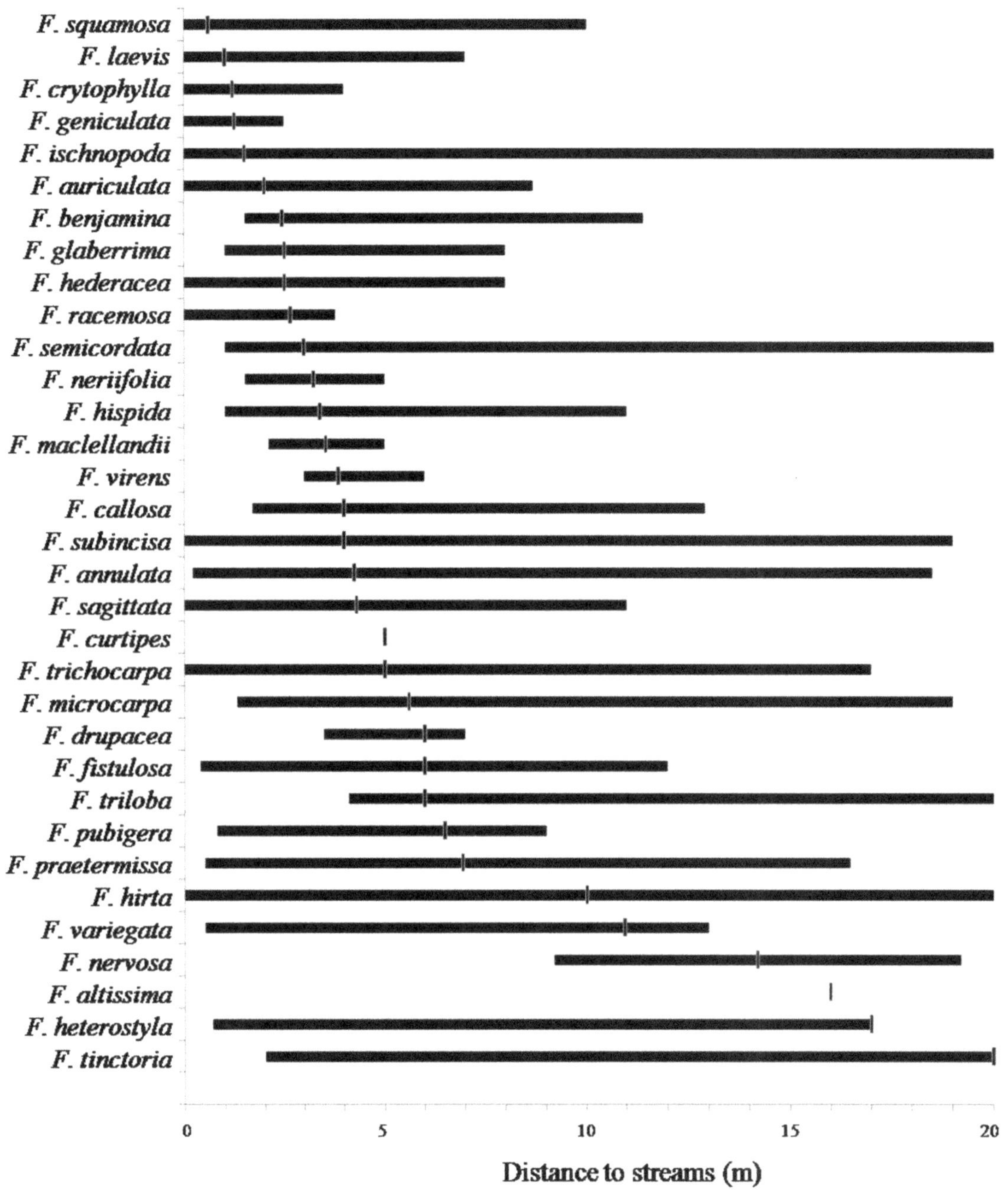

Figure 3. Median distance to streams (vertical solid lines) and ranges of occurrence of *Ficus* species (all four sites combined). See Table S2 in File S3 for number of stems of each species and where 75% of the total stems occur.

Table 3. Differences in fig tree mean species richness and density in different river distance bands (four study sites combined).

Distance strata from streams	Total species richness	Number of individuals	Species richness (Mean ± SD)	Density (stems/ha) (Mean ± SD)
1. (0–5 m)	24	937	4.100±2.17	234.25±114.81
2. (≥5–10 m)	17	128	1.375±1.75	32.00±13.49
3. (≥10–15 m)	11	43	0.525±0.78	10.75±8.26
4. (≥15–20 m)	8	61	0.350±0.74	15.25±24.62

Table 4. One-way ANOVA for number of stems, species richness and diversity between three elevation ranges of each quadrats (All four sites combined, n = 40).

	MS	*F*-Value	*p*
Number of stems	2118.1	8.731	0.0007*
Richness	4.306	1.098	0.344
Shannon (H')	0.618	3.007	0.062
Simpson	0.150	3.298	0.048
Evenness	0.160	4.162	0.023
Density	0.002	8.69	0.0008*

Asterisks* indicates significant differences.

limestone substrate and flows through hill evergreen forest, deciduous dipterocarp forest and agricultural land before joining the Ping River.

Species Diversity

The abundance of each *Ficus* species was estimated using the belt transect method [50]. Ten 20 m wide belt transects that extended 50 m from the stream edge were set at each site. A stratified random sampling design was employed, to cover all the elevation ranges and forest types in each study site. All *Ficus* species present in the transects were identified following *Flora Malesiana* and *Flora of Thailand* [10,51].

For each belt transect, *Ficus* species richness was taken as the total number of fig species present (all growth forms). Species diversity in each site was calculated using the Shannon-Weaver diversity index (H') which accounts for both abundance and evenness [52], in the package PRIMER [53]. The Shannon index ranges typically from 1.5 to 3.5 and rarely reaches 4.5 [54]. Other parameters such as species evenness (e) and Simpson's diversity index (1-D) were also derived [55].

Descriptive statistics were computed to compare the mean abundance, species richness, diversity indices and evenness values of fig trees at the four study sites. All variables were compared using one-way analysis of variance (one-way ANOVA), after verifying normality and homogeneity of variances. Tukey HSD tests were used when ANOVA resulted in significant differences [56].

Fig Tree Distributions

To reveal individual species distributions in relation to streams, we measured the distance of all individual trunks to the stream edge using a tape measure and then generated a median distance of each fig tree species from the stream. In addition to calculating total stem density versus distance from streams, each belt transect (20×50 m) was also divided into four contiguous 5×50 m subplots. Species richness and number of stems were calculated in each subplot. A two-way ANOVA tested how species richness and stem abundance differed between streams and in distance bands from the streams (0–5 m, $\geq$5–10 m, $\geq$10–15 m and $\geq$ 15–20 m). Linear regressions determined the relationship between mean species richness and number of stems in each distance band by using the minimum distance to the stream of each subplot. All tests were applied using R version 3.0.0 [56].

Fig Tree Composition Along Elevation Gradients

We classified all quadrats (n = 40) from four streams according to their elevations. The elevation gradient was divided into three ranges: 400–600 m asl (n = 16), 600–1,000 m asl (n = 11) and > 1,000 m asl. (n = 13) (Fig. S2). These ranges followed the forest classification schemes in Thailand; dry deciduous dipterocarp, mixed deciduous and evergreen, and hill evergreen respectively [47,49,57]. One-way ANOVA then compared the mean abundance, species richness, diversity indices and evenness values, and density (stems/1000 m^2) between the three elevation ranges. Normality and post hoc tests were conducted as before [56].

We used linear regressions to determine the relationship between number of stems, species richness, species diversity (H') and elevation by using the mean elevation of each quadrat. We generated similarity matrics (Sorensen's index) of species abundance per quadrat and all regressions using the Labdsv and Vegan packages in R. We performed Fuzzy Set Ordination (FSO) analysis on the similarity matrics. The FSO analysis creates an axis that is relativized (zero to one) using both a vector variable and the similarity matrix [58]. In this study, the environmental vector variable was the mean elevation of each quadrat. The resulting relativized axis is termed the "apparent elevation", based on the similarity matrix of species abundance, and mean elevation of the quadrat. In this way, the environmental vector of "elevation" was used to predict the species composition of the quadrats, and their variation in relation to the environmental variable, in this case, elevation.

Results

Species Richness

A total of 1169 individual fig trees were recorded in the surveys along the four streams with 174 (14.88%) individuals at MK, 258 (22.07%) at HK, 370 (31.65%) at MS and 367 (31.39%) at MKL. These *Ficus* communities are dominated by shrubs or small trees (Fig. 1A). Thirty-three fig tree species were recorded (Table S1 in File S3), representing 13 monoecious and 20 dioecious species from all six subgenera of *Ficus*. The largest taxonomic group was subgenus *Urostigma*, the strangler figs, with 10 species recorded. The second largest group consisted of eight species of free-standing trees (subgenus *Sycomorus*), followed by five species of creepers (subgenus *Synoecia*), with subgenera *Ficus* and *Sycidium* (both comprising small shrubs) each represented by four species, and free-standing *Pharmacosycea* by two species. Many species co-occurred at each site (17 species were present at MK, 20 at HK, 15 at MS and 17 at MKL) (Fig. 1B). Five species were common and occurred in all four sites including *F. auriculata* Lour, *F. hispida* L.f., *F.semicordata* Buch.-Ham. ex Sm., *F.squamosa* Roxb. and *F. microcarpa* L.f. Fourteen species were local and occurred at only one of four sites (Table S1 in File S3).

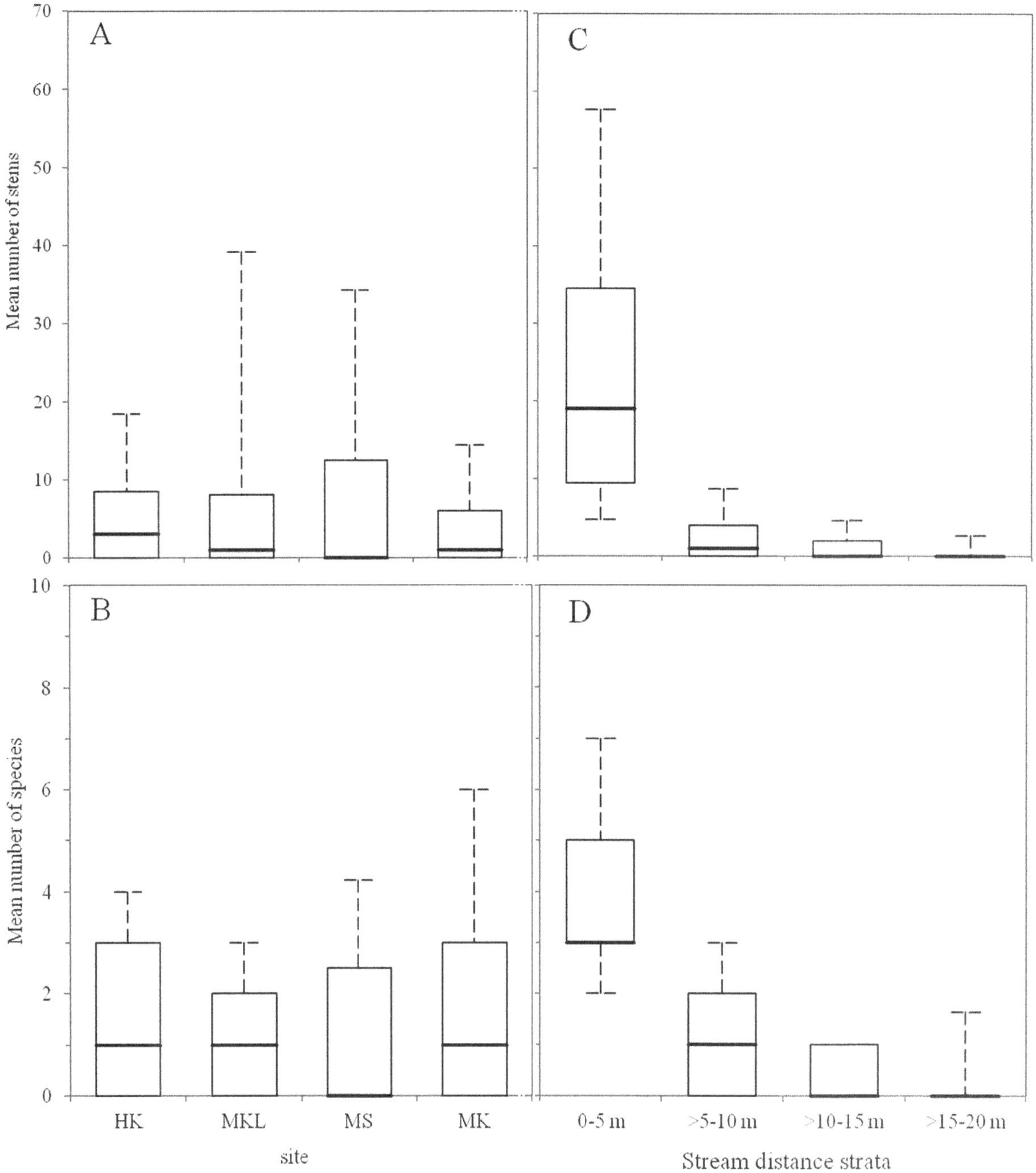

Figure 4. Mean number of stems (A, C) and mean number of species (B, D) in four stream sites (left) and at each stream distance stratum (right). The boxplots describe the relationship between mean number of stems and species richness in each stream and stream distance strata. The maximum and minimum extents of the boxes indicate 25th and 75th percentiles and thick solid lines indicate the medians.

From 1–11 fig tree species were recorded per belt transect. The mean fig tree density across all belt transects was 292.25±94.45 stems/ha. The highest density was recorded for *F. squamosa* (117±81.95 stems/ha), followed by *F. ischnopoda* Miq. (68.25±89.88 stems/ha), *F. hirta* Vahl (23.75±39.30 stems/ha), *F. semicordata* (9±5.94 stems/ha), *F. auriculata* (7±2.16 stems/ha), *F. praetermissa* Corner (6.5±13.0 stems/ha), *F. fistulosa* Reinw. ex Blume (6.25±6.24 stems/ha), and *F. sagittata* J. Konig ex Vahl (5.25±4.57 stems/ha). The remaining species had densities of less than 5 stems/ha. One-way ANOVA showed that among study sites, there were no significant differences between numbers of fig trees and sites, or species richness and sites (Table 2).

Fig Tree Diversity

Despite an absence of significant differences in species numbers, there were significant differences in fig tree diversity across the four sites (Table 2), as measured by Shannon's index (H'), Evenness index (E) and Simpson's diversity index (1-D) (one-way ANOVA, $F=5.039$, $P<0.001$), $F=5.642$, $P<0.001$ and $F=5.667$, $P<0.001$, respectively) (Figs. 2A, 2B). Tukey HSD tests revealed that all diversity indices and evenness values for MK

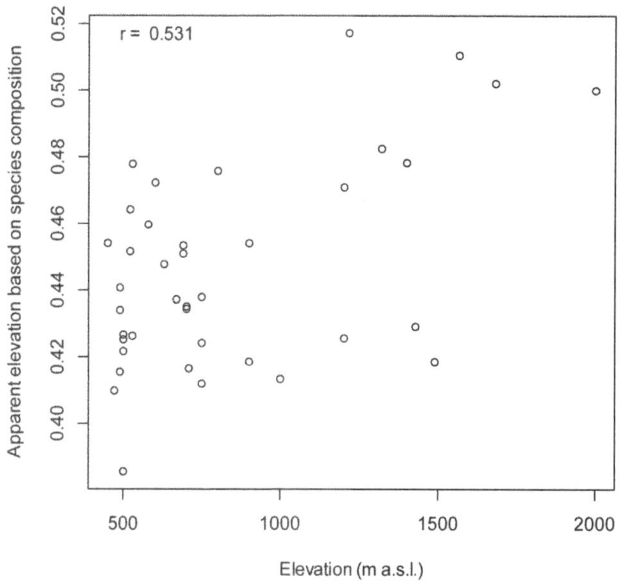

Figure 5. Plot of ordination analysis (Fuzzy Set Ordination; FSO) based on binary data, (presence/absence of species) and mean elevation in each quadrat. Sørensen indices were evaluated with rank and linear correlations with the apparent elevation gradient.

were significantly higher than at the other sites (all $P \leq 0.01$), whereas HK, MKL and MS did not differ significantly in their diversity indices (Table 2).

Fig Tree Distributions

The median distance of occurrence of individual fig stems from stream edges varied between 0.6 and 20 m. *F. auriculata*, *F. ischnopoda*, *F. laevis*, *F. squamosa*, *F. geniculata* and *F. crytophylla* grew closest to the streams, with median distances to the streams of between 0.5 and 2 m (Fig. 3). Twenty one species occurred at a median distances between 2 and 8 m. and the six species that were least closely associated with the streams, with no stems within 10 m of the stream edges. were *F. nervosa*, *F. hirta*, *F. variegata*, *F. altissima*, *F. heterostyla* and *F. tinctoria*.

In terms of percentiles of the total distributions of stems, fifty percent of *F. squamosa* were within 0.6m from the streams. Twelve species had seventy-five percent of their stems at distances less than 5 m away from the streams (*F. ischnopoda*, *F. nerrifolia*, *F. hederacea*, *F. laevis*, *F. crytophylla*, *F. auriculata*, *F. racemosa*, *F. geniculata*, *F. benjamina*, *F. glaberrima*, *F. maclellandii* and *F. curtipes*) (Table S2 in File S3).

Across sites, fig species richness varied with distance from the streams, with nearest the stream (0–5 m) having the highest number of species and the farthest distance (≥15–20 m) the lowest number (Table 3). The same pattern was evident for stem abundance. A two-way ANOVA showed that there was no significant difference between streams or sites ($F_{3,144} = 2.808$, $P = 0.042$) in the distributions of stems from the streams. There was also no significant difference in fig species richness distances between sites ($F_{3,144} = 1.153$, $P = 0.330$). However, the mean numbers of stems in the different distance bands were significantly different within sites ($F_{3,144} = 57.210$, $P < 0.001$), as were species richness ($F_{3,144} = 54.894$, $P < 0.001$) (Table 4). One-way ANOVA showed that the first distance bands (0–5 m) varied highly significantly in terms of stem abundance (HK; $F_{3,36} = 4.088$, $P < 0.05$, MKL; $F_{3,36} = 13.76$, $P < 0.001$, MS; $F_{3,36} = 42.34$, $P < 0.001$, MK; $F_{3,36} = 15.43$, $P < 0.001$). When compared between sites, MS had significantly ($P < 0.05$) greater number of stems in the streamside zone (0–5 m) than HK, MKL and MS (Tukey tests).

Comparisons of the box plots suggest that there was no significant difference between the numbers of stems ($F = 2.808$, $P = 0.042$) and species ($F = 1.1.53$, $P = 0.330$) between sites (Fig. 4A, 4B), but there appeared to be differences between sites and highly significant relationships with distance to the stream in terms of both the number of stems ($F = 57.210$, $P < 0.001$) (Fig. 4C) and species richness ($F = 54.894$, $P < 0.001$) (Fig. 4D). Two-way ANOVA showed that there was a significant interaction effect between site and stream distance strata ($F_{9,144} = 4.831$, $P < 0.001$) on the mean number of stems and no significant difference between mean number of species ($F = 1.1.68$, $P = 0.320$).

Within the forty quadrats, there was a significant relationship between species richness ($r^2 = 0.421$, $P < 0.001$) and distance from the streams. In addition, there was a significant positive relationship between number of stems ($r^2 = 0.311$, $P < 0.001$) and stem density ($r^2 = 0.308$, $P < 0.001$) and proximity to streams, with the overall number of stems increasing with proximity to the streams.

Fig Composition Along An Elevation Gradient

Fig tree abundance and density varied significantly between elevation ranges and forest types (Table 4). Tukey's tests indicated that the elevation range 600–1000 m. with mixed deciduous and evergreen forest types had significantly higher densities of fig trees than the others ($F = 8.89$, $P < 0.001$). In contrast, there were no statistically significant differences in species richness and diversity between the elevation ranges (Table 4). Linear regression analysis showed that there was also no relationship between mean elevation of quadrats and number of stems, species richness, or diversity (all values $r^2 < 0.08$, $P > 0.04$).

In the analysis of community composition using fuzzy set ordination (FSO) with binary (presence/absence) data (Fig. 5), we found a significant relationship between species abundance and composition as a function of elevation range ($F = 2.118$, $P < 0.001$). The FSO indicated that there was a relationship between species composition of quadrats and elevation ranges ($r^2 = 0.263$, $P < 0.001$). The relationship between elevation and"apparent elevation based on species composition" was moderate, with a correlation of 0.531, meaning that we can predict the elevation range to some degree of a site based on its species composition. At elevations lower than 1,000 m asl. we observed greater differences among quadrats in species composition (more turnover) than at higher elevations, where riparian fig tree species richness is more limited.

Discussion

High species richness is a frequently cited property of riparian zones [30,32,35,59] and may be related to factors including disturbance [60,61], productivity, flow-facilitated dispersal of propagules [62], and the diversity of physical conditions present at the interface between aquatic and terrestrial ecosystems [30]. Fig trees are commonly found in moderate to moist areas near rivers or streams [63,64] and riparian zones in northern Thailand were found to be extremely rich in *Ficus* species. We recorded 33 species in 40 plots that had a total area of only about 4 ha. This is about two thirds of all the *Ficus* species native to Northern Thailand [23] and more than one quarter of all native fig tree species in Thailand [51]. This high level of species richness is by no means unusual for fig trees and is comparable to other sites in the region [65]. Elsewhere in Thailand, 50 percent of the 38 *Ficus* taxa recognized by [8] were recorded at least occasionally from riparian situations, and several species were only recorded from this habitat. Similarly, Kong-ied [43] reported that *Ficus* L. was the most diverse genus in the riparian forest along the Sok Canal, Surat Thani Province, southern Thailand, though only 13 species were represented. The fig trees were also abundant at our study sites and these results emphasize that, as in some other habitats, riparian *Ficus* are likely to be keystone resources for threatened frugivores such as hornbills [66] as well as for many other species of birds and mammals.

Shrubs, creepers, stranglers and free-standing trees were all present in the riparian vegetation. A small number of the *Ficus* species were extreme riverine habitat specialists capable of growing on rocks in mid-stream (rheophytes), some other species were bank-side specialists largely restricted to riparian zones, and others were more widespread species that are also found routinely in other habitats [10]. In this study, the most common *Ficus* species at the four sites, and the species most closely located to the streams, was *F. squamosa*. It is a rheophytic shrub with rooting stolon-like stems, and is particularly associated with fast-flowing streams. Our field observations found that this species is tolerant of extreme disturbance during rainy season flooding events. It also displays anatomical adaptations for seed dispersal by water [10]. In addition to *F. squamosa, F. auriculata, F. geniculata, F. ischnopoda,* and *F.crytophylla* all had median distances to stream edges of less than 2 m. Environmental factors related to proximity to streams, such as disturbance, readily available water and increased light from the open canopy present above are likely to have been influencing their distributions [39]. Hemi-epiphytic strangler figs were found mostly at greater distances from the streams and their association with the riparian zone appears to be more casual. They germinate from seeds deposited in the canopy and are likely to be less responsive to the increased light levels at stream edges, and because they are epiphytic as young plants they also display adaptations for resistance to dehydration, so the elevated soil moisture near streams will be less important [11]. Consequently, the availability of suitable host trees may be more important than where the host trees are growing.

There was significant positive relationships between species richness, abundance and density of figs with proximity to streams. Higher light levels and the availability of water clearly play the important role in organizing the distribution of riparian figs. The relationship between hydrology and riparian plant composition has previously been identified as an important research area [59,61] that requires further interdisciplinary research.

Ecological studies of riparian zones have been predominantly in North America. Temperate research from there and elsewhere has demonstrated the importance of the disturbance caused by flood scour events for the maintenance of local biodiversity [67]. Community processes in tropical riparian zones are less well understood, but they are likely to show significant differences from temperate examples. Hydrological regimes may or may not be more highly variable in the tropics, depending on local rainfall regimes, and predictable seasonal events such as snow melt will usually be absent. The chemical composition of stream substrates, and of adjacent land, may also often be different. Tropical riparian zones also have a different plant and animal taxonomic composition and much higher species richness. Interactions between plants and animals may also be more significant, with greater dependence on animal pollination and seed dispersal. Chantarasuwan *et al.* [68] concluded that suitable habitats for riparian fig species was determined mainly by moisture, with aspect and slope of less importance. In South America, Banack [69] found that *F. insipida* occurs only along larger streams and rivers and concluded that it is more establishment-limited than disperser-limited, despite being dispersed mainly by bats and fish.

Ficus species richness and abundance (numbers of stems) did not differ between the four streams we examined. The species diversity of the Mae Ka stream (MK) was nonetheless higher than those of the others. It was situated at a lower elevation and had a different substrate to the other streams (with a combination of limestone bed rock and granite). This finding is consistent with that of Munishi [70], who revealed that there was difference in species richness, density and diversity of tree resources outside forest in river strata on the Southern side of Mount Kilimanjaro. This variation did not follow any particular pattern with regard to distances from river catchments. Topography and soil types are often important factors in the distribution of tree species [71] and the nature of the dipterocarp forest at the MK sites may also have had an influence on fig tree distribution.

There were more species and a higher density of fig trees between 400–600 m asl. The distribution of species richness along elevation gradients is governed by a series of interacting biological, climatic and historical factors [72] and elevation represents a complex gradient along which many environmental variables change simultaneously [73]. Other factors, such as soil fertility and topography may also affect the patterns of species richness along elevation gradients. In mountain regions, the pattern of different forest types and other communities often corresponds to elevation and topography. Variation in microclimate with topography and elevation is also a major factor of species distribution within a forest landscape [74]. While Harrison *et al.* [75] found that most hemi-epiphytic fig trees are widely distributed, the assemblages differed substantially between sites. They suggested that large-scale habitat associations may also influence fig tree species richness at higher elevations.

In conclusion, tropical riparian vegetation influences several important ecological functions such as stream bank stabilization, reduction of flood velocities, shading and the provision of food for animals. It is also a particularly vulnerable and threatened habitat in Thailand and elsewhere [30,32–38]. We found that in northern Thailand fig trees are a numerically important and diverse component of riparian vegetation, and that this habitat supports a high proportion of all figs in the region, comprising a mix of growth forms and both habitat specialists and generalists. Streams on limestone appear to differ in character to those on acid substrates, and there is some variation in community composition with altitude. This study emphasizes the likely conservation importance of fig trees in riparian areas of Thailand, but their role in broader ecosystem functioning remains largely unknown.

Supporting Information

Figure S1 The locations of four riparian fig tree study sites in Chiang Mai Province, Northern Thailand. 1. Mae Ka stream (MK) 2. Mae Sa stream (MS) 3. Huay Kaew stream (HK) 4. Mae Klang stream (MKL).

Figure S2 Distribution of plot sites along an elevation gradient followed the forest classification scheme in Thailand; dry deciduous dipterocarp, mixed diciduous and evergreen, and hill evergreen respectively. The elevation gradient was divided into three ranges i.e. 400–600 m asl ($n = 16$ plots), 600–1,000 m asl. ($n = 11$ plots) and the final one at >1,000 m asl. ($n = 13$ plots).

File S1 Tables S1 and S2. Table S1. Species list of riparian figs present along four streams in Chiang Mai, Northern Thailand. Table S2. Median distances to streams of *Ficus* species (all four sites combined). (N = total stems, Median = median distance to stream, SD = standard deviation, p75 = distance from streams at which 75% of stems occur, min = minimum distance, and max = maximum distance recorded from stream edges).

Acknowledgments

We are deeply grateful to Wattana Tanming who assisted with the maps and provided support in identifying fig species. We would like to offer our regards to all those who supported us in any way during the completion of this research.

Author Contributions

Conceived and designed the experiments: PP SC PW. Performed the experiments: PP. Analyzed the data: PP SC PW. Contributed reagents/materials/analysis tools: PP PW. Contributed to the writing of the manuscript: PP SC PW.

References

1. Corner EJH (1965) Check-list of *Ficus* in Asia and Australasia with keys to identification: Botanic Gardens Singapore. 186 p.
2. Berg CC (1989) Classification and distribution of *Ficus*. Experientia 45: 605–611. doi: 10.1007/BF01975677.
3. Berg CC (2003) Flora Malesiana precursor for the treatment of Moraceae 2: *Ficus* subgenus *Pharmacosycea* Section *Oreosycea*. Blumea - Biodiversity, Evolution and Biogeography of Plants 48: 289–301. doi: 10.3767/000651903X674991.
4. Harrison RD, Shanahan M (2005) Seventy-seven ways to be a fig: Overview of a diverse plant assemblage. Pollination ecology and the rain forest: Springer New York. 111–127.
5. Shanahan M, So S, Compton SG, Corlett R (2001) Fig-eating by vertebrate frugivores: a global review. Biological reviews of the Cambridge Philosophical Society 76: 529–572.
6. Whittaker RJ (1995) Surface and buried seed banks from Krakatau, Indonesia: implications for the sterilization hypothesis. Biotropica 27: 346–354.
7. Metcalff DJ, Turner IM (1998) Soil seed bank from lowland rain forest in Singapore: canopy-gap and litter-gap demanders. Journal of Tropical Ecology 14: 103–108. doi: 10.1017/S0266467498000091.
8. Corner EJH (1963) *Ficus* in the Pacific region. In: Gressitt JL, editor. In Pacific basin biogeography Honolulu: HI: Bishop Museum Press. 233–245.
9. Corner EJH (1969) The complex of *Ficus deltoidea*; A recent invasion of the Sunda Shelf. Philosophical Transactions of the Royal Society of London B, Biological Sciences 256: 281–317. doi: 10.1098/rstb.1969.0043.
10. Berg CC, Corner EJH, Nooteboom HP, Coode MJE, Kirkup DW, et al. (2005) Flora Malesiana Series 1 - Seed Plants: Vol. 17, part 2: Moraceae (Ficus). Leiden: Nationaal Herbarium Nederland.
11. Verkerke W (1989) Structure and function of the fig. Experientia 45: 612–622. doi: 10.1007/BF01975678.
12. Harrison RD, Yamamura N (2003b) A few more hypotheses for the evolution of dioecy in figs (*Ficus*, Moraceae). Oikos 100: 628–635. doi: 10.1034/j.1600-0706.2003.11829.x.
13. Cook JM, Rasplus J-Y (2003) Mutualists with attitude: coevolving fig wasps and figs. Trends in Ecology & Evolution 18: 241–248. doi: 10.1016/S0169-5347(03)00062-4.
14. Anstett MC, Hossaert-McKey M, Kjellberg F (1997) Figs and fig pollinators: evolutionary conflicts in a coevoled mutualism. Trends in Ecology & Evolution 12: 94–99. doi: 10.1016/S0169-5347(96)10064-1.
15. Ramírez BW (1969) Fig Wasps: Mechanism of Pollen Transfer. Science 163: 580–581. doi: 10.1126/science.163.3867.580.
16. Wiebes JT (1979) Co-evolution of figs and their insect pollinators. Annual Review of Ecology and Systematics 10: 1–12. doi: 10.1146/annurev.es.10.110179.000245.
17. Galil J, Eisikowitch D (1968) On the pollination ecology of *Ficus sycomorus* in east Africa. Ecology 49: 259–269.
18. Herre EA (1996) An overview of studies on a community of Panamanian figs. Journal of Biogeography 23: 593–607. doi: 10.1111/j.1365-2699.1996.tb00020.x.
19. Noort Sv, Compton SG (1996) Convergent evolution of agaonine and sycoecine (Agaonidae, Chalcidoidea) head shape in response to the constraints of host fig morphology. Journal of Biogeography 23: 415–424. doi: 10.1111/j.1365-2699.1996.tb00003.x.
20. Kerdelhué C, Rossi, Rasplus J-Y (2000) Comparative community ecology studies on old world figs and fig wasps. Ecology 81: 2832–2849. doi: 10.1890/0012-9658(2000)081[2832:CCESOO]2.0.CO;2.
21. Hossaert-McKey M, Bronstein JL (2001) Self-pollination and its costs in a monoecious fig (*Ficus aurea*, Moraceae) in a highly seasonal subtropical environment. American Journal of Botany 88: 685–692.
22. Weiblen GD (2002) How to be a fig wasp. Annual review of entomology 47: 299–330. doi: 10.1146/annurev.ento.47.091201.145213.
23. Database CH (2000) The trees of Doi Suthep-Pui National Park. Unpublished output from Chiang Mai University Herbarium Database.
24. Elliott S, Blakesley D (2003) Eden helps restore Thailand's tropical forests. Eden Project Friends 10: 33–35.
25. unit Frr (1998) Forests for the future: growing and planting native trees for restoring forest ecosystems. In: Elliott S, Blakesley D, Anusarnsunthorn V, editors. Thailand: Biology Department, Faculty of Science, Chiang Mai University. 60.
26. Janzen DH (1979) How to be a Fig. Annual Review of Ecology and Systematics 10: 13–51. doi: 10.1146/annurev.es.10.110179.000305.
27. Gautier-Hion A, Michaloud G (1989) Are figs always keystone resources for tropical frugivorous vertebrates? A test in Gabon. Ecology 70: 1826–1833. doi: 10.2307/1938115.
28. Lambert FR, Marshall AG (1991) Keystone characteristics of bird-dispersed *Ficus* in a Malaysian lowland rain forest. Journal of Ecology 79: 793–809. doi: 10.2307/2260668.
29. Nason JD, Herre EA, Hamrick JL (1998) The breeding structure of a tropical keystone plant resource. Nature 391: 685–687.
30. Naiman RJ, Décamps H (1997) The ecology of interfaces: riparian zones. Annual review of Ecology and Systematics: 621–658.
31. Petts GE (1990) Forested river corridors:a lost resource. In: Cosgrove DE, Petts GE, editors. Water, Engineering and Landscape: water and control and landscape transformation in the modern period. Belhaven, London. 12–34.
32. Gregory SV, Swanson FJ, McKee WA, Cummins KW (1991) An ecosystem perspective of riparian zones. Bioscience 41: 540–551.
33. Naiman RJ, Decamps H, Pollock M (1993) The role of riparian corridors in maintaining regional biodiversity. Ecological applications 3: 209–212.
34. Naiman RJ, Bilby RE, Bisson PA (2000) Riparian Ecology and Management in the Pacific Coastal Rain Forest. Bioscience 50: 996.
35. Naiman RJ, Decamps H, McClain ME (2010) Riparia: ecology, conservation, and management of streamside communities: Academic Press. 421 p.
36. Tang S, Montgomery D (1995) Riparian buffers and potentially unstable ground. Environmental Management 19: 741–749. doi: 10.1007/BF02471956.
37. Prach K, Bartha S, Joyce CB, Pyšek P, van Diggelen R, et al. (2001) The role of spontaneous vegetation succession in ecosystem restoration: A perspective. Applied Vegetation Science 4: 111–114. doi: 10.1111/j.1654-109X.2001.tb00241.x.
38. Hood WG, Naiman RJ (2000) Vulnerability of riparian zones to invasion by exotic vascular plants. Plant Ecology 148: 105–114.
39. Heartsill Scalley T, Crowl TA, Thompson J (2009) Tree species distributions in relation to stream distance in a mid-montane wet forest, Puerto Rico. Caribbean Journal of Science 45: 52–63.
40. Sabo JL, Sponseller R, Dixon M, Gade K, Harms T, et al. (2005) Riparian zones increase regional species richness by harboring different, not more, species. Ecology 86: 56–62. doi: 10.1890/04-0668.
41. Norton BDA, Miller CJ (2000) Some issues and options for the conservation of native biodiversity in rural New Zealand. Ecological Management & Restoration 1: 26–34. doi: 10.1046/j.1442-8903.2000.00005.x.
42. Mongkolsawat C (2008) Estimating Riparian Forest in the Watershed Areas of Northeast Thailand: Application of SPOT Data. Kasetsart University, Thailand.
43. Kong-ied K (2011) Floristic composition and plant community structure of the riparian forest along the Sok canal, Surat Thani province, Thailand.: Prince of Sungkla University. 166 p.
44. center C (2012) Meteorological data at Chiang Mai Province station. Unpublished output from Thai meteorological.
45. Taesombat W, Sriwongsitatanon N (2010) Flood investigation in the Upper Ping River Basin using mathematical models. Kasetsart Journal (Natural Science) 44: 152–166.
46. Wood SH, Ziegler AD (2008) Floodplain sediment from recent 50-year-recurrence floods of the Ping River in northern Thailand. Hydrology and Earth Systems Science 12: 959–973.
47. Santisuk T (1988) An account of the vegetation of northern Thailand. Stuttgart: Franz Steiner Verlag Wiesbaden GMBH. 101 p.
48. Smitinand T (1989) Floristic inventory of tropical countries: Thailand. In: Campbell DG, Hammond HD, editors. New York: The New York Botanical Garden. 63–82.
49. Maxwell JF, Elliot S (2001) Vegetation and vascular flora of Doi Sutep-Pui National Park, Northern Thailand: Thai studies in biodiversity 5. Biodiversity Research and Training Program. 205 p.
50. Buckland ST, Anderson DR, Burnham KP, Laake JL (2005) Distance Sampling. Encyclopedia of Biostatistics: John Wiley & Sons, Ltd.

51. Berg CC, Pattharahirantricin N, Chantarasuwan B (2011) Flora of Thailand, Volume 10, Part 4: Cecropiaceae and Moraceae; Santisuk T, Larsen K, editors. Forest Herbarium: Royal Forest Department. 240 p.
52. Clarke KR, Gorley RN (2006) PRIMER v6: User manual/tutorial. Plymouth UK: PRIMER-E. 192 p.
53. Shannon CE (1948) A mathematical theory of communication. Bell System Technical Journal 27: 379–423. doi: 10.1002/j.1538-7305.1948.tb01338.x.
54. Gaines WL (1999) Monitoring biodiversity quantification and interpretation; Harrod RJ, Lehmkuhl JF, editors. Portland: U.S. Dept. of Agriculture, Forest Service, Pacific Northwest Research Station. 38 p.
55. Simpson EH (1949) Measurement of diversity. Nature 163: 688. doi: 10.1038/163688a0.
56. Team RDC (2011) R: A Language and Environment for Statistical Computing.Austria: the R Foundation for Statistical Computing.
57. Maxwell JF (1988) The vegetation of Doi Suthep-Put National Park, Chiang Mai Province, Thailand. Tiger paper 15: 6–14.
58. Roberts D (1986) Ordination on the basis of fuzzy set theory. Vegetatio 66: 123–131. doi: 10.1007/BF00039905.
59. Nilsson C, Svedmark M (2002) Basic principles and ecological consequences of changing water regimes: riparian plant communities. Environmental Management 30: 468–480. doi: 10.1007/s00267-002-2735-2.
60. Pollock MM, Naiman RJ, Hanley TA (1998) Plant species richness in riparian wetlands-a test of biodiversity theory. Ecology 79: 94–105. doi: 10.1890/0012-9658(1998)079[0094:PSRIRW]2.0.CO;2.
61. Tabacchi E, Lambs L, Guilloy H, Planty-Tabacchi, Muller E, et al. (2000) Impacts of riparian vegetation on hydrological processes. Hydrological Processes 14: 2959–2976. doi: 10.1002/1099-1085(200011/12)14:16/17<2959::AID-HYP129>3.0.CO;2-B.
62. DeFerrari CM, Naiman RJ (1994) A multi-scale assessment of the occurrence of exotic plants on the Olympic Peninsula, Washington. Journal of Vegetation Science 5: 247–258. doi: 10.2307/3236157.
63. Corner EJH (1960) Taxonomic notes on *Ficus Linn.*, Asia and Australasia: U. S.Government Printing Office.
64. Gardner S, Sitthisunthorn P, Anusarnsunthorn W (2000) A field guide to forest trees of Northern Thailand. Thailand: Kobfai Pub. 545 p.
65. Laman T, Welblen G (1998) Figs of Gunung Palung National Park (West Kalimantan, Indonesia). Tropical Biodiversity 5: 245–297.
66. Balasubramanian P, Saravanan R, Maheswaran B (2004) Fruit preferences of Malabar pied hornbill *Anthracoceros coronatus* in Western Ghats, India. Bird Conservation International 14: S69–S79.
67. Wintle BC, Kirkpatrick JB (2007) The response of riparian vegetation to flood-maintained habitat heterogeneity. Austral Ecology 32: 592–599. doi: 10.1111/j.1442-9993.2007.01753.x.
68. Chantarasuwan B, Marod D, Pattanakiat S (2007) Species diversity and habitat suitability assessment for genus *Ficus* in Mae Klong watershed research station, Amphoe Thong Pha Phum, Changwat Kanchanaburi. The Thailand Natural History Museum Journal 2: 12.
69. Banack SA, Horn MH, Gawlicka A (2002) Disperser- vs. establishment-limited distribution of a riparian fig tree (*Ficus insipida*) in a Costa Rican tropical rain forest. Biotropica 34: 232–243. doi: 10.1111/j.1744-7429.2002.tb00534.x.
70. Munishi LK (2007) The distribution and diversity of tree resources outside forest in southern side of Mount Kilimanjaro. Discovery and Innovation 19: 36–44.
71. Whitmore TC (1984) Tropical rain forests of the Far East: Clarendon Press. 352 p.
72. Colwell RK, Lees DC (2000) The mid-domain effect: geometric constraints on the geography of species richness. Trends Ecol Evol 15: 70–76.
73. Austin MP, Pausas JG, Nicholls AO (1996) Patterns of tree species richness in relation to environment in southeastern New South Wales, Australia. Australian Journal of Ecology 21: 154–164. doi: 10.1111/j.1442-9993.1996.tb00596.x.
74. Mark AF, Dickinson KJM, Hofstede RGM (2000) Alpine vegetation, plant distribution, life forms, and environments in a perhumid New Zealand region Oceanic and tropical high mountain affinities. Arctic Antarctic & Alpine Research 32: 240–254.
75. Harrison RD, Hamid AA, Kenta T, Lafrankie J, Lee, et al. (2003a) The diversity of hemi-epiphytic figs (*Ficus*; Moraceae) in a Bornean lowland rain forest. Biological Journal of the Linnean Society 78: 439–455. doi: 10.1046/j.0024-4066.2002.00205.x.

11

Non-Stomatal Limitation to Photosynthesis in *Cinnamomum camphora* Seedings Exposed to Elevated O³

Junfeng Niu[1], Zhaozhong Feng[2], Weiwei Zhang[3], Ping Zhao[1], Xiaoke Wang[2]*

1 Key Laboratory of Vegetation Restoration and Management of Degraded Ecosystems, South China Botanical Garden, Chinese Academy of Sciences, Guangzhou, Guangdong, China, **2** State Key Laboratory of Urban and Regional Ecology, Research Center for Eco-Environmental Sciences, Chinese Academy of Sciences, Beijing, China, **3** Key Laboratory of Black Soil Ecology, Northeast Institute of Geography and Agroecology, Chinese Academy of Sciences, Harbin, Heilongjiang, China

Abstract

Ozone (O_3) is the most phytotoxic air pollutant for global forests, with decreased photosynthesis widely regarded as one of its most common effects. However, controversy exists concerning the mechanism that underlies the depressing effects of O_3 on CO_2 assimilation. In the present study, seedlings of *Cinnamomum camphora*, a subtropical evergreen tree species that has rarely been studied, were exposed to ambient air (AA), ambient air plus 60 [ppb] O_3 (AA+60), or ambient air plus 120 [ppb] O_3 (AA+120) in open-top chambers (OTCs) for 2 years. Photosynthetic CO_2 exchange and chlorophyll a fluorescence were investigated in the second growing season (2010). We aim to determine whether stomatal or non-stomatal limitation is responsible for the photosynthesis reduction and to explore the potential implications for forest ecosystem functions. Results indicate that elevated O_3 (E-O_3) reduced the net photosynthetic rates (P_N) by 6.0-32.2%, with significant differences between AA+60 and AA+120 and across the four measurement campaigns (MCs). The actual photochemical efficiency of photosystem II (PSII) in saturated light (F_v'/F_m') was also significantly decreased by E-O_3, as was the effective quantum yield of PSII photochemistry (Φ_{PSII}). Moreover, E-O_3 significantly and negatively impacted the maximum rates of carboxylation (V_{cmax}) and electron transport (J_{max}). Although neither the stomatal conductance (g_s) nor the intercellular CO_2 concentration (C_i) was decreased by E-O_3, P_N/g_s was significantly reduced. Therefore, the observed reduction in P_N in the present study should not be attributed to the unavailability of CO_2 due to stomatal limitation, but rather to the O_3-induced damage to Ribulose-1,5-bisphosphate carboxylase/oxygenase and the photochemical apparatus. This suggests that the down-regulation of stomatal conductance could fail to occur, and the biochemical processes in protoplasts would become more susceptible to injuries under long-term O_3 exposure, which may have important consequences for forest carbon and water budget.

Editor: Dafeng Hui, Tennessee State University, United States of America

Funding: Funding for this project was provided by the National Natural Science Foundation of China (No. 31170424, No. 71003092, No. 41030638 and No. 41275169), Tiantong National Station of Forest Ecosystems (No. XT200707) and South Botanical Garden, Chinese Academy of Sciences (No. 201229). The funders had no role in study design, data collection and analysis, decision to publish, or preparation of the manuscript.

Competing Interests: The authors have declared that no competing interests exist.

* E-mail: wangxk@rcees.ac.cn

Introduction

Of the known phytotoxic air pollutants, ozone (O_3) has the greatest potential to detrimentally impact forests [1]. The negative effects of O_3 on plants are commonly attributed to its highly oxidative properties that can damage cell membranes, denature critical enzymes, and give rise to various reactive oxygen species (ROS) [2,3,4]. One of the most consistent effects of O_3 on trees is the inhibition of photosynthesis [5,6,7]. Based on a quantitative meta-analysis of over 100 studies of O_3 effect on trees, it has been estimated that an 11% decrease in net photosynthetic rates (P_N) has resulted from the increase in O_3 levels that has occurred since the Industrial Revolution [8].

However, controversy exists concerning the mechanism that underlies the depressing effects of O_3 on CO_2 assimilation in plants [9]. While previous studies have frequently linked these effects to decreased stomatal conductance (g_s), many others have related the O_3-induced decline in photosynthesis to altered mesophyll activities, such as reduced maximum rates of carboxylation (V_{cmax}) and electron transport (J_{max}) [10,11,12,13]. Additionally, under high levels of O_3, changes in chlorophyll a fluorescence, including reductions in photochemical quenching (qP), actual photochemical efficiency (F_v'/F_m') and the effective quantum yield of photosystem II (PSII) in saturated light (Φ_{PSII}), have also been widely reported [14,15]. In fact, due to its high-resolution, real-time, and non-invasive nature, chlorophyll a fluorescence measurement has become an important technique parallel to gas exchange analyses for confirming the primary site of photosynthetic limitation [16]. Combined measurements of chlorophyll a fluorescence and gas exchange can provide valuable information regarding plant photosynthetic performance [17].

Field and experimental evidence suggests that broadleaf evergreen trees in Mediterranean areas are comparatively more O_3 tolerant than deciduous species in temperate and boreal regions [18,19]. This can be largely attributed to the sclerophyllous traits of the former, as well as the uniquely Mediterranean climate that concentrates high levels of O_3 during seasons of drought, which triggers stomatal closure [20,21,22,23]. In subtropical regions, trees also develop glossy and leathery leaves, however, the climate is generally characterized by four clearly demarcated seasons with rain and heat co-occurring in summers, and so far very little information is available concerning the impact of O_3 on the regional evergreen tree species.

In the present study, seedlings of *C. camphora*, a subtropical evergreen broadleaf tree species native to the Yangtze River Delta in eastern China, were exposed to ambient air (AA), ambient air plus 60 [ppb] O_3 (AA+60), or ambient air plus 120 [ppb] O_3 (AA+120) in open-top chambers (OTCs) for 2 years. During the second growing season (2010), gas exchange and chlorophyll a fluorescence were measured and analyzed. The aims of this experiment were to: (1) determine the extent to which photosynthesis is reduced in the experimental seedlings exposed to elevated O_3 (E-O_3, AA+60 or AA+120), (2) clarify whether stomatal or non-stomatal limitation is responsible for this reduction, and (3) explore the ecological meaning of our findings to broad-scale studies.

Materials and Methods

Experimental site and plant material

Permits and approvals for the work were obtained from East China Normal University and Tiantong Forest Farm, which are responsible for the protection of the Tiantong National Forest Park. The experiment was carried out within the park, at the Tiantong National Field Observation and Research Station for Forest Ecosystems (29°48′N, 121°47′E), Ningbo, Zhejiang province, China. One-year-old *C. camphora* seedlings, a typical subtropical evergreen broadleaf tree species widely distributed throughout eastern China and Japan, were planted in 5-[L] plastic pots and fumigated with E-O_3 for 2 years. These seedlings were purchased from a nearby commercial nursery and selected for phenotypic homogeneity. The potting soil consisted of a mixture of native yellowish-brown lateritic soil and litter collected from a fir forest at a 1:1 ratio. All seedlings were acclimated to OTC conditions for two weeks before O_3 fumigation. More information about the climate conditions of the experimental site and the plant cultivation prior to O_3 treatment were described in Niu *et al.*[24].

OTCs and treatments

Six OTCs (octagonal base, 5.5 [m^2] of basal area, and 2.6 [m] in height) were set up at the experimental field in early 2009. The rate of light transmittance of the OTCs was 98.3% and the average air velocity corresponded to a turnover rate of two complete air changes per minute. O_3 was generated from pure oxygen using an electrical discharge O_3 generator (*HY003, Chuangcheng Technology Co., Ltd.*, Jinan, China) and piped into four of the six OTCs in mixture with ambient air. O_3 flow was modulated using mass flow meters (*SY-9311, Beijing Shengye Science and Technology Development Co., Ltd.*, Beijing, China) in order to obtain the designated O_3 concentration within each OTC. The seedlings were fumigated from 9:00–17:00, 7 days per week, 25 May to 10 September 2009 and 1 May to 7 October 2010, except for rainy and mostly cloudy days.

Replicate AA, AA+60, and AA+120 chambers were randomly arranged in the experimental field. Seedling positions within each OTC were changed every 3-5 days. Every 10-15 days, all chambers were emptied and randomly reassigned a new O_3 level, and the seedlings were replaced according to their specified treatment levels. This allowed us to eliminate position and chamber effects, treating each plant as an independent experimental unit. Five seedlings within each OTC and a total of 30 plants (5 plants×3 groups×2 OTCs) were investigated.

Measurements

Temperature and relative air humidity were recorded at 30-minute intervals using thermo-hygrographs (*DSR-TH, ZOGLAB Microsystem Inc.*, Hangzhou, China) inside and outside the OTCs. O_3 concentrations at approximately 10 [cm] above the plant canopy were monitored using a UV-absorption O_3 analyzer (*Model 49i, Thermo Scientific Inc.*, Connecticut, USA).

Gas exchange and chlorophyll a fluorescence under light conditions were measured using an infrared gas analyzer (IRGA) fitted with a 6400-40 leaf chamber fluorometer (*LI-6400, LI-COR Inc.*, Lincoln, NE, USA). All measurements were made from 9:00–12:00 and recorded when the coefficient of variance (CV) was less than 3%. The photosynthetic photon flux density (PPFD) was fixed at a saturating intensity of 1200 [μmol m^{-2} s^{-1}]. CO_2 was supplied with pure CO_2 cylinders and maintained at 380 [μmol mol^{-1}]. Block temperature of the cuvette was set to 30±0.5°C, and relative humidity 60±5%. Maximum, minimum and steady state fluorescence under light conditions (F_m', F_o' and F_s) were measured, and Φ_{PSII}, qP and F_v'/F_m' were calculated as (F_m'-F_s)/F_m', (F_m'-F_s)/(F_m'-F_o') and (F_m'- F_o')/F_m', respectively. Six plants per treatment were analyzed, and only fully expanded upper leaves were screened. Tracking analyses of leaves in the same leaf position were carried out monthly (2 July, 7 August, 7 September, and 8 October 2010).

In order to determine the maximum photochemical efficiency of PSII (F_v/F_m), a field-portable chlorophyll fluorometer (*FMS 2, Hansatech Instruments Ltd.*, Norfolk, UK) was employed. The same leaves used for gas exchange analyses were screened. Leaves were adapted to dark for 30 minutes and the minimum fluorescence (F_o) was measured by switching on the modulating light (0.6 [kHz]). Then, the application of a saturating light pulse (8000 [m^{-2} s^{-1}] for 1 [s]) led to the rapid closure of PSII reaction centers, yielding the maximum fluorescence (F_m). F_v/F_m was calculated as (F_m-F_o)/F_m.

The response of carbon assimilation rates to changing CO_2 concentrations (*A*-C_i response curves) was determined by sequentially measuring the rates of photosynthesis at CO_2 concentrations of 380, 200, 150, 100, 50, 400, 600, 900, 1200 and 1500 [μmol mol^{-1}]. Light intensity, block temperature and relative humidity were set equal to those used in gas exchange analyses. V_{cmax} and J_{max} were determined and adjusted to 25°C according to Long and Bernacchi [25]. Four seedlings per treatment were screened, and two measurement campaigns (MCs) were carried out for the *A*-C_i response curves, on 22 July and 24 September 2010, respectively.

Data analysis

Data were analyzed using SAS software (*Version 9.1.3, SAS Institute*, Cary, NC, USA). Repeated measures analyses of variance (RANOVAs) were performed in order to analyze the overall effect of O_3 on the examined parameters throughout the growing season. Variances across MCs, as well as the interaction between O_3 and MCs were also investigated. For each MC, ANOVA model was applied to test the O_3 effect and Bonferroni methods were adopted for post-hoc multiple comparisons. Normality of distribution and homogeneity of variance were tested before all analyses. Differences between treatments were considered significant if $p \leq 0.05$.

Table 1. AOT40s and SUM60s under different O_3 exposure regimes during the 2010 growing season of *Cinnamomum camphora* seedlings.

	AOT40 [ppm h][b]				SUM60 [ppm h][c]			
O_3 regimes[a]	**2 July**	**7 Aug.**	**7 Sept.**	**8 Oct.**	**2 July**	**7 Aug.**	**7 Sept.**	**8 Oct.**
AA	3.3	4.8	5.5	6.7	3.9	5.7	6.7	8.7
AA+60	12.9	19.1	21.7	26.1	14.3	26.6	32.3	45.6
AA+120	23.1	42.8	47.3	56.3	35.4	48.2	56.9	74.9

[a]AA: ambient air; AA+60: ambient air plus 60 [ppb] O_3; AA+120: ambient air plus 120 [ppb] O_3.
[b]AOT40 [ppm h]: accumulated O_3 exposure over a threshold of 40 [ppb].
[c]SUM60 [ppm h]: sum of hourly O_3 concentration when the concentration is equal to or greater than 60 [ppb].

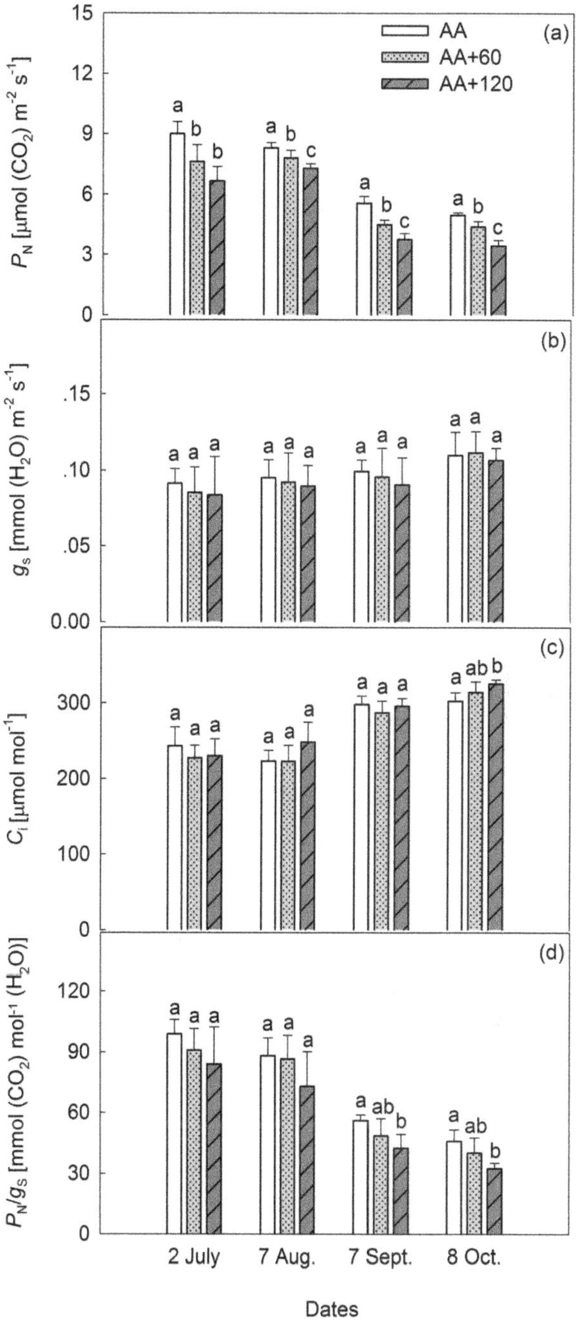

Figure 1. Effects of elevated O_3 on gas exchange parameters. Vertical bars represent average levels and distinct letters indicate significant differences among O_3 regimes ($n=6$) (AA: ambient air; AA+60: ambient air plus 60 [ppb] O_3; AA+120: ambient air plus 120 [ppb] O_3).

Results

OTC microclimate and O_3 monitoring

Table 1 shows AOT40s (accumulated O_3 exposure over a threshold of 40 [ppb]) and SUM60s (sum of hourly O_3 concentration when the concentration is equal to or greater than 60 [ppb]) during each MC of gas exchange and chlorophyll a fluorescence during the 2010 growing season. Because of persistent rain (19 days) from 7 August to 7 September, the accumulation of

Table 2. Repeated measures ANOVAs (RANOVAs) of the gas exchange and chlorophyll a fluorescence parameters of *Cinnamomum camphora* seedlings during the 2010 growing season (*P* values are shown, $n=4$ for V_{cmax}, J_{max} and J_{max}/V_{cmax}, n = 6 for other parameters).

	Parameters	O_3	MCs	$O_3 \times MCs^a$
Gas exchange	P_N [μmol (CO_2) m^{-2} s^{-1}]	**<.0001**	**<.0001**	**0.0169**
	g_s [mmol (H_2O) m^{-2} s^{-1}]	0.5738	**0.0001**	0.9956
	C_i [μmol mol^{-1}]	0.1755	**<.0001**	**0.0377**
	P_N/g_s [mmol (CO_2) mol^{-1} (H_2O)]	**0.0072**	<.0001	0.9435
	V_{cmax} [μmol m^{-2} s^{-1}]	**0.0031**	0.3435	0.1481
	J_{max} [μmol m^{-2} s^{-1}]	<.0001	**0.0007**	**0.0005**
	J_{max}/V_{cmax}	0.4994	0.1172	0.3419
Chlorophyll a fluorescence	F_v'/F_m'	**0.0310**	**0.0011**	0.1329
	Φ_{PSII}	**<.0001**	**<.0001**	0.6421
	qP	<.0001	<.0001	0.3827
	F_v/F_m	0.5042	**<.0001**	0.9998

[a]MCs: measurement campaigns.

AOT40 and SUM60 was lower during this period, as shown in Table 1. Under AA, the total dose of O_3 was 6.7 and 8.7 [ppm h] in the forms of AOT40 and SUM60, respectively. AOT40s were generally lower than values expressed as SUM60s in all O_3 regimes. Detailed descriptions of the average O_3 concentrations, as well as the OTC microclimate conditions throughout the 2 years of this experiment can be found elsewhere [24,26].

Gas exchange

Throughout the 2010 growing season, P_N was reduced, on average, by 13.0% and 25.3% under AA+60 and AA+120, respectively. Differences between these two treatment regimes were statistically significant, except on 2 July (Table 2 and Figure 1a). P_N/g_s was also significantly reduced, while C_i was significantly increased by AA+120 on 8 October. A negative O_3 effect on P_N/g_s was also observed on 7 September (Figure 1d). Variations of P_N, g_s, C_i, and P_N/g_s across the four MCs were statistically significant, and O_3 interacted significantly with MCs for P_N and C_i (Table 2).

V_{cmax} was significantly decreased by AA+60 and AA+120 on 24 September, but only by AA+120 on 22 July (Figure 2b). Both AA+60 and AA+120 exerted significantly negative effects on J_{max} (Figure 2a) across the two MCs. The difference in J_{max} between AA+60 and AA+120 was statistically significant on 22 July, but not on 24 September (Figure 2a). J_{max} varied significantly, while V_{cmax} maintained the same levels across the two MCs (Table 2). J_{max}/V_{cmax} was not significantly affected by E-O_3 in the present study (Figure 2c).

Chlorophyll a fluorescence

F_v'/F_m' and Φ_{PSII} were significantly decreased by E-O_3 (Table 2). Φ_{PSII} was significantly decreased by AA+120 across all four MCs, and also by AA+60 on 2 July and 8 October. For F_v'/F_m', only AA+120 exerted significant impact, on 2 July and 8 October (Figure 3a, c). Additionally, qP was significantly depressed by AA+120 across all four MCs, and also by AA+60 on 7 August and 8 October (Figure 3b). However, F_v/F_m was not significantly influenced by E-O_3 (Figure 3d). Differences across MCs were statistically significant for all fluorescence parameters. However, no interactions were found between O_3 and MCs (Table 2).

Discussion

E-O_3 significantly reduced P_N in *C. camphora* over the course of the present study. At the end of the 2010 growing season, AA+60, which corresponded to an AOT40 of 26.1 [ppm h], reduced P_N by 11.7%. A similar 11.4% decrease in P_N occurred with an AOT40 of 36.2 [ppm h] in the Mediterranean evergreen Satsuma mandarin (*Cirtus unshiu* [Mak.] Marc.) [15]. However, in deciduous *Quercus pyrenaica*, *Quercus robur* and *Quercus faginea*, an AOT40 of 26.2-28.8 [ppm h] decreased P_N by 64%, 38% and 33%, respectively [27]. Based on the comparisons with these results, our findings suggest that *C. camphora* is less resistant to O_3 than Mediterranean evergreen broadleaves, but more tolerant than deciduous species [7,19]. Reduction in P_N under E-O_3 has also been reported in other species, deciduous (*Quercus serrata*, *Populus tremuloides* Michx., *Betula pendula* Roth.) as well as evergreen (*Pinus taeda* L.) [28,29,30,31], and therefore this may represent a common response behavior to high levels of O_3 in woody plants [32].

Similar to observations in European beech (*Fagus sylvatica*) and black aspen (*Populus nigra*) [33], g_s was not affected by E-O_3 in the present study (Figure 1b), indicating that the significant reduction in P_N of *C. camphora* cannot be attributed to stomatal behavior. Additionally, C_i was not reduced, but in fact significantly increased by AA+120 on 8 October, confirming that CO_2 supply was not the limiting factor in reducing P_N. Moreover, significant decrease in P_N/g_s was detected under AA+120 on 7 September and 8 October, further suggesting that factors other than g_s should be considered when attempting to clarify the mechanism responsible for the reduction in P_N that results from O_3 stress. Increased C_i, as well as the negative relationship between P_N and C_i under elevated O_3, has also been documented in previous literature [34,35].

RuBisCO is commonly regarded as one of the primary targets of O_3-induced damage [36,37]. On both 22 July and 24 September, AA+120 significantly reduced V_{cmax}, which was also notably decreased by AA+60 on 24 September (Figure 2b). These findings concur broadly with those reported in aspen (*P. tremuloides*) and birch (*B. pendula* Roth.) [28,38,39]. Decreases in RuBisCO quantity and activity may be responsible for the decline of V_{cmax} under E-O_3 [40]. J_{max} was also significantly decreased by E-O_3 in the present study, while J_{max}/V_{cmax} was not affected (Figure 2c).

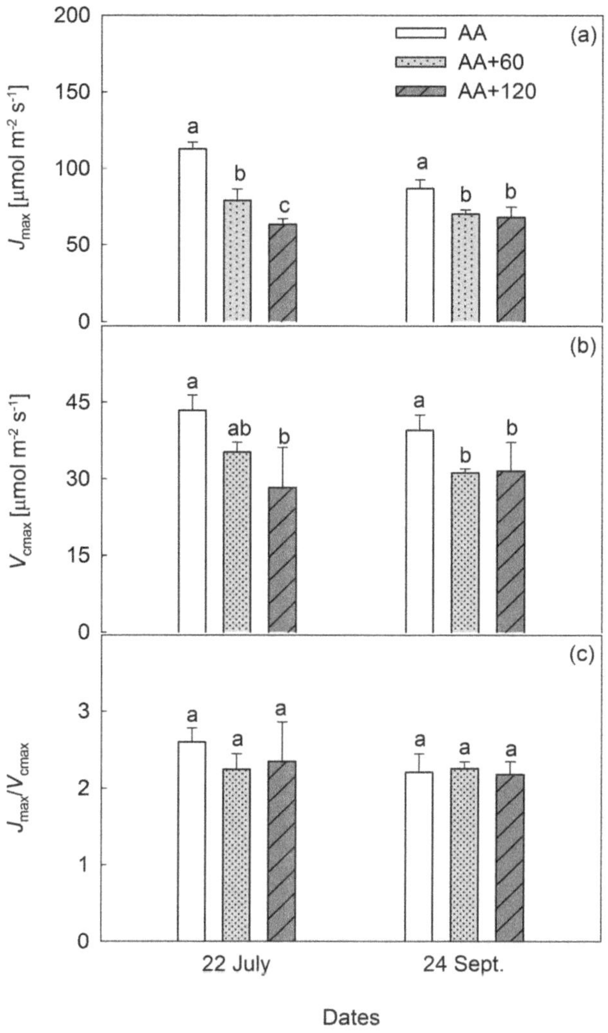

Figure 2. Effects of elevated O_3 on the maximum rates of carboxylation (V_{cmax}) and electron transport (J_{max}). Vertical bars represent average levels and distinct letters indicate significant differences among O_3 regimes ($n = 4$) (AA: ambient air; AA+60: ambient air plus 60 [ppb] O_3; AA+120: ambient air plus 120 [ppb] O_3).

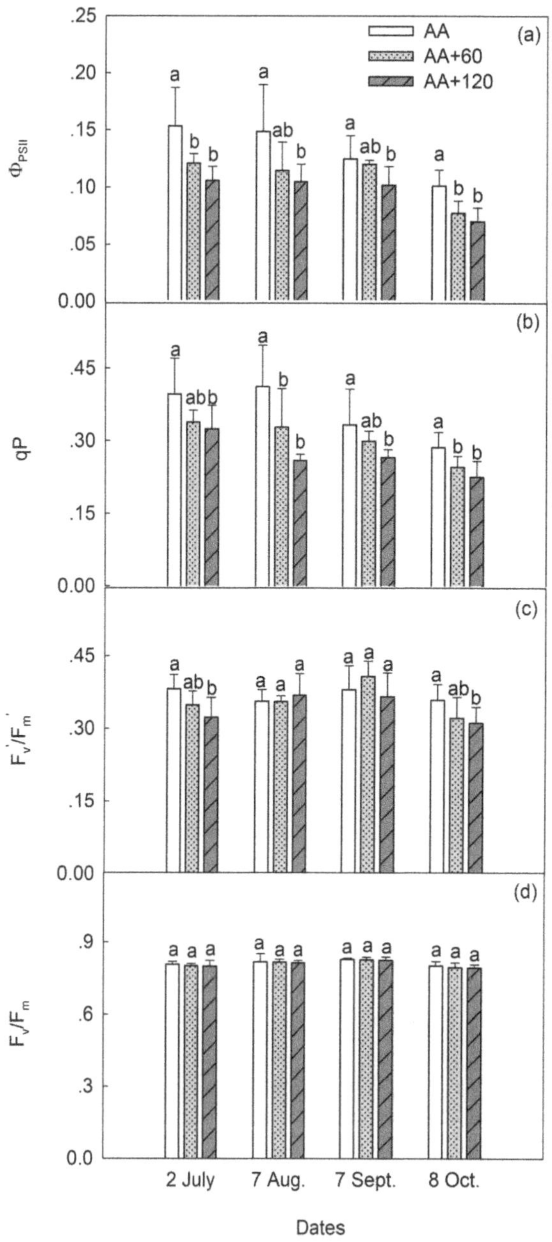

Figure 3. Effects of elevated O_3 on the chlorophyll a fluorescence parameters. Vertical bars represent average levels and distinct letters indicate significant differences among O_3 regimes ($n = 6$) (AA: ambient air; AA+60: ambient air plus 60 [ppb] O_3; AA+120: ambient air plus 120 [ppb] O_3).

The constant J_{max}/V_{cmax} ratio indicates the close coupling between RuBP carboxylation and light-driven electron transport. Activities of these two processes are commonly related, and vary in parallel with environmental conditions [41]; thus the observed decrease in J_{max} might have occurred in response to the declining V_{cmax} [42]. Therefore, O_3-induced degradation and deactivation of RuBisCO, as well as its feedback inhibitory effect on the electron transport system, might be a primary cause of the reduction in P_N.

Confirming the previous findings in evergreen Mediterranean species [43,44], F_v/F_m in *C. camphora* was not influenced by E-O_3 (Figure 3d). However, Fv′/Fm′ was significantly reduced by AA+120 on 2 July and 8 October, implying enhanced energy decay via non-radiative processes at the PSII reaction centers [45]. Meanwhile, in the present study, significant reductions of qP (notably under AA+120 at all four MCs and under AA+60 on 7 August and 8 October) and Φ_{PSII} (notably under AA+120 at all four MCs and under AA+60 on 2 July and 8 October) were also observed under E-O_3 (Figure 3a, b). Similar results have been reported in Scots pine (*Pinus sylvestris* L.) and Satsuma mandarin (*C. unshiu* [Mak.] Marc.) [15,17]. Reduction in qP and Φ_{PSII} may correlate with a decrease in the proportion of available excitation energy used in the photochemistry [11]. Decreased photochemistry but enhanced non-radiative decay suggests that more energy absorbed by PSII is dissipated as heat, and this could then lead to the overheating of PSII reaction centers. Therefore, direct oxidative damage, as well as indirect heat-related injuries to the photochemical apparatus, may also play an important role in mediating the down-regulation of P_N in *C. camphora* exposed to E-O_3.

Previous studies have frequently linked the O_3-induced decline in plant photosynthesis to its inhibitory effect on foliar stomatal conductance. Torsethaugen *et al.* found that acute O_3 exposure inhibited the guard cell K^+ channels, which mediate K^+ uptake that drives stomatal opening, and thus led to decreased photosynthesis in *Vicia faba* [46]. Zhang *et al.* reported that the reduction in photosynthesis (-27%) of *Liriodendron chinense* (Hemsl.) Sarg seedlings was accompanied by a significant decrease of stomatal conductance (-34.7%) after O_3 exposure for 40 days at a concentration of [150 ppb] [47]. During the first growing season (2009) of the present study, we also observed concurrent reduction in photosynthesis (-24.6%) and stomatal conductance (-34.2%) under [AA+60] [26], which was however not observed during the second growing season (2010). This suggests that the coupling between photosynthesis and stomatal conductance in plants could fail and the biochemical processes in protoplasts would become more susceptible to injuries under long-term O_3 exposure.

Decoupling between photosynthesis and stomatal conductance under elevated O_3 may have important implications for water use and carbon cycling of forest ecosystems [48]. One the one hand, failure or sluggishness of stomatal closure could give rise to excessive plant transpiration [49], resulting in unnecessary water loss, leading to regional water shortage, or even causing tree wilt and dieback if soil water supply is particularly tight, especially in arid and semi-arid areas. At the same time, increased exposure of mesophyll cells to O_3 through open stoma, on the other hand, could decrease the efficiency of light use by photosystem II for CO_2 assimilation [25], resulting in lower forest carbon sequestration and leading to a warmer atmosphere. Therefore, the potential impact of O_3 under both current and future enriched conditions should be considered adequately in carbon budget calculations, forest hydrology simulations and climate change predictions at regional and global scales.

It should be noted that the present study was conducted on just one subtropical evergreen species of 2 to 3 years age. The unique physiological characteristics of seedlings and the optimal water status under OTC conditions as well as the restriction of root growth in pots may bias tree performance [4,24]. To attain a comprehensive understanding of the effect of O_3 on forests, as well as forest responses and feedbacks to global changes, further investigations based on mature trees of a wider range of other species are critically needed.

Conclusions

E-O_3 (AA+60 or AA+120) significantly reduced P_N in *C. camphora.* Comparisons of this reduction with those observed in other species suggest that *C. camphora* is less tolerant to O_3 than Mediterranean evergreen trees, but more resistant than deciduous species. Reduction of stomatal conductance is not a reasonable explanation for the decline of P_N in the present study, as manifested by the increased C_i and decreased P_N/g_s. As with P_N, decreases in V_{cmax}, J_{max}, Φ_{PSII} and qP were detected, indicating that direct oxidative damage and indirect heat-related injuries to RuBisCO and photochemical apparatus were responsible for the reduction in P_N that was observed in *C. camphora* under E-O_3. This suggests that the biochemical processes in protoplasts will become more susceptible to injuries under long-term O_3 exposure, which may bear important implications for forest water use and carbon cycling.

Acknowledgments

For fieldwork assistance, we wish to thank Mr Wenzhi Sun. Great appreciation is also given to Professor Zongwei Feng for his constructive suggestions on the experimental design.

Author Contributions

Conceived and designed the experiments: JFN ZZF. Performed the experiments: JFN WWZ. Analyzed the data: JFN. Contributed reagents/materials/analysis tools: JFN. Wrote the paper: JFN. Helpful suggestions in data analyses and manuscript preparation: XKW PZ.

References

1. Bussotti F (2008) Functional leaf traits, plant communities and acclimation processes in relation to oxidative stress in trees: a critical overview. Global Change Biology 14: 2727-2739.
2. Karnosky DF, Pregitzer KS, Zak DR, Kubiske ME, Hendrey GR, et al. (2005) Scaling ozone responses of forest trees to the ecosystem level in a changing climate. Plant Cell and Environment 28: 965-981.
3. Matyssek R, Sandermann H, Wieser G, Booker F, Cieslik S, et al. (2008) The challenge of making ozone risk assessment for forest trees more mechanistic. Environmental Pollution 156: 567-582.
4. Samuelson L, Kelly JM (2001) Scaling ozone effects from seedlings to forest trees. New Phytologist 149: 21-41.
5. Bortier K, Ceulemans R, De Temmerman L (2000) Effects of ozone exposure on growth and photosynthesis of beach seedlings (Fagus sylvatica). New Phytologist 146: 271-280.
6. Karnosky DF, Skelly JM, Percy KE, Chappelka AH (2007) Perspectives regarding 50 years of research on effects of tropospheric ozone air pollution on US forests. Environmental Pollution 147: 489-506.
7. Paoletti E (2006) Impact of ozone on Mediterranean forests: A review. Environmental Pollution 144: 463-474.
8. Wittig VE, Ainsworth EA, Long SP (2007) To what extent do current and projected increases in surface ozone affect photosynthesis and stomatal conductance of trees? A meta-analytic review of the last 3 decades of experiments. Plant Cell and Environment 30: 1150-1162.
9. Degl'Innocenti E, Vacca C, Guidi L, Soldatini GF (2003) CO_2 photoassimilation and chlorophyll fluorescence in two clover species showing different response to O_3. Plant Physiology and Biochemistry 41: 485-493.
10. Darbah JNT, Kubiske ME, Nelson N, Kets K, Riikonen J, et al. (2010) Will photosynthetic capacity of aspen trees acclimate after long-term exposure to elevated CO_2 and O_3? Environmental Pollution 158: 983-991.
11. Feng Z, Pang J, Kobayashi K, Zhu J, Ort DR (2011) Differential responses in two varieties of winter wheat to elevated ozone concentration under fully open-air field conditions. Global Change Biology 17: 580-591.
12. He XY, Fu SL, Chen W, Zhao TH, Xu S, et al. (2007) Changes in effects of ozone exposure on growth, photosynthesis, and respiration of Ginkgo biloba in Shenyang urban area. Photosynthetica 45: 555-561.
13. Matyssek R, Gunthardtgoerg MS, Keller T, Scheidegger C (1991) Impairment of gas-exchange and structure in birch leaves (betula-pendula) caused by low ozone concentrations. Trees-Structure and Function 5: 5-13.
14. Bussotti F, Desotgiu R, Cascio C, Pollastrini M, Gravano E, et al. (2011) Ozone stress in woody plants assessed with chlorophyll a fluorescence. A critical reassessment of existing data. Environmental and Experimental Botany 73: 19-30.
15. Calatayud A, Iglesias DJ, Talon M, Barreno E (2006) Effects of long-term ozone exposure on citrus: Chlorophyll a fluorescence and gas exchange. Photosynthetica 44: 548-554.
16. Soja G, Pfeifer U, Soja AM (1998) Photosynthetic parameters as early indicators of ozone injury in apple leaves. Physiologia Plantarum 104: 639-645.
17. Wang KY, Kellomaki S (1997) Effects of elevated CO_2 and soil-nitrogen supply on chlorophyll fluorescence and gas exchange in Scots pine, based on a branch-in-bag experiment. New Phytologist 136: 277-286.
18. Bussotti F, Gerosa G (2007) Are the Mediterranean forests in Southern Europe threatened from ozone? Journal of Mediterranean Ecology 3: 23-34.
19. Nali C, Paoletti E, Marabottini R, Della Rocca G, Lorenzini G, et al. (2004) Ecophysiological and biochemical, strategies of response to ozone in Mediterranean evergreen broadleaf species. Atmospheric Environment 38: 2247-2257.
20. Gerosa G, Finco A, Mereu S, Vitale M, Manes F, et al. (2009) Comparison of seasonal variations of ozone exposure and fluxes in a Mediterranean Holm oak forest between the exceptionally dry 2003 and the following year. Environmental Pollution 157: 1737-1744.
21. Grulke NE, Paoletti E (2005) A field system to deliver desired O_3 concentrations in leaf-level gas exchange measurements: results for Holm Oak near a CO_2 spring. Phyton-Annales Rei Botanicae 45: 21-31.
22. Manes F, Vitale M, Fabi AM, De Santis F, Zona D (2007) Estimates of potential ozone stomatal uptake in mature trees of Quercus ilex in a Mediterranean climate. Environmental and Experimental Botany 59: 235-241.
23. Ribas A, Penuelas J, Elvira S, Gimeno BS (2005) Ozone exposure induces the activation of leaf senescence-related processes and morphological and growth changes in seedlings of Mediterranean tree species. Environmental Pollution 134: 291-300.

24. Niu J, Zhang W, Feng Z, Wang X, Tian Y (2011) Impact of elevated O_3 on visible foliar symptom, growth and biomass of Cinnamomum camphora seedlings under different nitrogen loads. Journal of Environmental Monitoring 13: 2873-2879.
25. Long SP, Bernacchi CJ (2003) Gas exchange measurements, what can they tell us about the underlying limitations to photosynthesis? Procedures and sources of error. Journal of Experimental Botany 54: 2393-2401.
26. Feng Z, Niu J, Zhang W, Wang X, Yao F, et al. (2011) Effects of ozone exposure on sub-tropical evergreen Cinnamomum camphora seedlings grown in different nitrogen loads. Trees-Structure and Function 25: 617-625.
27. Calatayud V, Cervero J, Calvo E, Garcia-Breijo F-J, Reig-Arminana J, et al. (2011) Responses of evergreen and deciduous Quercus species to enhanced ozone levels. Environmental Pollution 159: 55-63.
28. Noormets A, Kull O, Sober A, Kubiske ME, Karnosky DF (2010) Elevated CO2 response of photosynthesis depends on ozone concentration in aspen. Environmental Pollution 158: 992-999.
29. Oksanen E (2003) Responses of selected birch (Betula pendula Roth) clones to ozone change over time. Plant Cell and Environment 26: 875-886.
30. Richardson CJ, Sasek TW, Fendick EA, Kress LW (1992) Ozone exposure-response relationships for photosynthesis in genetic strains of loblolly pine seedlings. Forest Ecology and Management 51: 163-178.
31. Watanabe M, Yamaguchi M, Tabe C, Iwasaki M, Yamashita R, et al. (2007) Influences of nitrogen load on the growth and photosynthetic responses of Quercus serrata seedlings to O_3. Trees-Structure and Function 21: 421-432.
32. Bussotti F, Desotgiu R, Cascio C, Strasser RJ, Gerosa G, et al. (2007) Photosynthesis responses to ozone in young trees of three species with different sensitivities, in a 2-year open-top chamber experiment (Curno, Italy). Physiologia Plantarum 130: 122-135.
33. Bussotti F, Strasser RJ, Schaub M (2007) Photosynthetic behavior of woody species under high ozone exposure probed with the JIP-test: A review. Environmental Pollution 147: 430-437.
34. Novak K, Cherubini P, Saurer M, Fuhrer J, Skelly JM, et al. (2007) Ozone air pollution effects on tree-ring growth, delta C-13, visible foliar injury and leaf gas exchange in three ozone-sensitive woody plant species. Tree Physiology 27: 941-949.
35. Yan K, Chen W, He X, Zhang G, Xu S, et al. (2010) Responses of photosynthesis, lipid peroxidation and antioxidant system in leaves of Quercus mongolica to elevated O_3. Environmental and Experimental Botany 69: 198-204.
36. Fontaine V, Pelloux J, Podor M, Afif D, Gerant D, et al. (1999) Carbon fixation in Pinus halepensis submitted to ozone. Opposite response of ribulose-1,5-bisphosphate carboxylase/oxygenase and phosphoenolpyruvate carboxylase. Physiologia Plantarum 105: 187-192.
37. Noormets A, Sober A, Pell EJ, Dickson RE, Podila GK, et al. (2001) Stomatal and non-stomatal limitation to photosynthesis in two trembling aspen (Populus tremuloides Michx.) clones exposed to elevated CO_2 and/or O_3. Plant Cell and Environment 24: 327-336.
38. Eichelmann H, Oja V, Rasulov B, Padu E, Bichele I, et al. (2004) Photosynthetic parameters of birch (Betula pendula Roth) leaves growing in normal and in CO_2- and O_3-enriched atmospheres. Plant, Cell & Environment 27: 479-495.
39. Kets K, Darbah JNT, Sober A, Riikonen J, Sober J, et al. (2010) Diurnal changes in photosynthetic parameters of Populus tremuloides, modulated by elevated concentrations of CO_2 and/or O_3 and daily climatic variation. Environmental Pollution 158: 1000-1007.
40. Dann MS, Pell EJ (1989) Decline of activity and quantity of ribulose bisphosphate carboxylase oxygenase and net photosynthesis in ozone-treated potato foliage. Plant Physiology 91: 427-432.
41. Warren CR, Dreyer E, Adams MA (2003) Photosynthesis-Rubisco relationships in foliage of Pinus sylvestris in response to nitrogen supply and the proposed role of Rubisco and amino acids as nitrogen stores. Trees-Structure and Function 17: 359-366.
42. Rogers A, Humphries SW (2000) A mechanistic evaluation of photosynthetic acclimation at elevated CO_2. Global Change Biology 6: 1005-1011.
43. Ribas A, Penuelas J, Elvira S, Gimeno BS (2005) Contrasting effects of ozone under different water supplies in two Mediterranean tree species. Atmospheric Environment 39: 685-693.
44. Vitale M, Salvatori E, Loreto F, Fares S, Manes F (2008) Physiological responses of Quercus ilex leaves to water stress and acute ozone exposure under controlled conditions. Water Air and Soil Pollution 189: 113-125.
45. Strand M, Oquist G (1985) Inhibition of photosynthesis by freezing temperatures and high light levels in cold-acclimated seedlings of scots pine (pinus-sylvestris) .2. effects on chlorophyll fluorescence at room-temperature and 77-k. Physiologia Plantarum 65: 117-123.
46. Torsethaugen G, Pell EJ, Assmann SM (1999) Ozone inhibits guard cell K+ channels implicated in stomatal opening. Proceedings of the National Academy of Sciences of the United States of America 96:13577-13582.
47. Zhang WW, Niu JF, Wang XK, Tian Y, Yao FF, et al. (2011) Effects of ozone exposure on growth and photosynthesis of the seedlings of Liriodendron chinense (Hemsl.) Sarg, a native tree species of subtropical China. Photosynthetica 49: 29-36.
48. Lombardozzi D, Levis S, Bonan GB, Sparks JP (2012) Predicting photosynthesis and transpiration responses to ozone: Decoupling modeled photosynthesis and stomatal conductance. Biogeosciences 9: 3113-3130.
49. Sun GE, McLaughlin SB, Porter JH, Uddling J, Mulholland PJ, et al. (2012) Interactive influences of ozone and climate on streamflow of forested watersheds. Global Change Biology 18: 3395-3409.

Deforestation and Benthic Indicators: How Much Vegetation Cover Is Needed to Sustain Healthy Andean Streams?

Carlos Iñiguez-Armijos[1,2]*, Adrián Leiva[1], Hans-Georg Frede[2], Henrietta Hampel[3], Lutz Breuer[2]

1 Departamento de Ciencias Naturales, Universidad Técnica Particular de Loja (UTPL), Sección de Ecología, Loja, Ecuador, **2** Institute for Landscape Ecology and Resources Management (ILR), Research Centre for Bio Systems, Land Use and Nutrition (IFZ), Justus-Liebig-University Giessen, Giessen, Germany, **3** Universidad de Cuenca, Quinta Balzaín, Cuenca, Ecuador

Abstract

Deforestation in the tropical Andes is affecting ecological conditions of streams, and determination of how much forest should be retained is a pressing task for conservation, restoration and management strategies. We calculated and analyzed eight benthic metrics (structural, compositional and water quality indices) and a physical-chemical composite index with gradients of vegetation cover to assess the effects of deforestation on macroinvertebrate communities and water quality of 23 streams in southern Ecuadorian Andes. Using a geographical information system (GIS), we quantified vegetation cover at three spatial scales: the entire catchment, the riparian buffer of 30 m width extending the entire stream length, and the local scale defined for a stream reach of 100 m in length and similar buffer width. Macroinvertebrate and water quality metrics had the strongest relationships with vegetation cover at catchment and riparian scales, while vegetation cover did not show any association with the macroinvertebrate metrics at local scale. At catchment scale, the water quality metrics indicate that ecological condition of Andean streams is good when vegetation cover is over 70%. Further, macroinvertebrate community assemblages were more diverse and related in catchments largely covered by native vegetation (>70%). Our results suggest that retaining an important quantity of native vegetation cover within the catchments and a linkage between headwater and riparian forests help to maintain and improve stream biodiversity and water quality in Andean streams affected by deforestation. This research proposes that a strong regulation focused to the management of riparian buffers can be successful when decision making is addressed to conservation/restoration of Andean catchments.

Editor: Tomoya Iwata, University of Yamanashi, Japan

Funding: This research was funded through a scholarship of the Secretaria Nacional de Educación Superior, Ciencia, Tecnología e Innovación of Ecuador (SENACYT), a DFG (Deutsche Forschungsgemeinschaft) grant (FR615/28-1/-2) and by the Universidad Técnica Partucular de Loja (UTPL). The support from Henrietta Hampel was thanks to the PROMETEO Program of SENESCYT. The funders had no role in study design, data collection and analysis, decision to publish, or preparation of the manuscript.

Competing Interests: The authors have declared that no competing interests exist.

* Email: cainiguez@utpl.edu.ec

Introduction

Tropical deforestation is one of the most important topics worldwide associated with biodiversity loss and ecosystem degradation [1–3]. However, the effects of deforestation on ecological condition of tropical streams have been rarely investigated despite evidence that forest loss negatively affects these streams [4–7]. In Ecuador, deforestation for agriculture and/or urbanization is a critical threat for Andean ecosystems [8], reaching annual deforestation rates of 2.7% between 1989 and 2008 in the montane forests of southern Andes (Tapia Armijos, pers. communication). However many of these ecosystems provide essential environmental services including biodiversity conservation, carbon storage and water supply [9]. Several studies focused along the Ecuadorian Andean region have demonstrated that deforestation affects habitat structure and macroinvertebrate community assemblages in streams [10], litter breakdown rates [11] and hydrological processes [12]. Clearly the Ecuadorian montane streams are influenced by the surrounding landscape and several research groups have recently deployed considerable efforts to assess these relationships. Nonetheless, information gaps concerning the influence of land cover/use and spatial scale on stream condition remains, despite they affect water quality and aquatic diversity [13,14]. These data are crucial for management and conservation of developed catchments to insure long term sustainability [15].

In other regions, organizations such as the European Environment Agency (EEA), the United States Environmental Protection Agency (EPA) and national agencies in New Zealand have identified, and are currently considering within their monitoring programs, the impacts of land cover/use changes on stream ecological conditions based on responses of macroinvertebrate communities [16–19]. In addition, they have developed catchment and/or riparian rehabilitation projects in order to enhance stream condition [20]. In South and Central America such extensive monitoring or restoration programs by governmental initiatives are rare, despite their potential to reduce the impact of deforestation on stream condition [13]. In this regard, a Costa

Rican regulation has showed the effectiveness of protecting riparian buffers to reduce the impacts of deforestation on macroinvertebrate communities and sensitive taxa in tropical streams [4].

Several authors [10,21–23] have applied a variety of macroinvertebrate metrics (e.g. Fisher's index; EPT taxa) to assess the impact of land cover/use on the ecological condition of streams in South America focusing on the structure and composition of benthic communities. These studies have confirmed changes to aquatic biodiversity and water quality because of deforestation around the streams. Nevertheless, how does vegetation cover influences stream ecological condition? In streams running through forested catchments the macroinvertebrate communities are more heterogeneous, diversity of intolerant taxa to environmental stress is higher, and abundance and variety of habitats is greater [10,14,21]. Likewise, streams with forested riparian areas have low variation in water temperature and less nutrients and sediment loads along the entire length [4,13]. At the catchment scale, there is also an influence of vegetation cover on controlling hydrology and the entire diversity of the streams [12]. In this regard, estimating the extent of native vegetation cover in a catchment and the importance of riparian management and headwater forest protection to maintain and improve biodiversity and water quality of streams affected by deforestation is a pressing task for management and conservation [13,14,24].

Geographic Information Systems (GIS) allow researchers to explore correlations in spatial patterns with ecological features in the landscape. Allan [13] describes the ability to apply GIS tools with remote sensing data to assess the influence of land cover/use on stream ecological condition. Studies in temperate areas have described the influence of land cover and spatial scale on stream communities [25,26] and on stream water quality [14] by applying GIS based evaluation techniques to derive land cover data and to relate them with physical, chemical and biological data.

In this study, our objective was to investigate the influence of vegetation cover at three spatial scales (local, riparian buffer and catchment) on structure and composition of macroinvertebrate communities and on water quality of headwater streams in the southern Ecuadorian Andes. These streams are located in two contrasting regions of varying intensities of anthropogenic pressure (i.e. deforestation, population growth, and farming). The questions we addressed were: what spatial scale and how much vegetation cover are associated with healthy streams based on macroinvertebrate metrics? In order to address these questions, we calculated several diversity and water quality metrics from macroinvertebrate communities, derived spatial information on land cover from GIS, and related these metrics statistically to three spatial scales and stream parameters. Finally, we addressed management and conservation problems in order to help mitigating impacts of deforestation on stream ecosystems.

Methodology

For this study, the Ministry of Environment of Ecuador issued us a general research permit for every sampling site and to collect water samples and macroinvertebrate specimens. In the case of the Zamora upstream catchment (Zu), the Municipality of Loja granted a specific permission to work at sites Zu1–2, 4–5, 8 because these streams are used for the water supply of the city. To work at the San Francisco catchment (SF), the administration of the San Francisco Research Station and Nature and Culture International allowed us to access sites SF1–7. For all other sites we used the general research permit.

Most of macroinvertebrates collected in this study were identified at genus level and we do not know if any of these genera are endangered or protected, because of the lack of information for Ecuador.

Study area

We studied 23 streams of two headwater catchments of the Zamora River basin, located in the mountain forests in Ecuadorian southern Andes (Figure 1). The Zamora upstream catchment covers an area of 276.2 km^2 and ranges between 2,020 to 3,250 m a.s.l. It is the most populated catchment within the Zamora River basin with approximately 120, 000 inhabitants in the city of Loja. The San Francisco catchment is located in a sparsely populated area of the Zamora River basin with an area of 75.3 km^2 and an elevation range from 1,800 to 3,250 m a.s.l. The natural vegetation in both study catchments is evergreen mountain rain forests, humid shrub, cloud forest and páramo [27,28]. Anthropic activities in both catchments have substituted the original vegetation mainly to pasture, and to a lesser extent to cropland, pine (*Pinus*) and eucalyptus (*Eucalyptus*) plantations [29]. Urban development has only taken place in the Zamora upstream catchment and has spread along its main river. Both catchments have similar climatic conditions with a wet season from May to September and a dry season from October to April. The average annual precipitation and temperature in the Zamora upstream and San Francisco catchments are about 2,200 mm and 15.2°C respectively.

Stream measurements and macroinvertebrates sampling

We sampled 16 streams in the Zamora upstream catchment and 7 streams in the San Francisco catchment. Sampling focused on riffles with an altitudinal variation of 350 m among sites. Every site consisted of a 30–60 m reach placed at the outlet of each stream in order to capture the influences of land use upstream on macroinvertebrate communities. According to the sampling frequency applied in other studies with benthic macroinvertebrates both in Ecuador and in other Neotropical regions [4,10,11,30–34], we collected physical–chemical and macroinvertebrate data during both dry (November 2010; December 2012) and wet seasons (May 2011; April 2012) at the same stream reach. Physical–chemical parameters such as pH, temperature (*T*), specific conductance (SC), dissolved oxygen (DO), were assessed in–stream in each of the sampling periods (Table S1) with a portable YSI 556 Multi Probe System and were averaged from several measurements during the sampling campaigns. Grab samples were taken for measuring further water chemistry at the laboratory of the Universidad Técnica Particular de Loja (UTPL) using standard methods [35]: nitrate (NO_3^-), phosphate (PO_4^{3-}), biochemical oxygen demand during decomposition in a period of 5 days (BOD_5), total coliforms (TC) and turbidity (Table S1).

For sampling of macroinvertebrate communities several methods exist and each one provides different results, nevertheless a combination of sampling methods is preferable [30]. In this study, qualitative macroinvertebrates samples were taken along the stream reach applying two sampling methods: 1) collecting macroinvertebrates directly in their stream habitats from stones and leaves, and 2) using a D–frame net (500 µm mesh) for kick sampling at 3–6 points selected randomly. Collection method 1 was applied during 10 min along the reach to remove all visible macroinvertebrates from stones and collect leaves respectively. Selected stones collected were totally submerged and had an average diameter of 20 cm approximately. Leaves were stored in a plastic bag and rinsed with tap water to remove macroinvertebrates in the laboratory. For collection method 2, macroinverte-

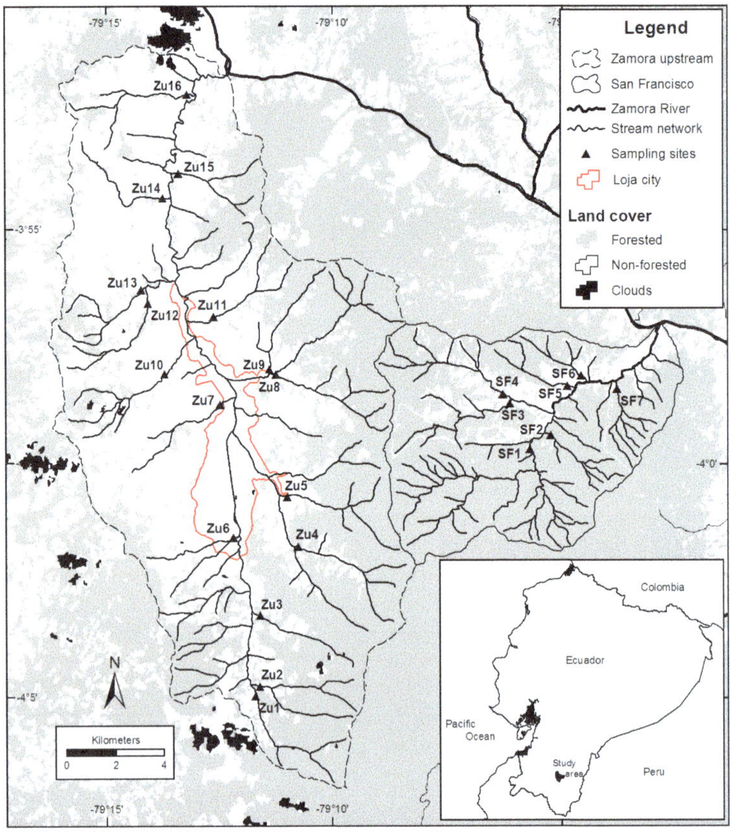

Figure 1. Study area and location of the sampling sites. Codes represent the sampling sites in the Zamora upstream (Zu1 to Zu16) and San Francisco (SF1 to SF7) catchments. Native vegetation cover was classified into forested and non-forested. Only the complete stream network for the study area is shown. Catchment characteristics are further shown in Table S1.

brates were sampled by kicking the substrate while holding the D–frame net downstream at each point. The application of this technique was based on its suitability for sampling in a variety of stream substrates and depths [36,37]. Both collections were combined into a single collection per site in order to avoid differences on macroinvertebrate communities due to seasonality of the Ecuadorian Andes [38,39]. All macroinvertebrates were preserved in 95% ethanol and brought to the laboratory to be sorted and identified to the lowest taxonomical level possible using identification keys for South American macroinvertebrates [40,41].

GIS data and vegetation cover

The Geographic Information System software ArcGIS 9.3 was used to evaluate spatial digital information. Catchments were delineated from a Digital Elevation Model (DEM) with a10 m resolution, available at the Research Unit 816 (RU 816) database (www.tropicalmountainforest.org/) of the Deutsche Forschungsgemeinschaft (DFG). Stream networks were obtained from the national cartography created by the National Mapping Agency (IGM; Instituto Geográfico Militar).

Based on Allan [13], the influence of vegetation cover (VC) on stream condition was evaluated at three spatial scales: 1) catchment scale representing the entire area upstream of the sampling site; 2) riparian scale, using a buffer of 30 m on each bank along the entire length of the stream; and 3) local scale, comprised by the same buffer width along a 100 m reach upstream of the sampling site. Buffer width chosen was based on regulations by local governments (called Municipal Ordinances) in Ecuador. These ordinances reflect several law articles of the Ecuadorian Government in terms of water resources protection, but it does not specify or recommend an amount of riparian vegetation.

A 2010 land use map provided by M.F. Tapia–Armijos (pers. communication) was used to quantify VC within each catchment. This map was generated for southern Ecuador from ASTER images (http://asterweb.jpl.nasa.gov/) with a minimum mapping unit of 1 ha and classified into eight main land use types: páramo, forest, shrub, pasture, crop, plantation, urban and un–vegetated. In Table 1 we show the land use and percentage area covered by each land use type within the 23 studied catchments, with forest and pasture classes as the most dominating ones. Urban land use was absent or very poorly represented. For this study we aggregated this classification to two land cover categories, i.e. forested and non–forested. Forest class included all types of native vegetation and non–forest class all types of disturbed areas (pasture, cropland, plantation, urban, un–vegetated). The forest class was used to determine the VC only at catchment scale, because resolution and accuracy may limit the interpretation of riparian land cover in mountainous areas [13,25,42]. In order to address the riparian and local scales, the VC within the 30 m buffer was digitized and interpreted from 2011 rectified colored orthophotographs (0.30 m resolution). These orthophotographs were obtained and provided by the Ecuadorian project SIG-TIERRAS (Ministerio de Agricultura, Ganadería, Acuacultura y

Table 1. Land use characteristics and area (%) of each land use type of the 23 studied catchments.

Catchment	Páramo	Forest	Shrub	Pasture	Plantation	Crop	Urban	Un-vegetated
Zu1	0.8	69.4	13.4	13.6	0.3	2.3	0.0	0.2
Zu2	9.3	61.1	14.4	12.5	0.5	2.1	0.0	0.0
Zu3	1.4	59.4	10.7	25.0	0.7	2.6	0.0	0.1
Zu4	2.8	46.9	19.3	25.2	0.5	5.3	0.0	0.2
Zu5	4.4	54.9	27.9	10.4	0.9	1.4	0.1	0.1
Zu6	0.3	8.8	17.6	66.0	2.2	2.9	1.5	0.7
Zu7	0.0	4.6	13.2	71.1	0.8	9.0	0.0	1.2
Zu8	2.7	39.0	25.6	30.9	0.7	1.0	0.0	0.2
Zu9	0.5	41.0	12.4	35.9	1.5	8.3	0.2	0.1
Zu10	0.4	8.0	28.0	47.1	2.5	13.4	0.0	0.6
Zu11	0.4	33.7	9.1	42.0	2.7	11.9	0.0	0.2
Zu12	0.2	17.6	4.7	57.3	2.9	16.3	0.0	0.9
Zu13	0.4	16.0	19.2	48.9	1.2	13.5	0.0	0.7
Zu14	0.0	6.5	0.1	73.2	1.9	15.0	0.0	3.3
Zu15	0.0	51.9	9.9	32.9	1.1	4.0	0.0	0.2
Zu16	0.0	15.6	0.2	72.4	2.5	7.0	0.0	2.3
SF1	0.0	82.3	5.3	5.6	0.0	0.0	0.0	6.8
SF2	2.0	93.5	4.0	0.0	0.0	0.0	0.0	0.5
SF3	0.0	78.0	5.0	15.8	0.3	0.0	0.0	0.9
SF4	1.9	87.6	3.7	6.3	0.1	0.0	0.0	0.3
SF5	0.4	81.8	3.0	14.1	0.4	0.0	0.0	0.3
SF6	0.4	76.3	2.8	18.9	1.2	0.0	0.0	0.4
SF7	2.0	92.0	3.9	2.0	0.0	0.0	0.0	0.1

Pesca; Proyecto Sistema Nacional de Información y Gestión de Tierras Rurales e Infraestructura Tecnológica).

Data analysis

We calculated nine indices (also referred as metrics) to describe ecological conditions of the streams. We calculated three structural, two compositional and four water quality indices (Table 2) that allowed us assessing stream conditions in response to increasing VC. The structural indices were richness, Fisher's index, and evenness (Pielou's index). For compositional indices we classified macroinvertebrates by functional feeding groups (FFG) according to [43,44], but we only used percentage of scrappers and shredders because they have an expected decreasing response to perturbation [36], in our case deforestation.

Water quality was assessed with the Water Quality Index (WQI), the EPT index, the Biological Monitoring Working Party index adapted to Colombia (BMWP/Col), and the %5 Dominant taxa. WQI is a single score composed of several measurements of various physical, chemical and bacteriological parameters, where low WQI values show high influence of environmental stress [45]. EPT index shows the taxa richness within the macroinvertebrate groups which are considered sensitive to stressors, and therefore should increase with increasing water quality [46]. BMWP/Col index provides a single value at the family level, representative of the organism's tolerance to stressors, where high tolerance to pollution corresponds to low BMWP scores [40]. %5 Dominant taxa is a measure of the abundance of the five most dominant taxa to the total number of organisms in the sample at each site [36]. When a community is dominated by few taxa it will have high dominance values, thus indicating the community is under the influence of environmental stress.

Two statistical approaches have been used to evaluate relationships between vegetation cover/land use and ecological indices; these are regression [14,42,47] and ordination analyses [4,10,48]. Prior to analyses, macroinvertebrate abundance data (x) were transformed by $\log_{10}(x)$ to reduce the effect of large numbers in single taxa (e.g. *Baetodes*). Rare taxa, i.e. taxa represented by three or less individuals in a sampling site, were excluded in order to avoid the influence of these taxa in the ordination analysis [49]. Because explanatory variables did not have a uniform scale, stream parameters were $\log_{10}(x+1)$ transformed in order to avoid the problem with 0 values and reduce the range of data. All of these variables were analyzed by pairs in scatterplot matrices, and through an exploratory Redundancy Analysis (RDA) to identify the least correlated and most informative variables (PO_4^{3-}, DO, TC and turbidity).

An important issue was to identify whether ecological condition of the streams was attributable to a land use type and/or a stream parameter. We used a stepwise multiple linear regression (SMLR) with calculated macroinvertebrate metrics as dependent variables against land use and PO_4^{3-}, DO, TC and turbidity to determine which of those independent variables affect our response variables. SMLR was applied with the 'stepAIC' function of 'MASS' package in R [50] using backward elimination of variables and Akaike Information Criterion for variables or predictors selection. In addition, we estimated the relative importance of predictors based on their contribution to R^2 using boostrap ('relaimpo' package). All p-values were adjusted with a Bonferroni correction for multiple comparisons during regression analyses. These analyses indicated that forest class (i.e. VC) was consistently more informative to explain macroinvertebrate metrics than other land use types or stream parameter (see Table S2 for SMLR results). Therefore, to examine the linear dependency of each of our response variables with increasing VC, macroinvertebrate metrics (dependent) were regressed against percentages of VC at each spatial scale (independent) using the method of least mean squares ('lm' function of 'stats' package).

We used a RDA ('rda' function of 'vegan' package) to identify macroinvertebrate community assemblages associated with percentages of VC, PO_4^{3-}, DO, TC and turbidity. RDA evaluates multivariate response data and is a powerful tool of constrained ordination in community ecology [51]. RDA works on a matrix Y of response data (e.g. macroinvertebrate data) and a matrix X of explanatory variables (e.g. water chemistry, VC). It identifies groups and shows the vector fitting of variables used in the analysis. While the vector fitting is commonly used for environmental interpretation in ordination diagrams, this implies a linear relationship between ordination scores and the environment

Table 2. Results of regression analysis between calculated macroinvertebrate metrics and percentage of vegetation cover at multiple spatial scales.

Spatial scale	Catchment		Buffer		Local	
Metric	R^2	p	R^2	p	R^2	p
Structural						
Richness	0.30	<0.01	0.34	<0.001	0.10	0.13
Fisher's index	0.50	<0.001	0.38	<0.001	0.01	0.30
Evenness	0.24	<0.01	0.11	0.11	0.00	0.93
Compositional						
% Scrappers	0.02	0.57	0.01	0.72	0.02	0.55
% Shredders	0.42	<0.001	0.46	<0.001	0.12	0.09
Water Quality						
WQI†	0.56	<0.001	0.50	<0.001	0.12	0.11
EPT index	0.53	<0.001	0.59	<0.001	0.03	0.39
BMWP/Col	0.44	<0.001	0.46	<0.001	0.17	0.05
%5 Dominant taxa	0.85	<0.001	0.60	<0.001	0.02	0.50

†WQI is the unique metric derived from water chemistry data showed in Table S1.

variables and may not always be appropriate if it is desired to understand the ordination of studied sites or species in relation to one or more variables. Thus, we fitted the surface of the best predictor variable, identified with an ANOVA of the RDA results, to detect how macroinvertebrate communities are grouped and spread in relation to this variable. We examined whether macroinvertebrate communities were different between those groups using an Analysis of Similarities (ANOSIM; 'anosim' function of 'vegan' package). ANOSIM is a nonparametric procedure that provides a test of whether there is a significant difference between two or more groups of sampling units.

Results

Stream condition and vegetation cover

Relationships between VC and all metrics are generally stronger at catchment and buffer scale than at local scale where associations are not apparent (Table 2). Significant correlations are found on catchment and buffer scale in 90% and 80% of all cases. Within the structural metrics, only evenness does not show significant relationship with the percentage of VC at buffer scale ($R^2 = 0.11$; $p = 0.11$). In general, correlations decrease with smaller spatial scales, disappearing completely at local scale.

As for compositional metrics, only one FFG shows a strong relationship with VC at the two larger scales ($p<0.001$). The percentage of shredders is more related to VC at catchment and buffer scales (Table 2). Scrappers show weaker regressions at all spatial scales ($R^2<0.02$). At local scale no significant correlation existed at all.

All water quality metrics have stronger relationships at catchment and buffer scale than at local scale ($R^2>0.44$ and $p<0.001$, Table 2). Within water quality metrics, the %5 Dominant taxa depicts the highest correlation across all scales with 0.85 on catchment scale. Regression slopes indicate significant differences in water quality when VC increases at both spatial scales. However, there are some streams with a large area covered by forest at catchment and buffer scale which present low values in water quality indices (Figure 2B–C; F–G). For example, in both spatial scales the EPT index increase with VC ($R^2 = 0.53$; $p<0.001$ and $R^2 = 0.59$; $p<0.001$), but there is one site with approximately 70% of forest cover whose index value is less than 50, indicating fairly poor water quality. In addition, the strongest relationship is inversely correlated between the %5 Dominant taxa and the percentage of VC in the catchment ($R^2 = 0.85$; $p = 0.001$). As a general pattern most of the macroinvertebrate metrics indicate good water quality when the percentage of native VC is higher than 70% within the catchment and the riparian buffer (see Figure 2).

Macroinvertebrate communities and vegetation cover

All structural indices are related with VC at catchment scale rather than at other spatial scales (Table 2). Thus, we decided to further analyze the community assemblage patterns related to catchment forest cover. A correlation analysis run for those variables showed scatterplot matrices and correlation values from 0 to 1.We defined a decision threshold of 0.50 to remove those variables which are correlated with more than one variable. Remaining variables include VC, DO, PO_4^{3-}, TC and turbidity. RDA reveal that 51% of the variation in the data can be explained by these five selected variables and the ANOVA identified that catchment VC is the best predictor for macroinvertebrate composition in the ordination ($p<0.001$). In Figure 3A, most variables are spread along negative axis 1 and negative axis 2, with VC, PO_4^{3-} and turbidity being the most influential factors for axis 1, while DO and TC are more associated with axis 2. Figure 3B depicts the fitted surface of the best predictor variable (i.e. VC) to detect how macroinvertebrate communities are grouped and spread in relation to VC. Thus the ordination plot illustrates two visibly distinguishable groups; one group associated to catchments with a forest cover above 70% which is even more aggregated than the other group related to lower VC. The ANOSIM indicates that both groups of macroinvertebrate assemblages were strongly dissimilar between both ranges of VC ($R = 0.55$; $p = 0.001$). Both RDA plot and Bray–Curtis distance of the ANOSIM suggest that macroinvertebrate assemblages differ in taxonomical composition when there is a decline in the percentage of VC within the catchments.

Overall, macroinvertebrate communities in the 23 study streams present specific patterns. They are composed of a total of 13,512 individuals belonging to 11 orders, 33 families and 53 genera (data available at PANGAEA – Data Publisher for Earth & Environmental Science; doi:10.1594/PANGAEA.831687). Only two taxa (taxa–1 = Empididae; taxa–2 = Pyralidae) were not identified to genera level.

Comparing occurrence and abundance patterns between both groups of catchments (i.e. less than 70% and greater than 70% VC) we detected several singularities. In catchments with a VC between 70 to 100% we collected 4,061 individuals of 11 orders. In contrast 9,451 individuals of nine orders were found in catchments with less VC (Basommatophora and Lepidoptera were absent). We found 33 families, all of them present in catchments highly forested; whilst 10 families were absent in catchments with less than 70% forest cover (Calamoceratidae, Ceratopogonidae, Coenagrionidae, Dolichopodidae, Gyrinidae, Odontoceridae, Physidae, Psephenidae, Ptilodactylidae, and Pyralidae). With regard to the 53 genera found, only *Scirtes* (Coleptera) and *Molophilus* (Diptera) were absent in catchments above 70% forest cover, and 25 genera were absent in catchments with lower forest cover showing the greatest difference between both groups of catchments at this taxonomic level. Here genera absent at <70% catchment VC were: *Physa* (Basommatophora); *Noelmis*, *Phanacerus*, *Dineutus*, *Anchytarsus*, *Psephenops*, *Elodes* (Coleoptera); *Probezzia*, *Aphrosytus*, *Hexatoma* (Diptera); *Andesiops*, *Moribaetis*, *Haplohyphes*, *Tricorythodes*, *Farrodes* (Ephemeroptera); taxa–2 (Lepidoptera); *Argia* (Odonata); *Phylloicus*, *Leptonema*, *Macrostemum*, *Grumichella*, *Nectopsyche*, *Triplectes*, *Marilia*, *Cernotina* (Tricoptera).

A trend can be observed here as well, confirmed by the ANOSIM, as there are differences in macroinvertebrate compositions between both groups of catchments (Table 3). We detect an increase for almost all metrics of the macroinvertebrate communities with an increase of the percentage of VC in the catchment, except for evenness where no changes are observed.

Discussion

Stream condition and vegetation cover

Our results show a stronger relationship between VC and the macroinvertebrate metrics at catchment and buffer scales. This suggests an important linkage between headwater and riparian forests to conserve diversity and water quality of the streams in Ecuadorian Andean catchments. We observe that stream condition begins to improve when native VC within the catchment is over 70% (see Figure 2). Therefore our suggestion is to retain at least 70% native VC within the catchment in order to keep sufficient ecological condition of the streams. Such a percentage of VC can contribute significantly to riparian management for protection or restoration of watersheds where forest has been

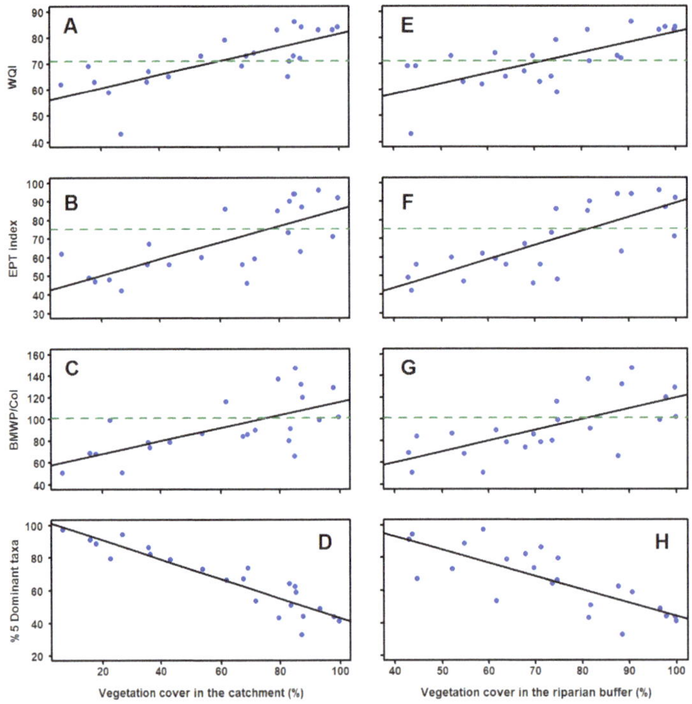

Figure 2. Water quality indices plotted against vegetation cover (%) at catchment and buffer scales. Horizontal green dashed lines represent thresholds of good water quality defined for each metric in the literature. Only the %5 Dominant index does not have ranges of tolerance, but high percentages indicate stress in the communities. For both spatial scales R^2 values are over 0.44 with a *p*-value <0.001, see Table 2 for specific values of each regression.

replaced by grasslands. Further, our findings show that local scale strategies (e.g. conserving small forest patches in riparian areas) do not improve water quality and are unlikely to improve the overall ecological condition of streams. According to [13,47], the maintenance of ecological conditions may require conservation of more than only a few native forest patches, and efforts should

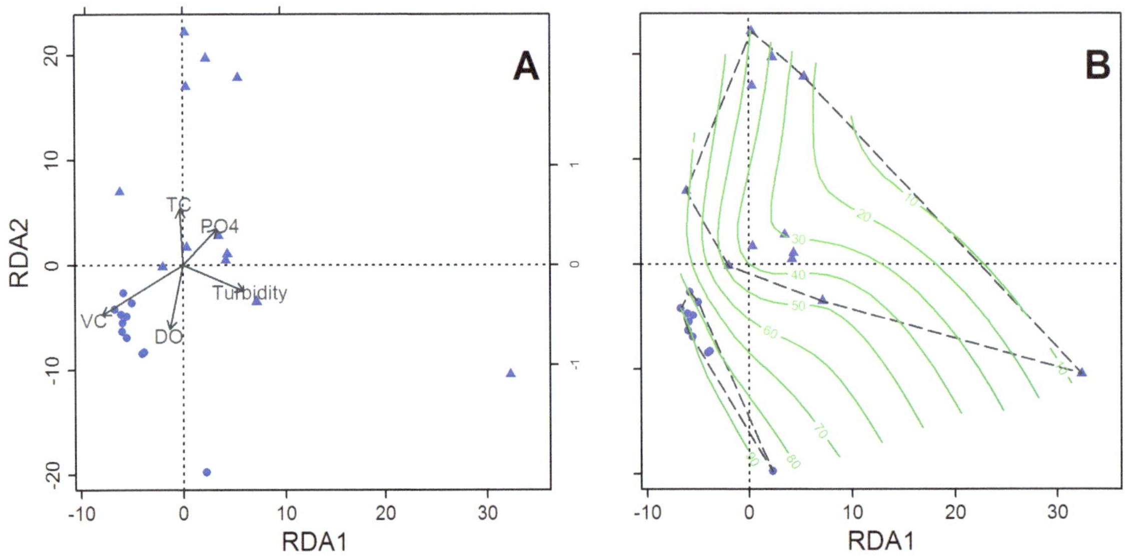

Figure 3. Redundancy Analysis (RDA) of macroinvertebrate community assemblages from catchments with different vegetation covers (VC). This ordination analysis detected two groups of macroinvertebrate assemblages: solid circles represent macroinvertebrate communities in catchments with 70–95% in VC and triangles indicate macroinvertebrate communities in catchments with less than 70% in VC. Each symbol represents the community of a single catchment, i.e. macroinvertebrate community at catchment SF3. A Shows the RDA ordination with vector fitting of five variables selected during analysis. DO = dissolved oxygen, TC = total coliforms, PO_4 = phosphate. B Displays in green curves the fitted surface only for the percentage of VC, which is the most explicative variable of the ordination (65%). Dashed lines enclose and separate the detected groups.

Table 3. Diversity patterns of macroinvertebrate communities in both groups of catchments identified in RDA ordination plot (Figure 3B). VC = catchment vegetation cover.

Vegetation cover	**<70% VC**		**>70% VC**		**Response to increasing VC**
Metric	**Mean**	**SD**	**Mean**	**SD**	
Structural					
Richness	14.3	2.1	20.1	6.2	Increase
Fisher's index	2.5	0.4	4.9	1.3	Increase
Evenness	0.7	0.1	0.7	0.1	Variable
Compositional					
% Scrappers	26.0	16.3	30.6	17.3	Increase
% Shredders	0.3	0.8	3.1	2.6	Increase

focus on considering larger spatial patterns that have more important consequences for stream conservation.

Macroinvertebrates diversity is more related to native VC at catchment and buffer scale than at a local scale. Many studies have documented the influence of upstream vegetation on stream ecological condition downstream, especially on water chemistry and habitat conditions [13,14,52]. For example, DO increases as the stream enters the forest [53]. Likewise, the similar relationships shown between water quality metrics and VC at catchment and buffer scale is not surprising given the strong longitudinal connectivity in a riverine landscape [13,54]. Riparian forests can influence water chemistry and habitat quality, which are directly linked to the macroinvertebrate metrics and help to minimize the effects of catchment deforestation on streams [4,42,55].

In this study, there were two particular sampling points in streams of catchments Zu4 and Zu8 (see Figure 1 and Table S1). Both had low values in water quality indices but a relatively high VC (60–68%). We assume that livestock and minor roads which are located parallel to the streams influence these metrics. For example, in [4] the authors found a stream affected by inputs of sediments and nutrients derived from roads and livestock, despite having a wide riparian buffer. We believe those elements are affecting the macroinvertebrate composition of very well forested catchments as discussed below.

Macroinvertebrate communities, indicators and vegetation cover

Analyzing the composition of macroinvertebrate communities of 23 Andean streams, we are able to identify direct relationships with the coverage of the catchment's native forest. The difference between macroinvertebrate assemblages increases when decreasing the percentage of native VC. We find a group with a more aggregated distribution when VC is large in the RDA plot. This pattern was also observed by [14] in a larger sample of streams in New Zealand. It supports the premise that deforestation is affecting the macroinvertebrate community assemblages in Ecuadorian Andean streams as in other Neotropical streams [4,21,56].

The difference between both groups of macroinvertebrate communities detected in the RDA is also revealed in the ANOSIM. Bücker et al. [10] found similar patterns in the San Francisco catchment having less variation at structural and compositional level and higher water quality metric values in forested than in deforested streams. Comparable results were also found in other Andean streams [11] and Amazon streams [21] of northern Ecuador. Nonetheless, among all metrics used here and mostly applied in similar studies, %5 Dominant taxa reveals a superior fitting to the regression slope despite it has not been used to reflect deforestation effects on streams before. In [36], authors denoted that high values of this metric indicate a community under environmental stress.

In terms of the FFG composition, we observe a direct response to increasing VC in both groups of macroinvertebrates (Table 3). In average, the abundance of shredders present in streams with > 70% VC in catchment is always higher than in less forested catchments. Shredders abundance is typically linked to leaf litter from riparian vegetation and it is sensitive to riparian zone impacts [43,57,58]. Shedders could be considered as an indicator of deforestation in Andean streams, despite proportion of shredders in macroinvertebrate communities in our catchments are always small (see Table 3). According to [4,5] this seems to be a general pattern in tropical streams. On average, scrapers abundance is slightly higher in more forested catchments in contrast to other Neotropical streams where their abundance is higher in less forested streams [4]. Additionally no association between scrappers and VC was detected.

Management and conservation

Allan et al. [13,47] have recommended that it is preferable to retain as much forest cover as possible in a catchment to preserve natural stream condition. However, in tropical regions like Ecuadorian Andes, deforestation has been increasing throughout the years affecting a large part of catchments along the Andes [29,59,60]. Deforestation has stopped in the upper Andean catchments because of topography and accessibility, preserving headwater forests and/or páramo [29]. However, this native vegetation in the headwaters is not enough to sustain healthy streams. In Andean countries, human activities are especially concentrated on midstream sections and valleys of catchments affecting water/habitat quality downstream and other rivers after the confluence. For example, within Guayas river basin in Ecuador there are several cities and in combination with intensive agricultural land use they are affecting the entire river network and its water quality [61]. Our results suggest that retaining over 70% native forest cover in Andean catchments now dominated by grasslands and keeping the riparian buffers, the water quality and stream biodiversity could be enhanced or at least preserved. However, this task is not so easy to achieve if there is not a strong regulation at national level.

The Ecuadorian constitution defines the right to water and promotes the protection, restoration and management of basins and water resources; but it does not mention how much forest needs to be retained in a catchment or a recommended width for riparian buffers. Best conservation actions that intend maintaining as much forest as possible have been focused on catchments for drinking water supply from initiatives of local governments (Municipalities) supported by non–governmental organizations, through the establishment of trust funds for catchment conservation. Also, local governments through Municipal Ordinances have decided to protect riparian strips with different buffer widths (15–30 m), depending if the stream is running in the city or outside the urban boundary. Riparian buffers could considerably reduce the effects of deforestation on benthic communities and water quality, but there is need for a proper management of riparian areas [4]. Unfortunately, Municipal Ordinances in Ecuador do not include the management of riparian buffers. Here, we assessed the effect of a 30 m buffer width on stream condition showing strong associations with the water quality metrics. Therefore, we also recommend protection and an effective management of native riparian forest buffers in Andean streams to assist headwater forests to sustain ecological condition of the streams.

We did not evaluate any management approaches of riparian areas, but we would like to recommend some strategies that governments can include in national regulations. We believe that the most important problems that managers will face in Andean streams are livestock density, residential areas and parallel roads along riparian zones which affect water quality and macroinvertebrate communities in forested catchments. Therefore, these problems must be addressed; for example, livestock exclusion jointly with a long–term research program to assess its effectiveness [62]; or an in–depth monitoring to further study road effects on stream and riparian vegetation [63]. Another important issue to be considered is the restoration of riparian vegetation, mainly because it enhances water quality, as our study indicates. Restoration positively affects habitat and macroinvertebrate community heterogeneity [48] and can be ecologically successful by using native plant species and avoiding the use of invasive plant species, which can affect structure and functionally of streams [64]. Finally, as mentioned by [4] we would like to emphasize that "riparian forest buffers are not a replacement for continuous forest, but can promote biodiversity and stream ecosystem integrity in tropical catchments affected by deforestation".

Supporting Information

Table S1 List of environmental variables considered in the 23 studied Andean streams. Data of water parameters are means obtained from several measurements in two sampling campaigns. Altitude was determined once at the sampling sites. Catchment area and percentage of vegetation cover (VC) at three spatial scales were calculated in ArcGIS 9.3 Geographic Information System software. T = Temperature, SC = Specific Conductance, DO = Dissolved Oxygen, NO_3^- = Nitrate, PO_4^{3-} = Phosphate, BOD_5 = Biochemical Oxygen Demand in 5 days. Turbidity in Nephelometric Turbidity Units or NTU and Total Coliforms (TC) as Colony Forming Units or CFU.

Table S2 Stepwise multiple linear regression analysis between macroinvertebrate metrics, land use and stream parameters for 23 Andean streams. Degrees of freedom for all regressions were 1, 20. AIC = Akaike Information Criterion used for selection of predictor variables. DO = Dissolved Oxygen, PO_4^{3-} = Phosphate, TC = Total Coliforms.

Acknowledgments

Special thanks are due to the personnel of the San Francisco Research Station, Nature and Culture International, Municipality of Loja and the Ministry of Environment for the provision of research facilities. Many thanks are due to C. Prieto to provide us additional information on stream parameters and macroinvertebrates of the Zamora catchment; and to L.M. Cun for his support with field sampling and stream measurements. The help of C. Romero, S. Aguilar and M. Villa with water analyses is appreciated. We would like to thank S. Bourret for proof reading, C. Espinosa for his assistance with R functions and routines, and P.F. Quintana–Ascencio and two anonymous reviewers for constructive comments on previous drafts of the manuscript.

Author Contributions

Conceived and designed the experiments: CIA HGF LB. Performed the experiments: CIA AL. Analyzed the data: CIA LB HH. Contributed reagents/materials/analysis tools: CIA HGF LB. Wrote the paper: CIA HH LB.

References

1. Laurance WF (1999) Reflections on the tropical deforestation crisis. Biol Conserv 91: 109–117.
2. Achard F, Eva HD, Stibig H-J, Mayaux P, Gallego J, et al. (2002) Determination of deforestation rates of the world's humid tropical forests. Science (80-) 297: 999–1002.
3. Lambin EF, Geist HJ, Lepers E (2003) Dynamics of l and-use and land -cover change in tropical regions. Annu Rev Environ Resour 28: 205–241.
4. Lorion CM, Kennedy BP (2009) Relationships between deforestation, riparian forest buffers and benthic macroinvertebrates in neotropical headwater streams. Freshw Biol 54: 165–180.
5. Wantzen KM, Wagner R (2006) Detritus processing by invertebrate shredders: a neotropical – temperate comparison. J North Am Benthol Soc 25: 216–232.
6. Iwata T, Nakano S, Inoue M (2003) Impacts of past riparian deforestation on stream communities in a tropical rain forest in Borneo. Ecol Appl 13: 461–473.
7. Ordóñez MV (2011) Influencia del uso del suelo y la cobertura vegetal natural en la integridad ecológica de los ríos altoandinos al noreste del Ecuador Universidad San Francisco de Quito.
8. Sierra R, Campos F, Chamberlin J (2002) Assessing biodiversity conservation priorities: ecosystem risk and representativeness in continental Ecuador. Landsc Urban Plan 59: 95–110.
9. Buytaert W, Cuesta-Camacho F, Tobón C (2011) Potential impacts of climate change on the environmental services of humid tropical alpine regions. Glob Ecol Biogeogr 20: 19–33.
10. Bücker A, Sondermann M, Frede H-G, Breuer L (2010) The influence of land-use on macroinvertebrate communities in montane tropical streams – a case study from Ecuador. Fundam Appl Limnol 177: 267–282.
11. Encalada AC, Calles J, Ferreira V, Canhoto CM, Graça MAS (2010) Riparian land use and the relationship between the benthos and litter decomposition in tropical montane streams. Freshw Biol 55: 1719–1733.
12. Crespo PJ, Feyen J, Buytaert W, Bücker A, Breuer L, et al. (2011) Identifying controls of the rainfall–runoff response of small catchments in the tropical Andes (Ecuador). J Hydrol 407: 164–174.
13. Allan JD (2004) Landscapes and riverscapes: the influence of land use on stream ecosystems. Annu Rev Ecol Evol Syst 35: 257–284.
14. Death RG, Collier KJ (2009) Measuring stream macroinvertebrate responses to gradients of vegetation cover: when is enough enough? Freshw Biol 55: 1447–1464.
15. Van Sickle J, Baker J, Herlihy A, Bayley P, Gregory S, et al. (2004) Projecting the biological condition of streams under alternative scenarios of human land use. Ecol Appl 14: 368–380.
16. Collier KJ, Hamer M (2012) The ecological condition of Waikato wadeable streams based on the Regional Ecological Monitoring of Streams (REMS) Programme. Waikato Regional Council, Technical Report, Waikato.
17. European Environment Agency (2012) European waters - assessment of status and pressures. EEA Report No 8. Copenhagen.
18. Environmental Protection Agency (2013) National Rivers and Streams Assessment 2008–2009: A Collaborative Survey. Draft February 28. Washington.
19. Perrie A, Morar S, Milne JR, Greenfield S (2012) River and stream water quality and ecology in the Wellington region: State and trends. Greater Wellington Regional Council, Publication No. GW/EMI-T-12/143, Wellington.
20. Parkyn SM, Davies-Colley RJ, Halliday NJ, Costley KJ, Croker GF (2003) Planted Riparian Buffer Zones in New Zealand: Do They Live Up to Expectations? Restor Ecol 11: 436–447.
21. Bojsen BH, Jacobsen D (2003) Effects of deforestation on macroinvertebrate diversity and assemblage structure in Ecuadorian Amazon streams. Arch für Hydrobiol 158: 317–342.

22. Miserendino ML, Casaux R, Archangelsky M, Di Prinzio CY, Brand C, et al. (2011) Assessing land-use effects on water quality, in-stream habitat, riparian ecosystems and biodiversity in Patagonian northwest streams. Sci Total Environ 409: 612–624.
23. Villamarín C, Rieradevall M, Paul MJ, Barbour MT, Prat N (2013) A tool to assess the ecological condition of tropical high Andean streams in Ecuador and Peru: The IMEERA index. Ecol Indic 29: 79–92.
24. Gergel SE, Turner MG, Miller JR, Melack JM, Stanley EH (2002) Landscape indicators of human impacts to riverine systems. Aquat Sci 64: 118–128.
25. Stewart JS, Wang L, Lyons J, Horwatich JA, Bannerman R (2001) Influences of watershed, riparian-cprridor and reach-scale characteristics on aquatic biota in agricultural watersheds. J Am Water Resour Assoc 37: 1475–1487.
26. Sutherland AB, Meyer JL, Gardiner EP (2002) Effects of land cover on sediment regime and fish assemblage structure in four southern Appalachian streams. Freshw Biol 47: 1791–1805.
27. Göttlicher D, Obregón A, Homeier J, Rollenbeck R, Nauss T, et al. (2009) Land-cover classification in the Andes of southern Ecuador using Landsat ETM+ data as a basis for SVAT modelling. Int J Remote Sens 30: 1867–1886.
28. Thies B, Meyer H, Nauss T, Bendix J (2014) Projecting land-use and land-cover changes in a tropical mountain forest of Southern Ecuador. J Land Use Sci 9: 1–33.
29. Torrachi S, Tapia MF, Escudero A, de la Cruz M (2013) Deforestación en una región montañosa megadiversa en los Andes: Dinámica del paisaje en el sur de Ecuador. In: de la Cruz M, Maestre FT, editors. Avances en el Análisis Espacial de Datos Ecológicos: Aspectos Metodológicos y Aplicados. Móstoles: ECESPA–Asociación Española de Ecología Terrestre. pp. 275–289.
30. Stein H, Springer M, Kohlmann B (2008) Comparison of two sampling methods for biomonitoring using aquatic macroinvertebrates in the Dos Novillos River, Costa Rica. Ecol Eng 34: 267–275.
31. De Cáceres M, Legendre P, Moretti M (2010) Improving indicator species analysis by combining groups of sites. Oikos 119: 1674–1684.
32. Sites RW, Willig MR, Linit MJ (2003) Macroecology of Aquatic Insects: A Quantitative Analysis of Taxonomic Richness and Composition in the Andes Mountains of Northern Ecuador. Biotropica 35: 226–239.
33. Jacobsen D (2004) Contrasting patterns in local and zonal family richness of stream invertebrates along an Andean altitudinal gradient. Freshw Biol 49: 1293–1305.
34. Tarras-Wahlberg N., Flachier A, Lane S., Sangfors O (2001) Environmental impacts and metal exposure of aquatic ecosystems in rivers contaminated by small scale gold mining: the Puyango River basin, southern Ecuador. Sci Total Environ 278: 239–261.
35. American Public Health Association (1995) Standard Methods for the Examination of Water and Wastewater. 19th ed. American Public Health Association, editor Washington DC.
36. Barbour MT, Gerritsen J, Snyder BD, Stribling JB (1999) Rapid Bioassessment Protocols for Use in Streams and Wadeable Rivers: Periphyton, Benthic Macroinvertebrates, and Fish. 2nd ed. Environmental Protection Agency.
37. Hauer FR, Resh VH (2007) Macroinvertebrates. In: Hauer FR, Lamberti GA, editors. Methods in Stream Ecology. San Diego CA: Academic Press. pp. 435–454.
38. Ríos-Touma B, Encalada AC, Prat Fornells N (2011) Macroinvertebrate Assemblages of an Andean High-Altitude Tropical Stream: The Importance of Season and Flow. Int Rev Hydrobiol 96: 667–685.
39. Jacobsen D, Encalada AC (1998) The Macroinvertebrate fauna of Ecuadorian high-land streams in the wet and dry season. Arch für Hydrobiol 142: 53–70.
40. Roldán Pérez G (2003) Bioindicación de la calidad del agua en Colombia. Uso del método BMWP/Col. 1st ed. Medellín: Editorial Universidad de Antioquia.
41. Fernádez HR, Domínguez E (2001) Guía para la determinación de los artrópodos bentónicos sudamericanos. 1st ed. Tucumán: Universidad Nacional de Tucumán.
42. Sponseller R, Benfield E, Valett H (2001) Relationships between land use, spatial scale and stream macroinvertebrate communities. Freshw Biol.
43. Cummins KW, Merritt RW, Andrade PCN (2005) The use of invertebrate functional groups to characterize ecosystem attributes in selected streams and rivers in south Brazil. Stud Neotrop Fauna Environ 40: 69–89.
44. Tomanova S, Goitia E, Helešic J (2006) Trophic levels and functional feeding groups of macroinvertebrates in neotropical streams. Hydrobiologia 556: 251–264.
45. Brown R (1970) A Water Quality Index – Do We Dare? Water Sew Work 11: 339–343.
46. Lenat DR (1988) Water quality assessment of streams using a qualitative collection method for benthic macroinvertebrates. J North Am Benthol Soc 7: 222–233.
47. Allan JD, Erickson DL, Fay J (1997) The influence of catchment land use on stream integrity across multiple spatial scales. Freshw Biol 37: 149–161.
48. Harding JS, Benfield EF, Bolstad P V, Helfman GS, Jones III EBD (1998) Stream biodiversity: The ghost of land use past. Proc Natl Acad Sci U S A 95: 14843–14847.
49. Legendre P, Gallagher E (2001) Ecologically meaningful transformations for ordination of species data. Oecologia 129: 271–280.
50. R Development Core Team (2011) R: A language and environment for statistical computing. R Foundation for Statistical Computing, Vienna. http://www.r-project.org/.
51. Bocard D, Gillet F, Legendre P (2011) Numerical Ecology with R. 1st ed. New York: Springer.
52. Harding JS, Claassen K, Evers N (2006) Can forest fragments reset physical and water quality conditions in agricultural catchments and act as refugia for forest stream invertebrates? Hydrobiologia 568: 391–402.
53. Suga CM, Tanaka MO (2012) Influence of a forest remnant on macroinvertebrate communities in a degraded tropical stream. Hydrobiologia 703: 203–213.
54. Shandas V, Alberti M (2009) Exploring the role of vegetation fragmentation on aquatic conditions: Linking upland with riparian areas in Puget Sound lowland streams. Landsc Urban Plan 90: 66–75.
55. Rios SL, Bailey RC (2006) Relationship between Riparian Vegetation and Stream Benthic Communities at Three Spatial Scales. Hydrobiologia 553: 153–160.
56. Couceiro SRM, Hamada N, Luz SLB, Forsberg BR, Pimentel TP (2006) Deforestation and sewage effects on aquatic macroinvertebrates in urban streams in Manaus, Amazonas, Brazil. Hydrobiologia 575: 271–284.
57. Lecerf A, Dobson M, Dang CK, Chauvet E (2005) Riparian plant species loss alters trophic dynamics in detritus-based stream ecosystems. Oecologia 146: 432–442.
58. Lecerf A, Richardson JS (2010) Litter decomposition can detect effects of high and moderate levels of forest disturbance on stream condition. For Ecol Manage 259: 2433–2443.
59. James LA, Marcus WA, Harden CP (2006) Human impacts on headwater fluvial systems in the northern and central Andes. Geomorphology 79: 249–263.
60. Molina A, Vanacker V, Balthazar V, Mora D, Govers G (2012) Complex land cover change, water and sediment yield in a degraded Andean environment. J Hydrol 472: 25–35.
61. Borbor-Córdova MJ, Boyer EW, McDowell WH, Hall CA (2006) Nitrogen and phosphorus budgets for a tropical watershed impacted by agricultural land use: Guayas, Ecuador. Biogeochemistry 79: 135–161.
62. Sarr D (2002) Riparian livestock exclosure research in the Western United States: A critique and some recommendations. Environ Manage 30: 516–526.
63. Jones JA, Swanson FJ, Wemple BC, Snyder KU (2000) Effects of roads on hydrology, geomorphology, and disturbance patches in stream networks. Conserv Biol 14: 76–85.
64. Hladyz S, Kajsa A, Giller PS, Woodward G (2011) Impacts of an aggressive riparian invader on community structure and ecosystem functioning in stream food webs. J Appl Ecol 48: 443–452.

13

Landscape Genetics for the Empirical Assessment of Resistance Surfaces: The European Pine Marten (*Martes martes*) as a Target-Species of a Regional Ecological Network

Aritz Ruiz-González[1,2,3]*, Mikel Gurrutxaga[2,4], Samuel A. Cushman[5], María José Madeira[1,2], Ettore Randi[3,6], Benjamin J. Gómez-Moliner[1,2]

1 Department of Zoology and Animal Cell Biology, University of the Basque Country, UPV/EHU, Vitoria-Gasteiz, Spain, **2** Systematics, Biogeography and Population Dynamics Research Group, Lascaray Research Center, University of the Basque Country, UPV/EHU, Vitoria-Gasteiz, Spain, **3** Conservation Genetics Laboratory, National Institute for Environmental Protection and Research, ISPRA, Ozzano dell'Emilia, Bologna, Italy, **4** Department of Geography, University of the Basque Country, UPV/EHU, Vitoria-Gasteiz, Spain, **5** U.S. Forest Service, Rocky Mountain Research Station, Flagstaff, AZ, United States of America, **6** Department 18/Section of Environmental Engineering, Aalborg University, Aalborg, Denmark

Abstract

Coherent ecological networks (EN) composed of core areas linked by ecological corridors are being developed worldwide with the goal of promoting landscape connectivity and biodiversity conservation. However, empirical assessment of the performance of EN designs is critical to evaluate the utility of these networks to mitigate effects of habitat loss and fragmentation. Landscape genetics provides a particularly valuable framework to address the question of functional connectivity by providing a direct means to investigate the effects of landscape structure on gene flow. The goals of this study are (1) to evaluate the landscape features that drive gene flow of an EN target species (European pine marten), and (2) evaluate the optimality of a regional EN design in providing connectivity for this species within the Basque Country (North Spain). Using partial Mantel tests in a reciprocal causal modeling framework we competed 59 alternative models, including isolation by distance and the regional EN. Our analysis indicated that the regional EN was among the most supported resistance models for the pine marten, but was not the best supported model. Gene flow of pine marten in northern Spain is facilitated by natural vegetation, and is resisted by anthropogenic landcover types and roads. Our results suggest that the regional EN design being implemented in the Basque Country will effectively facilitate gene flow of forest dwelling species at regional scale.

Editor: Jesus E. Maldonado, Smithsonian Conservation Biology Institute, United States of America

Funding: This study has been funded by the Basque Government through the Research group on "Systematics, Biogeography and Population Dynamics" (Ref. IT317-10; GIC10/76; IT575/13) and by the University of the Basque Country (UPV-EHU) and the Department of Environment, Territorial Planning, Agriculture and Fisheries (Basque Government) through IKT S.A under the University-Enterprise research program (Ref. UE07/02). Ruiz-González holds a Post doc fellowship awarded by the Dept. of Education Universities and Research of the Basque Government (Ref. DKR-2012-64). Several samples analysed in this study have been obtained in the framework of different carnivore surveys funded by regional or national administrations (Spanish Ministry of Environment, Regional Governments of Navarre and Aragon, Alava Provincial Council). The funders had no role in study design, data collection and analysis, decision to publish, or preparation of the manuscript.

Competing Interests: The authors have declared that no competing interests exist.

* Email: aritz.ruiz@ehu.es

Introduction

Long-term biodiversity conservation requires the preservation of ecological and evolutionary processes, such as gene flow, dispersal movements and population range shifts [1]. The ability of individuals to move across changing landscapes is crucial for maintaining regional populations [2,3]. The preservation of these processes requires, in turn, that landscape connectivity be preserved, especially when we take into account the synergetic effects of habitat fragmentation and climate change [1]. Landscape connectivity is defined as the degree to which landscape facilitates or impedes movement of organisms among resource patches [4]. Connectivity is species-specific and reflects the response of individuals to landscape features and the patterns of dispersal and gene flow that result from these individual responses [5]. Thus, landscape connectivity depends to a large extent on how the spatial configuration of habitat and land use interact with the movement ecology of particular species [6].

Ecological networks have been promoted as coherent systems composed of core areas linked by ecological corridors capable of facilitating the dispersal, migration and gene flow of wild species in landscapes and regions [7–9]. They are configured and managed with the objective of maintaining ecological functions and conserving biodiversity [7]. Although the development of ecological networks is based on the precautionary principle and on ecological theory [8], the absence of empirical evidence regarding their effectiveness and the difficulty in obtaining this evidence has been a focus of criticism about the extent to which they have in fact ensured landscape connectivity and increased biodiversity conservation [10,11].

In the design of ecological networks there is a need to predict regional ecological corridors and to quantify the degree of expected landscape connectivity between specific areas [3,9,11–13]. 'Least-cost modeling' is one commonly employed approach for designing ecological corridors [9,14], in which resistance values are assigned to distinct habitat or land use types and the least-cost paths (LCP) between specific locations are calculated using a geographical information system (GIS). How landscape influences effective distances between locations is calculated as the accumulated cost through the least cost paths [14,15]. However, for most organisms, setting the resistance values is a difficult process in which expert judgment and data available in the literature play an important role [16–19].

Accurate identification of the potential factors that drive gene flow in heterogenous landscapes and the scales at which they are acting is a foundation of reliable mapping of corridors [9,18]. Thus, reliable development of corridors must be based on a correct representation of the local resistance relative to the movement ecology of the organism of focus [9,18]. Landscape genetics, a research area that integrates landscape ecology, population genetics and spatial statistics, provides a valuable framework for testing the influence of landscape structure and composition on dispersal and gene flow [20,21]. It facilitates quantification of the resistance to gene flow a given landscape element poses [12,22]. Thus, one of the principal applications of landscape genetics in landscape planning and conservation biology is to empirically test and optimize resistance maps [23–26]. This facilitates the optimal design of ecological corridors [3,16,23], the detection of barriers to gene flow [27–29] and the identification of the landscape features which favour or impede dispersal [30–35].

Landscape genetics has shifted towards individual-based sampling and analysis, especially when organisms are continuously distributed [12,22]. However, sufficient sample collection for this purpose is a difficult task, especially in rare and elusive species in which sampling is a limiting factor [36]. In this context, non-invasive genetic sampling allows us to address studies of wildlife species without the need to capture or even observe them [37–40].

In 2005 a regional ecological network was established in the Basque Country (North Spain) by delimiting the ecological corridors linking forest protected areas [41]. A functional group of forest mammal species was selected to guide the development of a generic resistance map, which would, in turn, serve as a basis least-cost modeling of the network of ecological corridors linking these core areas. These mammals were considered suitable target species due to their sensitivity to recent fragmentation and homogenization dynamics in the regional landscape, such as road construction, urbanization and agrarian intensification [41,42]. The resistance map was parameterized through bibliographical review and expert opinion and was based on the assignment of different resistance levels to each land use [41]. The regional government of the Basque country incorporated that coherent ecological network as a reference for the environmental assessment of plans, programs and projects in 2005 [41]. In addition to its intrinsic internal relevance, the Basque country has been chosen for its crucial role in the regulation of biotic flows in south-western Europe [43]. This is because of its strategic location between two important biodiversity reservoirs in south-western Europe, the mountain chains of the Pyrenees and the Cantabrian Range [43–45]. Consequently, the preservation and restoration of connectivity in this transitional area between mountain ranges requires reliable knowledge about ecological responses of organisms to landscape composition and structure [45].

Among the set of functional forest mammals used in the design of the coherent regional ecological network, the European pine marten (*Martes martes*) is the most forest dependent species [46]. The pine marten is generally associated with forest habitats, mainly mature forests [46–48]. Deforestation and forest fragmentation limit the distribution and density of pine martens [48–50], which are believed to need a minimum woodland area to survive (ca. 2km^2) [47] and tend to avoid treeless areas [48,51,52]. Their occurrence patterns are affected by forest patch size, percentage of woodland cover, food abundance, sex, age class and habitat fragmentation levels [47,51]. Given their strong associations with high forest structural complexity, the species is particularly sensitive to human influences on their habitats, including habitat loss and landscape-scale effects of habitat fragmentation [48,51,53]. Nonetheless, they have also been recently reported in fragmented landscapes characterized by isolated, small forest fragments within an agricultural landscape matrix [48,51,54], suggesting they are not as obligately interior-forest dependent as previously described [51]. However, in such landscapes, linear features, such as hedgerows and small woods, play a key role to connect adjacent forest patches [48,54,55].

Consequently, the pine marten is a species which is well suited to studies focused on the effects of forest fragmentation on genetic structure and gene flow [56]. However, whether habitat characteristics that predict marten occupancy act as barriers to dispersal, influencing gene flow and population genetic structure across the landscape, is largely unknown [56].

The main objective of this research is to evaluate a large suite of alternative resistance hypotheses for the pine marten and compare the most supported empirical model with the expert-derived landscape resistance model used to parameterize the corridor network for the Basque Country. Specifically, we aim to evaluate (1) different binary landscape resistance maps which cover a gradient from greater to lesser preference of the pine marten for forest environments in order to identify which land uses favour or impede genetic interchange in the study area; and secondly (2) whether or not the resistance map with which the regional ecological network was originally designed in the Basque Country was correctly parametrized to reflect European pine marten gene flow.

If there is no effect of landscape structure on dispersal and gene flow in martens, then we expected either: (a) panmixia, where there is no genetic pattern, or (b) isolation-by-distance, where genetic differences increase with geographic distance [57]. If landscape structure influences marten dispersal, then we expected (c) isolation by landscape resistance [30]. Given that the most consistent marten-habitat relation appears to be a general association with forest habitats, and avoidance of open, non-forested habitats [47,48,51,52], we expected that open and human altered landscapes would act as a barrier for martens, and hence that landscape structure would have an effect on gene flow. In addition, we hypothesize that the intervening landscape features between forest patches (i.e., matrix) could also play a key role to substantially affect pine marten dispersal, and consequently the connectivity between forest environments [48,55,58].

Methods

Study area and spatial data

The region of the Basque Country is located in the northern Iberian Peninsula (Fig. 1) within the Atlantic and Mediterranean biogeographical regions. It comprises an area of 7,235 km^2 and has an average human population density of 298 inhabitants per square kilometer. Forests cover 28%, forestry plantations 29%,

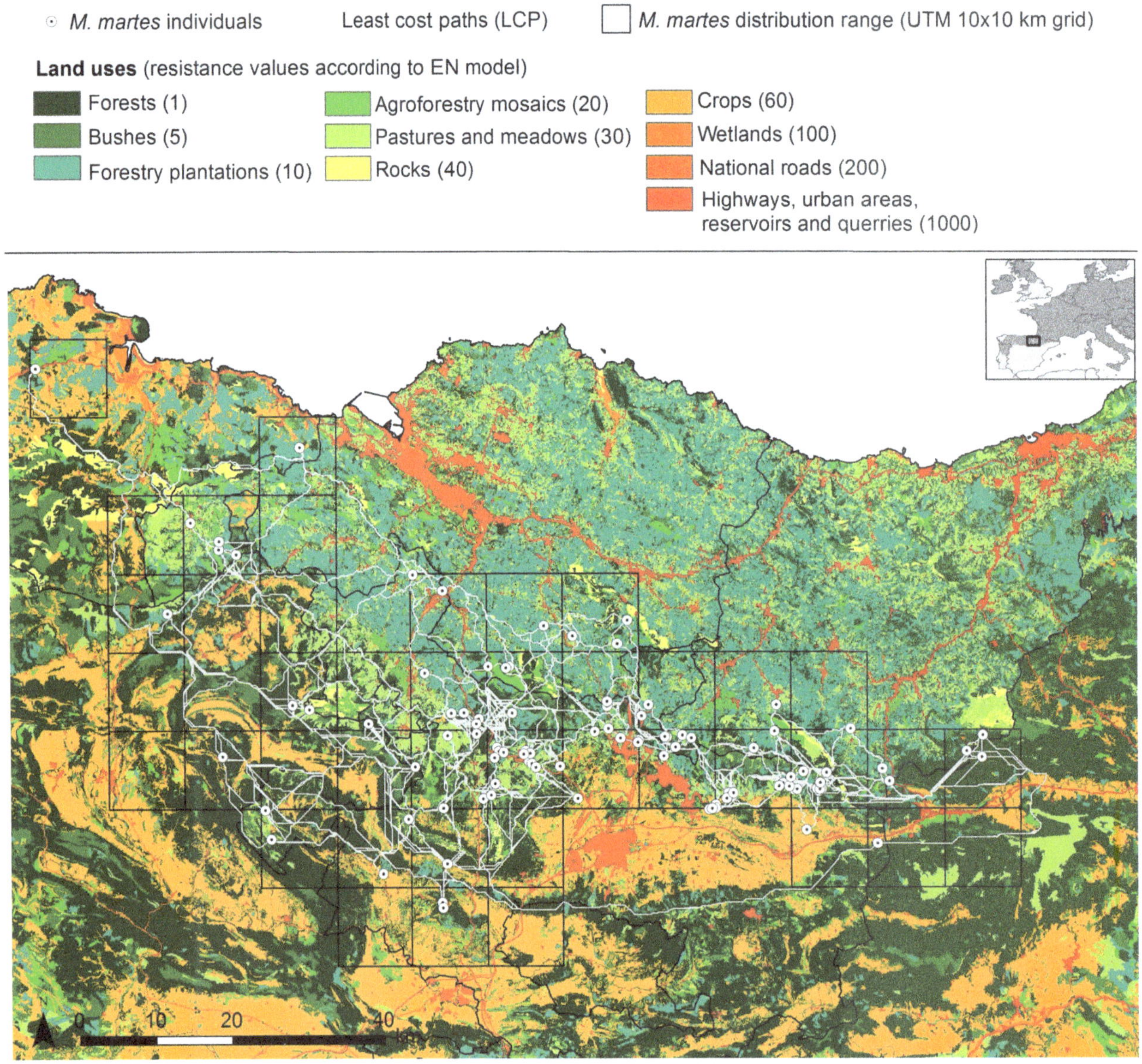

Figure 1. Ecological network resistance map (EN) and LCP analysis between European pine marten individuals in the study area. Least cost paths (LCP) obtained between the 101 pine marten individuals in accordance with the EN resistance map, analogous to that used in the design of the corridors in the ecological network of the Basque Country (North Spain) [41]. Resistance values for each land use are indicated in brackets.

non-wooded mountains 24%, cultivated land 14%, and urban land and infrastructures 5.7% of the land area, respectively.

Land use information was obtained in vector format from the most recent forest map of Spain [59] and from national road network maps [60].

Non-invasive genetic sampling and species identification

We used non-invasive scat sampling to collect genetic samples from the *Martes* sp. (*Martes martes* and *Martes foina*) in the study area between 2004 and 2010. Thus, no specific permissions were required for faecal sampling purposes, as the sampling was carried out without needing to intervene directly in the species in focus. We conducted a multi-stage sampling scheme, in which samples from a pilot study were used to assess the appropriateness of the sampling with respect to the research questions. Thus, two scat-based surveys were conducted between 2004 and 2010 across the sympatric range of both species in the study area. The first survey, conducted in 2004–2005, was used to initially estimate the distribution range of the two sympatric species of the genus *Martes* in the study area and isolate genetic samples of the focal species (*M. martes*). The second, conducted in 2006–2010, was used to refine species distribution information and to obtain a higher number of *M. martes* samples for microsatellite genotyping after a genetic species identification process [52] Aiming to homogenously cover the wide study area and obtain the highest number of different individuals we prioritized our sampling to faecal samples that were separated a minimum of 1km apart (i.e. potentially avoiding re-samplings of the same individual). We also prioritized fresh scat samples to increase genotyping success [61]. Additionally, fresh tissue specimens from road-killed pine martens were included in the data base, when possible. Tissue specimens were collected by authorized veterinarian personnel of the Wildlife

Rehabilitation Centre of Martioda (Alava Regional Council. Department of Environment. Biodiversity section), in line with the laws and ethical protocols governing wildlife management (Law 42/2007) and were submitted to Department of Zoology and Animal Cell Biology (UPV/EHU) for further DNA analyses. No animals were sacrificed for the only purposes of this study. Therefore, a formal approval by an Institutional Animal Care and Use Committee was not necessary. Universal Transversal Mercator (UTM) coordinates were recorded for all the samples collected using a global positioning system (Garmin eTtrex) [61]. The faecal samples were stored in autoclaved tubes containing ethanol 96% and frozen at −20°C until processed [52]. DNA was isolated from tissues and scat using the Qiagen DNeasy Tissue DNA (Qiagen, Hombrechtikon, Switzerland) and DNA Stool MiniKit (Qiagen, Hombrechtikon, Switzerland) according to the manufacturer's instructions, respectively. As pine marten faeces cannot be distinguished from those of the sympatric stone marten (*M. foina*), which is widespread in the study area, and can also be easily confused with those of other carnivores [62], a molecular technique was applied for the identification of faecal samples. Species identification was accomplished by a polymerase chain reaction – restriction fragment length polymorphism (PCR-RFLP) method, providing for an effective genetic identification of sympatric marten species following the method described in Ruiz-González et al. [52].

Microsatellite analyses and individual identification

Identification of individual pine martens used nuclear DNA following methods in Ruiz-González et al. [61]. All the faecal samples identified by the PCR-RFLP method [52] as pine marten were genotyped at 15 variable microsatellite loci (Table S1) using a multiplex protocol specifically designed for degraded faecal DNA analysis [61] and following a modified multitube-approach [63]. The multitube-approach of 4 independent replicates followed by a stringent criteria to construct consensus genotype (i.e. accepting heterozygotes if the two alleles were seen at least in two replicates and homozygotes if a single allele was seen at least in three replicates) is a commonly used approach in non invasive genetic studies leading to a low probability of retaining a false homozygote or false allele error (e.g. [64–66]). Briefly, DNA quality was initially screened by PCR-amplifying each DNA sample four times at four loci (Multiplex 1: MP0188; MP0059; Gg-7; Ma-1), since the results obtained for this four loci are indicative of the genotyping success for the full panel of 15 microsatellites [61].

Only samples showing> 50% positive PCRs were further amplified four times at the remaining 11 loci. Samples with ambiguous results after four amplifications per locus or with < 50% successful amplifications across loci were removed from further analysis as they were not considered reliable genotypes. Multiplex PCR products were run on an ABI (Foster City, CA) 3130XL automated sequencer (Applied Biosystems), with the internal size standard GS500 LIZ (Applied Biosystems). Fragment analyses were conducted using the ABI software Genemapper 4.0.

RELIOTYPE software [67] was used to assess genotype reliability obtained by 4 independent replicates. Samples that were not reliably typed at all loci after 4 replicates (at score threshold $R = 0.95$) were discarded from the analysis. GIMLET software v 1.3.4 [68] was used to calculate the probabilities of identity (PID and PID-sibs) so as to quantify the efficacy in discriminating the fifteen loci in combination. Consensus genotypes from four replicates were reconstructed using GIMLET, accepting heterozygotes if the two alleles were seen at least in two replicates and homozygotes if a single allele was seen at least in three replicates (e.g. [64–66]). GIMLET was also used to estimate genotyping errors: allelic dropout (ADO) and false alleles (FA) [63,69].

The raw microsatellite data and geographic coordinates of the 101 pine marten individuals are included in Table S2.

Genetic diversity and pairwise individual genetic distances

We summarized genetic variation through the number of alleles per locus (A), expected (HE) and observed (HO) heterozygosities using GENETIX v 4.05.2 [70]. Estimates of pairwise linkage disequilibria for each pair of loci and deviation from Hardy Weinberg equilibrium (HWE) genotypic proportions at each locus were tested using the exact test implemented in GenePop version 4.0 [71]. Statistical significance was evaluated by running a Markov Chain Monte Carlo (MCMC) consisting of 10,000 batches of 10,000 iterations each, with the first 10,000 iterations discarded before sampling [72]. Significance levels were adjusted with sequential Bonferroni correction in order to correct for the effect of multiple tests [73], (i.e. $\alpha = 0.05/\text{number markers}$). MICRO-CHECKER software [74] was used to check for potential scoring errors and the presence of null alleles. The Rousset's a_r inter-individual genetic distance [75] was computed using the program SPAGeDI [76] since this parameter of relatedness does not rely on a reference population [77] and has been successfully applied to infer the effect of landscape on genetic structure of continuously distributed vertebrates [32,78–80].

Construction of landscape resistance models

We produced different resistance maps representing 59 different hypotheses about the resistance of different land use types using ARCGIS version 9.3 [81], with a raster cell size set to 50 m (Table 1; File S1). As suggested by Anderson et al. [82] the sampling grain selected (i.e. 50×50 m) is adequate to infer landscape effects on gene flow as is smaller than the average home-range size of the study species (i.e.> 0.5Km^2, [47]). In addition, this resolution allows representation of small landscape patches, but also those smaller elements in the landscape that will be crucial for the resulting effective distances, including linear elements such roads and highways [14].

1) Isolation by distance: Our first hypothesis and null model was a test of isolation by distance across a uniform resistance landscape [57,83]. In this model we assumed movement could occur with equal facility in any direction, with all raster cell values equal in resistance (i.e. resistance value 1).

2) Binary Landscape resistance maps: Our second set of hypotheses propose that some land uses promote genetic connectivity for forest dependant species, such as the pine marten, that specialize in such habitats [47,48], while others resist gene flow. Thus, different binary resistance maps were developed to evaluate the specific land uses which were favourable and unfavourable to the dispersal movements of the martens (i.e. habitat *vs* non-habitat-model). As pine martens are believed to need a minimum woodland area to survive (ca. 2km^2) [47] and tend to avoid treeless areas [48,49,51], we expected to find a positive effect of closed-canopy forest habitats and negative effect of open areas and human transformed landscapes on gene flow. Thus, the different binary resistance maps created (Land_A, Land_B, Land_C, Land_D, Land_E, Land_F and Land_G) covered a gradient from greater to lesser preference of the focal species for forest environments, ranging from strictly forest land (Land_A) up to and including open spaces (Land_G) (Table 1, Fig. 2). Therefore, we classified land use data as habitat *vs.* non-habitat and parameterized the models according to a range of plausible resistance values. As there is not a general rule for the

Table 1. Resistance values corresponding to the resistance maps taken evaluated.

		Binary landscape resistance maps														Ecological Network	
		Land_A_x to Land_G_x							Land_A_x to Land_G_x								
Land uses	IBD	A_x	B_x	C_x	D_x	E_x	F_x	G_x	Ab_x	Bb_x	Cb_x	Db_x	Eb_x	Fb_x	Gb_x	EN	ENnb
Forests	1	1	1	1	1	1	1	1	1	1	1	1	1	1	1	1	1
Forestry plantations	1	X	1	1	1	1	1	1	X	1	1	1	1	1	1	10	10
Scrubland	1	X	X	1	1	1	1	1	X	X	1	1	1	1	1	5	5
Agroforestry mosaics	1	X	X	X	1	1	1	1	X	X	X	1	1	1	1	20	20
Pastures and meadows	1	X	X	X	X	1	1	1	X	X	X	X	1	1	1	30	30
Rocks	1	X	X	X	X	X	1	1	X	X	X	X	X	X	1	40	40
Crops	1	X	X	X	X	X	X	1	X	X	X	X	X	X	1	60	60
Wetlands	1	X	X	X	X	X	X	X	100	100	100	100	100	100	100	100	100
National roads	1	X	X	X	X	X	X	X	200	200	200	200	200	200	200	200	50
Highways, urban areas, reservoirs and quarries	1	X	X	X	X	X	X	X	1000	1000	1000	1000	1000	1000	1000	1000	50

Binary landscape resitance maps: 1) *Land_A-Land_G*: Binary resistance maps, on a gradient from greater to lesser preference of the focal species in relation to forest environment; 2) *Land_Ab-Land_Gb*: Maps with letter "b" include the barrier effect of national roads, highways, urban areas, reservoirs and quarries. All the models were evaluated for 4 different resistance values (X = 5, 25, 50, 100) (e.g. Land_A_5 correspond to Land_A model with resistance value of 5). *Ecological Network resistance map*: 1) *EN*: a resistance map analogous to that used in the design of the ecological network of the Basque country; *ENnb*: a variant of the latter which diminishes the barrier effect of national roads, highways, urban areas, water reservoirs and quarries.

Table 2. Results of causal modeling of landscape resistance on genetic distance in European pine marten according to Mantel and partial mantel tests.

	Model/ Resistance Values	1) Simple mantel		Rank	2) Partial mantel test		Rank	3) Partial mantel		CMS?		1) Simple mantel test		Rank	2) Partial mantel test		Rank	3) Partial mantel		CMS?
		*G*L*	*G*L*		*G*L/Dis*	*G*L/Dis*		*G*Dis/L*	G*Dis|L			*G*L*	*G*L*		*G*L/Dis*	*G*L/Dis*		*G*Dis/L*	*G*Dis/L*	
		R	p		R	p		R	p			R	p		R	*p*		R	p	
Binary Landscape Resistance Models (Land_A to Land G)	**Land_A** *Forest*										**Binary Landscape Resistance Models (Land_Ab to Land_Gb)**	**Land_Ab**								
	5	*0.221*	*0.0001*	49	0.057	0.202	54	*−0.007*	*0.877*	**N**		*0.243*	*0.0001*	28	*0.126*	*0.0005*	26	*−0.045*	*0.256*	**Y**
	25	*0.219*	*0.0001*	53	0.066	0.215	51	*0.047*	*0.317*	**N**		*0.243*	*0.0001*	29	*0.117*	*0.0138*	32	*0.008*	*0.861*	**Y**
	50	*0.217*	*0.0001*	56	0.074	0.170	49	*0.066*	*0.156*	**N**		*0.238*	*0.0001*	34	*0.112*	*0.0279*	36	*0.038*	*0.421*	**Y**
	100	*0.211*	*0.0001*	58	0.080	0.138	46	*0.089*	*0.053*	**N**		*0.229*	*0.0001*	42	*0.107*	*0.0445*	38	*0.067*	*0.159*	**Y**
	Mean (±SE)	**0.2169 (±0.0044)**			**0.0692 (±0.0099)**							**0.2381 (±0.0063)**			**0.1152 (±0.0078)**					
	Land_B *Forest + Forestry plantations)*											**Land_Bb**								
	5	*0.216*	*0.0001*	57	0.027	0.463	58	*0.011*	0.764	**N**		*0.242*	*0.0001*	30	*0.125*	*0.0001*	27	*−0.050*	*0.161*	**Y**
	25	*0.224*	*0.0001*	47	0.068	0.134	50	*0.009*	0.838	**N**		*0.245*	*0.0001*	23	*0.124*	*0.0022*	28	*−0.028*	*0.511*	**Y**
	50	*0.227*	*0.0001*	46	0.082	0.085	45	*0.024*	0.612	**N**		*0.245*	*0.0001*	25	*0.121*	*0.0083*	29	*−0.004*	*0.938*	**Y**
	100	*0.232*	*0.0001*	40	0.099	0.051	41	*0.040*	0.380	**N**		*0.241*	*0.0001*	32	*0.117*	*0.0195*	31	*0.028*	*0.539*	**Y**
	Mean (±SE)	**0.2247 (±0.0069)**			**0.0689 (±0.0309)**							**0.2432 (±0.0018)**			**0.1220 (±0.0037)**					
	Land_C *Forest + forestry plantations + Scrublands*											**Land_Cb**								
	5	*0.218*	*0.0001*	54	0.041	0.128	56	*−0.006*	0.863	**N**		*0.245*	*0.0001*	24	*0.135*	*0.0002*	21	*−0.059*	*0.105*	**Y**
	25	*0.228*	*0.0001*	45	*0.080*	*0.043*	47	*−0.009*	0.840	**Y**		*0.250*	*0.0001*	14	*0.140*	*0.0005*	18	*−0.045*	*0.305*	**Y**
	50	*0.232*	*0.0001*	41	*0.091*	*0.033*	43	*0.008*	0.868	**Y**		*0.251*	*0.0001*	11	*0.137*	*0.0022*	20	*−0.024*	*0.587*	**Y**
	100	*0.239*	*0.0001*	33	*0.111*	*0.018*	37	*0.021*	0.648	**Y**		*0.248*	*0.0001*	17	*0.129*	*0.0085*	25	*0.009*	*0.859*	**Y**
	Mean (±SE)	**0.2291 (±0.0088)**			**0.0806 (±0.0292)**							**0.2487 (±0.0028)**			**0.1349 (±0.0045)**					
	Land_D *Forest + forestry plantations + scrublands + agroforestry mosaics*											**Land_Db**								
	5	*0.220*	*0.0001*	52	0.052	0.161	55	*−0.019*	0.608	**N**		*0.246*	*0.0001*	22	*0.137*	*0.0001*	19	*−0.062*	*0.090*	**Y**
	25	*0.229*	*0.0001*	44	*0.083*	*0.041*	44	*−0.012*	0.801	**Y**		*0.251*	*0.0001*	12	*0.142*	*0.0006*	15	*−0.047*	*0.280*	**Y**
	50	*0.235*	*0.0001*	37	*0.100*	*0.038*	40	*−0.001*	0.981	**Y**		*0.253*	*0.0001*	8	*0.140*	*0.0024*	17	*−0.027*	*0.564*	**Y**
	100	*0.242*	*0.0001*	31	*0.116*	*0.026*	33	*0.017*	0.721	**Y**		*0.251*	*0.0001*	13	*0.134*	*0.0064*	22	*0.005*	*0.915*	**Y**
	Mean (±SE)	**0.2313 (±0.0095)**			**0.0877 (±0.0272)**							***0.2501 (±0.0031)***			***0.1382 (±0.0036)***					
	Land_E *Forest + forestry plantations + scrublands + agroforestry mosaics + pastures*											**Land_Eb**								
	5	*0.220*	*0.0001*	51	*0.062*	*0.049*	53	*−0.035*	0.265	**Y**		0.248	*0.0001*	19	*0.149*	*0.0001*	13	*−0.078*	*0.016*	**Y**

Table 2. Cont.

Model/ Resistance Values	1) Simple mantel		Rank	2) Partial mantel test		Rank	3) Partial mantel		CMS?	1) Simple mantel test		Rank	2) Partial mantel test		Rank	3) Partial mantel		CMS?
	*G*L*	*G*L*		*G*L/Dis*	*G*L/Dis*		*G*Dis/L*	G*Dis|L		*G*L*	*G*L*		*G*L/Dis*	*G*L/Dis*		*G*Dis/L*	*G*Dis/L*	
	R	p		R	p		R	p		R	p		R	*p*		R	p	
25	*0.234*	*0.0001*	39	*0.113*	*0.009*	35	*−0.061*	0.132	**Y**	0.255	*0.0001*	7	*0.167*	*0.0001*	6	*−0.089*	*0.015*	**Y**
50	*0.243*	*0.0001*	27	*0.133*	*0.004*	24	*−0.064*	0.142	**Y**	*0.259*	*0.0001*	**4**	*0.171*	*0.0001*	**2**	*−0.086*	*0.032*	**Y**
100	*0.253*	*0.0001*	10	*0.150*	*0.002*	11	*−0.061*	0.176	**Y**	0.261	*0.0001*	**2**	*0.170*	*0.0002*	**4**	*−0.074*	*0.097*	**Y**
Mean (±SE)	**0.2373 (±0.0138)**			**0.1146 (±0.0380)**						**0.2556 (±0.0060)**			**0.1638 (±0.0103)**					
Land_F *Forest + forestry plantations + scrublands + agroforestry mosaics + pastures + rocky areas*										**Land_Fb**								
5	*0.220*	*0.0001*	50	*0.062*	*0.044*	52	−0.035	0.266	**Y**	*0.248*	*0.0001*	20	*0.149*	*0.0001*	12	−*0.078*	*0.016*	**Y**
25	*0.234*	*0.0001*	38	*0.114*	*0.008*	34	*−0.062*	0.134	**Y**	*0.255*	*0.0001*	6	*0.167*	*0.0001*	**5**	*−0.089*	*0.014*	**Y**
50	*0.243*	*0.0001*	26	*0.134*	*0.004*	23	*−0.064*	0.140	**Y**	*0.259*	*0.0001*	**3**	*0.171*	*0.0001*	**1**	*−0.086*	*0.028*	**Y**
100	*0.253*	*0.0001*	9	*0.150*	*0.003*	10	*−0.061*	0.184	**Y**	*0.262*	*0.0001*	**1**	*0.170*	*0.0002*	**3**	*−0.074*	*0.088*	**Y**
Mean (±SE)	**0.2374 (±0.0138)**			**0.1150 (±0.0381)**						**0.2557 (±0.0060)**			**0.1640 (±0.0104)**					
Land__G *Forest + forestry plantations + scrublands + agroforestry mosaics + pastures + rocky areas + croplands*										**Land_Gb**								
5	*0.217*	*0.0001*	55	*0.041*	*0.027*	57	−0.018	0.351	**Y**	*0.249*	*0.0001*	16	*0.157*	*0.0001*	7	*−0.089*	*0.007*	**Y**
25	*0.223*	*0.0001*	48	*0.074*	*0.002*	48	−0.039	0.125	**Y**	*0.249*	*0.0001*	15	*0.157*	*0.0001*	8	*−0.087*	*0.009*	**Y**
50	*0.229*	*0.0001*	43	*0.096*	*0.000*	42	−0.051	0.077	**Y**	*0.248*	*0.0001*	18	*0.152*	*0.0001*	9	*−0.081*	*0.017*	**Y**
100	*0.237*	*0.0001*	36	*0.120*	*0.000*	30	−0.063	0.043	**Y**	*0.246*	*0.0001*	21	*0.142*	*0.0001*	16	*−0.071*	*0.036*	**Y**
Mean (±SE)	**0.2263 (±0.0082)**			**0.0827 (±0.0338)**						**0.2481 (±0.0014)**			**0.1519 (±0.0069)**					
ENnb										**EN**								
	0.237	*0.0001*	35	*0.103*	*0.032*	39	−*0.003*	0.947	**Y**	*0.256*	*0.000*	*5*	*0.146*	*0.0018*	14	*−0.030*	*0.519*	**Y**

Model definitions according to Table 1. There are 3 Mantel tests comprising causal modeling: (1) G*L—simple Mantel test between the candidate model and genetic distance; (2) G*L|Dis—partial Mantel test between the candidate model and genetic distance, partialling out Euclidean distance; (3) G*D|L—partial Mantel test between the Euclidean and genetic distance, partialling out the candidate model. For a candidate model to be supported tests (1) and (2) must be significant, while test (3) must be negative or non-significant. Mantel tests meeting each criterion are italicized. Ranking of each model according to Mantel and partial Mantel r values is included. CMS? Indicates if the model is supported within the causal modeling framework (Y) or not (N).

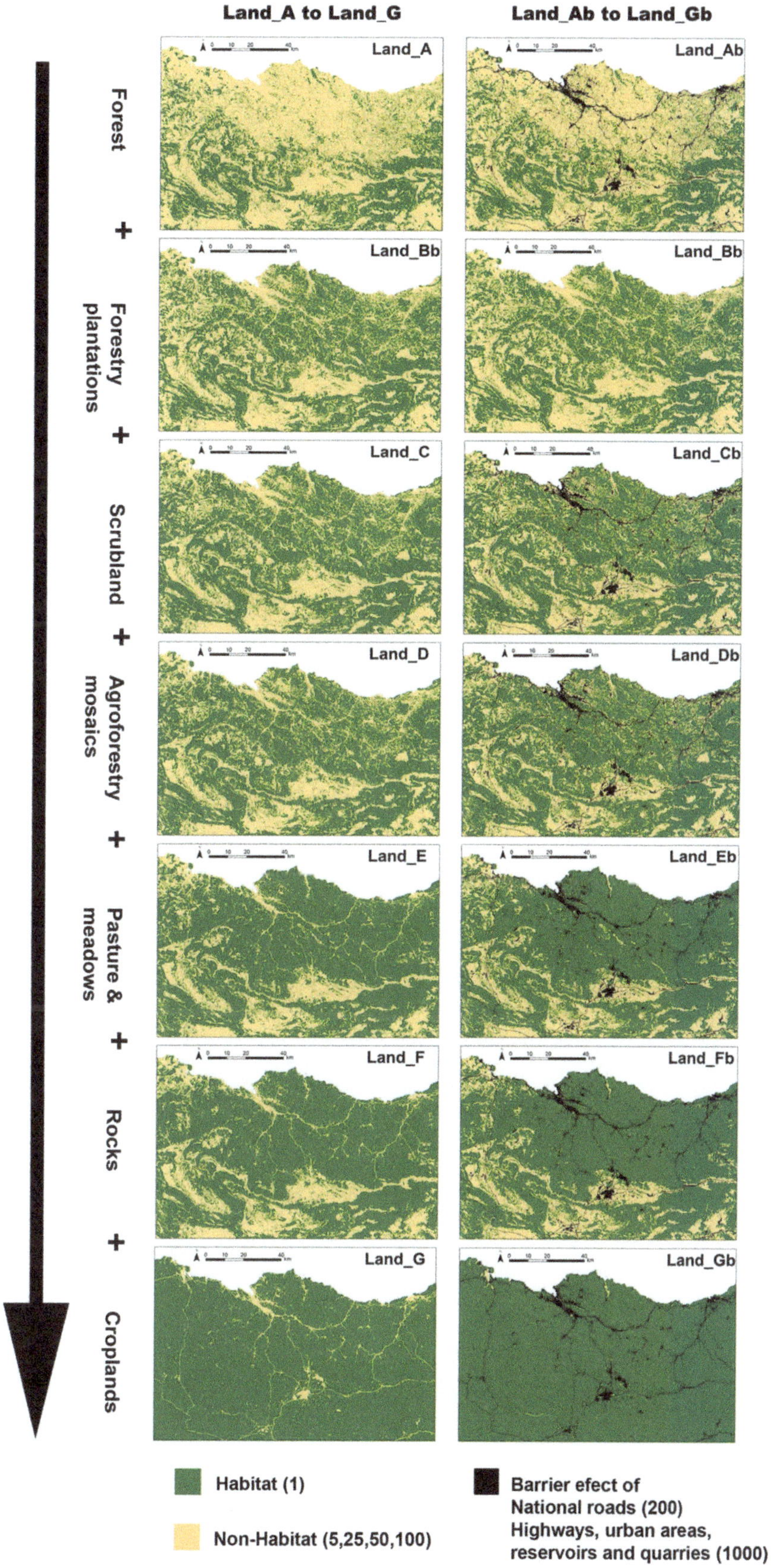

Figure 2. Binary landscape resistance maps on a gradient from greater to lesser preference of the focal species in relation to forest environment. Binary landscape resistance maps, on a gradient from greater to lesser preference of the focal species in relation to forest environment (Land_A to Land_G). Green-coloured cells represent "Habitat" (resistance value 1) and yellow-coloured cells "Non-Habitat" (resistance

values evaluated, 5, 25, 50,100). Models Land_Ab to Land_Gb additionally include black-coloured cells representing the barrier effect of national roads (resistance value 200), highways, urban areas, reservoirs and quarries (resistance value 1000).

assignment of resistance values to non-habitat, we explored 4 different resistance values (5, 25, 50, 100) of non-habitat relative to habitat to verify if it could affect the detectability of landscape genetic relationships, as it has been previously shown in both empirical [24] and simulation [84] studies. In this way, preferential land uses for dispersal were assigned a value of 1 (i.e. Habitat), while non-favourable to dispersal habitat (i.e. non-habitat) sites were assigned a value of 5, 25, 50 or 100, depending on the scenario.

For each of the resistance maps described above, we tested the effect of potential anthropogenic barriers, including national roads (resistance 200), highways, urban areas, reservoirs and quarries (resistance 1000) (Land_Ab to Land_Gb resistance maps; Table 1), using the same resistance values of the ecological network resistance map (see below), for comparative purposes [41]. As each resistance map (Land_A to Land_G; Land_Ab to Land_Gb) was explored for 4 different resistance values, we evaluated 56 different binary resistance maps (i.e. without barrier effect: Land_A_x to Land_G_x; With barrier effect: Land_Ab_x to Land_Gb_x; where X corresponds to the 4 evaluated resistances values per model: 5, 25, 50, and 100) (Table 1).

3) Ecological network resistance map: We evaluated the resistance map previously utilized in the design of ecological corridors linking forest Natura 2000 areas of the Basque Country [41]. The resistance map was based on the assignment of different resistance levels to each land use and parameterized through bibliographical review and expert opinion [41]. The resistance surface outlined in [41] was updated with the new available spatial data regarding land uses in the study area [59,60] (Table 1). Raster breaks in linear barrier elements were avoided by the reinforcement of the size of national roads and highways [14]. Sections of highways which run through viaducts or tunnels were assigned the resistance value corresponding to the land use of the surrounding area. Additionally, a second resistance map was used, with a view to testing the effect of noticeably decreasing the resistance value attributed to the potential barrier effects of national roads, highways, urban areas, reservoirs and quarries (ENnb map) (Table 1).

The raw ascii file of the EN is included in File S1. Following the resistance values outlined in Table 1, all the 59 resistance maps evaluated can be produced from the raw ascii EN resistance map (File S1).

The effective and Euclidean distances between each pair of individuals were calculated with PATHMATRIX 1.1 [15]. Pairwise effective distances between individuals were calculated as the accumulated cost through the least cost paths (LCP) throughout each resistance surface [14,15] (Fig. 1).

We proposed 59 alternative landscape models: 1) 56 binary landscape resistance maps; 2) two complex resistance maps based on the resistance surfaces used to develop the regional ecological network (EN and ENnb) (Table 1), and 3) the null model of Isolation by Distance.

Relationship between genetic and geographical distances within a reciprocal causal modeling framework

Mantel correlations between genetic distance and alternative resistance hypotheses. The pairwise genetic distances matrix (Rousset's a_r) was correlated with different matrices of geographical and (cost) distances encompassing a total number of 5151 pairwise comparisons, including: i) Euclidean distance, to determine whether the patterns of differentiation follow an isolation by distance pattern (null hypothesis) and ii) the effective distances calculated for each of the 58 resistance maps, to infer landscape structure effects on gene flow. The correlation between distance matrixes was calculated by means of the Mantel test [85] and partial Mantel tests [86] as implemented in the ECODIST package [86] in R version 2.7 (R Development Core Team 2008) with 10,000 permutations. Given the potential sensitivity of Mantel tests to non-linear relationships between genetic and cost-distances [87], we compared results between two sets of analyses, one log transforming the effective and Euclidean distances, and one using the original untransformed cost-distance matrices.

Factorial hypothesis cube randomization: Evaluation of the unimodality of support across landscape models. When hypotheses are constructed across a quantitative range of values for a parameter, it is possible to evaluate the degree to which the analysis indicates a unimodal peak of support for a global best model [30]. The degree of unimodality of model support in a factorial hypothesis cube is one measure of the reliability of model results [13]. This is done by computing the differences in support (in our case partial Mantel r values) among all neighbouring cells (i.e. different models) in the hypothesis cube and comparing the sum of those differences to the distribution of the sum of differences from a large number of randomizations of the hypothesis cube (e.g. [13]). We evaluated the unimodality of support across the 56 binary resistance hypotheses (i.e. without barrier effect: Land_Ax to Land_Gx; With barrier effect: Land_-Abx to Land_Gbx; where X corresponds to the 4 evaluated resistances values per model: 5, 25, 50, and 100) for the transformed and untransformed analyses using the randomization procedure introduced by Cushman et al. [13], in which the order of hypotheses in the hypothesis cube is randomized a large number of times and each time the difference in partial Mantel r (partialling out distance) is calculated between neighboring hypotheses in the cube. The sum of squared neighbor distances from the actual hypothesis cube (Actual Sum Differences, ASD) is then compared to the distribution of squared neighbor distances in the randomized hypothesis cubes (Mean Sum Randomized Differences, MSRD). We conducted this analysis with 1,000,000 randomizations of the hypothesis cube for both the untransformed and transformed analysis. If no randomizations produce a sum of squared neighbor distances as small as observed, it is strong evidence that the analysis has shown a strong peak of support (i.e. unimodal support).

Original causal modeling. In addition to the reciprocal causal modeling approach [84] (see below), we conducted the original causal modeling [30] as a comparative framework, in which the 58 alternative landscape resistance models are tested against the null model of isolation by distance (IBD) as described in Cushman et al. [30]. There were 3 sets of diagnostic Mantel and partial Mantel tests to complete the causal modeling. These included: (i) simple Mantel tests between genetic distance and landscape resistances; (ii) partial Mantel tests between genetic distance and landscape cost distances, partialling out the effects of Euclidean distance; (iii) partial Mantel tests between genetic distances and Euclidean distance, partialling out the effects of landscape resistance. To infer an effect of a landscape resistance scenario on dispersal, we expected (i) and (ii) to be significant, and we expected (iii) to be negative or non-significant if that scenario

a)

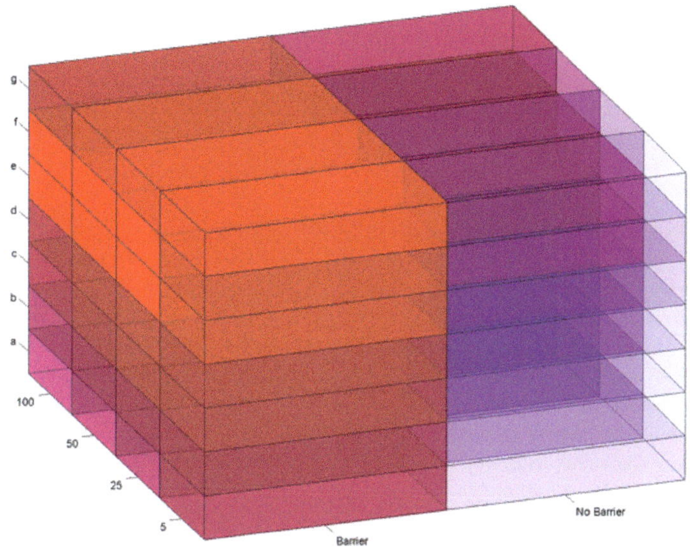

b)

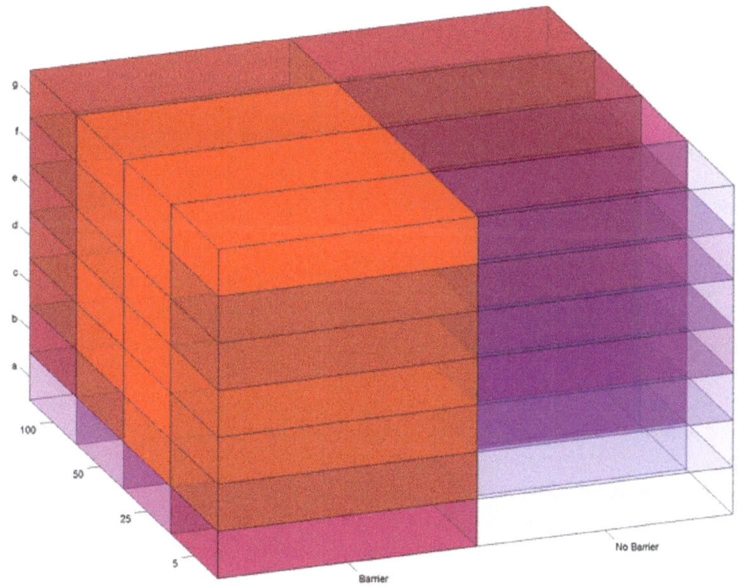

Figure 3. Factorial hypothesis cube randomization. Visualization of the 56 binary landscape-resistance hypotheses after the effects of geographical distance are partialed out on the a) log transformed and b) untransformed cost distances. The cubes each represent one of the 56 binary landscape-resistance models. The cubes are colored in a gradient from blue to red, with red being the most supported models based on the partial Mantel r value. The Mantel r values corresponding to each cube are found in Table 2 and Table S3 for the log transformed and the untransformed matrices, respectively.

'correctly' explained population connectivity in our study population [30].

Reciprocal causal modeling. We used (partial) Mantel tests in a reciprocal causal modeling framework [84] to analyse the influence of landscape structure on gene flow and to determine the extent to which possible landscape resistance models (i.e. resistance maps) explained the spatial pattern of genetic distance between individuals. Cushman and Languth [88] found that the inherent

Table 3. Factorial randomization of the hypothesis cube.

	Untransformed	Transformed
Rank	1	1
Actual Sum Differences (ASD)	19.6325	17.81
Mean Sum Randomized Differences (MSRD)	26.4738	24.17
SD error from MSRD	6.59E+04	6.97E+04

high correlation among alternative resistance models results in a high risk of spurious correlations using simple Mantel tests. Several refinements, including causal modeling [30], have been developed to reduce the risk of affirming spurious correlations and to assist model selection. However, Cushman et al. [84] showed these still suffer from elevated Type I error rates. Consequently, Cushman et al. [84] proposed "reciprocal causal modeling" which they showed greatly lessens Type I error rates in landscape genetic analysis [89]. In reciprocal causal modeling, each alternative resistance hypothesis is tested against all others with partial Mantel tests. A matrix of relative support is calculated by taking the difference between a) the partial Mantel r of each candidate model partialling out each alternative model, and b) the partial Mantel test of the alternative model partialling out the candidate model [84]. A fully supported hypothesis will have positive values of this difference with all alternative models, and no alternative models will have positive values compared to the supported model.

Results

Non-invasive sample collection and species identification

Out of 733 faecal samples collected from the entire study area, 141 were discarded because they were not fresh or because they presumably belong to the same individual (samples separated by $<$ 1km). 494 out of 592 analyzed samples were classified as *Martes* sp. (*M. martes* and *M. foina*) based on genetic species identification results. Thus, unequivocal species identification was possible in 83.45% of the samples. We effectively identified 232 faecal samples as stone marten and 262 as pine marten. Additionally, we obtained 57 tissue samples from road-killed pine martens.

Out of 262 faecal samples identified as pine marten, 108 were not included to the microsatellite genotyping procedure. These samples correspond to the sampling period from 2004–2005, which was used for a first distribution assessment of sympatric martens in the study area and were not potentially fresh enough for microsatellite analysis. Thus, 213 pine marten samples (154 faecal samples and 59 tissue samples) were used for microsatellite genotyping.

Individual identification, genotype checking and genetic diversity

The first quality-screening test, based on 4 replicates of four loci, was not passed by 73 non-invasive samples (47.40%), which were immediately discarded. The remaining 81 samples (52.59%) were amplified at the other 11 loci. After multiple-tubes genotyping, 27 samples from this sub-set (17.53% from the total analyzed samples) were then discarded because they showed $<$50% PCR success, or because of high failure rates. Full multilocus microsatellite genotypes were obtained for the remaining 54 samples (66.67% from the samples that passed the screening and 35.06% from the total samples analyzed) all showing reliability score$>$ 0.95 [67].

The observed average error rates across loci were: ADO = 0.188 and FA = 0.017. PID analysis showed that the set of 15 loci would produce an identical genotype with a probability of 1.69×10^{-10}, and with a probability of 4.45×10^{-5} for a full-sib, suggesting no "shadow effect" (i.e. all the genotypes identify distinct individuals; [90], and that matching genotypes were recaptures of the same individual).

After a regrouping procedure, we identified 42 individual genotypes from faecal samples. All of the 59 tissue samples were correctly genotyped at 15 loci and all provided new individuals. In total we identified 113 genotypes that corresponded with 101 different individuals. The number of times each individual was detected in the survey varied from 1 to 3, with a total number of 12 re-samplings. Complete genetic profiles and the geographic coordinates for the 101 pine marten individuals are included in Table S2.

The average observed (HO) and expected (HE) heterozygosity values were 0.53 and 0.58, respectively (Table S1). All 15 loci were variable with total numbers of alleles ranging between 3 and 8 per locus. The overall pine marten dataset showed a significant deficit of heterozygotes as compared to Hardy-Weinberg expectations ($p < 0.001$). Despite the broad scale of sampling, the majority of loci were in Hardy-Weinberg proportions (13 out of 15). Only loci Mp0188 and Lut-435 were out of Hardy-Weinberg proportions (Table S1). These results suggest signs of a Wahlund effect, due to the existence of an isolation by distance (Euclidean or effective) pattern in the study area. Linkage disequilibrium was not apparent for any pair of loci after performing Bonferroni corrections.

Correlation between genetic and effective distances

Factorial hypothesis cube randomization. We evaluated the unimodality of support across the 56 binary resistance maps for the log transformed and untransformed data to determine which form of the data should be used for subsequent analyses. After the effects of distance are partialled out, ranking the models by partial Mantel r value provides a means to determine which hypotheses have the greatest support and to identify the most related model to the genetic structure (Table 2, Fig. 3). According to the results outlined in Figure 3, there is a more coherent, unimodal pattern of support in the transformed analysis than the untransformed analysis. Additionally, factorial randomization of the hypothesis cube, in both the transformed and untransformed analyses, no instance of 1,000,000 randomizations produced a sum of squared differences between neighboring hypotheses (MSRD) as small as the actual sum of squared differences (ASD) in partial Mantel r values (Table 3), indicating very high unimodality in both forms of analysis. However, the transformed analysis had higher total support for optimal unimodal support of the best hypothesis as indicated by the larger number of standard errors of MSRD between neighboring hypotheses across the 1,000,000 randomizations (Table 3). Accordingly, all subsequent analyses are restricted to the log transformed resistance distances. As indicated

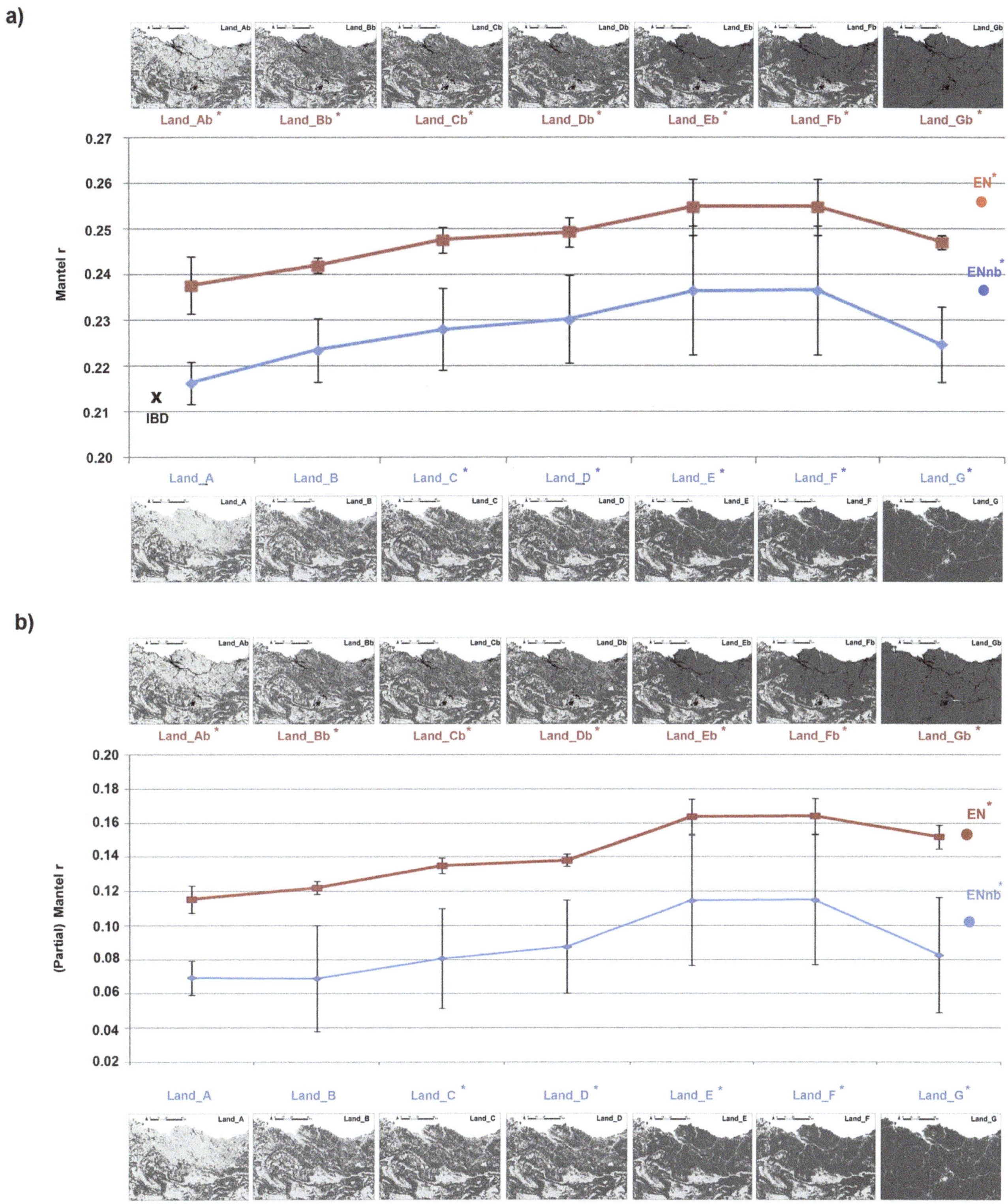

Figure 4. Mantel r results for the different landscape resistance maps evaluated (log transformed). a) Pearson correlation coefficients (Mantel r) between genetic distance and effective distance (log transformed) and **b**) Pearson correlation coefficients (Partial Mantel r) between genetic distance and effective distance (log transformed) after factoring out the effect of the Euclidean distance in the different landscape resistance maps examined. Models marked with an asterisk correspond to the models supported within the causal modeling framework [30].

by the hypothesis cube (Fig. 3), the different resistance values evaluated (5, 25, 50, 100) slightly modified the (partial) Mantel correlation results obtained for each model for both the log transformed (Table 2; Fig. S1 and Fig. S2) and the untransformed distances (Table S3; Fig. S3 and Fig. S4), but overall a consistent pattern was obtained.

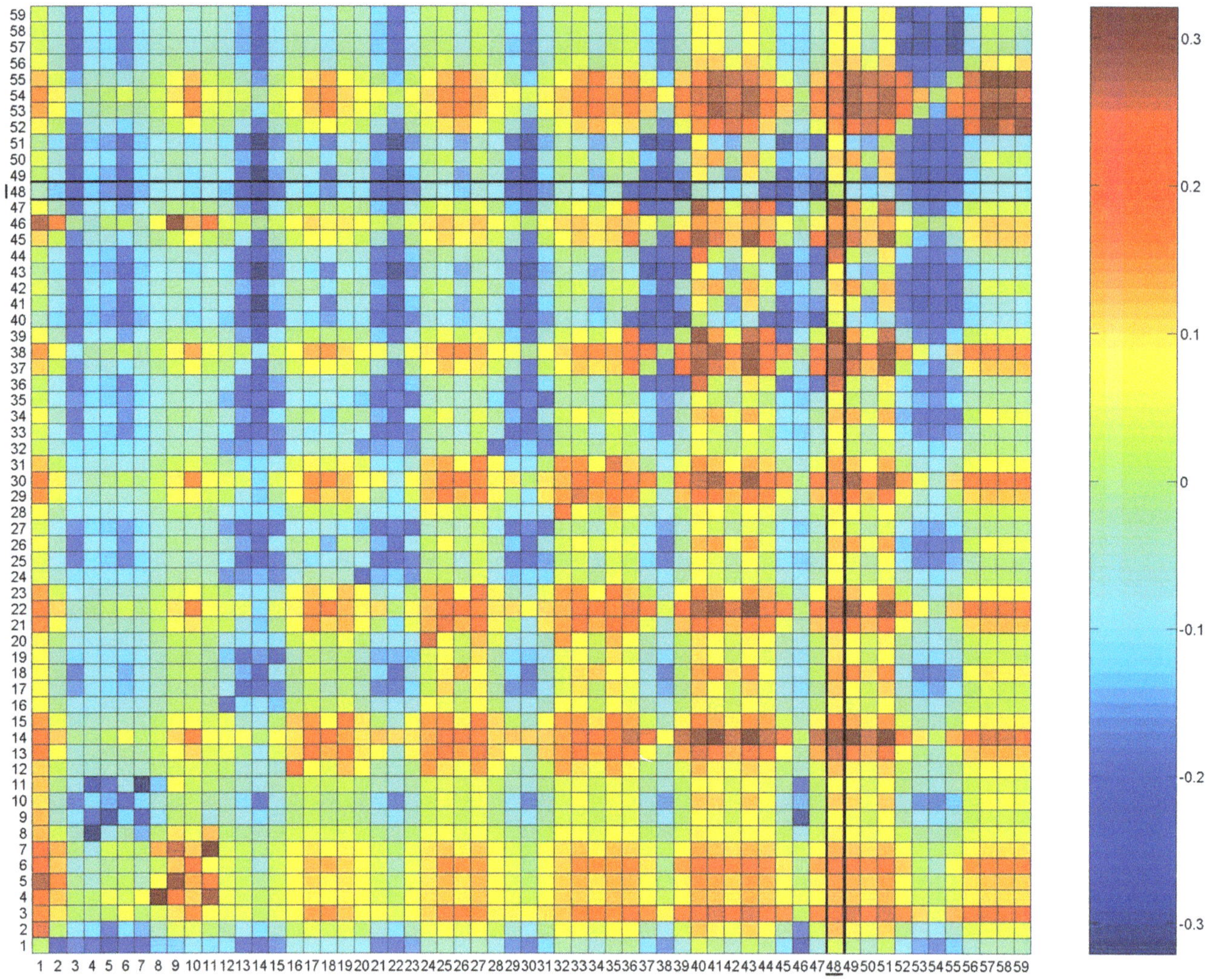

Figure 5. Reciprocal causal modeling results. Results of reciprocal causal modeling on the log transformed cost distances. A single resistance model (Model 48-Land_Fb100) is supported in analysis of the transformed cost distances. Columns indicate focal models, and rows indicate alternative models. The color gradient from blue to red indicates support for the focal model independent of the alternative model (e.g. focal model | alternative model – alternative model | focal model is positive). A fully supported model would have all positive values in the vertical dimension (e.g. that model is supported independently of all other models), and all negative values in the horizontal dimension (no other model is supported independently of the focal model). Model number and associated resistance map: 1 - EN, 2 - ENnb, 3 - Geo_dist, 4 - Land_A100, 5 - Land_A25. 6- Land_A5. 7 - Land_A50, 8 - Land_Ab100, 9 - Land_Ab25, 10 - Land_Ab5. 11 - Land_Ab50. 12 - Land_B100, 13 - Land_B25, 14 - Land_B5, 15 - Land_B50, 16 - Land_Bb100, 17 - Land_Bb25, 18 - Land_Bb5, 19 - Land_Bb50, 20 - Land_C100, 21 - Land_C25, 22- Land_C5, 23 - Land_C50, 24- Land_Cb100, 25 - Land_Cb25, 26 - Land_Cb5, 27 - Land_Cb50, 28 - Land_D100, 29 - Land_D25, 30 - Land_D5, 31 - Land_D50, 32 - Land_Db100, 33 - Land_Db25, 34 - Land_Db5, 35 - Land_Db50, 36 - Land_E100, 37 - Land_E25, 38 - Land_E5, 39 - Land_E50, 40 - Land_Eb100, 41 - Land_Eb25, 42 - Land_Eb5, 43 - Land_Eb50, 44 - Land_F100, 45 - Land_F25, 46 - Land_F5, 47 - Land_F50, 48 - Land_Fb100, 49 - Land_Fb25, 50 - Land_Fb5, 51 - Land_Fb50, 52 - Land_G100, 53 - Land_G25, 54 - Land_G5, 55 - Land_G50, 56 - Land_Gb100, 57 - Land_Gb25, 58 - Land_Gb5, 59 - Land_Gb50.

Simple Mantel correlations between genetic differentiation and alternative landscape models. *1) Isolation by distance:* A significant positive correlation was obtained between the genetic distances and Euclidean distances ($r = 0.214$; $p < 0.0001$), bearing clear witness to the existence of a pattern of isolation by distance (IBD) (Table 2, Fig. 3). However, when the models were ranked based on Mantel r, all the landscape models performed better than the null model (Table 2, Fig. 4).

2) Binary resistance maps: All of the simple Mantel tests were significant when analyzed in log-transformed form (Table 2). The correlation between genetic and effective distance gradually increased on including, in addition to natural forest (Land_A), forestry plantations (Land_B), scrublands (Land_C), agroforestry mosaics (Land_D), and pastures and meadows (Land_E) as environments favouring dispersal (Table 2, Fig. 4). This correlation did not change on including rocky areas as dispersal environments (Land_F), while it decreased on including cultivated land (Land_G). The same pattern was obtained with models which specifically increase the cost value of the main barrier features (national roads, highways, urban areas, reservoirs and quarries; Land_Ab to Land_Gb resistance maps), but with an increase of Mantel r values with respect to Land_A to Land_G models (Table 2, Fig. 4), indicating that including barrier effects due to linear features improves the resistance model. The correlation reached its maximum value on including barrier effects in resistance maps Land_E and Land_F (i.e. resistance maps

Land_Eb and Land_Fb; $r = 0.256 \pm 0.0060$; $p < 0.0001$) (Table 2, Fig. 4). The different resistance values (5, 25, 50, 100) slightly modified the correlation results obtained for each model (see Figure S2 and S4 for further details), with the highest correlations for Land_Fb100 (Table 2, Table S3; Fig. 4).

3) Ecological network resistance map: The effective distances calculated on the basis of the EN map were positively correlated with genetic distances and explained a slightly higher proportion of the observed genetic variance than the Euclidean distances (EN, $r = 0.256$; $p < 0.0001$; Table 2, Fig. 4). The degree of correlation when using the ENnb map was less than that obtained with EN, though still greater than that obtained using Euclidean distance ($r = 0.237$; $p < 0.0001$; Table 2, Fig. 4). However, the original model used in the design of the ecological network (EN), which included a higher barrier effect for national roads, highways, urban areas, reservoirs and quarries was better supported than the alternative model (ENnb).

Partial Mantel correlations between genetic differentiation and alternative resistance hypotheses. We found significant effects of nearly all of the landscape resistance models (46 out of 52), as the relationship between genetic distance and effective distance was always significant when Euclidean distance was factored out of the relationship ($p < 0.05$) (Table 2; Table S3).

1) Binary landscape resistance maps: The correlation values after factoring out the effect of Euclidean distance showed the same pattern of increase of that obtained by means of a simple Mantel test, with Land_Fb50 showing the highest partial Mantel r correlation (Table 2; Figure 4 and Fig. S2). However, Land_A_x, Land_B_x, Land_C5 and Land_D5 were not significant when the Euclidean distance was partialled out. Factorial support cubes indicate a clear unimodal peak of support in models Land_Fb50, Land_Eb50, Land_Fb100, Land_Eb100 (Figure 3) that are ranked from first to fourth according to partial r values. Similarly, these models have the highest simple Mantel r values of all of the evaluated models (Table 2; Figure 4). These results suggest that there is a strong peak of support for Land_Fb50 with a clear similarity with Land_Eb50, Land_Fb100 and Land_Eb100. Thus, the best supported models were associated with minimum resistance to movement on forest, forestry plantations, scrublands, agroforestry mosaics and pastures habitats and clear support for the barrier effect.

2) Ecological network resistance map: Both EN and ENnb models appeared better supported than the null model of IBD as this latter retained a significant positive relationship with a_r-based genetic distance after factoring out the effects of Euclidean distance, but less supported than the top resistance models (Land_Eb50, Land_Fb100, Land_Eb100; Table 2).

Original and reciprocal causal modeling. Using the original form of causal modeling approach [30] we found that all binary resistance hypotheses except Land_Ax, Land_Bx, Land_C5 and Land_D5 were supported (Table 2).

Using the novel Cushman et al. [84] method of "reciprocal causal modeling", only one resistance model out of the 59 candidate models was fully supported. The single supported model is model number 48, Land_Fb100. The reciprocal causal modeling method shows that the indexes of relative support of this model [i.e. calculated by taking the difference between 1) the partial Mantel r value of the candidate model partialling out each alternative model and 2) the partial Mantel r of the alternative model partialling out the candidate model (reciprocal partial Mantel test)] are all positive (Fig. 5). In addition to model Land_Fb100, several others are nearly perfectly supported. Land_Eb100 is supported independently of all but model Land_Fb100. Similarly, Land_Fb50 is supported independently of all except Land_Fb100 and Land_Eb100. Model Land_Eb50 is supported independently of all but Land_Fb100, Land_Fb50, and Land_Eb100. This verifies the peak of support seen in the hypothesis cube, with highest support for Land_Fb100, followed by Land_Eb100, Land_Fb50, and Land_Eb50 [i.e. models that included in addition to natural forest, forestry plantations, scrublands, agroforestry mosaics, pastures and meadows (Land Eb) and rocky areas (Land_Fb) as environments favouring dispersal and importance effect of roads as potential barriers to gene flow]. Importantly, the Ecological Network model EN was supported independently of all other models except Land_Fb100, Land_Eb100, Land_Fb50 and Land_Eb50, indicating that it is a highly effective surrogate for landscape resistance to pine marten gene flow.

Discussion

Recent studies suggested that individual-based landscape genetic analysis using partial Mantel tests in a causal modeling framework have high power to correctly identify landscape resistance as a driving process and reject spurious correlations with isolation by distance [84,88]. Cushman et al. [84] showed that the reciprocal causal modeling method we employed here substantially reduces the frequency of Type I errors. In this regards, Castillo et al. [91] recently showed that simulations support the reciprocal causal modeling with partial Mantel tests approach as an effective means to identifying the relationship between gene flow and landscape variables. Although there has been recent controversy over the use of Mantel tests in landscape genetics [92–94], a preferable alternative has yet to be identified that does not also suffer drawbacks [84]. There is no one-size- fits-all approach, and the most appropriate methodology will depend on the research question and landscape under investigation [92]. We use here the more robust modeling framework, proposed by Cushman et al. [84], that is based on the relative support of each candidate model and includes a reciprocal causal modeling step in the model optimization process. Moreover, we used the original causal modeling approach [30], factorial hypothesis cube randomization and ranking by simple Mantel test values as a comparative framework to further explore the performance of these several approaches.

Our results clearly indicate that a standard isolation-by-distance model is not sufficient to explain the observed genetic pattern, and including landscape variables through different resistance maps significantly improves the prediction of the target species gene flow. One model was supported in the transformed analysis using reciprocal causal modeling. This model, Land_Fb100, indicates that pine marten gene flow in northern Spain is facilitated by forests, forestry plantations, scrubland, agroforestry mosaics and pastures and meadows, and that crops have roughly 100 times higher resistance than optimal habitat. Further, this uniquely supported model indicates that anthropogenic barriers, such as national roads, highways, urban areas, reservoirs and quarries and wetlands likely pose much greater resistance to marten gene flow. This suggests that the population connectivity of pine martens in the study area may be vulnerable to habitat loss and fragmentation processes, due to the presence of anthropogenic barriers as has been previously suggested in other forest-dependant species [25,32,95].

The reciprocal causal modeling approach [84] clearly improved discrimination among the competing models, from only one fully supported model (i.e. Land_Fb100) versus more than 80% of models supported equivocally by the original form of causal modeling [30]. Indeed, recent studies showed that the novel reciprocal causal modeling approach is a strong improvement over other methods [91,96]. Even thought reciprocal causal modeling has a greater discrimination to detect the top model, the Cushman et al. [30] method (i.e. causal modeling + model rank + hypothesis cube) reached nearly identical conclusions. We found that ranking models based on simple or partial Mantel r gave consistent support. Likewise, our evaluation of unimodality of support in the hypothesis cube (e.g. [13,24,30]) verified the same peak of support. Reciprocal causal modeling resolved the Type I error problem that we saw in our results from the original form of causal modeling (evaluating support relative to isolation by distance), and was consistent with the peak of support seen in the factorial hypothesis cube. Thus, both forms of causal modeling are complementary and provided independent support to the obtained results, as did evaluating unimodality of support in the hypothesis cube.

In landscape genetics a large number of pairwise genetic relatedness measures have been applied to infer effects of landscape structure on gene flow [12,97], with Rousset's a_r or â [75] being one of the most widely used measures (e.g. [79,80,98,99]). Watts et al. [100] proposed a differentiation statistic (ê) that seems to improve Rousset's â performance for populations with large neighborhood-size (Dσ2) values (i.e. weak IBD pattern). However, differences in statistical performance usually highly depend on the data set used and the sampling scheme as well as how well the data set meets the underlying model assumptions [77]. Thus, and even if the results among different genetic distance measures have generally agreed [24,29,30,101] further studies based on empirical analyses of genetic patterns and simulation modeling are needed to properly evaluate the potential effect of different genetic distance estimator on disentangling landscape effects on gene flow.

Influence of land uses on gene flow: insights into pine marten ecology

The most consistent marten-habitat relation appears to be a general association with forest habitats, and avoidance of open, non-forested habitats [48,51,52]. Thus, the marten's unwillingness to cross open habitats may restrict the species' ability to disperse and colonise new forested areas [51,55]. Ruiz-González et al. [52] found that pine marten occurrence in the study area is highly dependent on the presence of forest and consequently sensitive to forest fragmentation as has been previously suggested in other studies across Europe [49,51]. Nevertheless, the presence of forest habitats is not the only factor which explains pine marten gene flow in the study area, indicating that the habitat selection and gene flow of pine martens may be driven by different factors [17,25,31].This may be because gene flow is driven by mating and dispersal events and habitat selection reflects the behaviour of individual organisms to maximize fitness within home ranges (e.g. [102]).

Our results suggest that it is not only forest masses which serve as favourable environments for dispersal. Scrubland, agroforestry mosaics and grassland habitats also potentially favour dispersal, since the correlation increases as, step by step, these environments are included as predictor variables of pine marten gene flow [Land_B(b) <Land_C(b) <Land_D(b) <Land_E(b)<Land_F(b)]. Original causal modeling identified the same pattern and suggested that Land_Eb and Land_Fb for resistance values of 50 and 100 are the most supported models. Likewise, novel reciprocal causal modeling highlighted similar results, identifying Land_Fb100 as the uniquely supported model, due to its greater discriminatory power [84].

These results are in consonance with recent ecological studies of European pine martens, based on radio tracking, which provide new data substantially differing from traditional descriptions in the scientific literature as strictly forest dependant species [48,51,55]. These studies show that martens are not exclusively confined to extensive forest patches but that they also use other patches including scrubland and agroforestry mosaics [48,51,54,55]. Indeed, the inclusion of scrub habitat in marten home ranges is likely to be related to its role in the connectivity of forest habitats [48,55]. In the same way, the improvement in correlation obtained by including pastures and meadows indicates that the species does not always avoid crossing these open spaces areas when there is forest habitat in the immediate vicinity as has been previously suggested by radiotracking data [48]. This is precisely the case in the area under study, where pastures and meadows are typically found in the immediate vicinity of forest. However, the inclusion of homogeneous croplands reduces the correlation between genetic distance and effective distance, suggesting that zones with intensive agriculture potentially impede species dispersal. This could be due to the scarcity of natural vegetation in these zones and the distance separating them from forest in the study area [41].

Additionally, models that increase the barrier effect of major roads and urban areas leads to a substantial improvement in the correlation between genetic distances and cost distances. The correlation with models Land_Ab-Land_Fb was greater than that obtained with models Land_A-Land_F, and the uniquely supported model in reciprocal causal modeling included the barrier effect (Land_Fb100). This suggests that the potential barrier effect of these land uses could have a synergic effect within a fragmented landscape, decreasing the gene flow due to road avoidance behaviour and/or road mortality [103–106].

Similar landscape genetics studies have also been conducted on other forest dependant *Martes* sp., providing contrasting results regarding landscape effects on gene flow [25,32,107–110]. Similar to the results found in this study, Broquet et al. [32], found that American marten (*Martes americana*) dispersal in Ontario is impeded by the loss and fragmentation of suitable habitat. Wasserman et al. [25] showed that gene flow in the Northern Idaho American marten population is driven by a gradient function of elevation, which was a proxy for snowpack, with marten avoiding lower elevations and dispersing in mid to high elevation montane forests. In contrast, Koen et al. [110] found that marten dispersal across Ontario can best be described as neighbour-mating with no directional bias caused by forest-management induced landscape structure, resulting in a pattern of isolation by distance, suggesting that Ontario landscape is well connected with respect to suitable marten habitat. These contrasting landscape hypotheses governing gene flow could be explained by the different limiting factors that could be acting in each of landscape under study [22].

Even though no previous individual-based landscape genetics data was available for the pine marten, Mergey et al. [58] found that genetic diversity is not associated with habitat fragmentation metrics in France, in spite of the existence of a high degree of forest fragmentation in the studied marten populations. However, this result does not demonstrate that the pine marten gene flow is not affected by forest fragmentation processes. Thus, a more detailed individual-based landscape genetics analysis (Larroque et al. Unpublished data), could provide better insights into the landscape processes governing gene flow and an interesting comparative framework with Spanish pine marten populations.

Empirical evaluation of ecological network resistance maps through landscape genetics

Maps of ecological corridors are commonly used in land use planning, but unfortunately are more often the product of expert opinion rather than empirical data [9,10]. Thus, using landscape genetic analysis, we could partially solve this limitation by studying the gene flow of a target species with regards to the resistance maps used to design the ecological networks [11,12]. Here, the parameterization found in the resistance map which was used to design the regional corridors linking forest protected areas of the Basque Country (north Spain) [41] was adequate to explain pine marten gene flow, with one of the highest partial Mantel r value ($r = 0.145$) of all the evaluated models. Based on reciprocal causal modeling only Land_Fb100, Land_Eb100, Land_Fb50 and Land_Eb50 were supported independently of EN. Even though Land_Fb100 better explains pine marten gene flow, the high mantel correlation value between the cost distances for Land_Fb100 and EN (Mantel $r = 0.9362$ $p < 0.001$) suggests EN is a good proxy. This indicates that the EN model used to develop regional connectivity networks among protected areas [41] likely performs very well as a surrogate for landscape resistance for pine marten.

Thus, the resistance map with which the regional ecological network was originally designed in the Basque Country (EN), appears to have high congruence with one of its official target species at regional scale [41]. This is a welcome finding, given that most past evaluations of expert-derived resistance values found that they performed poorly in comparison to empirically optimized models [18,24]. Given the importance of pine marten as a bio-indicator of species associated with natural vegetation [41], our results emphasize the importance of incorporating regional corridors into land use planning and management to preserve landscape connectivity for forest dwelling species.

The influence of resistance values and logarithmic transformations to detect landscape genetic relationships

Since our ability to detect the effects of landscape structure on genetic differentiation depends on both the landscape features used and the relative costs of each feature, different resistance values could provide different results [17,19,83]. Previous studies have found that the degree of contrast in resistance to gene flow in habitat as compared to non-habitat could affect whether or not a given landscape configuration will significantly affect genetic differentiation [22]. We found an increasing reliability of predictions as resistance contrast increased, with several models only supported by 25, 50 and 100 resistance values and unimodal peak of support for 50 and 100 resistance values. The hypothesis that was uniquely supported by reciprocal causal modeling indicated that non-habitat was 100 times more resistant than habitat, and anthropogenic barriers may impart additional resistance as high as 1000 times that of optimal habitat.

Some landscape genetic studies have found that the untransformed geographic distances perform better than logarithmic transformation (e.g. [24,30]), while most previously studies used transformed distances without any evaluation. However, the relationship between cost distances and genetic distances is highly dependant on the study area and the focal species. For example, at small extents the relationship between cost and genetic distances is nearly linear and the untrasformed correlations may fit the data better [24,30]. However, when the study area is large in extent relative to the dispersal ability of the species, as in the present study, the relationship between cost distance and genetic distance will be nonlinear and the logarithm transform will improve fit. Thus, taking into account the potential bias due to an incorrect use of transformations, we propose that future landscape genetics should evaluate the unimodality of support among the hypotheses as a means to determine the degree to which the transformation improves the analysis.

Conclusions

This paper presents a comprehensive individual-based landscape genetic analysis of the European pine marten, and the first formal use of landscape genetics to evaluate the effectiveness of regional ecological networks. We compared results from several methods of model selection and found that ranking based on Mantel r or partial Mantel r, the unimodality of support in the hypothesis cube, causal modeling and reciprocal causal modeling all identified the same best models of landscape resistance for European pine marten in northern Spain. Reciprocal causal modeling appeared to provide the strongest differentiation among hypotheses and enabled the identification of a single, independently supported model. Gene flow of European pine marten is facilitated by natural land cover, such as forest, scrublands and pastures and meadows, and is resisted by anthropogenic land uses and linear barriers such as major roads. We confirm that the resistance map used to develop the regional ecological network in the Basque Country is a close surrogate to the empirically optimized resistance model for marten.

Supporting Information

Figure S1 Mantel Correlation values for a) Land_A-Land_G and b) Land_Ab-Land_Gb models for the 4 different cost values evaluated on the log transformed cost distances.

Figure S2 Partial Mantel Correlation values for a) Land_A-Land_G and b) Land_Ab-Land_G models for the 4 different cost values evaluated on the log transformed cost distances.

Figure S3 Mantel Correlation values for a) Land_A-Land_G and b) Land_Ab-Land_G models for the 4 different cost values evaluated on the untransformed cost distances.

Figure S4 Partial Mantel Correlation values for a) Land_A-Land_G and b) Land_Ab-Land_G models for the 4 different cost values evaluated on the untransformed cost distances.

Table S1 Summary of the genetic variability. Genetic variability of the 15 microsatellite loci multiplexed used in this study. Number of alleles (N_A), Observed (H_O) and expected (H_E) heterozygosities for each locus and for whole data set.. Loci marked with an asterisk deviated from Hardy-Weinberg proportions.

Table S2 Sample locations and microsatellite data. Complete genetic profiles (15 microsatellite loci) and the geographic coordinates for the 101 pine marten individuals

Table S3 Results of causal modeling of landscape resistance on genetic distance in European pine marten according to Mantel and partial mantel tests for the untransformed distances.

File S1 Ecological network resistance map. Raw '.asc' file of the EN resistance map from which all the resistance maps evaluated can be produced following the resistance values outlined in Table 1.

Acknowledgments

The authors wish to thank the following persons and institutions for supplying some of the tissue and faecal samples used in this study: Patricia Lizarraga, Ricardo Gutierrez and Laura Elorza (CRF Martioda—Alava Provincial Council); Txema Fernandez and Javier Sesma (IKT S:A); the technical staff and rangers of the Natural Parks of Alava (Juan Carlos Ortíz, Ricardo Ortíz, María Elena García, Kepa García, Sonia Benítez, Elisabeth Cabanillas, Iker Ayala, Arantza Puente, Jesús Gómez, Iñaki Martínez, Lidia Lacha, Jokin Sáez de Camara); Rangers from Alava (Asier Martínez, Tomás Landa, Javier Pinedo, Mario Corral) and Bizkaia (Igor Aginako, Julio Ruiz Guijarro, Ignacio Martínez, Eneko Díaz, Juan Manuel Pérez de Ana, David Vado, Ander Eguia) Provincial Councils; Joseba Carreras (Alava Provincial Council); Jokin Larumbe (Navarre Government); Dr. Fermín Urra (GAVRN); Asun Gómez (TRAGSATEC); Dr. Madis Podra (Asoc. Visón Europeo); Nerea Ruiz de Azua (EKOS, S.L.); Gorka Belamendia (CEA); Felipe Canales and Miguel Ángel Campos (Consultora de Recursos Naturales, S.L); Enrique Arberas; Oskar Berdión; Javier López de Luzuriaga; Dr. Iñigo Zuberogoitia (ICARUS); Dr. Marta Barral (NEIKER); Rebeca Pérez; Haizea Aguirre; Pablo Pérez; Amaia García de Albéniz; Alvaro Osés; and Gerardo Domínguez. Especial thanks to Maria Vergara, Oihana Razkin, Luis Javier Chueca and Peio Lozano (UPV/EHU), the staff of IKT S.A and colleagues of the Conservation genetics laboratory (ISPRA) for their valuable help at different stages of the study.

Author Contributions

Conceived and designed the experiments: ARG MG SC ER BJGM. Performed the experiments: ARG MG MJM. Analyzed the data: ARG MG SC. Contributed reagents/materials/analysis tools: ER BJGM. Wrote the paper: ARG MG SC. Revised the manuscript: ARG MG SC MJM ER BJGM.

References

1. Opdam P, Wascher D (2004) Climate change meets habitat fragmentation: linking landscape and biogeographical scale levels in research and conservation. Biological Conservation 117: 285–297.
2. Fahrig L (2007) Non-optimal animal movement in human-altered landscapes. Functional Ecology 21: 1003–1015.
3. Cushman SA, McKelvey KS, Schwartz MK (2009) Use of Empirically Derived Source-Destination Models to Map Regional Conservation Corridors. Conservation Biology 23: 368–376.
4. Taylor PD, Fahrig L, Henein K, Merriam G (1993) Connectivity is a vital element of landscape structure. Oikos 68: 571–573.
5. Brooks CP (2003) A scalar analysis of landscape connectivity. Oikos 102: 433–439.
6. Cushman SA, Landguth EL (2012) Multi-taxa population connectivity in the Northern Rocky Mountains. Ecological Modelling 231: 101–112.
7. Bennett G, Wit P (2001) The development and application of ecological networks: a review of proposals, plans and programmes. Amsterdam: AIDEnvironment.
8. Jongman R, Pungetti G (2004) Ecological networks and greenways: concept, design, implementation. Cambridge: Cambridge University Press. pp. 368
9. Cushman S, McRae B, Adriansen F, Beier P, Shirley M, et al. (2013) Biological corridors and connectivity. In: MacDonald D, editor. Conservation in Theory and Practice. New York: Wiley. pp. 32.
10. Boitani L, Falcucci A, Maiorano L, Rondinini C (2007) Ecological networks as conceptual frameworks or operational tools in conservation. Conservation Biology 21: 1414–1422.
11. Luque S, Saura S, Fortin M-J (2012) Landscape connectivity analysis for conservation: insights from combining new methods with ecological and genetic data. Landscape Ecology 27: 153–157.
12. Segelbacher G, Cushman SA, Epperson BK, Fortin MJ, Francois O, et al. (2010) Applications of landscape genetics in conservation biology: concepts and challenges. Conservation Genetics 11: 375–385.
13. Cushman SA, Chase M, Griffin C (2010) Mapping landscape resistance to identify corridors and barriers for elephant movement in Southern Africa. In: Cushman SA, Huettmann F, editors. Spatial Complexity, Informatics and Wildlife Conservation. Tokyo: Springer. pp. 349–367.
14. Adriaensen F, Chardon JP, De Blust G, Swinnen E, Villalba S, et al. (2003) The application of 'least-cost' modelling as a functional landscape model. Landscape and Urban Planning 64: 233–247.
15. Ray N (2005) PATHMATRIX: a geographical information system tool to compute effective distances among samples. Molecular Ecology Notes 5: 177–180.
16. Beier P, Majka DR, Spencer WD (2008) Forks in the road: Choices in procedures for designing wildland linkages. Conservation Biology 22: 836–851.
17. Spear SF, Balkenhol N, Fortin MJ, McRae BH, Scribner K (2010) Use of resistance surfaces for landscape genetic studies: considerations for parameterization and analysis. Molecular Ecology 19: 3576–3591.
18. Zeller KA, McGarigal K, Whiteley AR (2012) Estimating landscape resistance to movement: a review. Landscape Ecology 27: 777–797.
19. Rayfield B, Fortin M-J, Fall A (2010) The sensitivity of least-cost habitat graphs to relative cost surface values. Landscape Ecology 25: 519–532.
20. Manel S, Schwartz MK, Luikart G, Taberlet P (2003) Landscape genetics: combining landscape ecology and population genetics. Trends in Ecology & Evolution 18: 189–197.
21. Manel S, Holderegger R (2013) Ten years of landscape genetics. Trends in Ecology & Evolution 28: 614–621.
22. Cushman SA, Shirk AJ, Landguth EL (2013) Landscape genetics and limiting factors. Conservation Genetics 14: 263–274.
23. Epps CW, Wehausen JD, Bleich VC, Torres SG, Brashares JS (2007) Optimizing dispersal and corridor models using landscape genetics. Journal of Applied Ecology 44: 714–724.
24. Shirk AJ, Wallin DO, Cushman SA, Rice CG, Warheit KI (2010) Inferring landscape effects on gene flow: a new model selection framework. Molecular Ecology 19: 3603–3619.
25. Wasserman TN, Cushman SA, Schwartz MK, Wallin DO (2010) Spatial scaling and multi-model inference in landscape genetics: *Martes americana* in northern Idaho. Landscape Ecology 25: 1601–1612.
26. Bull RAS, Cushman SA, Mace R, Chilton T, Kendall KC, et al. (2011) Why replication is important in landscape genetics: American black bear in the Rocky Mountains. Molecular Ecology 20: 1092–1107.
27. Zalewski A, Piertney SB, Zalewska H, Lambin X (2009) Landscape barriers reduce gene flow in an invasive carnivore: geographical and local genetic structure of American mink in Scotland. Molecular Ecology 18: 1601–1615.
28. Perez-Espona S, McLeod JE, Franks NR (2012) Landscape genetics of a top neotropical predator. Molecular Ecology 21: 5969–5985.
29. Blair C, Weigel DE, Balazik M, Keeley ATH, Walker FM, et al. (2012) A simulation-based evaluation of methods for inferring linear barriers to gene flow. Molecular Ecology Resources 12: 822–833.
30. Cushman SA, McKelvey KS, Hayden J, Schwartz MK (2006) Gene flow in complex landscapes: Testing multiple hypotheses with causal modeling. American Naturalist 168: 486–499.
31. Centeno-Cuadros A, Roman J, Delibes M, Antonio Godoy J (2011) Prisoners in Their Habitat? Generalist Dispersal by Habitat Specialists: A Case Study in Southern Water Vole (*Arvicola sapidus*). Plos One 6.
32. Broquet T, Ray N, Petit E, Fryxell JM, Burel F (2006) Genetic isolation by distance and landscape connectivity in the American marten (*Martes americana*). Landscape Ecology 21: 877–889.
33. Wang YH, Yang KC, Bridgman CL, Lin LK (2008) Habitat suitability modelling to correlate gene flow with landscape connectivity. Landscape Ecology 23: 989–1000.
34. Schwartz MK, Copeland JP, Anderson NJ, Squires JR, Inman RM, et al. (2009) Wolverine gene flow across a narrow climatic niche. Ecology 90: 3222–3232.
35. Blair C, Jimenez Arcos VH, Mendez de la Cruz FR, Murphy RW (2013) Landscape Genetics of Leaf-Toed Geckos in the Tropical Dry Forest of Northern Mexico. Plos One 8.
36. Beja-Pereira A, Oliveira R, Alves PC, Schwartz MK, Luikart G (2009) Advancing ecological understandings through technological transformations in noninvasive genetics. Molecular Ecology Resources 9: 1279–1301.
37. Taberlet P, Waits LP, Luikart G (1999) Noninvasive genetic sampling: look before you leap. Trends in Ecology & Evolution 14: 323–327.
38. Piggott MP, Taylor AC (2003) Remote collection of animal DNA and its applications in conservation management and understanding the population biology of rare and cryptic species. Wildlife Research 30: 1–13.
39. Waits LP, Paetkau D (2005) Noninvasive genetic sampling tools for wildlife biologists: A review of applications and recommendations for accurate data collection. Journal of Wildlife Management 69: 1419–1433.

40. Schwartz MK, Monfort SL (2008) Genetic and Endocrine Tools for Carnivore Surveys. In: Long R, MacKay P, Ray J, Zielinski W, editors. Noninvasive survey methods for North American carnivores. Washington D.C.: Island Press. pp. 228–250.
41. Gurrutxaga M, Lozano PJ, del Barrio G (2010) GIS-based approach for incorporating the connectivity of ecological networks into regional planning. Journal for Nature Conservation 18: 318–326.
42. Jongman RHG (2002) Homogenisation and fragmentation of the European landscape: ecological consequences and solutions. Landscape and Urban Planning 58: 211–221.
43. Jongman R, Bouwma I, Van Doorn A (2006) Indicative map of the Pan-European ecological network in Western Europe. Wageningen: Alterra. 104 p.
44. Gurrutxaga M, Rubio L, Saura S (2011) Key connectors in protected forest area networks and the impact of highways: A transnational case study from the Cantabrian Range to the Western Alps (SW Europe). Landscape and Urban Planning 101: 310–320.
45. Mallarach J, Rafa M, Sargatal J (2010) Cantabrian Mountains-Pyrénées-Massif Central-Western Alps great mountain corridor. In: Worboys GL, Francis WL, Lockwood M, editors. Connectivity conservation management A global guide London: Earthscan. pp. 269–279.
46. Proulx G, Aubry K, Birks J, Buskirk S, Fortin C, et al. (2004) World distribution and status of the genus Martes in 2000. In: Harrison D, Fuller A, Proulx G, editors. Martens and fishers (*Martes*) in human- altered environments: an international perspective. New York: Springer-Verlag. pp. 77–98.
47. Zalewski A, Jedrzejewski W (2006) Spatial organisation and dynamics of the pine marten *Martes martes* population in Bialowieza Forest (E Poland) compared with other European woodlands. Ecography 29: 31–43.
48. Pereboom V, Mergey M, Villerette N, Helder R, Gerard JF, et al. (2008) Movement patterns, habitat selection, and corridor use of a typical woodland-dweller species, the European pine marten (*Martes martes*), in fragmented landscape. Canadian Journal of Zoology-Revue Canadienne De Zoologie 86: 983–991.
49. Brainerd SM, Rolstad J (2002) Habitat selection by Eurasian pine martens *Martes martes* in managed forests of southern boreal Scandinavia. Wildlife Biology 8: 289–297.
50. Kurki S, Nikula A, Helle P, Linden H (1998) Abundances of red fox and pine marten in relation to the composition of boreal forest landscapes. Journal of Animal Ecology 67: 874–886.
51. Mergey M, Helder R, Roeder J-J (2011) Effect of forest fragmentation on space-use patterns in the European pine marten (*Martes martes*). Journal of Mammalogy 92: 328–335.
52. Ruiz-Gonzalez A, Rubines J, Berdion O, Gomez-Moliner BJ (2008) A non-invasive genetic method to identify the sympatric mustelids pine marten (*Martes martes*) and stone marten (*Martes foina*): preliminary distribution survey on the northern Iberian Peninsula. European Journal of Wildlife Research 54: 253–261.
53. Bright PW (2000) Lessons from lean beasts: conservation biology of the mustelids. Mammal Review 30: 217–226.
54. Balestrieri A, Remonti L, Ruiz-Gonzalez A, Gomez-Moliner BJ, Vergara M, et al. (2010) Range expansion of the pine marten (*Martes martes*) in an agricultural landscape matrix (NW Italy). Mammalian Biology 75: 412–419.
55. Caryl FM, Quine CP, Park KJ (2012) Martens in the matrix: the importance of nonforested habitats for forest carnivores in fragmented landscapes. Journal of Mammalogy 93.
56. Schwartz M, Ruiz-González A, Pertoldi C, Masuda R (2012) *Martes* conservation genetics: assessing within species movements, units to conserve and connectivity cross ecological and evolutionary time. In: Aubry K, Zielinski W, Raphael M, Proulx G, Buskirk S, editors. *Biology and conservation of marten, sables, and fisher: a new synthesis*. New York: Cornell University Press. pp. 398–428.
57. Wright S (1943) Isolation by distance. Genetics 28: 114–138.
58. Mergey M, Larroque J, Ruette S, Vandel J-M, Helder R, et al. (2012) Linking habitat characteristics with genetic diversity of the European pine marten (*Martes martes*) in France. European Journal of Wildlife Research 58: 909–922.
59. Environment SMft (2006) Mapa forestal de España 1:50.000. Madrid: Ministerio de Medio Ambiente.
60. Institut SNG (2008) Base cartográfica numérica BCN200. Ministerio de Fomento.
61. Ruiz-Gonzalez A, Jose Madeira M, Randi E, Urra F, Gomez-Moliner BJ (2013) Non-invasive genetic sampling of sympatric marten species (*Martes martes* and *Martes foina*): assessing species and individual identification success rates on faecal DNA genotyping. European Journal of Wildlife Research 59: 371–386.
62. Davison A, Birks JDS, Brookes RC, Braithwaite TC, Messenger JE (2002) On the origin of faeces: morphological versus molecular methods for surveying rare carnivores from their scats. Journal of Zoology 257: 141–143.
63. Taberlet P, Griffin S, Goossens B, Questiau S, Manceau V, et al. (1996) Reliable genotyping of samples with very low DNA quantities using PCR. Nucleic Acids Research 24: 3189–3194.
64. Frantz AC, Pope LC, Carpenter PJ, Roper TJ, Wilson GJ, et al. (2003) Reliable microsatellite genotyping of the Eurasian badger (*Meles meles*) using faecal DNA. Molecular Ecology 12: 1649–1661.
65. Stenglein JL, Waits LP, Ausband DE, Zager P, Mack CM (2010) Efficient, Noninvasive Genetic Sampling for Monitoring Reintroduced Wolves. Journal of Wildlife Management 74: 1050–1058.
66. Brzeski KE, Gunther MS, Black JM (2013) Evaluating River Otter Demography Using Noninvasive Genetic Methods. Journal of Wildlife Management 77: 1523–1531.
67. Miller CR, Joyce P, Waits LP (2002) Assessing allelic dropout and genotype reliability using maximum likelihood. Genetics 160: 357–366.
68. Valiere N (2002) GIMLET: a computer program for analysing genetic individual identification data. Molecular Ecology Notes 2: 377–379.
69. Pompanon F, Bonin A, Bellemain E, Taberlet P (2005) Genotyping errors: Causes, consequences and solutions. Nature Reviews Genetics 6: 847–859.
70. Belkhir K, Borsa P, Chikhi L, Raufaste N, Bonhomme F (2004) Genetix 4.02, Logiciel sous windows pour la génétique des populations. Université de Montpellier II, Montpellier, France.
71. Raymond M, Rousset F (1995) Genepop (Version-1.2) - Population-Genetics Software for Exact Tests and Ecumenicism. Journal of Heredity 86: 248–249.
72. Guo SW, Thompson EA (1992) Performing the Exact Test of Hardy-Weinberg Proportion for Multiple Alleles. Biometrics 48: 361–372.
73. Rice WR (1989) Analyzing Tables of Statistical Tests. Evolution 43: 223–225.
74. Van Oosterhout C, Hutchinson WF, Wills DPM, Shipley P (2004) MICRO-CHECKER: software for identifying and correcting genotyping errors in microsatellite data. Molecular Ecology Notes 4: 535–538.
75. Rousset F (2000) Genetic differentiation between individuals. Journal of Evolutionary Biology 13: 58–62.
76. Hardy OJ, Vekemans X (2002) SPAGEDi: a versatile computer program to analyse spatial genetic structure at the individual or population levels. Molecular Ecology Notes 2: 618–620.
77. Vekemans X, Hardy OJ (2004) New insights from fine-scale spatial genetic structure analyses in plant populations. Molecular Ecology 13: 921–935.
78. Coulon A, Cosson JF, Angibault JM, Cargnelutti B, Galan M, et al. (2004) Landscape connectivity influences gene flow in a roe deer population inhabiting a fragmented landscape: an individual-based approach. Molecular Ecology 13: 2841–2850.
79. Blair ME, Melnick DJ (2012) Scale-Dependent Effects of a Heterogeneous Landscape on Genetic Differentiation in the Central American Squirrel Monkey (*Saimiri oerstedii*). Plos One 7.
80. Dudaniec RY, Rhodes JR, Wilmer JW, Lyons M, Lee KE, et al. (2013) Using multilevel models to identify drivers of landscape-genetic structure among management areas. Molecular Ecology 22: 3752–3765.
81. ESRI (2009) ArcMap version 9.3. Redlands, CA, USA.: Environmental Systems Research Institute.
82. Anderson CD, Epperson BK, Fortin MJ, Holderegger R, James PMA, et al. (2010) Considering spatial and temporal scale in landscape-genetic studies of gene flow. Molecular Ecology 19: 3565–3575.
83. Jenkins DG, Carey M, Czerniewska J, Fletcher J, Hether T, et al. (2010) A meta-analysis of isolation by distance: relic or reference standard for landscape genetics? Ecography 33: 315–320.
84. Cushman SA, Wasserman TN, Landguth EL, Shirk AJ (2013) Re-Evaluating Causal Modeling with Mantel Tests in Landscape Genetics. Diversity 5: 51–72.
85. Mantel N (1967) Detection of disease clustering and a generalized regression approach. Cancer Research 27: 209–220.
86. Smouse PE, Long JC, Sokal RR (1986) Multiple-regression and correlation extensions of the mantel test of matrix correspondence. Systematic Zoology 35: 627–632.
87. Rousset F (1997) Genetic differentiation and estimation of gene flow from F-statistics under isolation by distance. Genetics 145: 1219–1228.
88. Cushman SA, Landguth EL (2010) Spurious correlations and inference in landscape genetics. Molecular Ecology 19: 3592–3602.
89. Castillo JA, Epps CW, Davis AR, Cushman SA (2014) Landscape effects on gene flow for a climate-sensitive montane species, the American pika. Molecular Ecology 23: 843–856.
90. Mills LS, Citta JJ, Lair KP, Schwartz MK, Tallmon DA (2000) Estimating animal abundance using noninvasive DNA sampling: Promise and pitfalls. Ecological Applications 10: 283–294.
91. Castillo JA, Epps CW, Davis AR, Cushman S (2014) Landscape effects on gene flow for a climate-sensitive montane species, the American pika. Molecular Ecology 23: 843–856.
92. Balkenhol N, Gugerli F, Cushman SA, Waits LP, Coulon A, et al. (2009) Identifying future research needs in landscape genetics: where to from here? Landscape Ecology 24: 455–463.
93. Guillot G, Rousset F (2013) Dismantling the Mantel tests. Methods in Ecology and Evolution 4: 336–344.
94. Graves TA, Beier P, Royle JA (2013) Current approaches using genetic distances produce poor estimates of landscape resistance to interindividual dispersal. Molecular Ecology 22: 3888–3903.
95. Coulon A, Guillot G, Cosson JF, Angibault JMA, Aulagnier S, et al. (2006) Genetic structure is influenced by landscape features: empirical evidence from a roe deer population. Molecular Ecology 15: 1669–1679.
96. Cushman SA, Max TL, Whitham TG, Allan GJ (2014) River network connectivity and climate gradients drive genetic differentiation in a riparian foundation tree. Ecological Applications 24:1000–1014.
97. Guillot G, Leblois R, Coulon A, Frantz AC (2009) Statistical methods in spatial genetics. Molecular Ecology 18: 4734–4756.

98. Latch EK, Boarman WI, Walde A, Fleischer RC (2011) Fine-Scale Analysis Reveals Cryptic Landscape Genetic Structure in Desert Tortoises. Plos One 6.
99. Quemere E, Crouau-Roy B, Rabarivola C, Louis EE, Chikhi L (2010) Landscape genetics of an endangered lemur (*Propithecus tattersalli*) within its entire fragmented range. Molecular Ecology 19: 1606–1621.
100. Watts PC, Rousset F, Saccheri IJ, Leblois R, Kemp SJ, et al. (2007) Compatible genetic and ecological estimates of dispersal rates in insect (*Coenagrion mercuriale*: Odonata: Zygoptera) populations: analysis of 'neighbourhood size' using a more precise estimator. Molecular Ecology 16: 737–751.
101. Row JR, Blouin-Demers G, Lougheed SC (2010) Habitat distribution influences dispersal and fine-scale genetic population structure of eastern foxsnakes (*Mintonius gloydi*) across a fragmented landscape. Molecular Ecology 19: 5157–5171.
102. Cushman SA, Lewis JS (2010) Movement behavior explains genetic differentiation in American black bears. Landscape Ecology 25: 1613–1625.
103. Balkenhol N, Waits LP (2009) Molecular road ecology: exploring the potential of genetics for investigating transportation impacts on wildlife. Molecular Ecology 18: 4151–4164.
104. Cushman SA, Compton BW, McGarigal K (2010) Habitat fragmentation effects depend on complex interactions between population size and dispersal ability: Modeling influences of roads, agriculture and residential development across a range of life-history characteristics. In: Cushman SA, Huettman F, editors. Spatial complexity, informatics and wildlife conservation. Tokyo: Springer. pp. 369–387.
105. Jackson ND, Fahrig L (2011) Relative effects of road mortality and decreased connectivity on population genetic diversity. Biological Conservation 144: 3143–3148.
106. Cushman SA, Lewis JS, Landguth EL (2013) Evaluating the intersection of a regional wildlife connectivity network with highways. Movement Ecology 1: 1–12.
107. Cushman SA, Raphael MG, Ruggiero LF, Shirk AS, Wasserman TN, et al. (2011) Limiting factors and landscape connectivity: the American marten in the Rocky Mountains. Landscape Ecology 26: 1137–1149.
108. Wasserman TN, Cushman SA, Littell JS, Shirk AJ, Landguth EL (2013) Population connectivity and genetic diversity of American marten (*Martes americana*) in the United States northern Rocky Mountains in a climate change context. Conservation Genetics 14: 529–541.
109. Garroway CJ, Bowman J, Wilson PJ (2011) Using a genetic network to parameterize a landscape resistance surface for fishers, *Martes pennanti*. Molecular Ecology 20: 3978–3988.
110. Koen EL, Bowman J, Garroway CJ, Mills SC, Wilson PJ (2012) Landscape resistance and American marten gene flow. Landscape Ecology 27: 29–43.

14

Bird Community Conservation and Carbon Offsets in Western North America

Richard Schuster[1]*, Tara G. Martin[1,2], Peter Arcese[1]

1 Department of Forest and Conservation Sciences, University of British Columbia, Vancouver, British Columbia, Canada, 2 Ecosciences Precinct, CSIRO Ecosystem Sciences, Brisbane, Queensland, Australia

Abstract

Conservation initiatives to protect and restore valued species and communities in human-dominated landscapes face huge challenges linked to the cost of acquiring habitat. We ask how the sale of forest carbon offsets could reduce land acquisition costs, and how the alternate goals of maximizing α or β-diversity in focal communities could affect the prioritization land parcels over a range of conservation targets. Maximizing total carbon storage and carbon sequestration potential reduced land acquisition costs by up to 48%. Maximizing β rather than α-diversity within forest and savannah bird communities reduced acquisition costs by up to 15%, and when these solutions included potential carbon credit revenues, acquisition cost reductions up to 32% were achieved. However, the total cost of conservation networks increased exponentially as area targets increased in all scenarios. Our results indicate that carbon credit sales have the potential to enhance conservation outcomes in human-dominated landscapes by reducing the net acquisition costs of land conservation in old and maturing forests essential for the persistence of old forest plant and animal communities. Maximizing β versus α-diversity may further reduce costs by reducing the total area required to meet conservation targets and enhancing landscape heterogeneity. Although the potential value of carbon credit sales declined as a fraction of total acquisition costs, even conservative scenarios using a carbon credit value of $12.5/T suggest reductions in acquisition cost of up to $235 M, indicating that carbon credit sales could substantially reduce the costs of conservation.

Editor: Mark S. Boyce, University of Alberta, Canada

Funding: The Natural Sciences and Engineering Research Council Canada Discovery Grant (PA, http://www.nserc-crsng.gc.ca/), W. and H. Hesse (RS, http://hesse-award.sites.olt.ubc.ca/) and the University of British Columbia (RS, PA, http://ubc.ca/) kindly funded our work. The funders had no role in study design, data collection and analysis, decision to publish, or preparation of the manuscript.

Competing Interests: The authors have declared that no competing interests exist.

* E-mail: mail@richard-schuster.com

Introduction

There is a pressing need to develop mechanisms to promote biodiversity conservation in the face of climate and land use change and the competing needs of humans [1–4]. This challenge is particularly severe in human-dominated landscapes, where private ownership prevails and the cost of purchasing properties or compensating land holders for lost opportunity incurred as a result of conservation can be substantial [5,6]. There are multiple routes available to land conservation, such as land purchase or private land conservation initiatives [7–10]. Here we focus on land purchase rather than conservation agreements on private lands, to avoid having to estimate the long-term costs of monitoring and enforcement or the probability that private conservation agreements are challenged in future [7,11,12]. One way of making conservation via land purchase more affordable is by offsetting those costs via payments for ecosystem services. The use of carbon markets to pay for carbon sequestration is an ecosystem service gaining global attention [13–15], in part because public concerns about the consequences of climate change have motivated 35 nations and 13 sub-national jurisdictions to put a price on carbon [16]. To the degree that carbon and biodiversity values overlap, carbon offsets could therefore be used to protect forests that would otherwise be logged [17,18] or to restore those still supporting valued old forest communities [19].

Several outstanding issues arise when considering the role of carbon markets in forest restoration. One issue is that biodiversity values may be lower in stands with the highest returns from carbon sequestration sales because sequestration rates typically peak in stands of intermediate age [20]. In contrast, older forests act as carbon sinks and continue to accumulate carbon over time, but at lower rates on average [21], resulting in high initial returns on the sale of carbon storage credits, where one carbon credit represents the offset of greenhouse gas emissions by one tonne of carbon dioxide equivalent (CO_2-e). A second issue is whether to develop conservation plans that maximize species richness (α-diversity) within habitats or maximize dissimilarities in community composition (β-diversity) to accommodate landscape complexity and species that utilize multiple habitats [22,23]. Under climate change, it has been suggested that an emphasis on community dissimilarity (β-diversity) may deliver more robust conservation plans than those based on species richness [23,24]. Here we examine the potential value of carbon credit sales to offset land acquisition costs by developing conservation area designs that maximize β or α-diversity in native old forest and savannah bird communities in relation to forest structure and human land use. Specifically, we ask how protecting forests with high carbon storage versus high carbon uptake is likely to affect conservation outcomes.

Carbon and Biodiversity in the Georgia Basin

The Coastal Douglas Fir (CDF) ecozone of the Georgia Basin (British Columbia, Canada [25]) is a classic example of an endangered but extraordinarily diverse region that has been rapidly converted to exclusive human use (≥60%) [26] and thus retains ≤0.3% of historic old forests (>250 years) [27] and ≤10% of oak woodland and savannah [28], both of which provide habitat for 117 species at risk of extirpation, which represents the highest density of species of global and provincial concern to conservation of any ecozone in BC [26]. Because regional, provincial and federal authorities own <20% of the region and only ~9% is already conserved, cost-efficient routes to conservation are urgently needed to help reduce the risk of extirpation for those species and related ecosystems.

Prior to European colonization the CDF occurred as uneven-aged forest (often >300 years) dissected by shallow and deep-soil meadow and woodland communities [25,29] maintained in part by aboriginal land management practices to enhance hunting opportunities and root and fruit harvests [30–33]. In addition to recent human-caused disturbances, oak woodland and savannah community distributions are predicted to shift under future climate conditions, and only a small fraction of the current protected areas have the potential to accommodate this shift [34]. The resulting land use heterogeneity within the region and potential for humans to directly or indirectly affect native species richness [19,35–37] make this system ideal for studying trade-offs involved when attempting to maximize α- versus β-diversity in conservation plans, while simultaneously maximizing ecosystem service values represented as total carbon stored or sequestration potential. To do so, we compared systematic conservation scenarios that maximized old forest and savannah bird biodiversity (α-diversity) or their dissimilarity (β-diversity), and then quantified their relative costs given alternate carbon markets, and in relation to increasing targets for the total area conserved [38] (Table 1).

Materials and Methods

Ethics Statement

Permits or permission for the use of bird point count locations were obtained from Parks Canada (locations in National Park Reserves), private land owners (locations on private land), or did not require specific permission as they occurred on public right of ways (e.g., roadsides, regional parks). As private land owners did not want their information posted publically please contact the authors for contact details. The field studies did not involve endangered or protected species. This study did not require approval from an Animal Care and Use Committee because it was a non-invasive observational field study, and did not involve the capture and handling of wild animals.

Biodiversity data

We used trained observers to conduct 1,770 point counts on mainland BC and 53 islands from 30 Apr–11 Jul, 2005–2011 (Fig.1, 48.7° N, 123.5° W) to record all birds detected in 10 min, 50 m radius counts between 5 AM–12 PM at 713 sample locations (>100 m apart). Locations were re-visited 1–12 times and geo-referenced via a GPS (GPS60, Garmin Ltd, Kansas, USA). We extended the approach of Schuster & Arcese [19] geographically (from 1560 km^2 to 2520 km^2) by adding 601 counts to create predictive distribution models for 47 bird species and 25 covariates based on remote-sensed data and models incorporating imperfect detectability [39]. To estimate detectability we used one site specific (crown closure) and three observation specific (time of date, Julian date and observer identity) covariates.

We associated bird species indicators with the habitats they were expected to occupy by using 11 experts to rank the likelihood of observing 47 species in 10 focal habitat types using photographic and text descriptions of herbaceous, shrub, woodland, wetland, four forest types (pole, young, mature and old), and 2 human-dominated habitats (rural, urban), to create two community metrics indicating Old Forest (OF, [19]) and Savannah (SAV) habitats standardized between 0 and 1 by dividing through the maximum value possible (details in Appendix S1), where:

$$OF = \frac{-2*Herb - 1*Shrub - 0.5*Pole + 0.5*YFor + 1*MFor + 2*OFor}{7}$$

$$SAV = \frac{2*Wood + 2*Herb + 1*Shrub}{5}$$

These metrics match our goals given the region's history and focus on Old Forest and Savannah community conservation (see Introduction). Specifically, each species contributed to the cumulative Old Forest or Savannah community score, weighted by its expert opinion score for the given sub-type, summed across species to create community specific association scores from 0 to 1, and corresponding to none versus all members of the community expected to be present. The metrics were then projected spatially as predictive maps of community occurrence over the entire study area (2520 km^2, Fig. 1) as 1 ha hexagonal polygons (Fig. S1–S2 in Appendix S1,see also [19]).

Table 1. Summary of diversity features, land cost metrics, conservation targets and carbon prices used in 144 Marxan scenarios.

Diversity features (n = 2)	Property cost metrics (n = 4)	Conservation Targets [%] (n = 9)	Carbon credit value (CC) (n = 3)
α-diversity (maximize Old Forest + Savannah individually)	Total Land value (TLV)	10 to 50 (in 5% steps)	9 $/T (lowest price PCT has paid for credits so far)
β-diversity (maximize β-score)	TLV - StC	-	12.5 $/T (half the cost PCT charges, as well as roughly the average price PCT is paying for credits)
'	TLV - SeqC		25 $/T (the price that PCT is charging for credits)
	TLV - TotC		

PCT= Pacific Carbon Trust; StC= Carbon Storage * CC; SeqC= Carbon. Sequestration potential * CC; TotC = StC+SeqC.

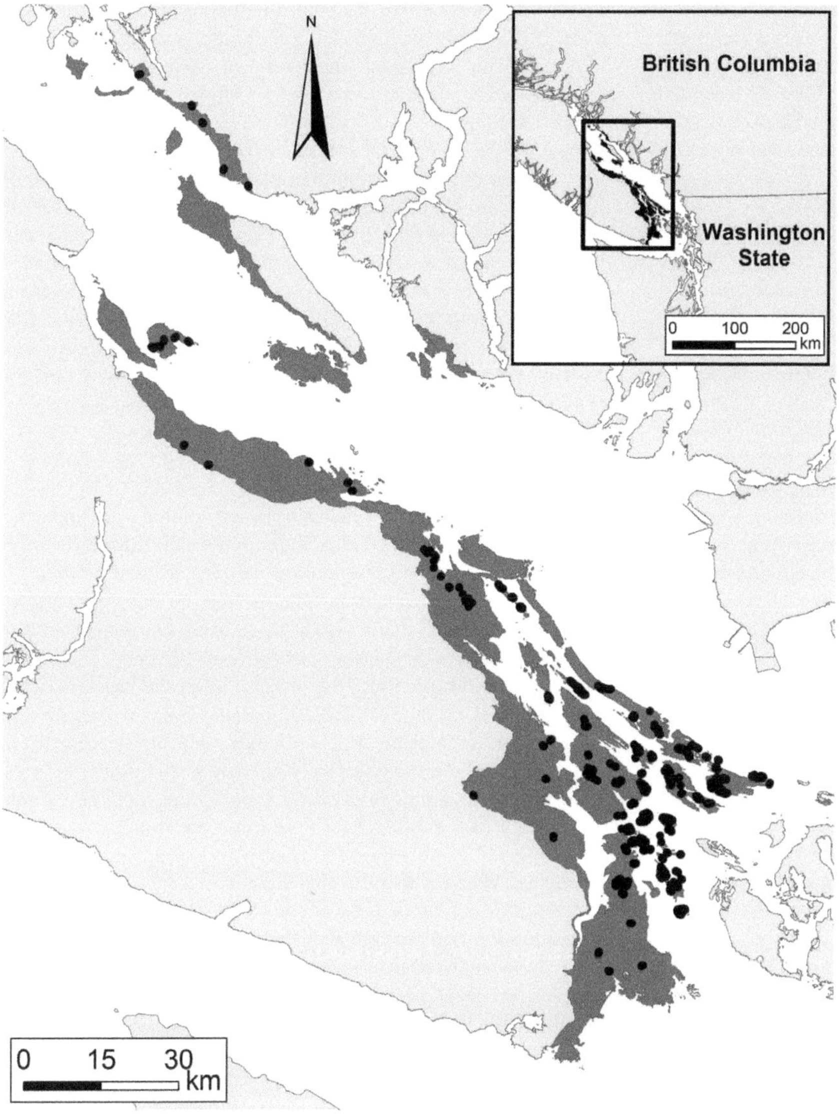

Figure 1. Georgia Basin study area including bird point count locations. Dark grey area indicates the extent of the study region and black dots represent bird point count locations.

Carbon estimates

Forest carbon storage and sequestration rates were estimated for all forested land in the study area using terrestrial ecosystem mapping (TEM; [27]) and FORECAST [40]. FORECAST is a stand-level forest ecosystem simulator that is one of two models approved by the BC Ministry of Forests for carbon budget assessments [41], and the only model calibrated for use in the CDF (Blanco et al. 2007) and linked to TEM [42]. To facilitate carbon analysis TEM polygons were stratified into homogenous analysis units based on site series. Net ecosystem carbon storage was limited to: above and below-ground tree biomass, deadwood biomass, and dead below-ground biomass. Each analysis unit was simulated for a period of 300 years with results reported for annual time steps to create carbon storage curves. FORECAST results were subsequently assigned to individual TEM polygons by estimating the age of each polygon subsection based upon the current assigned structural stage and estimated productivity class [42]. These age estimates were derived from ranges provided by Meidinger *et al.* [43] for regional forest ecosystems. Ages of old stands (structural stage 7) were set at 200 to be conservative. Age estimates were verified against a subset of TEM polygons (Southern Gulf Islands of Southwestern BC) for which direct age estimates were available (n = 254). For conservation prioritization analysis we used predicted net ecosystem carbon storage and net ecosystem carbon sequestration estimates for 20 years from now due to uncertainty about fire frequency in the future. Further details on this analysis are provided in Appendix S2.

Cadastral layer and property costs

We incorporated spatial heterogeneity in land values [5,44–46] in our plans by using cadastral data and 2012 land value assessments (Integrated Cadastral Information Society of BC, ICIS). However, because there is no centralized entity curating cadastral data for British Columbia, we combined data from ICIS, the BC Assessment agency and the Integrated Cadastral Fabric. Doing so required processing to remove stacked and overlapping polygons and slivers. The combined cadastral layer included 193,623 polygons. Current assessments were available for 187,139 polygons, but missing for 3,281 polygons or reduced relative to market value due to taxation or administrative reasons unrelated to our work (e.g farm or managed forest land, 3,203 ploygons). For these 6,484 polygons we applied an inverse distance weighted interpolation to estimate land values by splitting cadastral polygons into 10 groups based on polygon size to accommodate high size related heterogeneity in assessed cost using R v.2.15.2 [47] and packages gstat v.1.0-14 [48] and sp v.1.0-1 [49].

We used tax assessment land values to estimate acquisition costs because they are revised annually in the region, and because more 'realistic' strategies would require speculation on how purchase cost may be affected by location of existing reserves, evolving zoning plans, the willingness of owners to sell, or other effects. In particular, there is as yet no consensus on the effect of conservation agreements on land values [50–52].

Marxan inputs

We used Marxan [53] to prioritize cadastral polygons for inclusion in conservation area designs by using them as planning units (n = 193,623). We calculated biodiversity and carbon estimates for each planning unit using ArcGIS v.10.1 [54] and area weighted sums in Geospatial Modelling Environment v.0.7.2.1 [55].

To determine whether maximizing β- versus α- diversity affected conservation outcomes we created two sets of diversity features as inputs to Marxan. First we included the diversity features individually in the analysis and set conservation targets for Old Forest and Savannah scores as the percentage of total old forest or savannah habitat existing within the study region. The second approach we used was to pre-specify a β-diversity metric to combine biodiversity features with the goal to specifically maximize highly diverse habitat patches. For this purpose we created the following metric:

$$\beta - score = \frac{2 * OF * SAV}{OF + SAV}$$

This represents the Old Forest and Savannah community dissimilarity, using a scaling factor of 2 to create β-scores between 0 and 1 (Fig. S3 in Appendix S1). In Marxan analyses we set targets for the β-score, while still including Old Forest and Savannah metrics (without setting a target) to keep track of individual community representation. We used a total of four property cost metrics per diversity scenario: i) Total land value (TLV) for each property, which is the sum of the assessed property value and any improvement on that parcel; ii) TLV minus the current carbon storage (T) times the carbon credit value ($/T). Here we used $12.5 Canadian per credit, which is half the amount that Pacific Carbon Trust, a crown corporation established in 2008 to deliver greenhouse gas offsets in the province of British Columbia (http://pacificcarbontrust.com, date accessed: 2013-12-10), sells credits for and about the average amount they pay for credits. iii) TLV minus the amount of potential carbon sequestration over 20 years times the carbon credit value; iv) TLV minus ii and iii combined (Table 1).

Marxan scenarios

We used the two diversity scenarios α (Old Forest + Savannah) and β (β-score) in combination with the four cost scenarios (Table 1). An important consideration for this study was what level to set the required conservation target to, in order to ensure the study system will maintain viable populations of native species and be resilient to predicted environmental change in the future. As there is debate about what constitutes appropriate conservation goals [38] we used a range of conservation targets (10–50%) to investigate the potential trade-offs of different targets. We calibrated each diversity scenario to ensure robust analysis by initially setting the diversity target to 50% (the most costly to reach) and the number of restarts to 100, as we were not so much interested in the spatial representation of the reserve design but rather its cost effectiveness [56]. For the same reason we also refrained from setting boundary length modifiers. For each diversity scenario we created Marxan solutions for combinations of the following species penalty factors (SPF's): 1–10,15,20 and number of iterations: 10 k, 50 k, 100 k, 500 k, 1 M, 5 M, 10 M, 25 M, 50 M, 100 M, for a total of 65 calibration analyses per diversity scenario. We created cumulative distribution functions using number of solutions on the y-axis, solution cost on the x-axis for SPF and Marxan score for number of iterations [56]. Based on the results we used the following values for SPF and number of iterations respectively: Old Forest + Svannah (3/10 M); β-score (3/10 M). We also investigated summed solutions to make sure every restart met its targets, excluding ones that missed the target by >5%. For ease of computation we created an R function to batch run Marxan (Appendix S3).

We held the calibrated values constant in subsequent analyses and ran Marxan scenarios for the two diversity metrics in combination with the four cost metrics, using the baseline carbon credit value of $12.5. For each combination we further varied the conservation target from 10–50%. From each run we recorded the cost of the total reserve system averaged over the number or restart (100), while ensuring conservation targets were met. To examine the amount of remaining Old Forest and Savannah communities protected by maximizing β-diversity we compiled community scores as Marxan features in these scenarios without setting targets, allowing us to keep track of Old Forest and Savannah representation without affecting the analysis. We used the results from these analyses to compare the reserve prices within each diversity scenario as well as across scenarios. In addition we calculated the potential cost savings between fee simple acquisition scenarios (TLV only) and ones that utilize the sale of carbon credits. As market prices of carbon credits are highly variable we extended our approach to include variation in carbon credit value, by repeating the entire analysis for two additional carbon credit values: i) $9 per credit (the lowest rate PCT has ever paid for credits), and ii) $25 per credit (the price PCT sells credits for). In total 144 Marxan scenarios were investigated (Table 1). All results presented here relate to the baseline carbon credit value of $12.5 unless otherwise stated.

Results

Land acquisition cost, diversity and planning goals

Acquisition costs of conservation networks increased from $180 M to $2.45 B as targets increased from 10 to 50% of remaining Old Forest and Savannah bird communities when maximising α-diversity (Fig. 2), but reduced slightly when maximizing β-diversity ($172 M to $2.16 B, 10–50% target; Fig. 2), representing savings of 4–15% as compared to equivalent α-diversity scenarios depending on conservation target (Fig. 3a). Savings were due in part to a reduction in total area needed to reach a given target when maximising β versus α-diversity (mean = 7%, range = 5–11%; Fig. 3b). The amount of standing and sequestration potential carbon in conserved landscapes also declined slightly when maximizing β-diversity (mean = 2%, range = 0.7–5%; Fig. 3c). In contrast, representation of Old Forest communities slightly increased (1–2.5%) and representation of

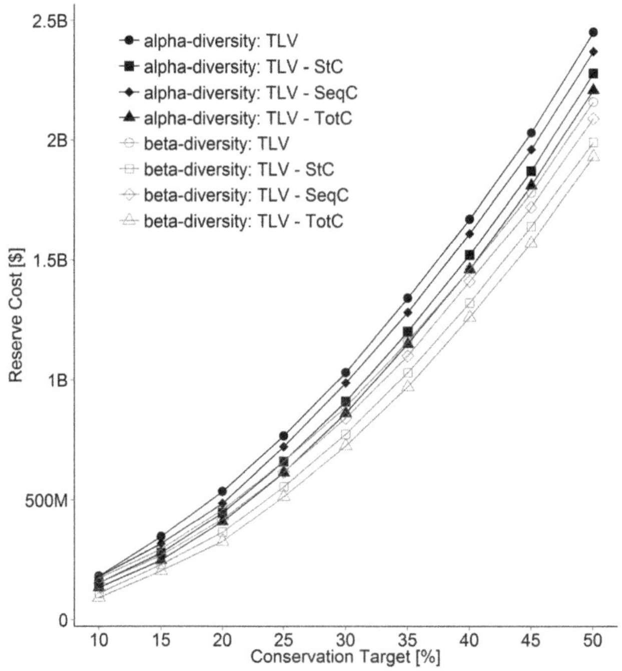

Figure 2. Reserve costs using alpha and beta diversity and a carbon credit value of $12.5 across a range of conservation targets (term definitions in Table 1).

Savannah declined (−2.0–−5.7%) when maximizing β versus α-diversity (Table 2).

Cost savings given carbon credits

Maximizing total (standing + sequestered) carbon resulted in the largest cost savings in both α and β-diversity scenarios aimed at protecting Old Forest and Savannah habitats. Acquisition costs increased from $133 M to $2.21 B as target increased from 10 to 50% when maximizing α-diversity, which represent potentials offset of $47 M–235 M, equivalent to a 10–28% cost reduction via carbon credit sales (Fig.2). In comparison, acquisition costs were lower for scenarios that maximised β-diversity ($90 M to $1.93 B), in part because implied carbon credit sales ($82–227 M) contributed slightly more to cost reduction (e.g., 11–48%; Fig. 2). Maximising carbon storage and carbon sequestration potential individually reduced acquisition costs to a smaller extent, but carbon storage offered superior savings (Fig. 2). Overall, maximizing total carbon returned networks that were 17.5% cheaper on average when maximizing β versus α-diversity compared to 12.3% without using carbon storage and sequestration values (Fig. 3a).

Conservation targets and carbon price

The cost of conservation networks increased exponentially with increasing targets for all scenarios (Fig. 2). In β-diversity scenarios the total area that needed to be acquired to reach a conservation target was 11–5% lower and acquisition costs 32–13% less than scenarios that maximized α -diversity (Fig. 3). The percent reduction in total acquisition costs due to carbon value also declined as conservation targets increased in α and β-diversity scenarios (Fig. 4a,b). The magnitude by which acquisition costs were reduced by carbon value was similar across prices considered but maximized at $25/T in most scenarios (Fig. 4a,b). Relative

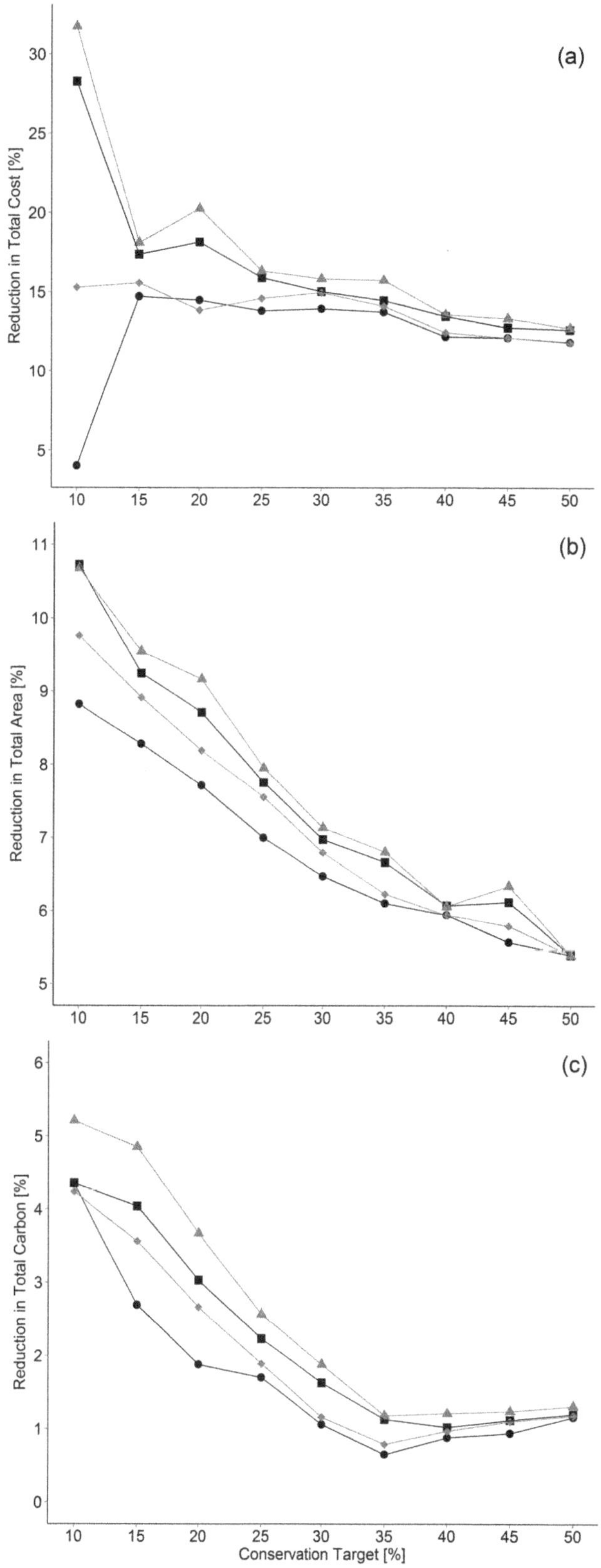

Figure 3. Comparison of α and β-diversity scenario results. Presented are the % reductions when using a β-diversity approach for: a) reserve network cost, b) reserve network area, c) total carbon included in the reserve networks. Circles represent TLV, squares StC, diamonds SeqC and triangles TotC (term definitions in Table 1) (c).

Table 2. Conservation targets (bold) and % actually included in β-diversity scenarios (BETA) using carbon offset value of \$12.5/T.

Scenario		Percent of Target Bird Community Protected [%]								
		10	15	20	25	30	35	40	45	50
Total Land Value (TLV)	BETA	10.00	15.00	20.00	25.00	30.00	35.00	40.00	45.00	50.00
	OF	11.12	16.31	21.60	26.81	32.07	37.01	42.18	47.29	52.41
	SAV	8.80	13.57	18.09	22.87	27.59	32.49	37.22	42.07	46.85
TLV - StC	BETA	10.00	15.00	20.00	25.00	30.00	35.00	40.00	45.00	50.00
	OF	11.18	16.40	21.73	26.89	32.07	37.13	42.23	47.51	52.47
	SAV	8.72	13.38	17.93	22.68	27.45	32.34	37.14	41.78	46.78
TLV - SeqC	BETA	10.00	15.00	20.00	25.00	30.00	35.00	40.00	45.00	50.00
	OF	10.98	16.31	21.67	26.83	32.04	37.10	42.14	47.35	52.43
	SAV	8.96	13.56	18.07	22.85	27.57	32.42	37.30	42.02	46.84
TLV - TotC	BETA	10.00	15.00	20.00	25.00	30.00	35.00	40.00	45.00	50.00
	OF	11.17	16.43	21.72	26.95	31.95	37.18	42.27	47.50	52.48
	SAV	8.72	13.36	17.94	22.70	27.63	32.26	37.10	41.79	46.73

β-targets were met in each case and Old Forest (OF) was generally over and Savannah (SAV) underrepresented. StC = carbon storage credits; SeqC = sequestration potential credits; TotC = StC+SeqC.

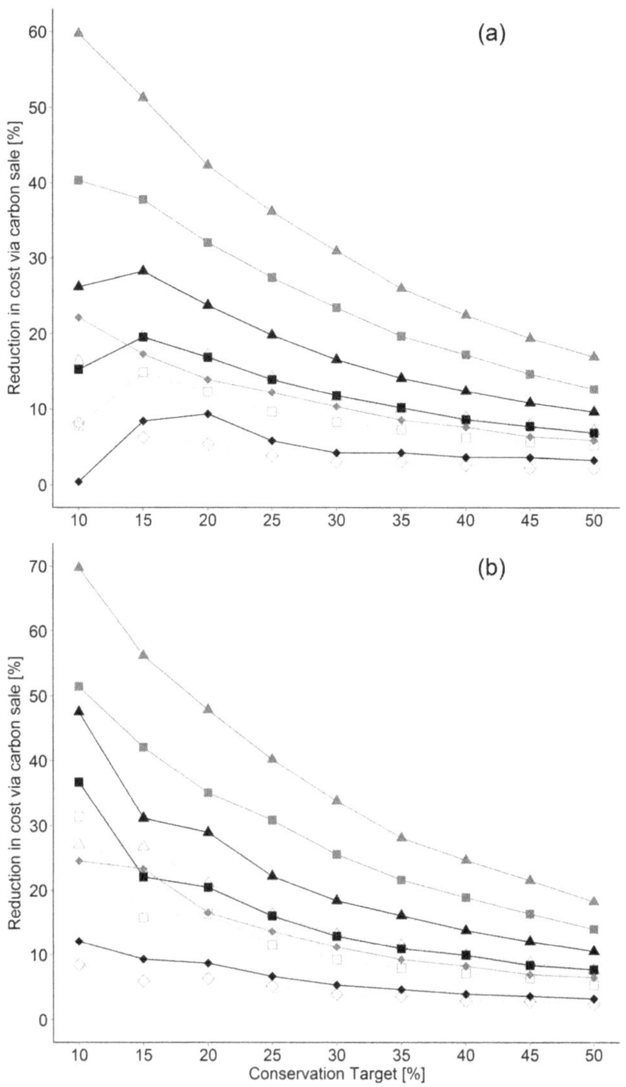

Figure 4. Cost savings when using carbon credit sales in relation to outright acquisition (TLV scenarios). Carbon credit values used are shown in parenthesis for a) α-diversity and b) β-diversity. Rectangles represent StC, diamonds SeqC and triangles TotC. Carbon credit values are as follows: dotted (\$9), black (\$12) and grey (\$25) (term definitions in Table 1).

reduction in cost due to carbon was maximized at the 10% target in all β-diversity scenarios (Fig. 4b).

Discussion

Carbon credit sales have the potential to reduce land acquisition costs by up to 48% in Coastal Douglas fir forest and woodland communities of western North America given values already paid in the region (\$12.5/T: [57]; Fig. 4b). The largest benefits were realized in scenarios that maximized total carbon and β-diversity in native woodland and old forest bird communities of the region, because those scenarios achieved their targets by selecting cheaper and slightly smaller networks than scenarios maximizing α-diversity in these communities independently. We now develop these points in light of literature on ecosystem services, land acquisition and conservation applied to threatened plant and animal communities in human-dominated landscapes of the Georgia Basin of western North America.

Maximizing β versus α diversity

Prioritizing β over α-diversity in Old Forest and Savannah bird communities reduced acquisition costs by up to 15%, or up to 32% including carbon values (Fig.3a). One reason for these savings is that the fraction of Old Forest bird habitat included in conservation networks was larger in β as compared to α-diversity scenarios (Table 2), resulting in more carbon stored per unit area conserved (Fig. 3b,c). However, over-representation of Old Forest relative to Savannah communities also reduced acquisition costs in scenarios not including carbon value, indicating that Old Forest habitat was on average less valuable than Savannah habitat in the region, perhaps due to high human amenity values (Table 2; e.g., [58]). Scenarios maximizing β-diversity also met conservation targets by protecting less total area and up to an 11% cost savings compared to scenarios maximizing α-diversity (Fig.3b), and this finding was largely independent of carbon value or conservation target (Fig.3b). Prior studies of the consequence of emphasizing α- versus β-diversity in conservation planning have concluded that a focus on β-diversity is likely to enhance long term persistence in diverse species assemblages and reserve networks [22,59,60]. Our results broaden these conclusions by showing that scenarios that maximize β-diversity may also reduce the cost of conservation by reducing the area required to meet realistic targets for land acquisition.

Conservation cost and carbon

Our results indicate that carbon credit sales should be considered as an option to maximize return on conservation investments in regions where land cost is high and old or growing forests offer habitat for valued focal communities. Our results therefore compliment suggestions that carbon credit sales have the potential to advance conservation and mitigate the impacts of climate change [13,61,62] but extend those suggestions by providing a spatially explicit, empirical example applied to a landscape with high conservation and cultural values [26,58,63]. The largest reductions in cost due to carbon credit sales were obtained by including carbon storage and carbon sequestration potential (Fig. 4), indicating that flexibility in carbon credit sales with respect to forest age can also increase economic efficiency. Although our results are based on a 20 year time-frame due to uncertainty about fire frequency, versus more typical 100 year time-frame for such projects [41,61], they could easily be revised with new data.

Our finding that carbon storage reduced costs more than sequestered carbon (Fig. 2,4) is partly a consequence of logging history, given that close to 30% of the region not converted to exclusive human use is covered by forest ≥80 years-old. The predominance of young forest has the potential to reduce adjacency between older, high-value forest and savannah habitat with rich and diverse native bird communities. However, young forest patches may also provide relatively low-cost opportunities to link high-value patches where acquisition costs can be offset by relatively high sequestration rate. Nevertheless, most scenarios preferentially included older stands with more carbon storage, but lower sequestration rates (Fig. S4,S5 in Appendix S2). Several other studies have suggested that carbon credits could be used to advance conservation, particularly on private land to compensate land owners for forgone opportunity costs [18,64,65]. We extended these suggestions by providing a particularly detailed example to demonstrate how land use planners might offset the costs of conservation area design by acquiring habitats that simultaneously maximize the diversity of valued vertebrate communities and realize the economic potential of carbon credit sales.

As an alternative to the sale of carbon credits to reduce conservation costs, land conservation and protection on private lands, accomplished through relationship building and alternative tax plans, could be an option [7,66,67]. But, one of the biggest challenges to realizing a theoretical approach of implementing private land conservation agreements on high value biodiversity landscapes is the need to work with landowners willing to put conservation agreements on their land [7]. In Canada, and British Columbia in particular, there are some incentives in place such as tax credits ([9], e.g. Natural Area Protection Tax Exemption Program), but none of the currently implemented compensation schemes would compensate land owners for the lost opportunity costs of developing their land or using it in other revenue-generating manners such as via agriculture or forestry, as proposed elsewhere [68–70]. Currently, a private conservation agreement approach depends largely on individuals wanting to create a legacy and see their property protected into the future [71]. It was beyond the scope of our study to address issues related to a landowners motivation to participate in conservation [7,72], nor did we want to speculate on the use of tax structure shifts.

Conservation targets

A key goal of our work was to demonstrate novel routes to protecting high-value, Old Forest and Savannah bird communities at landscape scales in western North America. However, the amount of habitat needed to achieve those goals remains uncertain. Policy-driven targets for biodiversity conservation place goals for terrestrial habitat conservation at 17% by 2020 [73], but recent reviews suggest much higher targets (25–75%; [38]). We used a range of targets to explore their influence on reserve design, carbon value and the conservation of Old Forest and Savannah ecosystems, but we found that carbon contributed proportionally less to acquisition costs as targets increased in all scenarios (Fig.4) because higher targets required the acquisition of more expensive parcels. Thus, although total carbon generally increased linearly with conservation target, acquisition costs increased exponentially, causing a decline in relative benefit (Fig.4). However, even for the largest targets (50%) in α and β-diversity scenarios, carbon values reduced acquisition cost by 9.6 and 10.5%, respectively ($235, 227 M; Fig. 4) at $12.5 per Ton.

A number of uncertainties in our study also have the potential to limit its interpretation. First, actual purchase costs may differ from assessed or predicted values [7,74]. Second, it may not be feasible to protect the areas offering the highest conservation value and least cost, particularly if regional representation or the augmentation of existing conservation areas is emphasized [75]. Third, although our results were robust over a range of carbon values, carbon markets remain unpredictable. Nevertheless, carbon markets are of substantial size, the European Union Emissions Trading System for example included 2.1 billion metric tons in 2011 [76]. In 2013 China, the largest national source of greenhouse gases (19.1% of total emissions), introduced pilot emission trading schemes [77,78], joining a growing number of countries with national emission trading schemes [16]. Voluntary carbon markets that are currently the biggest market place for forest carbon offset projects in countries like Canada had a market volume of $572 M in 2011 [76]. Assuming that carbon markets develop further, our results demonstrate that carbon value has the potential to substantially reduce land acquisition costs in human-dominated landscapes, particularly in the Georgia Basin of western North America, where diverse Old Forest and Savannah bird [19] and plant [79] communities still persist in relatively isolated, mature forest and woodland habitats.

Supporting Information

Appendix S1 Further details on our bird score modelling approach as well as residual spatial autocorrelation test results and community scores for individual bird species. We further present maps of the Old Forest, Savannah and β-scores.

Appendix S2 Summary results of carbon storage and carbon sequestration potential modelling and a technical report on the FORECAST modelling approach used.

Appendix S3 The R code we developed for Marxan calibration that can be used in combination with both the 32 and 64bit versions of Marxan. Function results can be used for Marxan post processing in R.

Appendix S4 The expert elicitation introduction document that expert birders used as guidelines for their input.

Acknowledgments

R. Butler, R. & R. Cannings, I. Criuckshank, K. Ferguson, S. Hannon, J. Kimm, A. Nightingale, S. Roias, A. Rousseau, R. Shortinghuis and A. Stewart offered expert advice. We also thank many private land owners, Parks Canada and BC Parks for facilitating access and are grateful for the helpful comments of V. LeMay, M. Boyce and two anonymous reviewers.

Author Contributions

Conceived and designed the experiments: RS TM PA. Performed the experiments: RS PA. Analyzed the data: RS. Contributed reagents/materials/analysis tools: RS PA. Wrote the paper: RS TM PA.

References

1. Ehrlich PR, Pringle RM (2008) Where does biodiversity go from here? A grim business-as-usual forecast and a hopeful portfolio of partial solutions. Proc Natl Acad Sci 105: 11579–11586. doi:10.1073/pnas.0801911105.
2. Bayon R, Jenkins M (2010) The business of biodiversity. Nature 466: 184–185. doi:10.1038/466184a.
3. Butchart SHM, Walpole M, Collen B, van Strien A, Scharlemann JPW, et al. (2010) Global biodiversity: indicators of recent declines. Science (80-) 328: 1164–1168. doi:10.1126/science.1187512.
4. Estes JA, Terborgh J, Brashares JS, Power ME, Berger J, et al. (2011) Trophic downgrading of planet Earth. Science (80-) 333: 301–306. doi:10.1126/science.1205106.
5. Naidoo R, Balmford A, Ferraro PJ, Polasky S, Ricketts TH, et al. (2006) Integrating economic costs into conservation planning. Trends Ecol Evol 21: 681–687. doi:10.1016/j.tree.2006.10.003.
6. Wunder S (2007) The efficiency of payments for environmental services in tropical conservation. Conserv Biol 21: 48–58.
7. Knight AT, Grantham HS, Smith RJ, McGregor GK, Possingham HP, et al. (2011) Land managers' willingness-to-sell defines conservation opportunity for protected area expansion. Biol Conserv 144: 2623–2630.
8. Armsworth PR, Sanchirico JN (2008) The effectiveness of buying easements as a conservation strategy. Conserv Lett 1: 182–189. Available: http://doi.wiley.com/10.1111/j.1755-263X.2008.00028.x.
9. Parker DP (2004) Land trusts and the choice to conserve land with full ownership or conservation easements. Nat Resour J 44: 483–518. Available: http://lawlibrary.unm.edu/nrj/44/2/08_parker_trusts.pdf.
10. Wilson KA, Underwood EC, Morrison SA, Klausmeyer KR, Murdoch WW, et al. (2007) Conserving biodiversity efficiently: what to do, where, and when. PLoS Biol 5: e223. Available: http://www.ncbi.nlm.nih.gov/pubmed/17713985.
11. Rissman AR, Butsic V (2011) Land trust defense and enforcement of conserved areas. Conserv Lett 4: 31–37.
12. Rissman AR (2013) Rethinking property rights: comparative analysis of conservation easements for wildlife conservation. Environ Conserv 40: 222–230.
13. Venter O, Laurance WF, Iwamura T, Wilson KA, Fuller RA, et al. (2009) Harnessing carbon payments to protect biodiversity. Science (80-) 326: 1368.
14. Agrawal A, Nepstad D, Chhatre A (2011) Reducing emissions from deforestation and forest degradation. Annu Rev Environ Resour 36: 373–396.
15. Phelps J, Webb EL, Adams WM (2012) Biodiversity co-benefits of policies to reduce forest-carbon emissions. Nat Clim Chang 2: 497–503.
16. Climate Commission (2013) The Critical Decade: Global Action Building On Climate Change (full report as pdf). Available: http://apo.org.au/node/33831.
17. Chan KMA, Hoshizaki L, Klinkenberg B (2011) Ecosystem Services in Conservation Planning: Targeted Benefits vs. Co-Benefits or Costs? PLoS One 6: e24378.
18. Douglass LL, Possingham HP, Carwardine J, Klein CJ, Roxburgh SH, et al. (2011) The Effect of Carbon Credits on Savanna Land Management and Priorities for Biodiversity Conservation. PLoS One 6: e23843.
19. Schuster R, Arcese P (2013) Using bird species community occurrence to prioritize forests for old growth restoration. Ecography (Cop) 36: 499–507. doi:10.1111/j.1600-0587.2012.07681.x.
20. Pregitzer KS, Euskirchen ES (2004) Carbon cycling and storage in world forests: biome patterns related to forest age. Glob Chang Biol 10: 2052–2077.
21. Luyssaert S, Schulze E-D, Borner A, Knohl A, Hessenmoller D, et al. (2008) Old-growth forests as global carbon sinks. Nature 455: 213–215.
22. Marsh CJ, Lewis OT, Said I, Ewers RM (2010) Community-level diversity modelling of birds and butterflies on Anjouan, Comoro Islands. Biol Conserv 143: 1364–1374.
23. Mokany K, Harwood TD, Overton JM, Barker GM, Ferrier S (2011) Combining α - and β -diversity models to fill gaps in our knowledge of biodiversity. Ecol Lett 14: 1043–1051. doi:10.1111/j.1461-0248.2011.01675.x.
24. Arponen A, Moilanen A, Ferrier S (2008) A successful community-level strategy for conservation prioritization. J Appl Ecol 45: 1436–1445.
25. Meidinger D, Pojar J (1991) Ecosystems of British Columbia. Victoria, BC: British Columbia Ministry of Forests.
26. Austin MA, Buffett DA, Nicolson DJ, Scudder GGE, Stevens V (2008) Taking Nature's Pulse: The Status of Biodiversity in British Columbia. Victoria, BC: Biodiversity BC. Available: www.biodiversitybc.org.
27. MES (2008) Terrestrial Ecosystem Mapping of the Coastal Douglas-Fir Biogeoclimatic Zone. Madrone Environmental Services LTD., Duncan, BC. Available: http://a100.gov.bc.ca/pub/acat/public/viewReport.do?reportId=15273.
28. Lea T (2006) Historical Garry oak ecosystems of Vancouver Island, British Columbia, pre-European contact to the present. Davidsonia 17: 34–50.
29. Mosseler A, Thompson I, Pendrel BA (2003) Overview of old-growth forests in Canada from a science perspective. Environ Rev 11: S1–S7. doi:10.1139/a03-018.
30. Turner NJ (2014) Ancient Pathways, Ancestral Knowledge: Ethnobotany and Ecological Wisdom of Indigenous Peoples of Northwestern North America. 2 vol. Montreal: McGill Queen's University Press.
31. MacDougall AS, Beckwith BR, Maslovat CY (2004) Defining conservation strategies with historical perspectives: a case study from a degraded oak grassland ecosystem. Conserv Biol 18: 455–465.
32. Dunwiddie PW, Bakker JD (2011) The future of restoration and management of prairie-oak ecosystems in the Pacific Northwest. Northwest Sci 85: 83–92.
33. McCune JL, Pellatt MG, Vellend M (2013) Multidisciplinary synthesis of long-term human—ecosystem interactions: A perspective from the Garry oak ecosystem of British Columbia. Biol Conserv 166: 293–300.
34. Pellatt MG, Goring SJ, Bodtker KM, Cannon AJ (2012) Using a down-scaled bioclimate envelope model to determine long-term temporal connectivity of Garry oak (Quercus garryana) habitat in western North America: implications for protected area planning. Environ Manage 49: 802–815.
35. Gonzales EK, Arcese P (2008) Herbivory more limiting than competition on early and established native plants in an invaded meadow. Ecology 89: 3282–3289. doi:10.1890/08-0435.1.
36. Martin TG, Arcese P, Scheerder N (2011) Browsing down our natural heritage: Deer impacts on vegetation structure and songbird populations across an island archipelago. Biol Conserv 144: 459–469.
37. Bennett JR, Vellend M, Lilley PL, Cornwell WK, Arcese P (2012) Abundance, rarity and invasion debt among exotic species in a patchy ecosystem. Biol Invasions 15: 707–716. doi:10.1007/s10530-012-0320-z.
38. Noss RF, Dobson AP, Baldwin R, Beier P, Davis CR, et al. (2012) Bolder Thinking for Conservation. Conserv Biol 26: 1–4.
39. Mackenzie DI, Nichols JD, Lachman GB, Droege SJ, Royle JA, et al. (2002) Estimating site occupancy rates when detection probabilities are less than one. Ecology 83: 2248–2255.
40. Kimmins JP, Mailly D, Seely B (1999) Modelling forest ecosystem net primary production: the hybrid simulation approach used in FORECAST. Ecol Modell 122: 195–224.
41. Ministry of Environment BC (2011) Protocol for the Creation of Forest Carbon Offsets in British Columbia. British Columbia Ministry of Environment, Victoria. Available: http://www.env.gov.bc.ca/cas/mitigation/fcop.html.
42. Seely B, Nelson J, Wells R, Peter B, Meitner M, et al. (2004) The application of a hierarchical, decision-support system to evaluate multi-objective forest management strategies: a case study in northeastern British Columbia, Canada. For Ecol Manage 199: 283–305.
43. Meidinger D, Trowbridge R, Macadam A, Tolkamp C (1998) Field Manual for Describing Terrestrial Ecosystems. Victoria, Canada: Crown Publications Inc.
44. Ando A, Camm J, Polasky S, Solow A (1998) Species Distributions, Land Values, and Efficient Conservation. Science (80-) 279: 2126–2128.

45. Polasky S, Camm JD, Garber-Yonts B (2001) Selecting Biological Reserves Cost-Effectively: An Application to Terrestrial Vertebrate Conservation in Oregon. Land Econ 77: 68–78.
46. Ferraro PJ (2003) Assigning priority to environmental policy interventions in a heterogeneous world. J Policy Anal Manag 22: 27–43.
47. R Development Core Team (2012) R: A language and environment for statistical computing 2.15.2, http://www.r-project.org.
48. Pebesma EJ (2004) Multivariable geostatistics in S: the gstat package. Comput Geosci 30: 683–691.
49. Bivand RS, Pebesma EJ, Gómez-Rubio V (2013) Applied Spatial Data Analysis with R. New York, NY: Springer. doi:10.1111/j.1541-0420.2009.01247_1.x.
50. Anderson K, Weinhold D (2008) Valuing future development rights: the costs of conservation easements. Ecol Econ 68: 437–446.
51. Schilling BJ, Sullivan KP, Duke JM (2013) Do Residual Development Options Increase Preserved Farmland Values? J Agric Resour Econ 38.
52. Nickerson CJ, Lynch L (2001) The effect of farmland preservation programs on farmland prices. Am J Agric Econ 83: 341–351.
53. Ball IR, Possingham HP, Watts ME (2009) Marxan and relatives: Software for spatial conservation prioritisation. In: Moilanen A, Wilson K, Possingham HP, editors. Spatial conservation prioritisation quantitative methods and computational tools. Oxford: Oxford University Press. pp. 185–195. Available: http://www.uq.edu.au/spatialecology/docs/Publications/2009_Ball_etal_MarxanAndRelatives.pdf.
54. ESRI (2012) ArcGIS 10.1 Economic and Social Reserach Institute Inc., Redlands, CA. http://www.esri.com/.
55. Beyer HL (2012) Geospatial Modelling Environment (Version 0.7.2.1). (software). URL: http://www.spatialecology.com/gme.
56. Ardron JA, Possingham HP, Klein CJ, editors (2010) Marxan Good Practices Handbook, Version 2. Victoria, BC, Canada: Pacific Marine Analysis and Research Association.
57. PCT (2013) Pacific Carbon Trust Carbon Offset Portfolio. Available: http://pacificcarbontrust.com/assets/Uploads/Pricing/PCTpricing2009-2012.pdf, date accessed: 2013-08-10. Accessed 10 August 2013.
58. Vellend M, Bjorkman AD, McConchie A (2008) Environmentally biased fragmentation of oak savanna habitat on southeastern Vancouver Island, Canada. Biol Conserv 141: 2576–2584.
59. Wiersma YF, Urban DL (2005) Beta Diversity and Nature Reserve System Design in the Yukon, Canada. Conserv Biol 19: 1262–1272. doi:10.1111/j.1523-1739.2005.00099.x.
60. Fairbanks D (2001) Species and environment representation: selecting reserves for the retention of avian diversity in KwaZulu-Natal, South Africa. Biol Conserv 98: 365–379. doi:10.1016/S0006-3207(00)00179-8.
61. Bradshaw CJA, Bowman DMJS, Bond NR, Murphy BP, Moore AD, et al. (2013) Brave new green world—Consequences of a carbon economy for the conservation of Australian biodiversity. Biol Conserv 161: 71–90.
62. Venter O, Meijaard E, Possingham H, Dennis R, Sheil D, et al. (2009) Carbon payments as a safeguard for threatened tropical mammals. Conserv Lett 2: 123–129.
63. Arcese P, Schuster R, Campbell L, Barber A, Martin TG (2014) Deer density and plant palatability predict shrub cover, richness and aboriginal food value in a North American island archipelago. Divers Distrib in press.
64. Crossman ND, Bryan BA, Summers DM (2011) Carbon Payments and Low-Cost Conservation. Conserv Biol 25: 835–845.
65. Evans MC, Carwardine J, Fensham RJ, Butler DW, Wilson KA, et al. (2014) The economics of biodiverse carbon farming: regrowth management as a viable mechanism for restoring deforested agricultural landscapes. Glob Environ Chang in press.
66. Fishburn IS, Kareiva P, Gaston KJ, Armsworth PR (2009) The Growth of Easements as a Conservation Tool. PLoS One 4: 6. Available: http://www.ncbi.nlm.nih.gov/pubmed/19325711.
67. Gordon A, Langford WT, White MD, Todd JA, Bastin L (2011) Modelling trade offs between public and private conservation policies. Biol Conserv 144: 558–566. Available: http://www.sciencedirect.com/science/article/pii/S0006320710004544.
68. Drechsler M, Wätzold F, Johst K, Bergmann H, Settele J (2007) A model-based approach for designing cost-effective compensation payments for conservation of endangered species in real landscapes. Biol Conserv 140: 174–186.
69. Klimek S, gen Kemmermann A, Steinmann H-H, Freese J, Isselstein J (2008) Rewarding farmers for delivering vascular plant diversity in managed grasslands: A transdisciplinary case-study approach. Biol Conserv 141: 2888–2897.
70. Bunn D, Lubell M, Johnson CK (2013) Reforms could boost conservation banking by landowners. Calif Agric 67.
71. Moon K, Cocklin C (2011) Participation in biodiversity conservation: motivations and barriers of Australian landholders. J Rural Stud 27: 331–342.
72. Knight A, Cowling RM, Difford M, Campbell BM (2010) Mapping Human and Social Dimensions of Conservation Opportunity for the Scheduling of Conservation Action on Private Land. Conserv Biol 24: 1348–1358. Available: http://dx.doi.org/10.1111/j.1523-1739.2010.01494.x.
73. Convention on Biological Diversity (2010) Strategic plan for biodiversity 2011–2020 and the Aichi targets. Secretariat of the Convention on Biological Diversity, Montreal.
74. Carwardine J, Wilson KA, Hajkowicz SA, Smith RJ, Klein CJ, et al. (2010) Conservation Planning when Costs Are Uncertain. Conserv Biol 24: 1529–1537.
75. Pressey RL, Cabeza M, Watts ME, Cowling RM, Wilson KA (2007) Conservation planning in a changing world. Trends Ecol Evol 22: 583–592. doi:10.1016/j.tree.2007.10.001.
76. Newell RG, Pizer WA, Raimi D (2013) Carbon markets 15 years after kyoto: Lessons learned, new challenges. J Econ Perspect 27: 123–146.
77. Lo AY (2013) Carbon trading in a socialist market economy: Can China make a difference? Ecol Econ 87: 72–74.
78. Wang Q (2013) China has the capacity to lead in carbon trading. Nature 493: 273.
79. Bennett JR, Arcese P (2013) Human Influence and Classical Biogeographic Predictors of Rare Species Occurrence. Conserv Biol 27: 417–421.

15

Community Turnover of Wood-Inhabiting Fungi across Hierarchical Spatial Scales

Nerea Abrego[1,2]*, Gonzalo García-Baquero[2], Panu Halme[1,3], Otso Ovaskainen[4], Isabel Salcedo[2]

1 Department of Biological and Environmental Science, University of Jyväskylä, Jyväskylä, Finland, 2 Department of Plant Biology and Ecology, University of the Basque Country (UPV/EHU), Bilbao, Spain, 3 Natural History Museum, University of Jyväskylä, Jyväskylä, Finland, 4 Department of Biosciences, University of Helsinki, Helsinki, Finland

Abstract

For efficient use of conservation resources it is important to determine how species diversity changes across spatial scales. In many poorly known species groups little is known about at which spatial scales the conservation efforts should be focused. Here we examined how the community turnover of wood-inhabiting fungi is realised at three hierarchical levels, and how much of community variation is explained by variation in resource composition and spatial proximity. The hierarchical study design consisted of management type (fixed factor), forest site (random factor, nested within management type) and study plots (randomly placed plots within each study site). To examine how species richness varied across the three hierarchical scales, randomized species accumulation curves and additive partitioning of species richness were applied. To analyse variation in wood-inhabiting species and dead wood composition at each scale, linear and Permanova modelling approaches were used. Wood-inhabiting fungal communities were dominated by rare and infrequent species. The similarity of fungal communities was higher within sites and within management categories than among sites or between the two management categories, and it decreased with increasing distance among the sampling plots and with decreasing similarity of dead wood resources. However, only a small part of community variation could be explained by these factors. The species present in managed forests were in a large extent a subset of those species present in natural forests. Our results suggest that in particular the protection of rare species requires a large total area. As managed forests have only little additional value complementing the diversity of natural forests, the conservation of natural forests is the key to ecologically effective conservation. As the dissimilarity of fungal communities increases with distance, the conserved natural forest sites should be broadly distributed in space, yet the individual conserved areas should be large enough to ensure local persistence.

Editor: Rick Edward Paul, Institut Pasteur, France

Funding: This study was partially funded by a PhD student fellowship to NA by the University of the Basque Country (UPV/EHU) (PIF10/2010/PIF10008), by Maj and Tor Nessling Foundation (a grant to PH), and by the Academy of Finland (grant no. 250444 to OO). The funders had no role in study design, data collection and analysis, decision to publish, or preparation of the manuscript.

Competing Interests: The authors have declared that no competing interests exist.

* Email: nerea.abrego@ehu.es

Introduction

Biodiversity is decreasing faster than ever in the history of the earth, largely as a result of human activity [1]. Consequently, conservation and management actions aiming at sustainable use of ecosystems and maintenance of biological diversity have become a priority worldwide. The most commonly applied conservation measures in terrestrial ecosystems have focused on preserving key habitats, species or communities of conservation interest. To date, most conservation strategies have focused on designating protected set aside areas or increasing the areas of existing protected areas. For example, regarding temperate forests, a considerable proportion of the remaining old-growth and semi-natural forests are already conserved, and the protected area is expected to continue increasing [2]. To make conservation efforts cost-effective in an evidence based manner, protection efforts need to be supported by research about the effects of habitat protection [2]. One factor potentially compromising the cost-effectiveness of biodiversity conservation plans is that the patterns and drivers of distribution and turnover of biodiversity across different spatial scales are still poorly understood for many organism groups [3].

A successful design of conservation area networks requires an understanding of the processes determining species composition and persistence at different spatial and temporal scales [4]. For example, if local processes are the most important factor determining the persistence and diversity of the focal species group (as has been observed for several taxa; [5–7]), it may be more effective to aim management actions or site selection procedures towards maintaining the heterogeneity and variability of habitats at the local scale, instead of e.g. increasing the connectivity among protected areas [5]. The influences of local versus regional processes in community assembly often interact with the traits of the species. For example, in a study of forest-dwelling beetles, Gering et al. [8] found that the spatial turnover of rare species was realised across large spatial scales (ecoregions), whereas the spatial turnover of common species was realised across smaller spatial scales. The most commonly used concept in measuring spatial species turnover is beta-diversity. However, this concept has been defined and used in multiple ways, making it difficult to synthesize studies based on this concept, and leading to a debate on its correct use [9,10].

Wood-inhabiting fungi are a diverse and functionally important group of organisms in forest ecosystems. In addition to their central role in nutrient cycling [11], they play a key role on the creation of new habitats and resources for other wood-dwelling species [12]. Furthermore, their dependence on dead wood makes them especially vulnerable to the changes caused by forest management (e.g. [13,14]). At the resource unit level, most species show preferences to a specific substrate type in terms of host species, log type (snag, log, stump…), decay stage and log size [15]. Therefore, the reduction of particular dead wood types (especially of coarse woody debris) in managed forests has directly led to declines and extinctions of certain species [16,17]. At the stand level, fungal communities are highly influenced by microclimatic conditions, which modify e.g. the water content of dead wood and thus can inhibit or facilitate the colonization and fructification of species [18,19]. Further, the size and isolation of forest stands can significantly influence fungal communities: specialist species have been found to be negatively affected by isolation while generalist species can even benefit from fragmented surroundings [15]. At the broadest scale, wood-inhabiting fungi are influenced by regional factors, related e.g. to climatic variation and management history [20].

Only a few studies have examined the importance of environmental factors at different spatial scales in determining the diversity of wood-inhabiting fungi [21,22]. Berglund et al. [23] applied a hierarchical study design and a hierarchical data analysis to show that habitat suitability determines the presence of polypore species at a small spatial scale. Bässler et al. [24] quantified the relative importance of the plot and forest management levels, showing that in spruce forests, species turnover is higher among plots than among management categories. Based on these previous studies, wood-inhabiting fungal species diversity is much affected by small scale processes, though also regional-scale processes seem important especially for specialized and threatened species. However, we are not aware of any previous studies in which the variation in species richness and community composition have been explicitly partitioned to different spatial scales. Such information would be critically needed to design effective conservation areas.

The overall objective of this study is to partition the variation in the assembly of wood-inhabiting fungal communities into different spatial levels. More specifically, our aim is to examine how community turnover of wood-inhabiting fungi is realised at the levels of forest management types, forest sites and plots, and how much of community variability is explained by variation in resource composition and spatial proximity. We utilize the conclusions of the present study to derive recommendations for the design of protection area networks for the conservation of wood-inhabiting fungi in temperate broadleaved forests.

Materials and Methods

Study permits were provided by the Government of Navarre (the local official institution for forest management) and the field studies did not involve endangered or protected species.

1. Study area

The study was carried out in the northern part of Navarre in northern Spain. The study area covers two biogeographical regions with temperate climate: the Atlantic region in the Northwest and the Alpine region in the Northeast. The predominant forest type is beech-dominated forest (with total area of ca. 61,000 ha, covering ca. 14% of the landscape), which also represents the most productive forest type for commercial harvesting in the area.

The northern part of Navarre consists of a valley-dominated landscape with a mountainous topography. Towns and farms are generally located in the lower parts of the valleys and the forests grow in the hillsides. The major part of the forested area has been intensively managed by thinning and selective cutting, but there is also a network of permanently protected areas (see Files S1 and S2). In this study, we examined the difference between two management types, called henceforth managed and natural forests. As natural forests we considered such protected areas in which logging activities have been prohibited, allowing the development of old-growth beech forests. As managed forests we considered such unprotected sites in which commercial harvesting has been continuously conducted (File S1).

2. Study design

2.1. Sampling design. We assembled a list of natural and managed forest sites based on the above definitions. From this list, eight managed and eight natural forest sites (File S1) were randomly selected. Within each site, fungal data (species abundances, defined as number of logs in which each species was observed) were recorded in five randomly placed 100 m^2 sampling plots (File S2). The sampling plots were separated from each other by tens of meters to hundreds of meters, whereas the study sites were separated by hundreds of meters to kilometres (the minimum, maximum and mean distances between plots were 0.07, 114 and 44 km respectively). The sites belonging to the two management types were distributed in space without a strong spatial autocorrelation, i.e. the managed and natural sites did not form their own clusters. The nested design has thus two factors: management (considered as fixed factor; with two levels) and forest site (considered as random factor; with eight levels, nested within management). Variation among study sites represents spatial variation at a regional scale, whereas the residual variation among the sampling plots represents spatial variation at a local scale.

2.2. Data collection. The fieldwork was carried out in the year 2011 during the main period of fungal fruit body production (from late September to early November). All woody debris (called henceforth resource units) with minimum diameter of 1 cm were measured for diameter and decay class, and examined for the presence of fungal species. The resource units were classified according to their diameter (three classes) and decay stage (five classes), thus forming altogether 15 classes. The diameter classes correspond to those used by Abrego & Salcedo [17]: very fine woody debris (VFWD), including resource units with diameter in the range 1–5 cm; fine woody debris (FWD), with 5 cm<diameter ≤10 cm; and coarse woody debris (CWD), with diameter >10 cm. The decay stages were classified into 5 categories based on a modified version of the classification by Renvall [25].

The presence of all saproxylic macromycetes (fungi with visible fruit bodies) was registered at the level of the resource units. Microscopic identification was carried out in the laboratory when necessary. The literature consulted for identifying fungal fruit bodies is summarized in File S3. Each species found on one resource unit was considered as one record, regardless of the number of fruit bodies. Thus, the abundances of the species correspond to the number of woody debris pieces in which each species was found in each sampling plot.

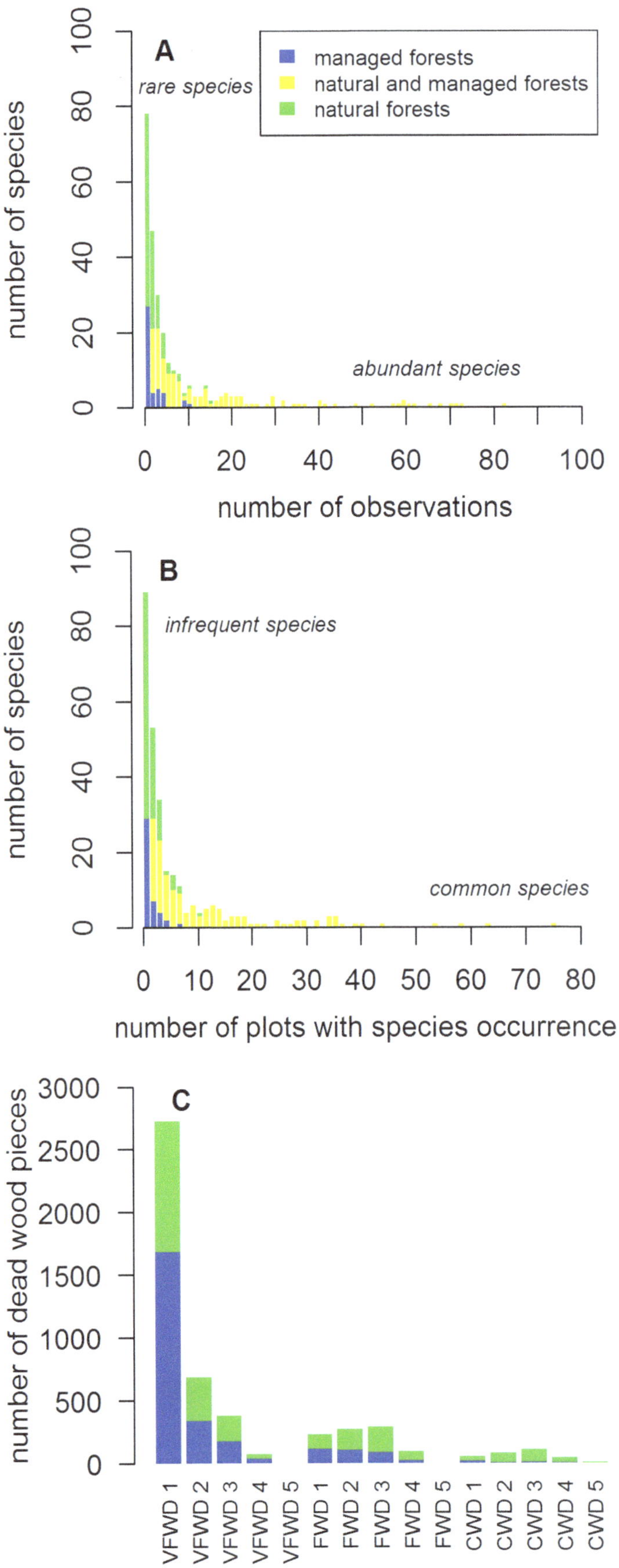
A
rare species
managed forests
natural and managed forests
natural forests
abundant species
number of species
0
20
40
60
80
100
number of observations
B
infrequent species
common species
0
10
20
30
40
50
60
70
80
number of plots with species occurrence
C
number of dead wood pieces
0
500
1000
1500
2000
2500
3000
VFWD 1
VFWD 2
VFWD 3
VFWD 4
VFWD 5
FWD 1
FWD 2
FWD 3
FWD 4
FWD 5
CWD 1
CWD 2
CWD 3
CWD 4
CWD 5

Figure 1. The abundance distributions of species and dead wood types. The total number of occurrences of each species (Panel A), the number of plots in which each species was recorded (Panel B) and the number of occurrences of each dead wood type (Panel C). The species and dead wood pieces that were recorded in managed forests are represented with blue colour, those that were recorded in natural forests with yellow colour, and those species recorded in both forest types with green colour.

3. Statistical analyses

We first conducted a descriptive analysis examining how species richness in fungal communities varies across the three hierarchical scales included in this study. To do so, we constructed accumulation curves in which the total number of species was computed as a function of the number of surveyed plots, assuming that the plots were selected randomly either a) within the same randomly selected managed forest site, b) within the same randomly selected natural forest site, c) among all sites in managed forests, d) among all sites in natural forests, or e) among all sites. We estimated the mean number of species and the interquartile (25%–75%) range for each of these five cases and for each number of sampling plots using 1000 random replicates.

We supplemented the accumulation curves by an additive partitioning in species richness at different hierarchical scales [26,27]. In this methodology, the total diversity γ (here, species

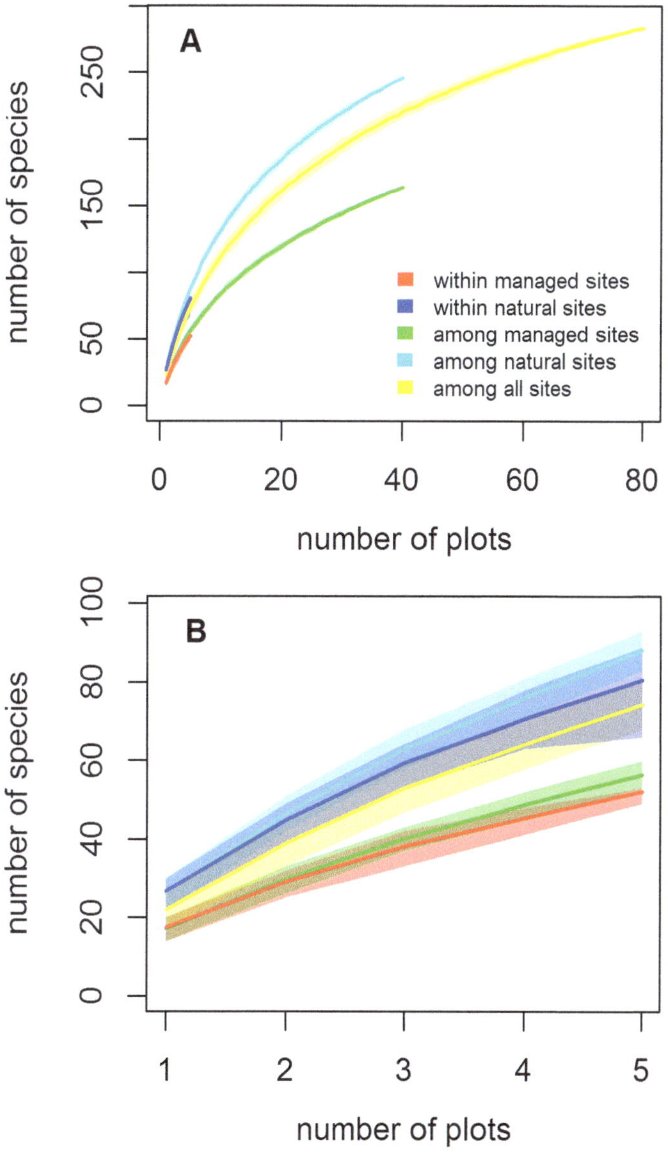

Figure 2. Species accumulation curves showing the total number of species in a given number of plots. The thick lines show means and the shaded regions the interquartile (25%–75%) ranges. The different lines correspond to the five cases which assume that the plots were selected randomly either within the same randomly selected managed forest site (red), the same randomly selected natural forest site (dark blue), among all sites in managed forests (green), among all sites in natural forests (light blue), or among all sites (yellow). Note that the maximum number of plots for which the number of species could be computed is either 5, 40 or 80, depending on the category. Panel A shows the full data, panel B a zoom, in which the number of plots varies from 1 to 5.

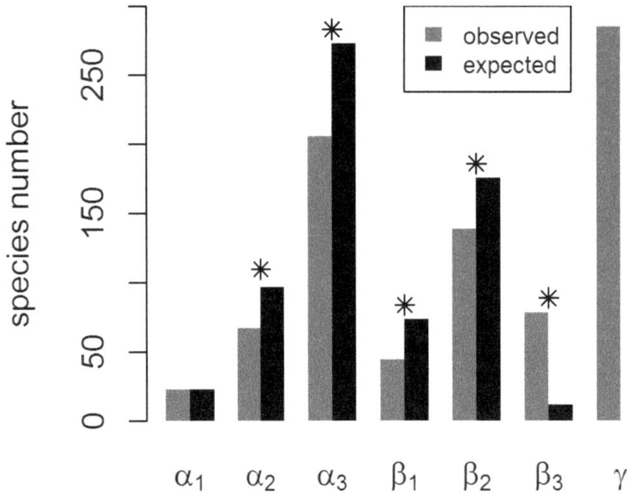

Figure 3. Total species richness (γ) explained by alpha and beta components of diversity at three spatial scales: the mean species number per plot (α_1), the mean species number per site (α_2), the mean species number per management category (α_3), the average addition of species from plots to sites (β_1), the average addition of species from sites to management categories (β_2), the average addition of species from management categories to the total species number (β_3) and the total species number (γ). The grey bars correspond to the observed values and the black bars to the expected values based on a null model. The asterisks indicate the levels of statistical significance (* corresponding to $p<0.001$) for a difference between observed and expected values.

richness) is modelled as a sum of the average diversities within sampling units (α) and among sampling units (β), the latter across the scales included in the hierarchical design. In our case, the partitioning of species richness γ was thus expressed as $\gamma = \alpha_{1(plots)} + \beta_{1(plots)} + \beta_{2(sites)} + \beta_{3(management)}$, where α_1 is the mean species number in the sampling plots, β_1 is the turnover (beta-diversity) between sampling plots, β_2 is the turnover between forest sites, and β_3 is the turnover between managed and natural forests. The additive partitioning and the related significance tests were performed under 999 permutations of the null model r0 of the function *adipart* in the R-package *vegan* [28]. The null model r0 is suited to binary (presence/absence) data, and it randomizes the community matrix in a way that preserves the number of species observed in each site.

We then analysed variation in community composition at the different levels of our hierarchical study design by calculating Bray-Curtis similarity measures [29] for all pairs of sampling plots. We constructed similarity matrices both for the fungal data (square root transformed species abundances, organised as a matrix for 80 sampling plots by 285 species) and for the dead wood data (square root transformed counts of resource unit types, organised as a matrix of 80 sampling plots by 15 resource unit types). We examined how much of the similarity in dead wood composition (SDC) was explained by the two hierarchical factors of management and site, as well as spatial distance (D). We examined how much of the similarity in fungal community composition (SFC) was explained by management and site, as well as by similarity in dead wood composition (SDC) and spatial distance (D). To do so, we constructed matrices measuring similarity in site (SS; with value 1 if the plots belonged to the same site and 0 otherwise), similarity in management (SM; with value 0 if the plots belonged to a different management category, and values 1 or 2 if both plots belonged to the managed or to the natural category, respectively) and log-transformed spatial distance (D) between the sampling plots. We treated SS and SM as fixed factors and SDC and D as continuous covariates. In the models (SDC = SS+SM+D and SFC = SS+SM+SDC+D) we considered also all two-way interactions (except for D). We used backward variable selection to include in the final model only the statistically significant variables. To account for the non-independence of data points that relate to pairs of plots rather than to individual plots, we tested for the statistical significance by comparing the *F-values* related to each factor to the null distribution of *F-values*. To compute the null distribution, we constructed 999 permutations of the species community, in each of which we permuted randomly the identity numbers of the sampling plots. As a complementary approach to quantify the variation on species and resource similarity explained by each scale, we carried out a permutational multivariate analysis of variance (PERMANOVA; [30]). The *pseudo-F* values were used to evaluate which components of the models explained more variability than expected by a null model. This analysis was performed using 999 permutations of residuals under a reduced model with the function *nested.npmanova* of the R-package *BiodiversityR*.

All statistical analyses were performed using R software [31], the code and the original data (the fungal species data matrix, UTM locations of all sampling plots and resource type data matrix) being available in the Supporting Information section (Files S4–S7).

Results

1. Variation in species richness and resource types

We recorded in total 3241 occurrences of 285 species in 2724 dead wood pieces. The species abundance distribution was dominated by rare species (Fig. 1A), with 27% of the recorded species (2.4% of the occurrences) being represented only once in our data. The five most abundant species were simultaneously the five species with highest prevalence at the plot level, comprising over 25% of all occurrences. In general, the species with high abundances were found in both (managed and natural) forest types (Fig. 1, A and B; black bars). Nevertheless, note that in the tail of Fig. 1 (A) there are also some species that were found only in managed forests. These species are *Steccherinum fimbriatum* and *Hyphodermella rosae*, which were recorded in 7 managed plots, 18 and 11 times respectively.

Table 1. Linear models for similarity in fungal and dead wood composition.

Response variable: Similarity in fungal composition (SFC)						
Source of variation	df	Estimate	SSq	MS	F	p
Management (SM)	2	MM: 0 MU: −0.060 UU: −0.047	1.93	0.96	152	<0.001
Site (SS)	1	Different: 0 Same: 0.019	0.71	0.71	112	<0.001
Similarity in dead wood composition (SDC)	1	0.11	1.11	1.11	175	<0.001
Spatial distance (D)	1	−0.02	0.46	0.46	73	<0.001
Residual	3154		20.02	0.01		
Response variable: Similarity in dead wood composition (SDC)						
Source of variation	df	Estimate	SSq	MS	F	p
Management (SM)	2	MM: 0 MU: −0.067 UU: −0.016	3.07	1.54	192	<0.001
Site (SS)	1	Different: 0 Same: 0.019	0.05	0.05	6.58	<0.001
Residual	3156		25.26	0.01		

df = degrees of freedom, SSq = sum of squares, MS = mean squares, MM = pairs of plots from managed forests, MU = pairs of plots belonging to a different management types, UU = pairs of plots from natural (unmanaged) forests. Statistical significance (*p*) based on permutation test.

The dead wood profile was numerically strongly dominated by small diameter classes in early decay stages (Fig. 1C). The coarser dead wood classes in the latest decay stages were the most uncommon, and were mostly found in natural forests.

The species accumulation curves indicate that natural forests were substantially more species rich than managed forests (Fig. 2A). Further, the number of species increased much faster if the sampling plots were located only in natural forests (Fig. 2A; light blue line) than if they were located both in natural and managed forests (Fig. 2A; yellow line), suggesting that the species assemblages in managed forests were largely a subset of those present in natural forests. The total number of species increased only marginally faster if the sampling plots were spread across forest sites than if they were located within a forest site, both in managed (Fig. 2B; green and red lines) and natural (Fig. 2B; light blue and dark blue lines).

The additive partitioning of total species richness (Fig. 3) was in line with the patterns visible in the species accumulation curves. The increase in species number when moving from the plot level to the forest site level (β_1) was lower than would expected by chance, and consequently so was also the number of species within a site, averaged across the sites (α_2). Similarly, the increase in species number when moving from the site level to the management level (β_2) was lower than would expected by chance, and consequently so was also the mean number of species within a management category, averaged across the two management categories (α_3). Both of these results reflect the fact that the most important part of species diversity was realized among the management categories (β_3). This is consistent with Fig. 2, which shows that natural forests were much more species rich than management forests, and that managed forests largely contained a subset of the species found from natural forests. Note that α_3 corresponds to the mean of the light blue and green lines in Fig. 2, which indeed is below the yellow line, which corresponds to a null expectation.

2. Variation in community and resource similarities

After applying the model selection procedure, the model for similarity in fungal community composition (SFC) included as explanatory variables management type similarity (SM), forest site similarity (SS), similarity in dead wood composition (SDC) and spatial distance (D) (Table 1). The model for similarity in dead wood composition (SDC) included management type similarity (SM) and forest site similarity (SS) but not spatial distance (Table 1). In both models none of the two-way interactions were significant (Permutation comparison of *F-values*, $p>0.05$). The variability explained by both models was low, as only the 17% of the total variability was explained in the SFC model and 11% in the SDC model. According to the predicted values of these models, the plots within managed forests showed higher fungal and dead wood similarities than the plots within natural forests, and the lowest fungal and dead wood similarities were found among pairs of plots out which one belonged to a natural and the other one to a managed forest (Fig. 4A). The effect of forest site similarity (SS) on the similarity of both fungal composition (SFC) and dead wood was small but yet significant (Fig. 4, Table 1). Fungal community similarity decreased with increasing spatial distance (Fig. 4B).

The results from the Permanova analysis (Table 2) were in line with the results of the linear model described above: for both fungal and dead wood species composition, both forest management type and forest site level explained a significant amount of variation. For dead wood composition, the variation explained by forest management type was much higher than that explained by forest site level.

Discussion

We found that the assembly of fungal communities was influenced by all scales that we considered in this study. Indeed, the similarity of fungal communities was higher within sites and within management categories than among sites or between the two management categories; it decreased with increasing distance

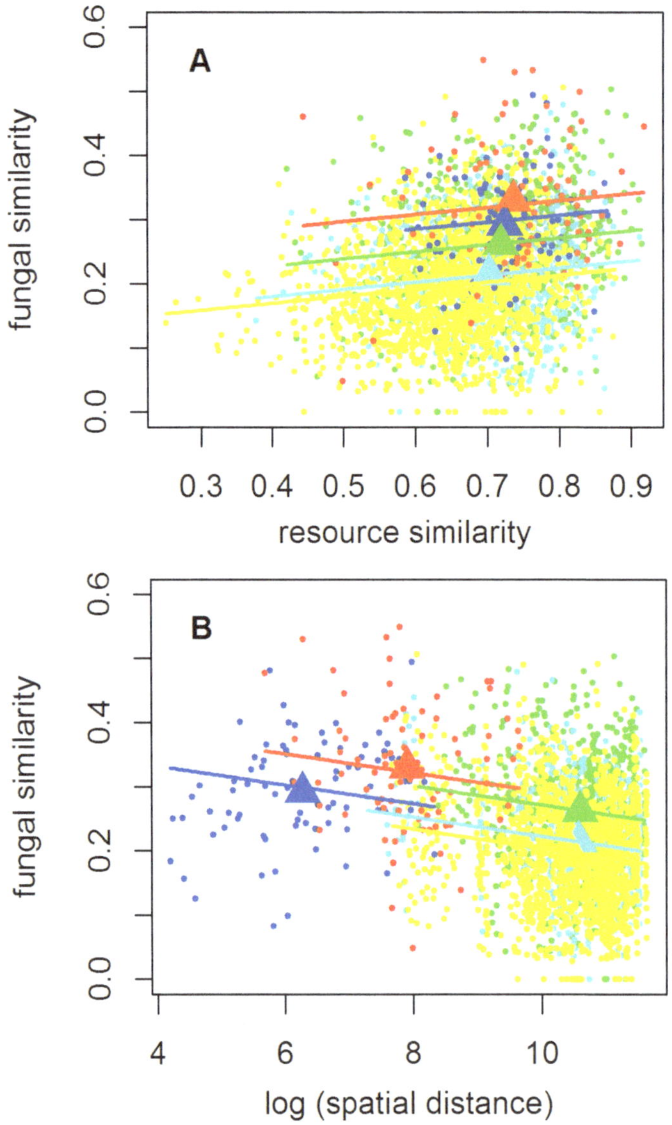

Figure 4. Predicted similarity of the fungal composition (SFC), among sampling plots. When two sampling plots are located either within the same managed forest site are represented by red colour, when located within the same natural forest site are represented by dark blue colour, when located in two different sites in managed forests are represented by green colour, when located in two different sites in natural forests are represented by light blue colour, and when plots are located in sites from different management types are represented by yellow colour. The dots show raw data, the triangles the mean values and the lines the model predictions. When constructing the model prediction as a function of resource similarity (panel A), we normalized spatial distance to its mean value within each category. When constructing the model prediction as a function of spatial distance (panel B), we normalized resource similarity to its mean value within each category.

among the sampling plots and increased with increasing similarity of dead wood resources. However, the majority of the variation remained unexplained, suggesting that biological interactions, neutral mechanisms such as inherent randomness in dispersal, establishment and fruiting [32], as well as environmental constraints not considered here are likely to play an important role in determining local assemblies of wood-inhabiting fungi. This result partly reflects the fact that the wood-inhabiting fungal communities studied here were dominated by locally rare and infrequent species, which directly leads to a high species turnover from one plot to another. This result is in line with a previous study in a coniferous forest where a high variability of wood inhabiting fungal species at small scales was also detected [24].

Bässler et al. [24] partitioned wood-inhabiting fungal richness between plots and management categories, with the result that community turnover was higher between study plots than between management categories. Thus Bässler et al. [24] concluded that the stand level is the most relevant scale for the conservation the wood-inhabiting diversity. In contrast, we found that fungal communities were significantly influenced by all scales that we considered, and in particular by the management category, as natural forests were significantly more diverse than managed forests. The difference between our results and those of Bässler et al. [24] can be at least partially attributed to a difference in the study design, as the study of Bässler et al. [24] involved only a single forest site for each management category.

Table 2. Output from the Permanova analysis for fungal species and dead wood composition models.

		Response variable: Similarity in fungal composition (SFC)				*Response variable: Similarity in dead wood composition (SDC)*			
Source of variation	**df**	**SS**	**MS**	***Pseudo-F***	***p***	**SSq**	**MS**	***Pseudo-F***	***p***
Management	1	17737	17737	3.53	0.001	8549.2	8549.2	12.45	0.001
Site	14	70343	5024.5	2.08	0.001	9612.1	686.58	1.74	0.002
Residual	64	154670	2416.7			25288	395.13		

df = degrees of freedom, SSq = sum of squares, MS = mean squares.

In general, wood-inhabiting fungal species show preferences for certain resource types in terms of tree species, size and decay class [15], so from a niche differentiation perspective [33], variation in resource type is expected to explain a major part of species variation at a small scale. In our study, although increasing similarity in resource composition (as measured by dead wood decay class and size) significantly increased the similarity of wood-inhabiting fungal communities, the trend was weak. Hence, at a small spatial scale, wood-inhabiting fungal communities seem to be largely influenced by stochastic processes related to species' idiosyncrasy and spatial proximity, rather than habitat factors. Some hierarchically designed studies have reported similar results, and showed that the variation in the occurrence of species among different forest stands [23] or study plots [34] remains largely unexplained even when the most important explanatory variables were considered.

The dispersal capabilities of species can be considered as a species-specific factor limiting potentially their occurrence on suitable habitats [35,36]. One of the covariates that significantly influenced fungal species similarity in our study was the spatial distance, with the result that similarity in fungal composition decreased with increasing distance between the study plots. This result suggests that dispersal limitation is one major factor shaping the local assembly of fungal communities.

Other factors creating unexplained variation between study units are related to the low detectability of wood-inhabiting fungi [37], partly due to their fruiting phenology and its dependence on microclimatic variation [38]. Many fungal species have ephemeral, short living fruit bodies, which makes it difficult to assess fungal diversity by a single fruit body based survey [37]. Further, a large proportion of fungal species can never be detected as fruit bodies [39]. With presence-absence data, low detectability generates a larger amount of random variation on small spatial scales because on larger scale the increasing number of potentially detectable individuals increases the likelihood that at least one is detected [40]. Thus, also in our study, the low general detectability of fungi as well as the climatic dependence of their phenology may have inflated the variation realized at the smallest spatial scale.

Due to the overall high number of species and the large proportion of rare species, the species accumulation curves we constructed were far from saturated. The only way to achieve saturated accumulation curves for hyperdiverse organism groups consists of sampling intensively over a limited area during several years [8,41]. Nevertheless, in spatially extensive studies even unsaturated accumulation curves provide a suitable method for the assessment of the heterogeneity of species diversity [8].

It is well documented that wood-inhabiting fungal communities in natural forests are significantly more species rich and different from those in managed forests (e.g. [16,42,43]), including the communities in beech forests [17]. The main factors behind this difference are related to resource availability and connectivity at forest stand level [15,42]. Our previous study [17] showed that species specialised in utilizing certain dead wood types were absent from managed forests, due to the absence of suitable resource type. Based on the previous [17] and present results, the community differences between different management categories are not related to qualitative differences in species compositions, but to quantitative differences, in the sense that the species from managed forests are a subset from those in natural ones (because their required resource has disappeared). Therefore we cannot say that the natural and managed forests have different species compositions, but rather that natural forests have more species. Indeed, the general trend reported in the literature is that the reduction of particular dead wood pieces caused by forestry practices in managed forests causes extinctions of habitat specialist saproxylic species (e.g. [44–46]). Thus, forest management significantly affects resource composition, and managed forests are more homogeneous in their dead wood and fungal species compositions. Nevertheless we found some species to be more abundant in managed forests than in natural ones, in particular *Steccherinum fimbriatum* and *Hyphodermella rosae*. Consistently with our results, earlier studies have shown that there are species adapted to disturbances, such as those caused by forest management [15,47].

Conservation implications and conclusions

Identifying the spatial scales at which species vary in their diversity and composition is an important research goal in conservation ecology [4,48]. Additive partitioning of diversity is one of the most commonly applied statistical approaches for quantifying the fraction of diversity explained by different scales and for concluding which scales are critical for species conservation (e.g. [5–8]). We found that fungal communities show a high level of turnover at all the studied scales, and that only a small part of the turnover can be explained by factors such as spatial proximity or resource type similarity. The large number of random variation and high number of rare species suggests that a large total area of protected forests is needed to conserve a substantial part of fungal diversity. As managed forests contributed only little to the total number of species, an ecologically effective way to conserve fungal communities necessarily requires conserving natural forests. Nevertheless, leaving dead wood in managed forests can be seen an additional supplementary method, partly because some rare species only appear in managed forests, and partly because increasing the quality of the matrix (here, managed forests) can be essential for ensuring the long-term viability of populations in fragmented landscapes [49]. We found the similarity of fungal communities to decrease with increasing spatial distance, suggesting that protected areas should be distributed thorough the landscape. However, the individual protected areas or local networks of protected areas must simultaneously be kept large enough to ensure the persistence of local populations, which process we did not however consider in the present study.

Supporting Information

File S1 Principal characteristics of the sampled beech forest sites.

File S2 A map of the sampled forest sites and plots in Navarre.

File S3 Literature consulted for the identification of wood-inhabiting species.

File S4 Code for performing the analyses in R.

File S5 Fungal species data.

File S6 Geographical UTM coordinates.

File S7 Dead wood composition data.

Acknowledgments

We are grateful to the Government of Navarre for giving the sampling permissions. We also thank A. Chiarucci and an anonymous reviewer for their valuable comments on the manuscript.

Author Contributions

Conceived and designed the experiments: NA GG-B IS. Performed the experiments: NA. Analyzed the data: NA GG-B OO. Contributed to the writing of the manuscript: NA GG-B PH OO IS.

References

1. Barnosky AD, Matzke N, Tomiya S, Wogan GOU, Swartz B, et al. (2011) Has the earth/'s sixth mass extinction already arrived? Nature 471: 51–57. Available: http://dx.doi.org/10.1038/nature09678.
2. Götmark F. (2013) Habitat management alternatives for conservation forests in the temperate zone: Review, synthesis, and implications. For Ecol Manage 306: 292–307. Available: http://dx.doi.org/10.1016/j.foreco.2013.06.014.
3. Chiarucci A, Bacaro G, Rocchini D. (2008) Quantifying plant species diversity in a natura 2000 network: Old ideas and new proposals. Biol Conserv 141: 2608–2618. Available: http://dx.doi.org/10.1016/j.biocon.2008.07.024.
4. Summerville KS, Boulware MJ, Veech JA, Crist TO. (2003) Spatial variation in species diversity and composition of forest lepidoptera in eastern deciduous forests of north america. Conserv Biol 17: 1045–1057. 10.1046/j.1523-1739.2003.02059.x.
5. Wagner H, Wildi O, Ewald K. (2000) Additive partitioning of plant species diversity in an agricultural mosaic landscape. Landscape Ecol 15: 219–227. Available: http://dx.doi.org/10.1023/A%3A1008114117913.
6. Summerville KS, Crist TO. (2003) Determinants of lepidopteran community composition and species diversity in eastern deciduous forests: Roles of season, eco-region and patch size. Oikos 100: 134–148. 10.1034/j.1600-0706.2003.11992.x.
7. Müller J, Goβner MM. (2010) Three-dimensional partitioning of diversity informs state-wide strategies for the conservation of saproxylic beetles. Biol Conserv 143: 625–633. Available: http://dx.doi.org/10.1016/j.biocon.2009.11.027.
8. Gering JC, Crist TO, Veech JA. (2003) Additive partitioning of species diversity across multiple spatial scales: Implications for regional conservation of biodiversity. Conserv Biol 17: 488–499. 10.1046/j.1523-1739.2003.01465.x.
9. Tuomisto H. (2010) A consistent terminology for quantifying species diversity? yes, it does exist. Oecologia 164: 853–860. Available: http://dx.doi.org/10.1007/s00442-010-1812-0.
10. Anderson MJ, Crist TO, Chase JM, Vellend M, Inouye BD, et al. (2011) Navigating the multiple meanings of β diversity: A roadmap for the practicing ecologist. Ecol Lett 14: 19–28. 10.1111/j.1461-0248.2010.01552.x.
11. Boddy L, Frankland JC, van West P. (2008) Ecology of saprotrophic basidiomycetes. London, United Kingdom: Elsevier Academic Press. Available: http://dx.doi.org/10.1016/S0275-0287(08)80001-X.
12. Stokland JN, Siitonen J, Jonsson BG. (2012) Biodiversity in dead wood. Cambridge: Cambridge University Press.
13. Penttilä R, Siitonen J, Kuusinen M. (2004) Polypore diversity in managed and old-growth boreal picea abies forests in southern finland. Biol Conserv 117: 271–283. 10.1016/j.biocon.2003.12.007.
14. Hottola J, Ovaskainen O, Hanski I. (2009) A unified measure of the number, volume and diversity of dead trees and the response of fungal communities. J Ecol 97: 1320–1328. 10.1111/j.1365-2745.2009.01583.x.
15. Nordén J, Penttilä R, Siitonen J, Tomppo E, Ovaskainen O. (2013) Specialist species of wood-inhabiting fungi struggle while generalists thrive in fragmented boreal forests. J Ecol 101: 701–712. 10.1111/1365-2745.12085.
16. Küffer N, Senn-Irlet B. (2005) Influence of forest management on the species richness and composition of wood-inhabiting basidiomycetes in swiss forests. Biodivers Conserv 14: 2419–2435. 10.1007/s10531-004-0151-z.
17. Abrego N, Salcedo I. (2013) Variety of woody debris as the factor influencing wood-inhabiting fungal richness and assemblages: Is it a question of quantity or quality? For Ecol Manage 291: 377–385. 10.1016/j.foreco.2012.11.025.
18. Heilmann-Clausen J. (2001) A gradient analysis of communities of macrofungi and slime moulds on decaying beech logs. Mycol Res 105: 575–596. Available: http://dx.doi.org/10.1017/S0953756201003665.
19. Siitonen P, Lehtinen A, Siitonen M. (2005) Effects of forest edges on the distribution, abundance, and regional persistence of wood-rotting fungi. Conserv Biol 19: 250–260. 10.1111/j.1523-1739.2005.00232.x.
20. Ódor P., Heilmann-Clausen J., Christensen M., Aude E., Dort K., et al. (2006) Diversity of dead wood inhabiting fungi an bryophytes in semi-natural beech forest in Europe. Biol Conserv. 131: 58–71. Available: http://dx.doi.org/10.1016/j.biocon.2006.02.004.
21. Junninen K, Komonen A. (2011) Conservation ecology of boreal polypores: A review. Biol Conserv 144: 11–20. 10.1016/j.biocon.2010.07.010.

22. Sverdrup-Thygeson A, Gustafsson L, Kouki J. (2014) Spatial and temporal scales relevant for conservation of dead-wood associated species: Current status and perspectives. Available: http://dx.doi.org/10.1007/s10531-014-0628-3.
23. Berglund H, Hottola J, Penttilä R, Siitonen J. (2011) Linking substrate and habitat requirements of wood-inhabiting fungi to their regional extinction vulnerability. Ecography 34: 864–875. 10.1111/j.1600-0587.2010.06141.x.
24. Bässler C, Müller J, Svoboda M, Lepšová A, Hahn C, et al. (2012) Diversity of wood-decaying fungi under different disturbance regimes–a case study from spruce mountain forests. Biodivers Conserv 21: 33–49. 10.1007/s10531-011-0159-0.
25. Renvall P. (1995) Community structure and dynamics of wood-rotting Basidiomycetes on decomposing conifer trunks in northern Finland. Karstenia 35: 1–51.
26. Lande R. (1996) Statistics and partitioning of species diversity, and similarity among multiple communities. Oikos 76: 5–13. Available: http://www.jstor.org/stable/3545743.
27. Crist TO, Veech JA, Gering JC, Summerville KS. (2003) Partitioning species diversity across landscapes and regions: A hierarchical analysis of α, β, and γ diversity. Am Nat 162: 734–743. Available: http://www.jstor.org/stable/10.1086/378901.
28. Oksanen J, Blanchet FG, Roeland R, Legendre P, Minchin PR, et al. (2013) Community ecology package "vegan", R package version 2.0–7.
29. Bray JR, Curtis JT. (1957) An ordination of the upland forest communities of southern wisconsin. Ecol Monogr 27: 325–349.
30. Anderson MJ. (2001) A new method for non-parametric multivariate analysis of variance. Austral Ecol 26: 32–46. 10.1111/j.1442-9993.2001.01070.pp.x.
31. R Development Core Team. (2013) R: A language and environment for statistical computing. Vienna, Austria: R Foundation for Statistical Computing.
32. Hubbell SP. (2001) The unified neutral theory of biodiversity and biogeography. Princeton, N. J.: Princeton Univ. Press.
33. Chase JM, Leibold MA. (2003) Ecological niches: Linking classical and contemporary approaches. Chicago, Illinois, USA: University of Chicago Press.
34. Abrego N, Salcedo I. (2014) Response of wood-inhabiting fungal community to fragmentation in a beech forest landscape. Fungal Ecol 8: 18–27. Available: http://dx.doi.org/10.1016/j.funeco.2013.12.007.
35. Edman M, Gustafsson M, Stenlid J, Ericson L. (2004) Abundance and viability of fungal spores along a forestry gradient - responses to habitat loss and isolation? Oikos 104: 35–42. 10.1111/j.0030-1299.2004.12454.x.
36. Norros V, Penttilä R, Suominen M, Ovaskainen O. (2012) Dispersal may limit the occurrence of specialist wood decay fungi already at small spatial scales. Oikos 121: 961–974. 10.1111/j.1600–0706.2012.20052.x.
37. Halme P, Kotiaho JS. (2012) The importance of timing and number of surveys in fungal biodiversity research. Biodivers Conserv 21: 205–219. 10.1007/s10531-011-0176-z.
38. Moore D, Gange AC, Gange EG, Boddy L. Chapter 5 fruit bodies: Their production and development in relation to environment. In: Anonymous British Mycological Society Symposia Series: Academic Press. pp.79–103. Available: http://dx.doi.org/10.1016/S0275-0287(08)80007-0.
39. Halme P, Heilmann-Clausen J, Rämä T, Kosonen T, Kunttu P. (2012) Monitoring fungal biodiversity – towards an integrated approach. Fungal Ecol 5: 750–758. Available: http://dx.doi.org/10.1016/j.funeco.2012.05.005.
40. MacKenzie DI, Nichols JD, Royle JA, Pollock KH, Bailey LL, et al. (2006) Occupancy estimation and modeling: Inferring patterns and dynamics of species occurrence. San Diego, USA: Elsevier.
41. Novotný V, Basset Y. (2000) Rare species in communities of tropical insect herbivores: Pondering the mystery of singletons. Oikos 89: 564–572. 10.1034/j.1600-0706.2000.890316.x.
42. Penttilä R, Lindgren M, Miettinen O, Rita H, Hanski I. (2006) Consequences of forest fragmentation for polyporous fungi at two spatial scales. Oikos 114: 225–240. 10.1111/j.2006.0030-1299.14349.x.
43. Juutilainen K, Mönkkönen M, Kotiranta H, Halme P. (2014) The effects of forest management on wood-inhabiting fungi occupying dead wood of different diameter fractions. For Ecol Manage 313: 283–291. Available: http://dx.doi.org/10.1016/j.foreco.2013.11.019.
44. Ódor P, Standovár T. (2001) Richness of bryophyte vegetation in near-natural and managed beech stands: The effects of management-induced differences in dead wood. Ecol Bull: 219–229. Available: http://www.jstor.org/stable/20113278.
45. Siitonen J. (2001) Forest management, coarse woody debris and saproxylic organisms: Fennoscandian boreal forests as an example. Ecol Bull: 11–41. Available: http://www.jstor.org/stable/20113262.
46. Stokland J, Kauserud H. (2004) Phellinus nigrolimitatus–a wood-decomposing fungus highly influenced by forestry. For Ecol Manage 187: 333–343. Available: http://dx.doi.org/10.1016/j.foreco.2003.07.004.
47. Stokland JN, Larsson K. (2011) Legacies from natural forest dynamics: Different effects of forest management on wood-inhabiting fungi in pine and spruce forests. For Ecol Manage 261: 1707–1721. 10.1016/j.foreco.2011.01.003.
48. Dray S, Pélissier R, Couteron P, Fortin M-, Legendre P, et al. (2012) Community ecology in the age of multivariate multiscale spatial analysis. Ecol Monogr 82: 257–275. 10.1890/11-1183.1. Available: http://dx.doi.org/10.1890/11-1183.1.
49. Jules ES, Shahani P. (2003) A broader ecological context to habitat fragmentation: Why matrix habitat is more important than we thought. J Veg Sci 14: 459–464. 10.1111/j.1654-1103.2003.tb02172.x.

Aboveground Tree Growth Varies with Belowground Carbon Allocation in a Tropical Rainforest Environment

James W. Raich[1]*, Deborah A. Clark[2], Luitgard Schwendenmann[3], Tana E. Wood[4,5]

1 Department of Ecology, Evolution & Organismal Biology, Iowa State University, Ames, Iowa, United States of America, **2** Department of Biology, University of Missouri-St. Louis, St. Louis, Missouri, United States of America, **3** School of Environment, The University of Auckland, Auckland, New Zealand, **4** United States Department of Agriculture Forest Service, International Institute of Tropical Forestry, Rio Piedras, Puerto Rico, **5** Fundación Puertorriqueña de Conservación, San Juan, Puerto Rico

Abstract

Young secondary forests and plantations in the moist tropics often have rapid rates of biomass accumulation and thus sequester large amounts of carbon. Here, we compare results from mature forest and nearby 15–20 year old tree plantations in lowland Costa Rica to evaluate differences in allocation of carbon to aboveground production and root systems. We found that the tree plantations, which had fully developed, closed canopies, allocated more carbon belowground - to their root systems - than did mature forest. This increase in belowground carbon allocation correlated significantly with aboveground tree growth but not with canopy production (*i.e.*, leaf fall or fine litter production). In contrast, there were no correlations between canopy production and either tree growth or belowground carbon allocation. Enhanced allocation of carbon to root systems can enhance plant nutrient uptake, providing nutrients beyond those required for the production of short-lived tissues such as leaves and fine roots, and thus enabling biomass accumulation. Our analyses support this deduction at our site, showing that enhanced allocation of carbon to root systems can be an important mechanism promoting biomass accumulation during forest growth in the moist tropics. Identifying factors that control when, where and for how long this occurs would help us to improve models of forest growth and nutrient cycling, and to ascertain the role that young forests play in mitigating increased atmospheric carbon dioxide.

Editor: Dafeng Hui, Tennessee State University, United States of America

Funding: Funding for measurements of the mature forest plots was provided by the U.S. National Science Foundation (http://www.nsf.gov) grants DEB-9629245, BCE-0421178, EAR-0421178 and DEB-0841872; the U.S. Department of Energy (http://science.energy.gov/funding-opportunities/find-funding/) grant DE-FG02-96ER62289; the Andrew W. Mellon Foundation (http://www.mellon.org/grant_programs/programs); and Conservation International's TEAM Initiative (http://www.conservation.org). Plantation studies were funded by U.S. National Science Foundation grants DEB-0236502 and DEB-0703561. Leaf fall N studies were funded by the University of Virginia. The funders had no role in study design, data collection and analysis, decision to publish, or preparation of the manuscript.

Competing Interests: The authors have declared that no competing interests exist.

* E-mail: jraich@iastate.edu

Introduction

Forests provide a variety of products and services to human societies, sustain diverse flora and fauna and, through their interception and processing of solar energy and precipitation, influence climate and the composition of the atmosphere [1], [2]. Continued loss of forest cover, and accelerating rates of loss in the tropics [3], impart importance to the understanding of processes that promote forest growth and of recovery of the benefits they provide [4]. For example, biomass accrual in secondary forests and plantations partially mitigates the carbon dioxide (CO_2) emissions that result from land use changes [5]. Biomass accumulation during forest stand development is one of the most paradigmatic processes in ecology [6], [7] and, on one level, is readily understood: the growth of a cohort of small trees to larger sizes through time reflects biomass accrual at the stand level. However, the processes that control the strength and duration of carbon (C) accumulation in growing forests remain poorly defined. Rates of biomass C accumulation can be particularly rapid in young tropical forests and plantations [8], [9], [10], but sustaining rapid tree growth requires substantial supplies of nutrients [11]. This implies, and we propose, that rates of aboveground biomass accumulation and belowground C allocation (BCA) in growing tropical forests are coupled, such that both processes vary together and independently of canopy production. We tested these hypotheses, and alternatives, using evidence provided by extensive measurements of key aspects of C cycling in young plantations and mature forest stands growing on the same soils in lowland Costa Rica.

Forest growth depends upon the capture of aerial resources, primarily sunlight and CO_2, by leaves, and of soil resources, primarily water and nutrients, by fine-root systems (we use this terminology to emphasize that mycorrhizas are specifically included). Increased rates of leaf production and tree growth must be balanced by increased uptake of soil resources by root systems. Plants may adjust their allocation of available resources to balance the uptake of multiple limiting resources [12], [13]. However, there is a fundamental difference between allocation of assimilated C to leaves versus to fine-root systems. Leaves absorb, reflect and transmit incoming solar radiation and, in so doing, reduce the amount of light reaching lower leaves. Once full canopy coverage has developed in a growing forest stand, the production of additional leaf area provides a diminishing return of photosynthetic products [14]. Photosynthesis saturates at high leaf area index (LAI, the area of leaves per unit ground area), although other factors such as leaf nitrogen (N) content and display angle are also involved [15]. Maximum LAI may develop relatively quickly in growing forests but does not thereafter increase, and may decrease [16].

Does a similar limit exist belowground? It is theoretically possible. Fine roots and mycorrhizas absorb nutrients from the soil and deplete soil nutrient concentrations within the rhizosphere, with the extent of depletion diminishing with distance from an absorbing root or hyphal surface [17]. At high densities, fine-root systems could efficiently capture available soil nutrients such that growth and maintenance of more extensive root systems would have diminishing values. However, in contrast to many theoretical and empirical studies that support the existence of a maximum sustainable leaf area, there is no parallel evidence that forest soils frequently are saturated with fine-root systems. Young secondary forests and plantations in the moist tropics can grow very rapidly [9], [10]. However, tree biomass contains nutrients and biomass accumulation entails nutrient sequestration in relatively long-lived woody tissues. Rapidly growing tropical forests may therefore be particularly prone to developing nutrient limitations. Increased allocation of newly fixed C to the support, regeneration, and expansion of fine-root systems can enhance plant nutrient uptake. By increasing the amount of absorbing surface area per volume of soil, more extensive fine-root systems may more effectively intercept available soil nutrients [17]; more effectively exploit nutrient-rich hot spots and transient pulses of nutrient availability [18]; and generally compete more successfully with soil microbes for nutrients. Some tree species produce root nodules that support mutualistic N-fixing bacteria [19]. Almost all trees enhance their nutrient uptake capacities by supporting mycorrhizas [20]. Many plant roots release organic compounds into the soil that promote soil organic matter turnover and nutrient availability to plants [21], [22], [23]. Via these and other processes, greater plant allocation of available substrates to fine-root systems may improve plant nutrition and thereby enhance forest growth. Greater allocation of C belowground could be a common feature of growing forests (*e.g.*, [24]).

The word 'allocation' has been used variably within plant physiology and forest ecology literatures [25]: we herein use 'allocation' to indicate a flux of organic substrates to specific forest components, with units of mass area^{-1} time^{-1}. Aboveground, we distinguish canopy production and tree growth, with allocation to canopy production being empirically determined from measurements of fine litterfall including leaves, and aboveground tree growth being equivalent to aboveground biomass increment (ABI). We also use leaf fall, the leaf component of fine litterfall, as a proxy for leaf production. We sum canopy production and tree growth to provide a widely used [26], [27] estimate of aboveground net primary productivity (ANPP). Belowground, we consider only total BCA, which includes both belowground NPP and respiratory CO_2 from roots and their rhizospheres. We do so because cleanly distinguishing the specific pathways of C flux belowground remains fraught with uncertainty: this does not influence the comparisons we make between specific fluxes measured across sites. We do not use ratios or proportions to express allocation without explicitly stating so, to avoid confusion with 'partitioning', which refers to the proportion of total photosynthesis (GPP) that is utilized by a particular forest component or process [25]. We recognize that canopy production implies more than light capture for photosynthesis by leaves: it includes leaves but also twigs ($\leq$ 1 cm diameter), meristems, and flowers and fruits. In addition to being the site of photosynthesis, plant canopies perform nitrate and sulfate reduction, and supply nutrients to roots [28]. They also produce and sense plant growth hormones that affect C allocation to roots and shoots [29].

Analysis of compiled data from a range of primarily mature and moist tropical forests indicated that NPP was distributed among canopy, wood, and fine root components in mean proportions of 34, 39 and 27%, respectively, and that variations from these mean proportions reflected a trade-off between fine roots and wood [27]. A similar conclusion was derived with a data-tested model of C allocation strategies of trees in old-growth forests [30]. Our proposition of a positive relationship between BCA and tree growth in young tropical forests contradicts these ideas. Nevertheless, based on the theoretical considerations described above, we suggest that growing forests may necessarily be more flexible in their allocation of production, to successfully balance nutrient uptake with aboveground growth [10], [11], [18].

We used a case-study approach to test alternative scenarios of forest C allocation (Fig. 1), which served as multiple working hypotheses *sensu* [31], by directly comparing data from concurrent studies of mature forest and replicated 15–20 year old plantations of native tree species located in close proximity to one another on the same highly weathered soils in lowland Costa Rica. Our primary objective was to assess whether between-site differences in tree growth rates (*i.e.*, wood productivity) could be ascribed to (I) differences in canopy production (scenario B in Fig. 1), (II) differences in BCA (scenario C in Fig. 1), or (III) proportionally equivalent differences in both processes (scenario A in Fig. 1). As illustrated in Figure 1, allocation of substrates to canopy production (flow 1) can generate a positive feedback to NPP (flow 6). In that case (Fig. 1B) a positive correlation between tree growth and canopy production is expected (*e.g.*, [26]). Similarly, preferential allocation of available substrates to root systems (BCA, flow 3 in Fig. 1) may stimulate soil nutrient turnover and uptake (flow 4), also producing a positive feedback to NPP (flow 5), in which case (Fig. 1C) tree growth and BCA should correlate positively. If available substrates are allocated in consistent proportions to leaves, to wood, and to root systems (Fig. 1A), *i.e.*, with 'fixed allocation coefficients' [27], [32], then canopy production, tree growth, and BCA will all correlate with one another across both growing and mature forest plots.

To fulfill our objective we tested multiple hypotheses that together made it possible to distinguish among the alternative allocation scenarios depicted in Figure 1. First, to assess possible differences in canopy attributes related to productivity ('canopy processes' in Fig. 1), we hypothesized that (H1a) total leaf area, (H1b) leaf litter production (*i.e.*, leaf fall), (H1c) N concentrations in leaf fall, and (H1d) leaf fall N fluxes were similar between the plantations and mature forest. Second, to determine if the two sites differed in aboveground productivity, we tested the hypotheses that (H2a) litterfall, (H2b) aboveground tree growth, and their sum, (H2c) ANPP, were similar. To identify belowground differences, we tested the hypothesis (H3) that BCA (flow 3 in Fig. 1) was greater in the plantations than in the mature forest. Belowground C allocation is an empirically based assay of the amount of photosynthetically produced substrate that is transported from the canopy through the phloem to root systems in forests [33]. Lastly, to determine which of the three alternative allocation scenarios (Fig. 1A, B, C) best describes observed differences among mature forest and plantation plots, we applied correlation analyses.

Methods

Ethics Statement

All field studies were conducted on land owned by the Organization for Tropical Studies, who gave permission for this research. The research also was approved and permitted by the

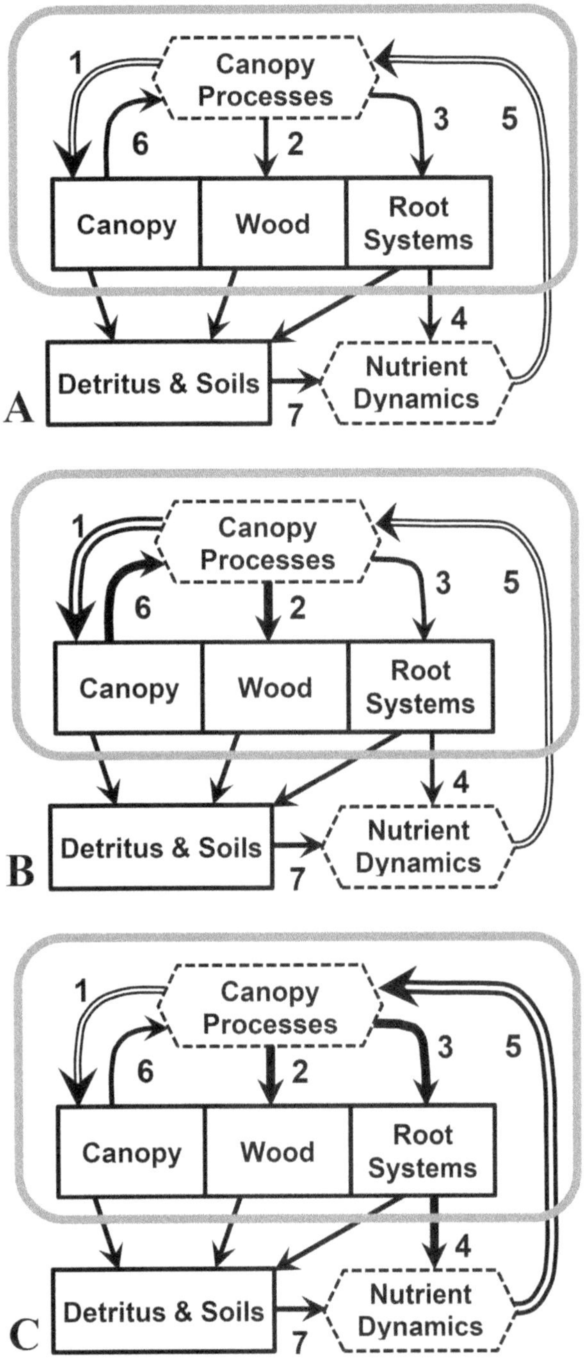

Figure 1. Schematic representation of alternative scenarios of organic matter flows through forest vegetation. Plant processes are encircled in gray, below which are soil processes: CO_2 uptake (photosynthesis) and release (respiration) are not depicted. Solid lines are organic matter flows; the two hollow arrows represents represent fluxes of organic matter and nutrients; solid boxes represent organic matter stocks (mass area^{-1}); and the two hexagons represent material transformations. Numbered flows represent: 1, allocation of photosynthates to the canopy; 2, allocation of NPP to woody tissues; 3, belowground carbon allocation (BCA); 4, root-system effects on soil nutrient availability; 5, root-system fluxes to canopy processes, *e.g.*, nutrient uptake; 6, feedbacks from canopy processes to NPP; and 7, transformations of detrital and soil nutrients into plant-available forms. Additional arrows show fluxes of plant detritus to soils; and transformation of detrital and soil nutrients into plant-available forms (flow 7). Three alternative scenarios are shown, with thicker arrows representing relatively larger fluxes. **A)** Fixed-allocation scenario, with organic matter fluxes to the canopy (flow 1), to wood production (flow 2), and to root systems (flow 3) varying uniformly. **B)** Canopy-feedback scenario, with relatively greater allocation of photoassimilates to the canopy (flow 1) and back (flow 6) providing a positive feedback to ANPP. **C)** Belowground-feedback scenario, with increased BCA stimulating nutrient uptake (flow 5) and wood NPP (flow 2).

Costa Rican Ministerio de Ambiente y Energía. We confirm that the field studies did not involve sampling of any endangered or protected species.

Study Area

Field studies were conducted at La Selva Biological Station of the Organization for Tropical Studies, in northeastern Costa Rica (10°26′N, 83°59′W). Annual rainfall from 1997–2009 averaged 4537 mm, air temperatures averaged 25.1°C, and elevations of the study sites ranged from 37–150 m [10], [34]. The driest season at La Selva normally extends from February through April, but every month averages >150 mm of rain. The vegetation of the mature forest was species-rich broad-leaved lowland evergreen forest with abundant palms and a relatively high abundance of the canopy tree *Pentaclethra macroloba* (Fabaceae). The forest contained an average of 550 trees ha^{-1} (stems ≥10 m diameter) with a basal area of 24 m^2 ha^{-1} [35]. The principal mature forest study sites are described in more detail elsewhere [34], [35]; we utilized data from the slope and plateau plots (twelve 0.5-ha plots) in the CARBONO plot network.

The experimental plantations were established on sites that previously were mature forest that was cut, burned and converted to pasture in the 1950's, and that remained in low-maintenance pastureland until the cattle were removed and the trees were planted, from June 1988 through February 1989. The 50-m×50-m plantation plots were established as a complete randomized block experiment having four blocks and one tree species per plot, and now have diverse and luxuriant understories [10]. The blocks originally contained 12 plots each, 11 planted to individual tree species and one left as an unplanted control. Each block draped over a single hill and thus included all hilltop and slope topographies. Seedlings were planted at 3-m×3-m spacing and plots of the fastest growing species were thinned at age four [36],

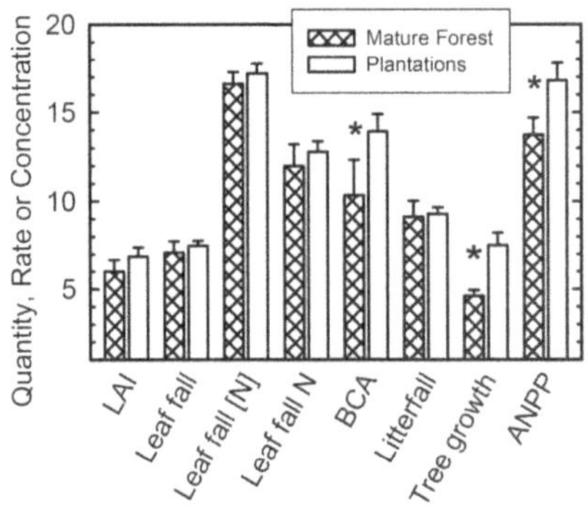

Figure 2. Comparisons of attributes of mature forests and plantations of native tree species at La Selva, Costa Rica. Variables are, left to right: total leaf area index (LAI, m^2 m^{-2}); annual leaf fall (Mg ha^{-1}); the mean N content of leaf fall (mg g^{-1}); the N flux in leaf fall (10^4 g ha^{-1} year^{-1}); belowground carbon allocation (BCA, Mg ha^{-1} year^{-1}); litterfall (Mg ha^{-1} year^{-1}); annual aboveground tree growth (Mg ha^{-1}); and aboveground net primary productivity (ANPP, Mg ha^{-1} year^{-1}). Statistically significant differences based on bootstrapped confidence intervals are identified with an asterisk (*). Additional information is provided in Table S1.

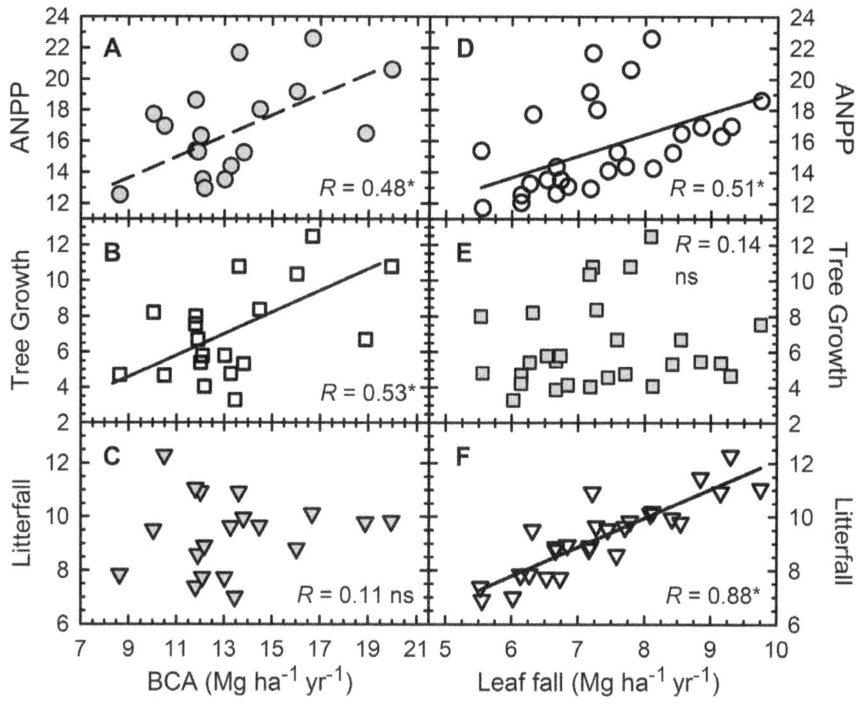

Figure 3. Belowground C allocation correlated positively with ANPP (A) and with tree growth (B) but not with litterfall (C). Leaf fall, in contrast, correlated with ANPP (D) and with litterfall (F), but not with tree growth (E). Lines show linear trends for significant relationships. Values are annual means in Mg ha^{-1} of organic matter, except that BCA is in units of C.

[37]. We utilize data from 16 plantation plots, four replicates each of four locally native tree species: *Hieronyma alchorneoides* Allemão (Phyllanthaceae), *Pentaclethra macroloba* (Willd.) Kuntze (Fabaceae), *Virola koschnyi* Warb. (Myristicaceae), and *Vochysia guatemalensis* Donn. Sm. (Vochysiaceae). Plantation understories were manually cleared of competing vegetation over the first four years [37] and saplings of *Vochysia guatemalensis*, which were seeding into other plots, were cut out in early 2006. To minimize edge effects the outer two rows of trees were not sampled; measurements were restricted to the central 0.13-ha subplot that originally contained 12 rows of 12 trees each. During this study, stem density in the plantations averaged 314 trees ha^{-1}, basal area averaged 18 m^2 ha^{-1}, and the planted trees were 15–21 years old.

Both the mature forest and plantation plots were located on the same soil type, a Typic Tropohumult according to [38], a Haplic Haploperox according to [39]. This soil is commonly referred to as residual soil in literature from La Selva, and we use that descriptor hereafter. The residual soils are deep, highly weathered, well-drained clays derived from very old basaltic lava flows. They are acidic, have high amounts of exchangeable acidity and low base saturation, are relatively high in organic matter, and are free of stones and gravel in the surface 3 m [38], [39]. The plantation experiment was designed to investigate the capacities of different tree species to improve soil conditions in degraded pasturelands [40]. Direct comparisons of soils from different plantation tree species at age 15 years [41] documented significant tree-species effects on soil properties, and demonstrated lower soil N contents in the plantations than in an adjacent mature forest site on residual soil. No differences in Olsen-extractable P in surface soils were observed between mature forest and plantation soils [41]. Based on those data from the plantation plots [41] and comparable measurements of soils from the mature forest plots [42], there were no significant differences between the mature forest and plantations in surface soil pH (range 4.0–4.8 in water) or in SOC stocks to 30 cm depth (range 83–90 Mg ha^{-1}) (Table S1). However, there was significantly less soil N (95% CI 6.6–7.03 versus 7.05–8.3 Mg ha^{-1}) and higher soil C: N (12.4–12.9 versus 11.8–12.2) in the plantations than in the mature forest (0–30 cm depth, Table S1). Soil P stocks may also differ between the sites, but the different analytical techniques used in the two studies preclude direct comparison.

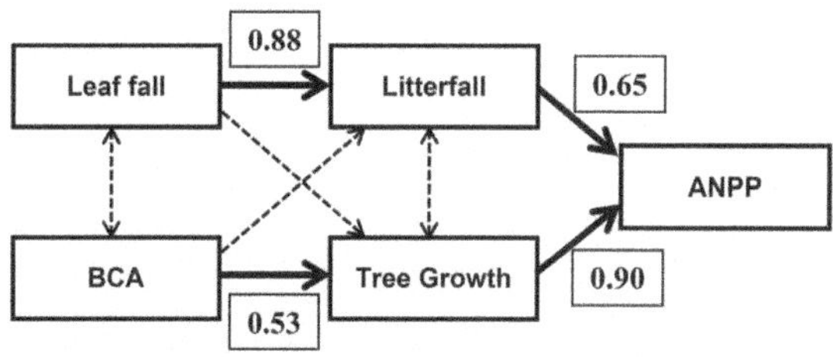

Figure 4. Correlations among carbon-flux variables used to distinguish among the alternative allocation scenarios depicted in Figure 1. Dashed arrows indicate insignificant relationships.

Field and Analytical Methods

To test our hypotheses, we integrated previously published and unpublished data from our field studies of the mature forest and plantations (Table S1). Most comparisons were based on concurrent studies conducted during 2003–2009, although exact time-matching was not always possible. Comparable methods were used in the two sites in most cases, as described below. Further methodological details are provided in Table S1 and in the source publications cited therein.

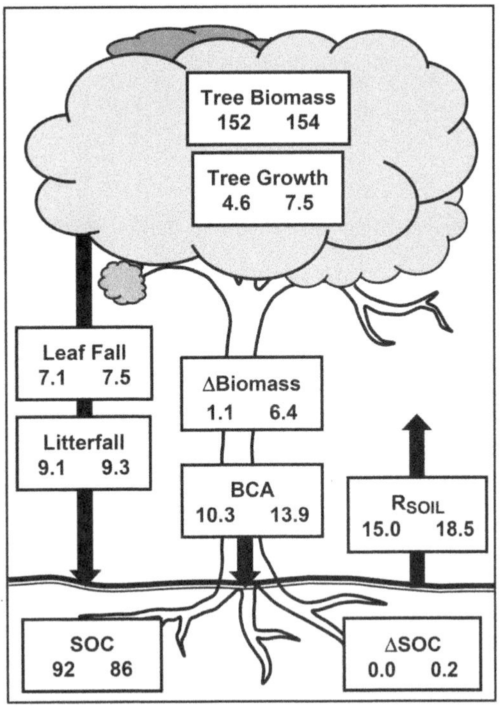

Figure 5. Carbon-cycling characteristics of mature rainforest and nearby tree plantations at La Selva, Costa Rica. In each box the value at the left is the mean annual value for the mature forest, and the value at right is the mean annual value of 15–20 year old plantations. BCA is belowground carbon allocation; R_{SOIL} is soil respiration; and ΔBiomass refers to the net annual increment in total aboveground tree biomass from 2003–2009; root biomass in excluded. Units are Mg ha^{-1} of biomass except for BCA and R_{SOIL}, which are in units of carbon. All fluxes are annual; data sources are in Table S1.

The aboveground biomass of each tree ≥10 cm diameter in each mature forest plot was estimated annually from tree diameter measurements, using the tropical wet forest regression of [43]. In the plantations, the diameters (*d*) and heights (*h*) of each tree in each plot were measured annually, and total aboveground biomass of each tree was estimated from d^2h based on species-specific regressions that were derived from the harvest of nine trees of each of the planted species from the experimental plots (Table S2). The harvested trees encompassed the entire size range of each species, including the largest individuals. An average of 8% of the tree biomass in the plantations was comprised of native tree species that naturally colonized the plots; the aboveground biomass of each of those trees (≥10 cm diameter) was similarly estimated with allometric equations derived from harvests of 49 trees from seven species (Table S2).

Annual aboveground tree growth was determined for each surviving tree each year as the difference in estimated biomass between years. The net annual change in aboveground biomass in each plot of each site was determined by difference between the plot-level biomass estimates derived from the 2003 and 2009 inventories. Aboveground fine-litter production (*i.e.*, litterfall) was measured in each plot with litter traps that were emptied at *ca.* two-week intervals from 2003–2009 (Table S1). Litterfall excluded branches >1 cm diameter and, in the mature forest, frass. We estimated aboveground NPP in each plot by summing tree growth and litterfall. Litterfall was sorted into components and the N flux in leaf fall was determined by multiplying its mass by its N content. Mature forest leaf fall samples from three collection dates per year (4–5 month intervals over 1997–2001, [44]) from all 12 forest plots were analyzed using a modified Kjeldahl digestion followed by colorimetric analysis on an Alpkem Flow Solution IV Autoanalyzer (OI Analytical, College Station, Texas, USA). Monthly leaf fall samples from each of the 16 plantation plots were analyzed with an elemental analyzer (Flash EA1112 CN Analyzer, CE Elantech, Lakewood, New Jersey, USA) over three years (2004, 2005, 2009).

Leaf area index was measured differently in the mature forest and plantations. In the mature forest, all leaves within 45 randomly located 1-m^2 vertical transects within old-growth forest were harvested and measured via repeated assembly and disassembly of a scaffolding tower [45]. In the plantations, the leaf areas of each of the planted trees, of all other trees ≥10 cm in diameter, and of saplings 2.5–10 cm in diameter were determined allometrically, based on regressions of total-tree leaf area from stem diameter (Table S2), with total-tree leaf area being determined from leaf biomass multiplied by the specific leaf mass (g m^{-2}) of each harvested tree. Leaf area of understory vegetation, including all non-woody plants and woody plants with diameters <2.5 cm diameter, was measured by direct harvests of four 0.5-m^2 quadrats per plot in 2005 [10]. Annual measurements of the diameters of trees and saplings by species were used to determine their total plot-level leaf area by year, and these were added to the 2005 understory leaf area (average 0.7 m^2 m^{-2}) to quantify total leaf area of each plot [10].

Total CO_2 emissions from the soil surface, *i.e.*, soil respiration rates (R_{SOIL}), were determined at approximately monthly intervals using dynamic closed chamber systems attached to LI-COR gas analyzers [10], [46]. In the mature forest, eight aluminum chambers (0.2 m in diameter, 0.15 m tall) were installed to about 0.02 m into the soil along four parallel transects in each of three 0.5-ha plots. Once inserted, the chambers were left in place and kept free of vegetation throughout the whole study period. Air was circulated at a flow rate of about 0.8 L min^{-1} between an infrared CO_2 gas analyzer (LI-800, Li-Cor Inc., Lincoln, NE, USA) and the flux chambers for about 5 min. To prevent pressure differences between chamber and atmosphere, chambers were vented to the atmosphere through a 0.025-m long stainless-steel tube. CO_2 concentrations were recorded at 5 s intervals with a datalogger (CR800, Campbell Scientific, Logan, UT, USA). Soil respiration rate was calculated from the linear change in CO_2 concentration multiplied by the density of air and the ratio of chamber volume to soil surface area. The infrared gas analyzer was calibrated in the lab using nitrogen as zero standard and a CO_2 standard (450 ppm). Measurements were conducted between 8 AM and 2 PM local time. For each of the sites the average soil respiration rate was calculated from the eight chamber flux measurements on a sampling day. Daily mean soil respiration for each site was calculated by linear interpolation between sampling dates. Daily CO_2 flux rates were then cumulated to estimate annual flux rates. In the plantations, soil respiration over 2004 through 2005 was measured in 3–4 locations per plot with an LI-8100 automated soil CO_2 flux system and 8100-102 (0.10-m diameter) chamber (LI-COR Biosciences, Lincoln, Nebraska, USA) [10], [47]. From 2007–2010 soil respiration was measured over two full years at three-week intervals in four locations per plot using the same system attached to an 8100-103 (0.20-m diameter) chamber. Mean fluxes were not affected by this change, but within-plot variability was reduced. Soil collars were 0.12 m tall and were inserted about 0.02 m into the surface soil. Living plants were removed from inside the collars, but surface litter and branches <5 cm diameter were left. Within-chamber CO_2 concentrations were measured every second for 3–5 minutes.

The resulting CO_2-concentration time series were analyzed using FV8100 software from LI-COR. Collar locations were moved annually or whenever they were disturbed. The LI-8100 was factory calibrated annually. Data from the plantations were averaged within months, and those monthly means were summed to estimate the mean annual flux, because sampling there did not include all months every year: in the mature forest each month was sampled each year. Belowground C allocation was estimated as R_{SOIL} minus litterfall-C. This approach is likely to underestimate total BCA because it excludes woody-root biomass accumulation, which is particularly important in growing forests [48]. For C budgeting, biomass components in the mature forest were assumed to be 48% C, similar to those in the plantations.

Belowground C allocation (flow 3 in Fig. 1) is a focus of our study (Fig. 1a) and so deserves further explication. The approach taken to estimate BCA included only the major belowground C fluxes that were directly measured in both sites, thus allowing for a robust comparison. As in the case of ANPP, however, several minor fluxes were not measured for this study. Measurements of some of these other minor fluxes have been made at La Selva, in forests on residual soils (Table S3). The mass of fine litter on the soil surface (*i.e.*, the forest floor) was similar in the mature forest and the plantations [49], so net accumulations of forest floor C were negligible. Net increments of soil organic C (SOC) stocks averaged 0.23 Mg ha^{-1} $year^{-1}$ in the plantations [10] but were miniscule in the mature forest [50]; we also assumed those to be negligible in both sites. In the mature forest, inputs of dissolved organic and inorganic C to the soil in canopy throughfall were small, and were balanced by equivalent (within measurement error) leaching losses (Table S3); a similar situation was presumed to exist in the plantations. We have no estimates of dissolved CO_2 fluxes in xylem sap water [51], but similarities in canopy characteristics between the forests being compared, which were at the same location on similar soils and landscape positions, minimize the likelihood of there being quantitatively important differences among the sites. It is likely that soils in both sites consumed some atmospheric methane, but at a very low rate [52]. Ignoring these minor and unmeasured soil C fluxes does not bias the comparison we make between BCA in the mature forest and plantations. We did not include estimates of net belowground coarse root accumulation within our estimates of BCA because coarse roots (including belowground stumps and stems) have not been measured in both sites, but we provide estimates of the possible magnitude of that exclusion (Table S3) based on root-shoot biomass ratios derived from other sites [53]. Inclusion of that flux would likely increase the difference in BCA between the mature forest and plantations, by perhaps 0.5 Mg ha^{-1} $year^{-1}$ (Table S3).

Statistical Analyses

To compare data from mature forest plots to those from plantation plots, we determined the arithmetic means of the plot-level data and determined their 95% confidence intervals (*CI*) using bootstrapping. Specifically, for each variable of interest within each site we used resampling with replacement to generate 5000 equal-sized populations of plots, confirmed that their average mean matched the empirically determined value, and then excluded the smallest and largest 2.5% to characterize the *CI*. In the plantations, bootstrapped populations were equally distributed among the four planted tree species. Bootstrapping is particularly appropriate for estimating confidence intervals when some sample sizes are low and heterogeneity among variances may exist [54]. For most variables sample sizes were 12 mature forest plots and 16 plantation plots (Table S1). To distinguish among the alternative scenarios of allocation of production (Fig. 1) we performed correlation analyses using the Restricted Maximum Likelihood (REML) method of JMP (Version 10.0.0, SAS Institute, Inc.). Bootstrapping was performed using the data analysis toolpack of Microsoft Excel 2010.

Results

We identified both similarities and significant differences between plantation and mature forest plots (Fig. 2, Table S1). Leaf fall was similar in the two sites ($P>0.1$), averaging (±1 *SEM*) 7.3±0.2 Mg ha^{-1} $year^{-1}$ ($n = 28$). Leaf fall N contents were measured in different years in the two sites but, based on those data, neither N concentrations in leaf fall ($n = 28$, 17.0±0.4 mg g^{-1}) nor leaf litter N fluxes ($n = 28$, 124±4 kg ha^{-1} $year^{-1}$) varied significantly between sites. Total leaf area index over 2003–2005, when the mature forest data were collected, averaged 6.0±0.3 m^2 m^{-2} ($n = 45$) in the mature forest and 6.9±0.5 m^2 m^{-2} ($n = 16$) in the plantations, but these were not statistically different due to high variability among plots ($0.5<P<0.10$). Aboveground NPP was greater in the experimental plantations (*CI* 15.8–17.8 Mg ha^{-1} $year^{-1}$) than in the mature forest (12.9–14.7 Mg ha^{-1} $year^{-1}$, Fig. 2). This difference was due to greater tree growth in the plantations: total fine litterfall did not differ significantly between the sites ($n = 28$, 9.2±0.3 Mg ha^{-1} $year^{-1}$). Tree growth, in contrast, averaged 60% faster in the plantations (mean 7.5, *CI* 6.8–8.2 Mg ha^{-1} $year^{-1}$) than in the mature forest (mean 4.6, *CI* 4.3–4.9 Mg ha^{-1} $year^{-1}$, Fig. 2). Belowground C allocation also was greater in the plantations (13.0–14.9 Mg ha^{-1} $year^{-1}$, $n = 16$) than in mature forest (8.3–12.3 Mg ha^{-1} $year^{-1}$, $n = 3$), which had less R_{SOIL} (Table S1).

Across all plots regardless of stand age, both tree growth and ANPP increased with increasing BCA, but litterfall did not (Fig. 3). Both litterfall and ANPP increased with increasing leaf fall, but tree growth did not (Fig. 3). Correlation analyses (Table S4) indicated two distinct pathways of influence. Leaf fall comprised an average of 79% of fine litterfall and so those two variables were strongly correlated ($r = 0.88$), but leaf fall was not related to either tree growth or BCA (Fig. 4). Belowground C allocation, in contrast, correlated significantly with aboveground tree growth ($r = 0.53$) but not with either leaf fall or litterfall ($P>0.1$).

Discussion

Rapid growth and biomass accumulation by tropical secondary forests and plantations partially mitigate C losses to the atmosphere that accompany deforestation [5] and promote restoration of ecosystem services on deforested lands [8], [55]. However, rapid forest growth requires nutrients. A case example is provided by rapidly growing plantations of four native tree species in lowland Costa Rica. There, C sequestration rates in biomass averaged >5 Mg ha^{-1} $year^{-1}$ over the first 16 years of tree growth and were still high, nearly 4 Mg C ha^{-1} $year^{-1}$, at age 16 [10]. This rapid C sequestration was accompanied by very rapid rates of N uptake by the vegetation, which averaged 350 kg N ha^{-1} $year^{-1}$, much of which was derived from soil N stocks [56]. It may be difficult to sustain such rapid tree growth and nutrient uptake. Nutrient limitations are widespread [57], [58] and potentially depress tree growth and stand-level biomass accumulation rates [11], [59]. We asked "What processes promote nutrient uptake by, and alleviate nutrient limitation to, these rapidly growing tropical forest stands?" We theorized that additional allocation of C to root systems in growing forests could promote the uptake of soil nutrients needed to support tree growth. We specifically conjectured that rates of aboveground biomass accumulation and belowground C allocation were linked, such that both processes would vary together and independently of leaf or canopy production, other factors being similar.

We tested these ideas by putting forth three alternative scenarios of C allocation (Fig. 1) and testing their predictions based on combining all relevant plot-level data from both sites. In the fixed-allocation scenario (Fig. 1A) BCA, tree growth and canopy production would all correlate with one another across both plantation and mature forest plots. However, neither BCA nor tree growth were significantly related to canopy production, measured as either leaf fall or total fine litter production (Fig. 4). In the canopy-feedback scenario (Fig. 1B) positive correlations between canopy production (flow 1 in Fig. 1) and tree growth (flow 2 in Fig. 1) are predicted. We found that leaf fall and litterfall provided virtually equivalent measures of canopy production ($r = 0.88$), but tree growth was not related to either variable (Fig. 4, Table S4). The belowground-feedback scenario (Fig. 1C), in contrast, predicts a positive correlation between BCA and tree growth, but no correlation between BCA and canopy production. This scenario is consistent with our findings: aboveground biomass increments (*i.e.*, tree growth) increased with increasing BCA whereas there was no correlation between BCA and canopy production ($P > 0.4$).

Direct comparisons of measured variables within the mature forest to those of the plantations also are consistent with a belowground-feedback scenario. Although ANPP was greater in the plantations than in the mature forest, no significant differences in (H1) leaf area, leaf fall or total fine litterfall were observed between the sites, which also had similar leaf fall N contents (mg g^{-1}) and fluxes based on the data available (Fig. 2, Table S1). Aboveground NPP (H2) averaged 22% higher in the plantations than in the mature forest, primarily due to 63% faster tree growth (Fig. 5). Aboveground biomass accumulated nearly six times faster in the plantations than in the mature forest (Fig. 5). Our estimate of BCA (H3) also was significantly greater in the plantations than in the mature forest (Fig. 2) despite the fact that we ignored probable differences in coarse-root biomass accumulation rates (Table S3), which likely mirrored the six-fold differences in net aboveground biomass accumulation. This finding is consistent with data from forests from different biomes summarized by [24], who found that forests <45 years old had relatively greater belowground C fluxes than did older forests. Overall, our findings are consistent with Fig. 1C in that both BCA and tree growth, but not canopy production, were significantly greater in the plantations than in the mature forest (Fig. 5).

Across a range of primarily mature tropical forests in Amazonia, ~34% of estimated NPP was allocated to canopy production, with indications of a possible trade-off in allocation of production to woody stems versus fine roots [27], [60]. Such a tradeoff was inferred to be evolutionarily advantageous based on an individual-based forest growth model that was compared against empirical data [30]. Our results from mature forests and 15–20 year old plantations of native tree species in lowland Costa Rica were not consistent with such a trade-off: we found a significantly positive relationship between BCA and aboveground tree growth (Fig. 3, 4). We suggest that young forest plots that are accumulating biomass [10] and are sequestering nutrients from the soil [56] reflect disequilibrium conditions, and thus are likely to have different allocation patterns than do mature forests, if allocation patterns are flexible (*e.g.*, Fig. 1B, 1C) and not fixed (Fig. 1A). Our data suggest flexibility in allocation of available growth substrates to canopies versus to root systems, with tree growth (wood production) in the plantations benefiting, in our case, from enhanced BCA. Clearly, healthy foliar canopies and root systems both contribute to ANPP: we presume that greater BCA in the experimental plantations at our site was related to soil degradation that occurred during pasture establishment and over the subsequent decades of low-level pasture management that preceded tree planting [40], [61]. This would be consistent with classical root-shoot allocation theory [12]. It also is generally consistent with a recent tree growth model [62] that posited that available C and N in trees were utilized to maximize annual wood production, with enhanced C fluxes to deep roots promoting N uptake in support of wood production. We infer that increasing allocation of C to root systems can be an important mechanism supporting high rates of tree growth and C sequestration in rapidly growing tropical forest plantations. Similar investigations of young forests at other sites are needed: identifying when, where and for how long enhanced allocation of C to root systems occurs during forest stand development would improve models of forest growth and forest-atmosphere C exchanges.

Acknowledgments

We thank Brian Wilsey for comments on an earlier draft of this manuscript. We thank the Organization for Tropical Studies for logistical support and for its continuing protection of the study sites.

Author Contributions

Conceived and designed the experiments: JWR DAC. Performed the experiments: JWR DAC LS TEW. Analyzed the data: JWR. Wrote the paper: JWR DAC LS TEW.

References

1. Foley JA, DeFries R, Asner GP, Barford C, Bonan G, et al. (2005) Global consequences of land use change. Science 309: 570–574.
2. Bonan GB (2008) Forests and climate change: Forcings, feedbacks, and the climate benefits of forests. Science 320: 1444–1449.
3. Hansen MC, Potapov PV, Moore R, Hancher M, Turubanova SA, et al. (2013) High-resolution global maps of 21st-century forest cover change. Science 342: 850–853.
4. Lamb D, Erskine PD, Parrotta JA (2005) Restoration of degraded tropical forest landscapes. Science 310: 1628–1632.
5. Pan Y, Birdsey RA, Fang J, Houghton R, Kauppi PE, et al. (2011) A large and persistent carbon sink in the world's forests. Science 333: 988–993.
6. Odum EP (1969) The strategy of ecosystem development. Science 164: 262–270.
7. Wardle DA, Walker LR, Bardgett RD (2004) Ecosystem properties and forest decline in contrasting long-term chronosequences. Science 305: 509–513.
8. Brown S, Lugo AE (1990) Tropical secondary forests. J Trop Ecol 6: 1–32.
9. Silver WL, Ostertag R, Lugo AE (2000) The potential for carbon sequestration through reforestation of abandoned tropical agricultural and pasture lands. Restor Ecol 8: 394–407.

10. Russell AE, Raich JW, Bedoya R, Valverde-Barrantes O, González E (2010) Impacts of individual tree species on carbon dynamics in a moist tropical forest environment. Ecol Appl 20: 1087–1100.
11. Evans J, Turnbull JW (2004) Plantation forestry in the tropics, 3rd edition. Oxford: Oxford University Press. 467 p.
12. Thornley JHM (1972) A balanced quantitative model for root:shoot ratios in vegetative plants. Ann Bot 36: 431–441.
13. Rastetter EB, Ågren GI, Shaver GR (1997) Responses of N-limited ecosystems to increased CO_2: A balanced-nutrition, coupled-element-cycles model. Ecol Appl 7: 444–460.
14. Monsi M, Saeki T (1953) Über den Lichtfaktor in den Pflanzengesellschaften und seine Bedeutung für die Stoffproduktion (On the factor light in plant communities and its importance for matter production). Ann Bot 95: 549–567.
15. de Pury DGG, Farquhar GD (1997) Simple scaling of photosynthesis from leaves to canopies without the errors of big-leaf models. Plant Cell Environ 20: 537–557.
16. Kira T, Shidei T (1967) Primary production and turnover of organic matter in different forest ecosystems of the western Pacific. Jpn J Ecol 17: 70–87.
17. Tinker PB, Nye PH (2000) Solute movement in the rhizosphere. Oxford: Oxford University Press. 444 p.
18. Johnson DW (2006) Progressive N limitation in forests: Review and implications for long-term responses to elevated CO_2. Ecology 87: 64–75.
19. Soltis DE, Soltis PS, Morgan DR, Swensen SM, Mullin BC, et al. (1995) Chloroplast gene sequence data suggest a single origin of the predisposition for symbiotic nitrogen fixation in angiosperms. Proc Natl Acad Sci USA 92: 2647–2651.
20. Marschner H, Dell B (1994) Nutrient uptake in mycorrhizal symbiosis. Plant Soil 159: 89–102.
21. Jones DL, Nguyen C, Finlay RD (2009) Carbon flow in the rhizosphere: carbon trading at the soil-root interface. Plant Soil 321: 5–33.
22. Fu S, Cheng W (2002) Rhizosphere priming effects on the decomposition of soil organic matter in C_4 and C_3 grassland soils. Plant Soil 238: 289–294.
23. Herman DJ, Johnson KK, Jaeger CH III, Schwartz E, Firestone MK (2006) Root influence on nitrogen mineralization and nitrification in *Avena barbata* rhizosphere soil. Soil Sci Soc Am J 70: 1504–1511.
24. Davidson EA, Savage K, Bolstad P, Clark DA, Curtis PS, et al. (2002) Belowground carbon allocation in forests estimated from litterfall and IRGA-based soil respiration measurements. Agric For Meteorol 113: 39–51.
25. Litton CM, Raich JW, Ryan MG (2007) Carbon allocation in forest ecosystems. Glob Chang Biol 13: 2089–2109.
26. Clark DA, Brown S, Kicklighter DW, Chambers JQ, Thomlinson JR, et al. (2001) Measuring net primary production in forests: concepts and field methods. Ecol Appl 11: 356–370.
27. Malhi Y, Doughty C, Galbraith D (2011) The allocation of ecosystem net primary productivity in tropical forests. Philos Trans R Soc Lond B Biol Sci 366: 3225–3245.
28. Marschner H, Kirkby EA, Cakmak I (1996) Effect of mineral nutritional status on shoot-root partitioning of photoassimilates and cycling of mineral nutrients. J Exp Bot 47: 1255–1263.
29. Wareing PF, Phillips IDJ (1979) The control of growth and differentiation in plants, 2nd edition. Oxford, New York, Toronto, Sydney, Paris, Frankfurt: Pergamon Press. 347 p.
30. Dybzinski R, Farrior C, Wolf A, Reich PB, Pacala SW (2011) Evolutionarily stable strategy carbon allocation to foliage, wood, and fine roots in trees competing for light and nitrogen: An analytically tractable, individual-based model and quantitative comparisons to data. Am Nat 177: 153–166.
31. Chamberlin TC (1890) The method of multiple working hypotheses. Science 15: 92–96.
32. Ryan MG, Binkley D, Fownes JH (1979) Age-related decline in forest productivity: pattern and process. Adv Ecol Res 27: 213–262.
33. Raich JW, Nadelhoffer KJ (1989) Belowground carbon allocation in forest ecosystems: Global trends. Ecology 70: 1346–1354.
34. Clark DA, Clark DB, Oberbauer SF (2013) Field-quantified responses of tropical rainforest aboveground productivity to increasing CO_2 and climatic stress, 1997–2009. J Geophys Res Biogeosci 118: 783–794.
35. Clark DB, Clark DA (2000) Landscape-scale variation in forest structure and biomass in a tropical rain forest. For Ecol Manage 137: 185–198.
36. González JE, Fisher RF (1994) Growth of native forest species planted on abandoned pasture land in Costa Rica. For Ecol Manage 70: 159–167.
37. Haggar J, Wightman K, Fisher RF (1997) The potential of plantations to foster woody regeneration within a deforested landscape in lowland Costa Rica. For Ecol Manage 99: 55–64.
38. Sollins P, Sancho FM, Mata RC, Sanford RL Jr (1994) Soils and soil process research. In: McDade LA, Bawa KS, Hespenheide HA, Hartshorn GS, editors. La Selva: Ecology and natural history of a Neotropical rain forest. Chicago: University of Chicago Press. 34–53.
39. Kleber M, Schwendenmann L, Veldkamp E, Rößner J, Jahn R (2007) Halloysite versus gibbsite: Silicon cycling as a pedogenetic process in two lowland Neotropical rain forest soils of La Selva, Costa Rica. Geoderma 138: 1–11.
40. Fisher RF (1995) Amelioration of degraded rain forest soils by plantations of native trees. Soil Sci Soc Am J 59: 544–549.
41. Russell AE, Raich JW, Valverde-Barrantes OJ, Fisher RF (2007) Tree species effects on soil properties in experimental plantations in tropical moist forest. Soil Sci Soc Am J 71: 1389–1397.
42. Espeleta JF, Clark DA (2007) Multi-scale variation in fine-root biomass in a tropical rain forest: A seven-year study. Ecol Monogr 77: 377–404.
43. Brown S (1997) Estimating biomass and biomass change of tropical forests: A primer. Food and Agriculture Organization Forestry Paper 134, Food and Agriculture Organization of the United Nations, Rome. Available: http//www.fao.org/docrep/W4095E/w4095e06.htm.
44. Wood TE, Lawrence D, Clark DA (2006) Determinants of leaf litter nutrient cycling in a tropical rain forest: Soil fertility versus topography. Ecosystems 9: 700–710.
45. Clark DB, Olivas PC, Oberbauer SF, Clark DA, Ryan MG (2008) First direct landscape-scale measurement of tropical rain forest leaf area index, a key driver of global primary productivity. Ecol Lett 11: 163–172.
46. Schwendenmann L, Veldkamp E, Brenes T, O'Brien JJ, Mackensen J (2003) Spatial and temporal variation in soil CO_2 efflux in an old-growth Neotropical rain forest, La Selva, Costa Rica. Biogeochemistry 64: 111–128.
47. Valverde-Barrantes OJ (2007) Relationships among litterfall, fine root growth, and soil respiration for five tropical tree species. Can J For Res 37: 1954–1965.
48. Giardina CP, Ryan MG (2002) Total belowground carbon allocation in a fast growing *Eucalyptus* plantation estimated using a carbon balance approach. Ecosystems 5: 487–499.
49. Raich JW, Russell AE, Bedoya-Arrieta R (2007) Lignin and enhanced litter turnover in tree plantations of lowland Costa Rica. For Ecol Manage 239: 128–135.
50. Veldkamp E, Becker A, Schwendenmann L, Clark DA, Schulte-Bisping H (2003) Substantial labile carbon stocks and microbial activity in deeply weathered soils below a tropical wet forest. Glob Chang Biol 9: 1171–1184.
51. Levy PE, Meir P, Allen SJ, Jarvis PG (1999) The effect of aqueous transport of CO_2 in xylem sap on gas exchange in woody plants. Tree Physiol 19: 53–58.
52. Keller M, Reiners WA (1994) Soil-atmosphere exchange of nitrous oxide, nitric oxide, and methane under secondary succession of pasture to forest in the Atlantic lowlands of Costa Rica. Global Biogeochem Cycles 8: 399–409.
53. Mokany K, Raison RJ, Prokushkin AS (2006) Critical analysis of root:shoot ratios in terrestrial biomes. Glob Chang Biol 12: 84–96.
54. Adams DC, Gurevitch J, Rosenberg MS (1997) Resampling tests for meta-analysis of ecological data. Ecology 78: 1277–1283.
55. Chazdon RL (2008) Beyond deforestation: Restoring forests and ecosystem services on degraded lands. Science 320: 1458–1460.
56. Russell AE, Raich JW (2012) Rapidly growing tropical trees mobilize remarkable amounts of nitrogen, in ways that differ surprisingly among species. Proc Natl Acad Sci USA 109: 10398–10402.
57. Lebauer DS, Treseder KK (2008) Nitrogen limitation of net primary productivity in terrestrial ecosystems is globally distributed. Ecology 89: 371–379.
58. Vitousek PM, Porder S, Houlton BZ, Chadwick OA (2010) Terrestrial phosphorus limitation: mechanisms, implications, and nitrogen-phosphorus interactions. Ecol Appl 20: 5–15.
59. Goll DS, Brovkin V, Parida BR, Reick CH, Kattge J, et al. (2012) Nutrient limitation reduces land carbon uptake in simulations with a model of combined carbon, nitrogen and phosphorus cycling. Biogeosciences 9: 3547–3569.
60. Aragão LEOC, Malhi Y, Metcalfe DB, Silva-Espejo JE, Jiménez E, et al. (2009) Above- and below-ground net primary productivity across ten Amazonian forests on contrasting soils. Biogeosciences 6: 2759–2778.
61. Reiners WA, Bouwman AF, Parsons WFJ, Keller M (1994) Tropical rain forest conversion to pasture: Changes in vegetation and soil properties. Ecol Appl 4: 363–377.
62. McMurtrie RE, Dewar RC (2013) New insights into carbon allocation by trees from the hypothesis that annual wood production is maximized. New Phytol 199: 981–990.

17

Production of Carbon Occluded in Phytolith Is Season-Dependent in a Bamboo Forest in Subtropical China

Zhang-Ting Huang[1], Pei-Kun Jiang[1,2]*, Scott Xiaochuan Chang[3]*, Yan Zhang[1], Yu-Qi Ying[1]

1 School of Environment and Resources, Zhejiang A&F University, Lin'an, Zhejiang, China, **2** Zhejiang Provincial Key Laboratory of Carbon Cycling in Forest Ecosystems and Carbon Sequestration, Zhejiang A&F University, Lin'an, Zhejiang, China, **3** Department of Renewable Resources, University of Alberta, Edmonton, Alberta, Canada

Abstract

Carbon (C) occluded in phytolith (PhytOC) is a stable form of C; when PhytOC is returned to the soil through litterfall it is stored in the soil which can be an effective way for long-term C sequestration. However, few estimates on the rate of PhytOC input to the soil are available. To better understand the seasonal dynamics of PhytOC production and the annual rate of stable C sequestration through PhytOC input, we quantified the monthly litterfall, phytolith and PhytOC return to the soil over a year in a typical Lei bamboo (*Phyllostachys praecox*) forest in subtropical China. The monthly litterfall ranged between 14.81 and 131.18 g m^{-2}, and the phytolith concentration in the monthly litterfall samples ranged between 47.21 and 101.68 g kg^{-1} of litter mass, with the PhytOC concentration in the phytolith ranged between 29.4 and 44.9 g kg^{-1} of phytolith, equivalent to 1.8–3.6 g kg^{-1} of PhytOC in the litterfall (based on litterfall dry mass). The amount of phytolith input to the soil system was 292.21±69.12 (mean±SD) kg ha^{-1} yr^{-1}, sequestering 41.45±9.32 kg CO_2−e ha^{-1} yr^{-1} of C in the studied Lei bamboo forest. This rate of C sequestration through the formation of PhytOC found in this study falls within the range of rates for other grass-type species reported in the literature. We conclude that return of C occluded in phytolith to the soil can be a substantial source of stable soil C and finding means to increase PhytOC storage in the soil should be able to play a significant role in mitigating the rapidly increasing atmospheric CO_2 concentration.

Editor: Shibu Jose, University of Missouri, United States of America

Funding: The authors wish to acknowledge the funding from the Natural Science Foundation of China (No. 31270667), and the key Fund of Science and Technology of Zhejiang Province (No. LZ12C16003). The funders had no role in study design, data collection and analysis, decision to publish, or preparation of the manuscript.

Competing Interests: The authors have declared that no competing interests exist.

* Email: jiangpeikun@zafu.edu.cn (PKJ); scott.chang@ualberta.ca (SXC)

Introduction

Phytolith, or plantstones or plant opal, is a noncrystalline form of mineral (SiO_2) that deposits in the intra- and extra-cellular structures of different plant tissues after the absorption of soluble silica by plant roots in the form of monosilicic acid ($Si(OH)_4$) [1–3]. Phytolith is essentially a silicified feature that forms as a result of biomineralization that occurs within plants and during the process some organic C can become occluded in the phytolith (PhytOC) during its formation in plant tissues [2,4]. Phytolith typically is highly resistant to decomposition [5,6], being stable in both aerobic and anaerobic conditions and in soils with pH ranging between 3.5 and 9.8 [7,8]. As such, the PhytOC is much more stable than organic C fractions in the soil that are not occluded in phytolith [2,9]. When plant litter decomposes after it falls onto the soil, PhytOC as a stable form of C will accumulate with phytolith and remain in the soil for a long period of time [2]. The accumulation of PhytOC in the soil is an important millennium scale mechanism for long-term C sequestration in terrestrial ecosystems, which could potentially contribute 15–37% of the estimated global soil organic C (SOC) sequestration of 24 kg C ha^{-1} yr^{-1}; this form of C sequestration could play a positive role in mitigating global climate change [2,5,10,11].

Despite the potential importance of phytolith in SOC stabilization and long-term C storage, few estimates exist on the concentration of phytolith in different plant species and ecosystem types [12–14]. Measurements of phytolith and PhytOC concentrations in different ecosystem types are urgently needed to improve our understanding of the potential contribution of PhytOC in global C sequestration [15]. Such data would also help us design management practices, such as the selection of species or genotypes that have high phytolith concentrations, to take advantage of this mechanism for SOC sequestration to reduce atmospheric CO_2 concentration [16].

Past research indicates that plants in the Poaceae family, a large and nearly ubiquitous family consisting of monocotyledonous flowering plants, have relatively high phytolith concentrations [3,17]. Among the more than 10,000 species in the Poaceae family, bamboo species have particularly high phytolith concentrations in their tissues [3]. Plants in the Poaceae family thus have an advantage in sequestering organic C in the form of PhytOC accumulation in the soil [2,5,11,12,15,18]. Through the determination of phytolith and PhytOC concentrations in different bamboo species, combined with the published bamboo litterfall rate [19,20], Parr et al. [15] estimated the organic C sequestration potential through PhytOC accumulation in bamboo forests of ten economically important bamboo species. They concluded that the global contribution to C sequestration on an annual basis is equivalent to 11% of the anthropogenic CO_2 release [15]. The

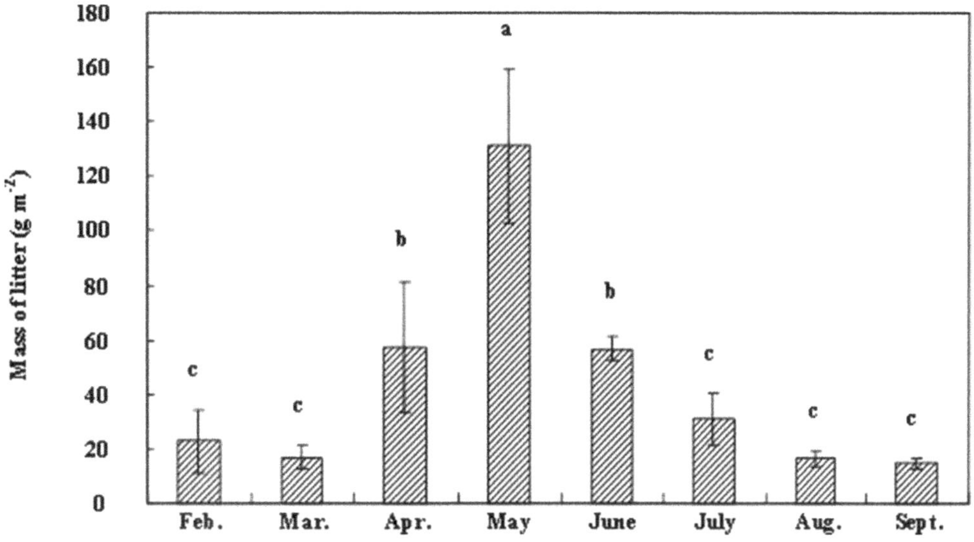

Figure 1. Seasonal dynamics of litter fall mass (oven-dry basis) in a Lei bamboo forest in subtropical China. Vertical error bars are standard deviations. Means followed by different lowercase letters are not significantly different.

work of Parr et al. [15] and others [16] illustrated the advantage of bamboo forests in long-term C sequestration through the formation of PhytOC. However, one potential deficiency with the work of Par et al. [15] is that the PhytOC contribution was estimated by its content in bamboo leaves rather than through their existence in the litterfall or in the soil. In reality, only PhytOC returned to the soil through litterfall and accumulated in the soil can be considered a contribution to SOC sequestration. The most reliable estimate of the rate of PhytOC input is through quantifying litterfall rate and the PhytOC concentration in the litterfall [16]. Since PhytOC return in litterfall constitutes PhytOC return to the soil, the seasonal dynamics of PhytOC return in litterfall should be understood but that has not been studied.

In this study, we determined the annual rate of PhytOC input to the soil in a Lei bamboo forest in subtropical China through measuring the amount of litterfall and the PhytOC concentration in the litterfall to provide an estimate of the potential contribution of C sequestration in the form of PhytOC input to the soil, as well as the seasonal dynamics of PhytOC input to the soil. Lei bamboo belongs to the Bambusoideae subfamily in the Poaceae family and is a species with high economic value due to the production of bamboo shoot, a widely used delicacy in Southeast Asia. This species is widely planted in southern China and the area planted to this bamboo species has increased dramatically in the last several decades [21]. As such, studying the rate of PhytOC production in Lei bamboo forests can improve our understanding of the potential contribution of this form of C sequestration in Lei bamboo forests and should have implications for estimating PhytOC input to the soil in forests of other bamboo species. The objectives of this study were therefore to determine the amount of litterfall and the associated PhytOC input to a Lei bamboo forest on a seasonal basis to understand the dynamics of both litterfall and PhytOC input through monthly sampling of the studied bamboo forest and to understand the contribution of PhytOC input from other bamboo components and obtain ecosystem level PhytOC storage and compare that with PhytOC storage in other ecosystems reported in the literature. Specifically, we hypothesize that 1) PhytOC production through litterfall in bamboo forests has a strong seasonal pattern and 2) the annual PhytOC return to the soil is a substantial contribution to soil C sequestration particularly as compared to rates of soil C sequestration in different ecosystems reported in the literature.

Materials and Methods

Ethics statement

No specific permits were required for the described field studies involving the collection of litterfall samples from the studied bamboo forest. The bamboo forest used in our sampling was privately owned and was not protected in any way and thus a specific permit for not-for-profit research is not required. The field studies did not involve endangered or protected plant or animal species in the study area.

Study area

The study site was located in Congkeng Village in Shankou Township (30°14′N, 119°42′E), Lin'an City, Zhejiang Province, in southeast China. The elevation of the site was 150 m above sea level and the general local landform was dominated by low hills. The study area had a monsoonal subtropical climate with four distinct seasons with hot and humid summers and cold winters. The average annual temperature and average annual precipitation of the study site were16°C and 1414 mm, respectively, based on the climate data between 2000 and 2009, with most of the precipitation occur between April and October. The site had an average of 236 frost-free days and 1946 day-light hours. The soils of the experimental area were classified as Anthrosols in the FAO soil classification system [22].

The site selected for this study had the bamboo forest established 10 years ago, with Lei bamboo density at 2,045 stems ha^{-1} and a diameter at breast height (1.3 m from the ground) of 3.9 cm. Those bamboo forests were managed for bamboo shoot production, a form of bamboo forest management that yields a high economic return [21]. The studied bamboo forest was converted from rice paddy in 2001, 10 years before this study was initiated in 2011. The bamboo forest was intensively managed as described in Li et al. [21].

Table 1. Carbon, SiO_2, phytolith and PhytOC concentrations (or stocks), and PhytOC fluxes through litter fall in a Lei bamboo stand in different months.

Months	Litter biomass (g m^{-2})	C concentration (g kg^{-1})	SiO_2 concentration (g kg^{-1})	Phytolith concentration (g kg^{-1})	PhytOC in phytolith (g kg^{-1})	PhytOC in litter (g kg^{-1})	Phytolith stock (kg ha^{-1} yr^{-1})	PhytOC stock (kg ha^{-1} yr^{-1})	PhytOC fluxes (kg CO_2-e ha^{-1} yr^{-1})
February	22.95±11.56 c	457.27±5.53 bc	76.55±8.91 bc	77.16±7.48 c	29.4±9.1 b	2.2±0.6 c	18.21±10.11 cd	0.51±0.32 c	1.86±1.19 c
March	17.00±4.17 c	451.76±2.16 cd	88.74±5.13 ab	88.09±3.42 bc	35.9±11.5 ab	3.2±1.1 ab	14.99±3.75 cd	0.52±0.15 c	1.90±0.55 c
April	57.25±23.81 b	442.13±3.32 e	91.93±6.98 a	101.68±6.99 a	34.6±4.3 ab	3.5±0.5 a	58.00±23.92 b	1.95±0.67 b	7.15±2.45 b
May	131.18±28.24 a	444.43±7.51 de	83.29±14.51 ab	90.37±15.03 ab	40.4±3.2 a	3.6±0.6 a	121.59±48.45 a	4.89±1.88 a	17.94±6.88 a
June	56.89±4.45 b	446.98±11.41 de	67.96±2.75 c	80.78±3.28 bc	43.6±2.4 a	3.5±0.2 a	45.88±2.47 bc	2.00±0.15 b	7.33±0.54 b
July	31.01±9.64 c	463.03±5.53 b	39.24±11.07 d	53.61±7.49 d	44.9±3.8 a	2.4±0.5 bc	17.00±7.06 cd	0.77±0.36 c	2.81±1.31 c
August	16.52±2.72 c	476.42±4.34 a	45.69±2.44 d	47.21±3.81 d	38.7±8.0 ab	1.8±0.2 c	7.86±1.75 d	0.30±0.05 c	1.09±0.17 c
September	14.81±2.29 c	472.52±3.49 a	47.93±4.70 d	58.68±7.89 d	43.9±5.7 a	2.5±0.2 bc	8.70±1.88 d	0.38±0.06 c	1.38±0.21 c
Total/Average	347.61±41.16	/	/	/	38.9±1.4	2.9±0.2	292.21±69.12	11.30±2.54	41.45±9.32

Values are means and SD. Means followed by the same lowercase letters in a column are not different among the month of litter collection.

Experimental design and sampling

Four randomly selected sampling sites were established in the bamboo forest described earlier. One 10×10 m plot was set up within each site and within each plot we set up four 1×1 m litter traps. Litter samples were collected on a monthly basis between January and December 2011. At each litter sample collection, the sample in each litter trap was transferred to a plastic bag, brought back to the laboratory, weighed, dried and weighed again. Water content was determined from those weights. The dried samples were ground and stored for chemical analysis. The main objective of this part of the study was to understand the seasonal dynamics of PhytOC return to the soil through bamboo litter fall.

In the second part of this research, we determined the biomass of standing bamboo in the bamboo stand we studied and compared that to the bamboo leaf litter production determined in the first part of the research described above. The main objective was to understand the partition of PhytOC in different bamboo components and compare that with bamboo leaf litter. This would also allow us to compare the ecosystem level PhytOC content with other ecosystems reported in the literature. In this part of the research, four sites were selected in 2012 in the studied stand after the new bamboo shoots produced in the spring were fully grown to a bamboo plant in late June. In each site, one 10×10 m plot was established. Within each plot, all bamboo plants were surveyed, with their average diameter and density, and grouped based on their age (all bamboo plants were from one to four years of age). After the field measurement was complete, three bamboo plants of each of the four ages representing a diameter gradient were harvested to determine the bamboo leaf biomass for each of those bamboo plants. The bamboo leaf samples from the three bamboo plants of each age were then mixed and 1 kg of each of those samples was collected and brought back to the laboratory. Once in the laboratory, the bamboo leaf samples were washed with deionized water, dried at 105°C for about an hour to quickly remove the majority of the water, and the samples were then dried at 65°C to constant weight to determine the water content of the leaf samples. The samples were ground and the phytolith and PhytOC concentrations determined using the methods described below.

The C and nitrogen (N) concentrations in the litter samples were determined based on a dry combustion method using an Elementar Vario MAX CN analyzer (Elementar Analysensysteme GmbH, Hanau, Germany). A microwave digestion method was used in this study to isolate phytolith from the litter samples [23]. After the microwave digestion, the samples were treated with a Walkley-Black type digestion [24] to ensure that extraneous organic materials in the samples were completely removed in recovering the phytolith sample [15]. The isolated phytolith samples were oven-dried at 75°C for 24 h in a centrifuge tube of known weight and the quantity of phytolith recovered was determined. The C concentration in the phytolith samples was then analysed on a Thermo Finnigan Flash EA 1112 CHNS-O Analyser (Thermo Finnigan, San Jose, CA). A soil standard sample (GBW07405) and a plant standard sample (GBW07602) were included in the analysis as part of the quality control measures. Precisions of better than 5% in the measurement of phytolith and better than 8% in the measurement of C concentration in phytolith were achieved through analysis of duplicate samples.

Data analysis

The phytolith concentration in litterfall samples, the C concentration in the phytolith, and the PhytOC concentration in litterfall samples were calculated using the following formulas:

Table 2. Biomass, phytolith and PhytOC concentrations and content (or stock), and PhytOC fluxes in fresh leaves in a Lei bamboo stand.

Age of leaf (yr)	Biomass ($t\ ha^{-1}\ yr^{-1}$)	Phytolith Concentration ($g\ kg^{-1}$)	PhytOC in phytolith ($g\ kg^{-1}$)	Phytolith stock ($kg\ ha^{-1}\ yr^{-1}$)	PhytOC stock ($kg\ ha^{-1}\ yr^{-1}$)	PhytOC fluxes ($kg\ CO_2\text{–}e\ ha^{-1}\ yr^{-1}$)
1	0.06±0.02 c	9.91±3.10 d	60.03±1.7 a	0.54±0.07 b	0.03±0.003 c	0.12±0.01 c
2	2.09±0.63 a	21.58±3.95 c	56.03±5.2 a	44.53±13.43 a	2.46±0.59 a	9.03±2.18 a
3	0.94±0.51 b	31.29±5.68 b	57.30±17.5 a	27.46±10.64 a	1.49±0.40 b	5.45±1.48 b
4	0.70±0.29 bc	41.33±8.38 a	51.18±4.4 a	29.32±13.82 a	1.54±0.82 b	5.64±3.02 b
Total	3.79±1.01	/	/	101.86±22.67	5.52±1.36	20.24±4.99

Values are means and SD. Means followed by same lowercase letters in a column are not different among the ages of leaves.

1) *Litterfall phytolith concentration* ($g\ kg^{-1}$)

$$= \frac{\textit{Phytolith weight}\ (g)}{\textit{Litterfall weight}\ (kg)}$$

2) *C concentration in phytolith* ($g\ kg^{-1}$)

$$= \frac{\textit{C content in phytolith}\ (g)}{\textit{Phytolith weight}\ (kg)}$$

3) *PhytOC concentration in litterfall samples* ($g\ kg^{-1}$)

$$= \frac{\textit{C content in phytolith}\ (g)}{\textit{Litterfall weight}\ (kg)}$$

The PhytOC addition rate from litterfall was calculated by multiplying the litterfall weight, the phytolith concentration and the C concentration in phytolith or by multiplying the litterfall weight and the PhytOC concentration in litterfall samples.

The normality and homogeneity of variance were tested before performing the ANOVA analysis and data were log-transformed if the assumptions for ANOVA were not met. An one-way analysis of variance (ANOVA) was conducted to test the effect of sampling month on the amount of litterfall, SOC concentration in the litterfall, and phytolith concentration, C concentration in phytolith and the PhytOC concentration in the litterfall. When the ANOVA analysis indicated a significant treatment effect, the Duncan's new multiple range test (MRT) was used to separate the means. Correlation analysis was performed to evaluate relationships between the studied parameters. An alpha level of 0.05 for significance was used in all statistical analyses, unless mentioned otherwise. All of the statistical analyses were performed using the SPSS software (SPSS 18.0 for windows, SPSS Inc., Chicago, USA).

Results

The amount of litterfall collected in each month was quite variable, with litterfall rates ranging from 14.81 to 131.18 $g\ m^{-2}$. The rates of litterfall were greater between April and June than in other months of the year, with the greatest litterfall rate obtained in May (Fig. 1). No litterfall was collected between October and

Table 3. PhytOC fluxes associated with different plant species based on data from the literature.

Plant species	PhytOC in phytolith ($g\ kg^{-1}$)	PhytOC in biomass ($g\ kg^{-1}$)	Biomass ($t\ ha^{-1}\ yr^{-1}$)	PhytOC fluxes ($t\ CO_2\text{–}e\ ha^{-1}\ yr^{-1}$)	Reference
Lei bamboo (litter)	38.9±1.4	2.9±0.2	3.48±0.41	0.042±0.009	This study
Sugarcane	38.8~192.6	0.9~2.5	40	0.123~0.364	[13]
Economic bamboo species	16.0~40.2	2.4~5.2	1~37	0.008~0.709	[12]
Foxtail millet	25.1±12.7	1.4±0.7	4.09	0.020±0.010	[8]
Common millet	19.2±14.3	1.3±0.9	4.79	0.023±0.015	[8]
Wheat	13.0~129.0	0.6~6.0	2.10~11.20	0.006~0.246	[14]
Rice	14.2~33.6	0.4~2.8	9.30~18.60	0.030~0.130	[24]
Riparian plants	4.9~39.7	0.1~1.6	0.01~11.50	0.052~23.562	[25]
Shallow-water emergent plants	11.0~23.3	0.1~1.5	16.00~34.06	3.775~159.359	[25]
Floating-leaf aquatic plants	9.1~19.4	0.2~0.4	0.46~99.18	0.320~87.278	[25]

Values represent the range or mean±SD.

January. The total amount of litterfall in the study year was 347.61 g m^{-2}.

The C concentration in the litterfall samples did not change dramatically between the sampling months but tended to be higher in late fall than in spring or summer (Table 1). The SiO_2 concentration in the litterfall samples markedly changed with the month of litterfall sample collection, with the highest concentrations found in late winter/early spring and the lowest in the summer growing season (Table 1). The seasonal change in the PhytOC concentration in the phytolith was not consistent with the seasonal change of the phytolith concentration in the litter samples, with a non-significant negative correlation between the two ($R^2 = 0.11$, $P > 0.05$; data not shown). When the PhytOC contained in the litter sample was expressed on the basis of the litter biomass (Table 1), the PhytOC concentration was positively correlated with the phytolith concentration ($R^2 = 0.51$, $P < 0.01$; data not shown).

The monthly amount of phytolith and PhytOC returned to the soil through litterfall (Table 1) was highly correlated ($R^2 = 0.98$, $P < 0.01$; data not shown). The total amount of PhytOC returned to the soil through litterfall was 11.30 kg ha^{-1} yr^{-1}, the equivalent of removing 41 kg of CO_2 ha^{-1} yr^{-1}.

The total bamboo green leaf biomass in the studied bamboo stand was 3.79 ton ha^{-1} (Table 2), which was slightly greater than the 3.48 ton ha^{-1} of the annual bamboo leaf litter determined (Table 1). The phytolith contained in the green bamboo leaf was only 101.86 kg ha^{-1} (Table 2), far less than the 292.21 kg ha^{-1} that can be returned to the soil through litterfall (Table 1). Consistent with the phytolith content in the standing bamboo leaves, the amount of PhytOC that was contained in the green bamboo leaf was also substantially less than that contained in the bamboo leaf litter.

Discussion

With the continued increase of greenhouse gas concentrations in the atmosphere and the related increase of the climatic forcing [25], the need to enhance C sequestration through biological means to mitigate the increasing greenhouse gas concentrations is becoming more urgent [26]. One of the main ways to biologically increase C sequestration is to increase C storage in the soil. The soil stores a vast amount of organic C and has the capacity to store more. However, organic C storage in the soil has different residence times and some of the SOC is very vulnerable to loss through mineralization under the actions of microbial populations in the soil [27,28]. Mechanisms for long-term SOC preservation in terrestrial ecosystems include physical protection of chemically recalcitrant organic matter within organo-mineral complexes [29], formation of charcoal [30], and organic C occlusion in phytolith [2]. The last mechanism has recently been promoted as a potential means to increase the long-term storage of SOC as PhytOC is considered a highly stable C that can be stored in the soil for hundreds of years [13–16].

The strong seasonal variation in PhytOC concentration in litterfall samples (Table 1) in the studied bamboo forest supported our first hypothesis. Such seasonal variations are to be expected as the litterfall may be consisted of leaves formed in different years and PhytOC concentrations in leaves change with leaf age [31–33]. Our own work also supports the literature data that PhytOC concentrations are dependent on the age of the leaf (Table 2). The implication of this is that we need to collect litter samples throughout the year to obtain accurate information about PhytOC production in bamboo forest ecosystems.

This is one of the first papers that report PhytOC concentrations in litter samples and the contribution of litterfall to the return of PhytOC to the soil [16]. Most earlier determinations on the contribution of plant species to ecosystem C sequestration through PhytOC production was based on estimates of PhytOC concentrations in green leaf samples [9,13,14,34]. This paper demonstrates that the input of PhytOC to the soil through litterfall is much greater than estimates derived from leaf biomass and PhytOC concentrations in green leaf (Tables 1 and 2) [15]. Our data indicate that the PhytOC input to the soil estimated using the method in Parr et al. [15] using phytolith concentrations in green leaves can be quite different from PhytOC input measurements based on leaf litter input and PhytOC concentrations in leaf litters. Those differences are likely caused by the different sampling times of the plant samples used for analysis of PhytOC concentrations, as phytolith concentrations increase with plant age [31–33]; as a result of that, the PhytOC concentrations in the leaf litter (the oldest leaves on a plant) should be higher than that in the green leaves, as was demonstrated in this study (Tables 1 and 2). Therefore, determining the amount of litterfall and the PhytOC concentration in the litterfall is a more accurate method to quantifying the PhytOC return to the soil in bamboo forest ecosystems.

The large difference in the PhytOC concentration between the bamboo leaf and bamboo leaf litter resulted in the two-time difference in the amount of C that can be sequestered in the form of PhytOC that were estimated through the two methods. This indicates that the amount of C sequestration in the form of PhytOC in previous publications (e.g., [15]) that were based on PhytOC concentrations in green leaves would represent an underestimate. If we follow the same approach used in Parr et al. [15] and consider that the global potential land that can be cultivated (4.1×10^9 ha) is planted to bamboo or plant species with similar PhytOC production potential, the global potential to sequester atmospheric CO_2 is going to be greater than the 11% of the current global CO_2 emissions suggested by Parr et al. [15].

We reviewed the literature and summarized the silicon and phytolith concentrations, biomass production, and the C sequestration potentials (expressed as CO_2 equivalent) of various plant species known to have high abilities to form phytolith (Table 3). The literature data indicate that silicon concentrations in plant biomass are variable, with the highest silicon concentrations reported in sugarcane plants. Silicon concentrations in wheat, bamboo, rice and some aquatic plants can also be high. The highest PhytOC concentration is also reported in sugarcane, followed by wheat, bamboo and rice. The 11.30 kg ha^{-1} yr^{-1} of PhytOC returned to the soil through litterfall (equivalent to the removal of 41 kg of CO_2 ha^{-1} yr^{-1}) supported our second hypothesis that the amount of annual PhytOC return to the soil is a substantial contribution to soil C sequestration. The C sequestration potential through PhytOC formation of the bamboo forest we studied falls within the range reported in other grass-type species [11–15,18]. The literature data indicate that the C sequestration potential through PhytOC formation was high in some aquatic plant species [14]. Despite that the PhytOC concentrations in different plant species in the Poaceae family are not dramatically different, the PhytOC production potential can be quite different largely because of the large differences in biomass production (Table 3).

For most economically important bamboo species, only bamboo leaf litter will be returned to the soil while bamboo stem, which would constitute the majority of biomass production in a bamboo forest, would be exported out of the ecosystem as an economic product that can be used for making furniture or other products.

Eventually, the PhytOC contained in those bamboo parts would be returned to the broader ecosystem and still contribute to C removal from the atmosphere [35]. Therefore, by only assessing the PhytOC content in the soil [36] and the contribution of PhytOC through litterfall [15] will substantially underestimate the potential for C sequestration in bamboo forests. Full accounting of the PhytOC sequestration potential of bamboo forests at the ecosystem level will be required to fully understand their C sequestration potential.

In summary, the production of litterfall and the concentration of phytOC in the litterfall should be used to calculate the potentials of bamboo forests for stable C sequestration in the form PhytOC. The PhytOC concentration in litterfall had seasonal variations and thus representative seasonal litter samples should be collected for analysis to provide more accurate estimates of the potential for PhytOC formation and climate change mitigation. PhytOC production can make a substantial contribution to long-term C sequestration in the studied bamboo forest. In estimating the contribution of PhytOC production for C sequestration in bamboo forests, PhytOC removal in the form of bamboo products from the ecosystem should also be considered.

Acknowledgments

Assistance from Professor Yongfu Li in the preparation of this manuscript is acknowledged.

Author Contributions

Performed the experiments: ZTH PKJ. Analyzed the data: ZTH PKJ. Contributed to the writing of the manuscript: ZTH PKJ SXC. Supervised the project: PKJ. Assisted in the interpretation of the data: SXC. Participated in data collection: YZ YQY. Discussed the results and commented on the contents of the manuscript: ZTH PKJ SXC YZ YQY.

References

1. Piperno DR, Pearsall DM (1993) Phytoliths in the reproductive structures of maize and teosinte - implications for the study of maize evolution. J Archaeol Sci 20: 337–362.
2. Parr JF, Sullivan LA (2005) Soil carbon sequestration in phytoliths. Soil Biol Biochem 37: 117–124.
3. Wang YJ, Lu HY (1993) Phytolith Study and Its Application. Beijing: Ocean Press.
4. Jones L, Milne AA (1963) Studies of silica in the oat plant. Plant Soil 18: 207–220.
5. Wilding LP (1967) Radiocarbon dating of biogenetic opal. Science 156: 66–67.
6. Santos GM, Alexandre A, Coe HHG, Reyerson PE, Southon JR, et al. (2010) The phytolith ^{14}C puzzle: a tale of background determinations and accuracy tests. Radiocarbon 52: 113–128.
7. Wilding LP, Drees LR (1974) Contributions of forest opal and associated crystalline phases to fine silt and clay fractions of soils. Clays Clay Min 22: 295–306.
8. Wang YJ, Lu HY, Lu HB, Tan DR, Huang L (1992) Phytolith and its importance in casting molds and braised soil from ancient China. Chin Sci Bull: 37: 342–345.
9. Song ZL, Liu HY, Si Y, Yin Y (2012) The production of phytoliths in China's grasslands: implications to the biogeochemical sequestration of atmospheric CO_2. Glob Change Biol 18: 3647–3653.
10. Wilding LP, Brown RE, Holowaychuk N (1967) Accessibility and properties of occluded carbon in biogenetic opal. Soil Sci 103: 56–61.
11. Zuo XX, Lü HY (2011) Phytoliths analysis for the discrimination of foxtail millet (*Setaria italica*) and common millet (*Panicum miliaceum*). Chin Sci Bull 56: 3451–3456.
12. Parr JF, Sullivan LA (2011) Phytolith occluded carbon and silica variability in wheat cultivars. Plant Soil 342: 165–171.
13. Li ZM, Song ZL, Parr JF, Wang HL (2013) Occluded C in rice phytoliths: implications to biogeochemical carbon sequestration. Plant Soil 370: 615–623.
14. Li Z, Song Z, Jiang P (2013) Biogeochemical sequestration of carbon within phytoliths of wetland plants: A case study of Xixi wetland, China. Chin Sci Bull 58: 2480–2487.
15. Parr JF, Sullivan LA, Chen BH, Ye GF, Zheng WP (2010) Carbon bio-sequestration within the phytoliths of economic bamboo species. Glob Change Biol 16: 2661–2667.
16. Huang ZT, Li YF, Jiang PK, Chang SX, Song ZL, et al. (2014) Long-term mulching increased carbon occluded in phytolith (PhytOC) in soils of intensively managed bamboo forests. Sci Rep 4: 3602.
17. Hodson MJ, White PJ, Mead A, Broadley MR (2005) Phylogenetic variation in the silicon composition of plants. Ann Bot 96: 1027–1046.
18. Parr JF, Sullivan LA, Quirk R (2009) Sugarcane phytoliths: Encapsulation and sequestration of a long-lived carbon fraction. Sugar Tech 11: 17–21.
19. Kleinhenz V, Midmore DJ (2001) Aspects of bamboo agronomy. Adv Agron. 74: 99–153.
20. He DJ, Hong W, Wu CT, Lan B, Huang H, et al. (2002) Study on biomass and energy distribution of natural *Phyllostachys heterocycla* cv. *Pubescens* in Wuyi Mountains and its comparison with high-yield forest. Acta Botanica Boreali-occidentalia Sinica 23: 291–296.
21. Li YF, Zhang JJ, Chang SX, Jiang PK, Zhou GM, et al. (2013) Long-term intensive management effects on soil organic carbon pools and chemical composition in Moso bamboo (*Phyllostachys pubescens*) forests in subtropical China. For Ecol Manage 303: 121–130.
22. WRB (World Reference Base for Soil Resources) (2006) A Framework for International Classification, Correlation and Communication. Food and Agriculture Organization of the United Nations, Rome.
23. Parr JF, Dolic V, Lancaster G, Boyd WE (2001) A microwave digestion method for the extraction of phytoliths from herbarium specimens. Rev Palaeobot Palynology 116: 203–212.
24. Walkley A, Black IA (1934) An examination of the Degtjareff method for determining soil organic matter, and a proposed modification of the chromic acid titration method. Soil Sci 37: 29–38.
25. Nisbet EG, Dlugokencky EJ, Bousquet P (2014) Methane on the rise–again. Science 343: 493–495.
26. Avinash J, Negi MS (2010) Adaptation and mitigation options in the forestry sector and climate change. J Trop For 26: 18–22.
27. Sollins P, Swanston C, Kramer M (2007) Stabilization and destabilization of soil organic matter–a new focus. Biogeochemistry 85: 1–7.
28. Lal R (2009) Challenges and opportunities in soil organic matter research. European Journal of Soil Science 60: 158–169.
29. Post WM, Kwon KC (200) Soil carbon sequestration and land-use change: processes and potential. Glob Change Biol 6: 317–327.
30. Skjemstad JO, Clarke P, Taylor JA, Oades JM, McClure SG (1996) The chemistry and nature of protected carbon in soil. Aust J Soil Res 34: 251–271.
31. Ma JF, Takahashi E (2002) Functions of silicon in plant growth. In: Soil, fertilizer, and Plant Silicon Research in Japan, 1st edn. Elsevier Science, Amsterdam, The Netherlands. 107–180.
32. Ma JF, Yamaji N (2008) Functions and transport of silicon in plants. Cell Mol Life Sci 65: 3049–3057.
33. Hodson MJ, Sangster AG (1998) Mineral deposition in the needles of white spruce [*Picea glauca* (Moench.) Voss]. Ann Bot 82: 375–385.
34. Song ZL, Liu HY, Li BL, Yang XM (2013) The production of phytolith-occluded carbon in China's forests: implications to biogeochemical carbon sequestration. Glob Change Biol 19: 2907–2915.
35. Jansson C, Wullschleger SD, Kalluri UC, Tuskan GA (2010) Phytosequestration: Carbon biosequestration by plants and the prospects of genetic engineering. BioScience 60: 685–696.
36. Zuo XX, Lu HY, Gu ZY (2014) Distribution of soil phytolith-occluded carbon in the Chinese Loess Plateau and its implications for silica–carbon cycles. Plant Soil 374: 223–232.

Contrasting Regeneration Strategies in Climax and Long-Lived Pioneer Tree Species in a Subtropical Forest

Haiyang Wang[1], Hui Feng[2], Yanru Zhang[2], Hong Chen[1,2]*

1 Institute of Landscape Ecology of Montane Horticulture, Southwest University, Chongqing, 400716, China, **2** Department of Botany, College of Horticulture and Landscape Architecture, Southwest University, Chongqing, 400716, China

Abstract

1: This study investigated 15 coexisting dominant species in a humid subtropical evergreen broad-leaved forest in southwest China, consisting of long-lived pioneers and climax species occurring in natural and disturbed regimes. The authors hypothesized that there would be non-tradeoff scaling relationships between sprouting and seed size among species, with the aim of uncovering the ecological relationship between plant sprouting and seed characteristics in the two functional groups.

2: The sprouting variations of the species were initially examined using pairwise comparisons between natural and disturbed habitats within and across species and were noted to show a continuum in persistence niches across the forest dominants, which may underlie the maintenance of plant diversity. Second, a significantly positive, rather than tradeoff, relationship between sprout number and seed size across species within each of the two functional groups was observed, and an obvious elevational shift with a common slope among the two groups in their natural habitat was examined. The results indicate the following: 1) the relationship of seed size vs. sprouts in the natural habitat is more likely to be bet-hedging among species within a guild in a forest; 2) climax species tend to choose seeding rather than sprouting regeneration, and vice versa for the long-lived pioneers; and 3) the negative correlation between sprouting and seed dispersal under disturbed conditions may imply a tradeoff between dispersal and persistence *in situ* during the process of plant regeneration.

3: These findings may be of potential significance for urban greening using native species.

Editor: Norman W. H. Mason, Landcare Research, New Zealand

Funding: This research was funded by Special Research Program for Public-Welfare Forestry (201004064) and Southwest University (SWU110032). The funders had no role in study design, data collection and analysis, decision to publish, or preparation of the manuscript.

Competing Interests: The authors have declared that no competing interests exist.

* Email: chenh.swu@gmail.com

Introduction

Plant regeneration is an important ecological process for a forest ecosystem in which the sprouting and seeding of woody species is involved [1]. As a mechanism for shaping community dynamics [2], plant regeneration is important for the succession of forest communities and for the stability and restoration of vegetation in a forest ecosystem following various disturbances. As new architectural units, species sprouts [3] have been widely perceived as key functional traits [4–6] linked to plant life-history strategies [7,8], functional types [9,10] or population persistence [11,12]. Using sprouts, some damaged trees can occupy their original space niche rapidly and eventually reach the forest canopy again [13].

Furthermore, sprouts, as one of the two plant regenerative traits [4], have been reported to correlate in part to seed size. For example, sprouters would most likely be expressed in seed fecundity [14]. Moreover, it has been argued that there is a tradeoff between sprouts and seed size [15,16] due to competition for resources between vegetative and reproductive growth [14] relative to starch-tissue content [17]. Plants that sprout vigorously tend to be poorer seed recruiters than non-sprouters [8], and non-sprouters produce more seeds than congeneric sprouters growing at the same sites [8,17]. Nevertheless, others have found no correlation between the two variables (e.g., [18]). In brief, few significant associations have been reported thus far.

In contrast, because both seeding and resprouting are all ascribed to plant regeneration [4], we may infer that seed size correlates with sprouting [19,20]. In addition, in fluctuating environments, seeding shows bet-hedging [21,22], which remains correlated with seed size among community species [23–25] during the process of seed germination as an evolutionarily stable strategy to control the plant population (or ESS, e.g., [26]). With respect to resprouting, Nzunda *et al.* [16] recently suggested a bet-hedging model between occasional events and fixed interference in the storage reserve in good resprouting species. Based on these reports, we may hypothesize that plant sprouting may show a pattern of bet hedging in response alternating seeding, which eventually results in a positive scaling relationship, instead of a tradeoff, between plant sprout and seed size.

Moreover, interspecific and intraspecific differences in sprouting ability have been observed [2,7] that were due to various ecological factors [27–30] or the biological properties of the plants themselves [31–33]. For example, as an example of the latter, light-demanding species generally show enhanced resprouting abilities compared with shade-tolerant species [18], and species tend to be more widely distributed in less-productive sites [17]. It is reasonable to infer that the connections between sprouting and seed traits may be intrinsically variable in different functional plants (i.e., guilds). An examination of these properties may be vital to understand the regenerative ecology of particular species.

However, direct evidence on the relationship between sprouts and seed size in woody species remains insufficient. Furthermore, much of the existing research on regeneration following disturbances has focused on the role of pioneer species (e.g., [2,34–36]). There have been few comparative studies among long-lived pioneers (LP) and climax species (CS; however, see [37]), which may be of practical value for forestry (see the summary in [38]).

The current study employed a subtropical evergreen broad-leaved forest in Mt. Jinyun Nature Reserve in China. We monitored the sprouting performance of 15 dominant species with varying seed size living in both natural and disturbed areas, and we analyzed intraspecific variations in sprouting responses to disturbances for the 15 species by examining the relationship between species individual sprout number and seed size among the two groups. Finally, we examined the associations among life-history attributes of the species, such as plant height, starch amount in current-year shoots, individual seed size, and appendages per seed carried (a parameter characterizing seed-dispersal ability that may be correlated in part with sprouting, see [39]). The primary objectives of this study were the following: 1) compare the sprouting performance within species across disturbance gradients; 2) test the expected ecological relationships between species' sprouting behaviors and seed traits among CS and LP.

Materials and Methods

Study site and vegetation

The study area is located in the Jinyunshan National Nature Reserve (29°50′N, 106°24′E) in southwestern China, covering about 14 km^2 in total. The elevation within the reserve ranges from 180 to 951.5 m. The climate is of a typical subtropical monsoon type, characterized by mid-subtropical monsoons with a rainy, hot summer and a dry, warm winter. The mean annual rainfall is 1143.1 mm, which primarily falls in summer; thus leading to a relatively high mean annual relative humidity of *c.* 85%. The annual mean temperature is 17°C, with annual effective accumulated temperature (>10°C) of *c.* 6000°C, a maximum annual frost-free period of 334 d, and a mean annual sunshine time of only 1160 h. The most common soil type is acid yellow (mean pH value is 4.36) developed from Triassic quartz sandstone, carbonaceous shale and argillaceous sandstone, which contains a total content of 0.0985% nitrogen, 0.040% phosphorus, 1.413% kalium and 2.099% organic matter [40,41]. In addition, Mt. Jinyun is both a nature reserve and a famous scenic area where there are many signs of human habitation, such as roads, farmhouses, and hotels. The plants near these sites are greatly influenced by human disturbance.

With respect to zonal vegetation, there is a humid subtropical evergreen broad-leaved forest with secondary succession of vegetation in varying stages that primarily result from a forest fire (~300 years ago), mining (~50 years ago), and an abandoned tea garden (~20 years ago). The forest shows an obvious vertical structure, primarily comprising Fagaceae, Lauraceae, Theaceae, Symplocaceae, and Elaeocarpaceae [42]. The first tree layer is generally 15–20 m high with a few trees taller than 25 m. This tree layer is chiefly dominated by Fagaceae and Theaceae species that average 15 m in height and range from 25–35 cm in diameter at breast height, interspersed with large Elaeocarpaceae, Lauraceae and Symplocaceae trees that also influence the forest community. The second shrub layer is typically dominated by large shrubs or saplings that are often less than 3 m in height [40] and include species that are common in the sunny bare soil or shaded understory.

Materials and study design

Fifteen dominant or co-dominant species in the tree layer and shrub layer were selected as target species that represented, to some extent, the characteristics of the evergreen broad-leaved forest. The 15 species included three Symplocaceae species, four Theaceae species, three Elaeocarpaceae species, two Lauraceae species, two Fagaceae species, and one Rubiaceae species. They are well-layered in the forest community, and *Castanopsis fargesii*, *Castanopsis carlesii*, *Elaeocarpus duclouxii*, *Machilus nanmu*, *Symplocos laurina* and *Gordonia acuminata* belong to the first tree layer, *Symplocos setchuanensis*, *Adinandra bockiana*, *Neolitsea aurata*, *Adinandra bockiana* and *Sloanea leptocarpa* belong to the second, and *Eurya loguiana*, *Camellia tsofuii*, *Aidia cochinchinensis* and *Symplocos lancifolia* belong to the third [41,43]. We designated these selected species as either climax species or long-lived pioneers based on their role in the forest community; the dividing approach for climax species and long-lived pioneers followed criteria used in the literature [43,44] and personal observations.

The field investigations were performed between April 2011 and May 2011 using stratified random sampling to arrange five sample plots (covering about 15 ha in total that were separated from each other by 400–600 m) in a sampling area of about 100 ha. The habitats of individual plants were divided into two types: disturbed habitat (DH; defined as a forest edge within 10 m of buildings, squares, or main roads), which was more strongly affected by human activities, and natural habitat (NH; greater than 50 m from the sources of interference), which was little influenced by human. The plots in the two types of habitats were kept as uniform as possible with respect to altitude, slope position and aspect based on target plants and plant habitat category.

This paper used sprout number to measure sprouting capability, which is the most commonly adopted method (c.f. [45,46]). Twenty to forty trees per plot were selected randomly to record sprout number per individual in the field. For the purposes of the census, we recorded "distinguished sprout" by specifically referring to the living shoots originating from the trunk [46], and the main stems branching from the trunk were not included.

Ethics statement

The field studies for each site were permitted by the staff of the Jinyunshan National Nature Reserve, and this study did not involve endangered or protected species.

Data analyses

In view of the variation in sprout data of interspecifics and intraspecifics under varying habitats, in prior to analyses of the measured traits we conducted nested ANOVAs to test variability. The results showed that both species and habitat influences sprout number markedly ($F = 45.630$, $p < 0.001$ and $F = 6.078$, $p = 0.014$, respectively). Moreover, as a variable, species contributed the largest variance (87.7%) of sprouting data, following by habitat (11.5%) and the individual plant of the same species under the same habitat contributed the least to variation (0.8%). These results showed that sprout data may be compared among species, regardless of their intraspecific variations in different habitats. Data were meaned across individuals of the same species living in

Table 1. The species properties and the mean sprout number per individual (mean±SE) in natural habitat and in disturbed habitat for the 15 species studied.

Species	Family	Height	Group	Abbr.	NH	DH	DH/NH
Castanopsis carlesii	Fagaceae	Mt	CS	Cc	3.37±0.39 ***	7.05±2.06	2.09
Eurya loguiana	Theaceae	Ls	LP	El	3.61±0.75 ***	6.11±1.89	1.70
Neolitsea aurata	Lauraceae	St	CS	Na	3.24±0.41 ***	5.44±1.08	1.68
Elaeocarpus japonicus	Elaeocarpaceae	Mt	CS	Ej	3.00±0.46 ***	5.00±0.92	1.66
Aidia cochinchinensis	Rubiaceae	Ls	CS	Ac	1.67±0.29 *	2.56±0.67	1.52
Sloanea leptocarpa	Elaeocarpaceae	Mt	CS	Slc	2.80±0.75 ***	4.02±3.00	1.43
Castanopsis fargesii	Fagaceae	Lt	CS	Cf	2.80±0.42 ***	3.82±0.71	1.36
Adinandra bockiana	Theaceae	St	LP	Ab	6.35±1.54 ***	8.57±2.72	1.35
Symplocos lancifolia	Symplocaceae	St	LP	Slf	6.56±2.12 **	8.64±3.33	1.32
Symplocos setchuanensis	Symplocaceae	Mt	CS	Ss	3.23±1.51 ns	4.07±1.61	1.26
Elaeocarpus duclouxii	Elaeocarpaceae	Lt	CS	Ed	5.48±1.14 ns	5.29±2.33	0.96
Camellia tsofuii	Theaceae	Ls	CS	Ct	3.46±0.93 ns	3.24±0.36	0.94
Gordonia acuminata	Theaceae	Mt	LP	Ga	10.50±2.18 ***	5.83±1.21	0.55
Machilus nanmu	Lauraceae	Lt	LP	Mn	9.70±3.77 ***	5.17±1.89	0.53
Symplocos laurina	Symplocaceae	Mt	LP	Sla	9.44±2.78 ***	2.16±0.31	0.23

* denotes the p-level of student's t-tests for the species in the two habitats. * = $p<0.05$, ** = $p<0.01$, *** = $p<0.001$, ns = non-significant. Ss = small shrub (≤0.5 m), Ms = middle shrub (0.5–2 m), Ls = large shrub (2–5 m), St = small tree (5–8 m), Mt = middle tree (8–25 m), Lt = large tree (≥25 m); NH & DH = sprout number in natural & in disturbed habitat, respectively; Abbr. = the abbreviation of the species studied. CS denotes climax species and LP does long-lived pioneer. The species are sorted according to the column of DH/NH.

the same habitat and log_{10}-transformed to improve normality prior to analyzing for the scaling relationships.

The scaling relationships between sprout number and seed size (i.e., seed mass, referring to individual seed dry mass, as in [47]) were analyzed using a Model Type II regression method, with allometric slopes of particular interest calculated as Standardized Major Axes (SMA; [48]). The heterogeneity of regression slopes and the common slopes were tested following the methods of [49], and the shifts between lines fitted to groups sharing a common slope (y-intercept) were examined using ANOVAs. The above allometric parameters were conducted using (S)MATR [48,50].

Although sprouting capability showed little phylogenetic conservatism [5], we nevertheless performed phylogenetic independent contrasts (PICs) analyses for the species used in the current study to determine whether a correlation between seed size and sprouts was biased by phylogenetic signal. The phylogenetic trees were constructed following Phylomatic (version 3; [51]), and PICs were conducted using the phylogenetic comparative methods of COMPARE (4.6b; [52]) for the LP and CS species.

In addition, to assess differences within species in the two habitat conditions, t-tests were conducted using Statistica 6.0 (Statsoft I 2001: Tulsa, Oklahoma). To detect broad correlations in variation among sprouting traits and plant seed traits, we performed a principal component analysis (PCA). Six statistics were employed to perform the PCA for 15 species. With the exception of two sprouting variables, the remaining four parameters were seed size (a life-history trait related to successional trade-offs), plant height (a parameter that has been shown to be strongly associated to the coexistence of pioneer and shade-tolerant tree species by [37]; sprouters are typically short plants, such as shrubs or bushes, whereas non-sprouters are commonly tall trees [11,53]), seed appendage per seed carried (taken from [39]), and starch content in shoot pith (estimated roughly using a scale of 1–5, denoting the least to the most amount based on anatomical sections of current-year shoots observed under a microscope). DCA (detrended correspondence analysis) was performed prior to PCA to determine the ordination model (i.e., unimodal or linear) to be adopted following the requirement of [54], using Canoco for Windows 4.5 software. Because all lengths of the gradient were less than 3 (the largest being 0.863), a linear multivariate PCA approach was chosen to examine the associations among the multiple variables including sprout number in NH and DH, seed size, seed appendage per seed carried and starch content in shoot pith and plant height. Of these variables, the first two were assigned as environmental variables and the remaining as species variables for the PCA. For the PCA, the species data were log-transformed, and the focus of scaling was on inter-species correlations. The species scores were divided by the standard deviations, and sample data were centered and standardized, whereas species data were centered by species. The PCA ordination biplot was created using CanoDraw 4.0.

Results

Sprouting characteristics under disturbed and natural habitats

The mean numbers of sprouts per trunk for the 15 species in DH and in NH are summarized in Table 1. Five of the species (*Gordonia acuminate*, *Machilus nanmu*, *Symplocos laurina*, *Camellia tsofuii*, *Elaeocarpus duclouxii*) showed more sprouts in NH than in DH; in particular, the former three were significantly different (t-test, $p<0.001$) and showed sprouting numbers that were 1.8-, 1.9-, and 5.9-fold greater than in the DH. The remaining 10 species showed fewer sprouts in NH than in DH. The sprout numbers for different tree species within the same habitat were also different. The biggest sprout number in the DH was that of *Symplocos lancifolia* (8.64), followed by *Adinandra bockiana* (8.57); the lowest two were *Symplocos laurina* and *Aidia cochinchinensis*, with only 1.6 and 2.56, respectively.

With the exception of 3 species (t-test, $p>0.05$; *Elaeocarpus duclouxii*, *Camellia tsofuii* and *Symplocos setchuanensis*), there were notable differences in paired sprouts for the 12 species between NH and DH (t-test, $p<0.05$; Table 1). Therefore, human activities affect sprouting capacity of trees in general; however, the effect of this interference differs with species.

Scaling relationship between sprouting capability and species seed size

Species sprout number in NH did not generally correlate with seed size in pooled data across all 15 species ($p>0.05$). However, when dividing the 15 species into two groups (CS group and LP group; Table 1), the sprout number closely scaled with seed mass across congeners within a group ($r^2=0.714$, $p=0.034$, slope = 5.884 [2.959 11.702] for the LP group; $r^2=0.907$, $p<0.001$, slope = 6.609 [5.049 8.653] for the CS group) with a significant elevation increase ($\text{y-intercept}_{LP}=-4.157$; $\text{y-intercept}_{CS}=-1.479$) fitting a common slope (= 6.476; Figure 1). Second, the sprout number in DH did not correlate with seed mass across species ($p>0.05$) even when the 15 species were divided into CS and LP ($p>0.05$). In addition, the above tendencies were maintained in PICs, with the exception of a weak correction in pooled data across the 15 species in DH (Table 2).

These results show that the relationship between sprouting capability and species seed size varies considerably between the two species groups in natural habitats and remained when phylogeny signals were removed. Furthermore, the apparent difference in y-intercept (increase) between the CS and LP groups indicates that the species in the CS group produced larger seeds than those of the LP group for given sprouts.

The broad trends among sprouting capability, stem and seed traits, and habitats

The PCA graph summarizes the general correlations (Figure 2). The PCA revealed that the first and second PCA axes explained 58.4% and 33.2% of the variability in species data, respectively; 91.6% in total. The variability of NH sprouts paralleled the direction of seed size (Figure 2) but opposed the direction of height and starch content; however, the DH variable was orthogonal to seed size but conversely parallel to the direction of seed appendage. Thus, the NH variable was more closely and positively associated with seed size and correlated negatively with height and starch content. The DH variable correlated strongly with seed appendage per seed carried but was independent of seed size.

Discussion

We performed intraspecific comparisons of sprouting performance for the dominant species in a subtropical community and tested our predictions on the interspecific relationships between sprouting and seed traits for LP and CS. We observed consistently significant relationships between species sprouting and seed traits; however, the details of these relationships varied based on species and habitat type.

1) Contrasting sprouting patterns within and across species between natural and disturbed habitats

The species sprouting patterns differed in conspecifics in different habitat types (consistently showed using t-tests) and also

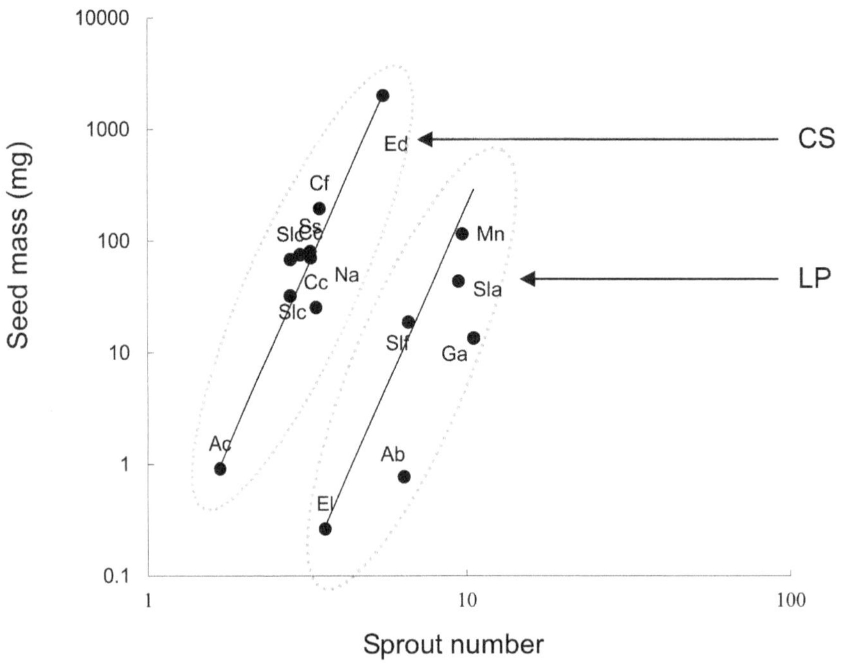

Figure 1. The relationship of seed size vs. sprouts. The bivariate relationship between seed size and trunk sprouting number among climax species (CS) and long-lived pioneers (LP) in natural habitats in a subtropical broad-leaved forest in southwestern China. The two lines in the graph denote the fitted lines to the two functional groups (slope = common slope shared by the two functional groups) using the standardised major axis (SMA); broken line, CS species group; solid line, LP group.

in interspecifics. Some species sprouted vigorously in natural habitats; however, others showed enhanced sprouting in disturbed habitats. The intraspecific sprouting pattern shows that, among the species measured, the majority of species were sensitive to disturbance, indicating that disturbance affects sprouting in plants.

Moreover, the differentiation among dominant species in a forest may be of particular significance due to their leader functions in a forest when considering spatial coexistence among species [55] and the dynamics of the community structure [56]. Some species sprouted more in DH whereas others performed better in NH, as mentioned above, we believe this diverse sprouting property in dominant species forms a vegetative regenerative spectrum, which lays the foundation for them to cope with various natural- or human-derived disturbances, and presumably, it also reflects a bet-hedging sprouting strategy at the species level within a community (c.f. [57]). Thus, the marked differences in sprouting performance may be critical for the dominant species to colonize gaps [58] in a complementary way and may enable maintenance of tree diversity following disturbances, based on the "alternate trait axis" argument [2].

Nevertheless, it is difficult to explain the origin of the difference in sprouting behaviors of some species that sprouted less in disturbed habitats, as observed here in this study. Further research such as investigation for underground clone structure is suggested.

2) Relationship between seed size and sprout number

Larger-sized species tend to produce more sprouts among species within a functional group in an undisturbed habitat; however, this relationship did not reach significance when all species were pooled, indicating that natural guild species may evolve the potential to reproduce vegetatively (bet-hedge rather than reproduce sexually) in response to environmental cues [57]. We believe that this result is reasonable based on two lines of reasoning. First, the observed sprouting continuum across dominant species along the disturbed gradient provided a necessity for species hedging their evolutionary bets in varying environments [59]. Second, woody plants are perennial and not annual. The sprouts on trunks were the sprouts saved for years that eventually become the biological "bet" and enable a positive relationship between seed size and sprouts; hence, there is no need

Table 2. Summary of phylogenetically independent comparative analysis (PICs) for the relationships between sprout number and seed size of the climax species (CS) and long-lived pioneers (LP) studied, and for the pooled data (Pooled) of the two groups.

Habitat	Group	r^2	p	*a*	*b*
NH	CS	0.807	0.002	6.395	0.057
NH	LP	0.722	0.002	3.326	−0.284
NH	Pooled	0.702	<0.001	4.508	−0.125
DH	Pooled	0.365	0.029	−2.187	−0.670

Letters *a* and *b* are the regressive coefficient and the intercept of the linear regressive equation $Y = ax+b$, where Y and x are seed mass and sprout number, respectively.

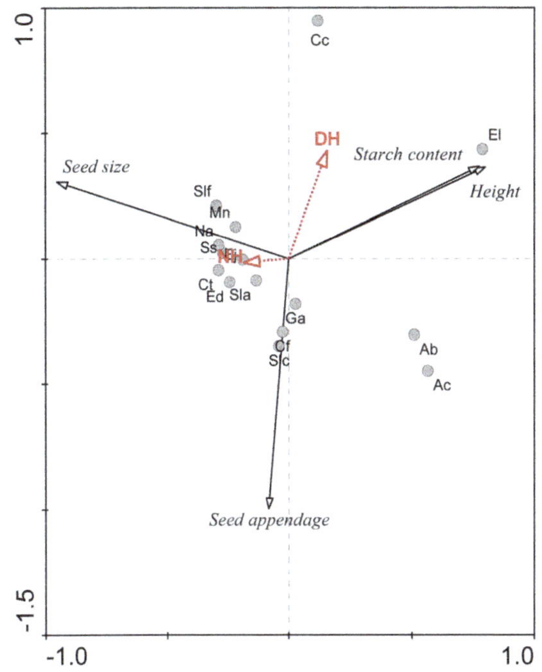

Figure 2. The PCA biplot. Principal component analysis (PCA) based on trunk sprouts in natural sites (NH) and disturbed sites (DH), seed size, seed appendage, starch content in shoot and plant height variables. Circles show sampled species, with abbreviations being as for Table 1.

for to trade-off their limited resources accumulated within one year. Thus, this strategy may be viewed as the integration of bet-hedging between sprouting and seeding during long-term regenerative evolution.

Furthermore, based on the equal sprouting number on trunks, climax species tended to show larger individual seeds than long-lived pioneers, revealing a difference between the two species groups with respect of life history strategy under natural conditions. This tendency held true in PIC analyses, revealing its ecological significance. The CS guild may win out in a forest by evolving larger seeds, enabling seedlings to settle and establish in a darker habitat. In contrast to CS, the LP species tended to adopt a vegetative means to enable plant regeneration in brighter sites. They produce more sprouts and smaller seeds under natural conditions; thus, enhanced sprouting ability but weaker seeds.

On one hand, this result is inconsistent with authors who argue that vegetative sprouting should require a tradeoff against sexual reproduction in general and consequently, that poor resprouters should show higher seed mass compared with good resprouters (e.g., [15,60]). This inconsistency may result from the differences in the type of environment studied (e.g., natural vs. disturbed habitat), the species type (the species focused upon were woody CS & LP) or the variation in seed size due to seed size/number tradeoff (e.g., [39]). On the other hand, the greater mean number of sprouts in LP compared with CS under natural conditions, as observed in our study, indicates a pattern of smaller-seeded species with more sprouts [the LP group in NH showed markedly more sprouts (= 7.695) than the CS group (= 3.232; t-test, $t = -4.661$, $p<0.001$)], which may also represent a tradeoff to some extent between seed size and sprouts. In addition, this relationship did not hold true for the species in disturbed habitats, as shown both in the PCA graph (Figure 2) and in regression analysis ($p>0.05$), which is consistent with a previous report [18].

3) Associations among NH & DH sprouts, seed dispersal, carbon storage and plant height

PCA showed a clear positive correlation of sprouts in NH with seed size, further verifying the relationship between these two variables (Figure 2). NH sprouting was also negatively correlated with starch content and plant height, showing that good sprouters in natural sites may store starch less and include relatively more short species. The DH sprouts showed contrasting results, supporting some previous reports, such as [60], demonstrating that sprouters show higher starch concentrations in roots.

The DH sprouting variable was inversely correlated with seed appendage (Figure 2), suggesting that good sprouters in DH may be weak seed dispersers and vice versa. This relationship may imply a tradeoff between the space occupied by the plant between far (i.e., seed dispersal) and near (i.e., sprouting). The ecological relationship between seed dispersal and plant sprouting may be an interesting topic that deserves further discussion.

4) Ecological and practical implications of the sprouting strategy

The sprouts of dominant species in the natural ecosystem may be very useful when managing forests for products including timber, firewood, edible fruits, and landscape plants, which are some of the different uses of the species observed. Thus, our results are likely to be helpful for silvicultural treatments. For example, the difference in seed size between CS and LP implies different regeneration niches in a subtropical forest and lays a good foundation for a more resilient forest ecosystem apart from human disturbances using complementary mechanisms of responding to a more or less disturbed habitat. Still, long-lived pioneers show some excellent features, such as their fast growth and relative shade tolerance, compared with pioneers, and they have light but strong timber (characteristics proposed as part of the long-lived pioneer syndrome by [38] (pp246), which is superior not to the slow-growing climax species but to the strongly r-selected pioneer. Therefore, LP is an issue of particular concern to foresters when applying "Miyawaki's Method" for reestablishing an urban forest [61].

Conclusions

We tested and confirmed our hypothesis on the ecological correlation of sprout vs. seed size. Dominant forest species differ in their sprouting capability, build divergent sprouting patterns along disturbed gradients and display contrasting relationships of sprouts vs. seed size among guilds, manifesting in diverse sprouting strategies across woody plants and suggesting a strong selection for species life-histories under varying conditions. These results provide evidence for the evolutionary mechanism and consequences of variations in plant sprouting behavior, which may contribute to the ecology of regeneration in a forest community.

Acknowledgments

We thank Yanfang Liu for analysis of the anatomy of plant shoots in the laboratory and Danrong Wang for preparing the images used in this paper. We also appreciate the two anonymous reviewers for their constructive comments on early versions of this manuscript.

Author Contributions

Conceived and designed the experiments: HW. Performed the experiments: HF HW. Analyzed the data: HC. Contributed reagents/materials/analysis tools: HC. Wrote the paper: HC YZ.

References

1. Pratt RB, Jacobsen AL, Hernandez J, Ewers FW, North GB, et al. (2012) Allocation tradeoffs among chaparral shrub seedlings with different life history types (Rhamnaceae). American Journal of Botany 99: 1464–1476.
2. Dietze MC, Clark JS (2008) Changing the gap dynamics paradigm: vegetative regeneration control on forest response to disturbance. Ecological Monographs 78: 331–347.
3. Turnbull CG (2005) Plant architecture and its manipulation: Blackwell.
4. Cornelissen J, Lavorel S, Garnier E, Diaz S, Buchmann N, et al. (2003) A handbook of protocols for standardised and easy measurement of plant functional traits worldwide. Australian Journal of Botany 51: 335–380.
5. Vesk PA, Westoby M (2004) Sprouting ability across diverse disturbances and vegetation types worldwide. Journal of Ecology 92: 310–320.
6. Clarke PJ, Lawes MJ, Midgley JJ, Lamont BB, Ojeda F, et al. (2013) Resprouting as a key functional trait: how buds, protection and resources drive persistence after fire. New Phytologist 197: 19–35.
7. Bellingham PJ, Sparrow AD (2000) Resprouting as a life history strategy in woody plant communities. Oikos 89: 409–416.
8. Bond WJ, Midgley JJ (2003) The evolutionary ecology of sprouting in woody plants. International Journal of Plant Sciences 164: S103–S114.
9. Poorter H, Niklas KJ, Reich PB, Oleksyn J, Poot P, et al. (2012) Biomass allocation to leaves, stems and roots: meta-analyses of interspecific variation and environmental control. New Phytologist 193: 30–50.
10. Weiher E, Werf A, Thompson K, Roderick M, Garnier E, et al. (1999) Challenging Theophrastus: a common core list of plant traits for functional ecology. Journal of Vegetation Science 10: 609–620.
11. Hodgkinson KC (1998) Sprouting success of shrubs after fire: height dependent relationships for different strategies. Oecologia 115: 64–72.
12. Houssard C, Escarré J (1995) Variation and covariation among life-history traits in *Rumex acetosella* from a successional old-field gradient. Oecologia 102: 70–80.
13. Bellingham PJ, Kohyama T, Aiba S (1996) The effects of a typhoon on Japanese warm temperate rainforests. Ecological Research 11: 229–247.
14. Lamont BB, Wiens D (2003) Are seed set and speciation rates always low among species that resprout after fire, and why? Evolutionary Ecology 17: 277–292.
15. Nzunda E, Lawes M (2011) Costs of resprouting are traded off against reproduction in subtropical coastal dune forest trees. Plant Ecology 212: 1991–2001.
16. Nzunda EF, Griffiths ME, Lawes MJ (2014) Resource allocation and storage relative to resprouting ability in wind disturbed coastal forest trees. Evolutionary Ecology 28: 735–749.
17. Bond WJ, Midgley JJ (2001) Ecology of sprouting in woody plants: the persistence niche. Trends in Ecology & Evolution 16: 45–51.
18. Shibata R, Shibata M, Tanaka H, Iida S, Masaki T, et al. (2014) Interspecific variation in the size-dependent resprouting ability of temperate woody species and its adaptive significance. Journal of Ecology 102: 209–220.
19. West GB, Brown JH, Enquist BJ (1999) The fourth dimension of life: fractal geometry and allometric scaling of organisms. Science 284: 1677–1679.
20. Lohier T, Jabot F, Meziane D, Shipley B, Reich PB, et al. (2014) Explaining ontogenetic shifts in root-shoot scaling with transient dynamics. Annals of Botany 114: 513–524.
21. Cohen D (1966) Optimizing reproduction in a randomly varying environment. Journal of theoretical biology 12: 119–129.
22. Philippi T (1993) Bet-hedging germination of desert annuals: variation among populations and maternal effects in *Lepidium lasiocarpum*. American Naturalist: 488–507.
23. Pake CE, Venable DL (1996) Seed banks in desert annuals: implications for persistence and coexistence in variable environments. Ecology 77: 1427–1435.
24. Fenner M (2000) Seeds: the ecology of regeneration in plant communities: CABI Publishing.
25. Fenner M, Thompson K (2005) The ecology of seeds: Cambridge University Press.
26. Clauss M, Venable D (2000) Seed germination in desert annuals: an empirical test of adaptive bet hedging. American Naturalist 155: 168–186.
27. Salk CF, McMahon SM (2011) Ecological and environmental factors constrain sprouting ability in tropical trees. Oecologia 166: 485–492.
28. Van Bloem SJ, Murphy PG, Lugo AE (2007) A link between hurricane-induced tree sprouting, high stem density and short canopy in tropical dry forest. Tree Physiology 27: 475–480.
29. Zhu W, Xiang J, Wang S, Li M (2012) Resprouting ability and mobile carbohydrate reserves in an oak shrubland decline with increasing elevation on the eastern edge of the Qinghai Tibet Plateau. Forest Ecology and Management 278: 118–126.
30. Moreira B, Tormo J, Pausas JG (2012) To resprout or not to resprout: factors driving intraspecific variability in resprouting. Oikos 121: 1577–1584.
31. Burrows GE (2008) *Syncarpia* and *Tristaniopsis* (Myrtaceae) possess specialised fire-resistant epicormic structures. Australian Journal of Botany 56: 254–264.
32. Meier AR, Saunders MR, Michler CH (2012) Epicormic buds in trees: a review of bud establishment, development and dormancy release. Tree Physiology 32: 565–584.
33. Wu L, Shinzato T, Chen C, Aramoto M (2008) Sprouting characteristics of a subtropical evergreen broad-leaved forest following clear-cutting in Okinawa, Japan. New Forests 36: 239–246.
34. Lusk CH (1999) Long-lived light-demanding emergents in southern temperate forests: the case of *Weinmannia trichosperma* (Cunoniaceae) in Chile. Plant Ecology 140: 111–115.
35. Lasso E, Engelbrecht BMJ, Dalling JW (2009) When sex is not enough: ecological correlates of resprouting capacity in congeneric tropical forest shrubs. Oecologia 161: 43–56.
36. Miura M, Yamamoto S (2003) Structure and dynamics of a *Castanopsis cuspidata* var. *sieboldii* population in an old-growth, evergreen, broad-leaved forest: the importance of sprout regeneration. Ecological Research 18: 115–129.
37. Gutiérrez AG, Aravena JC, Carrasco-Farías NV, Christie DA, Fuentes M, et al. (2008) Gap-phase dynamics and coexistence of a long-lived pioneer and shade-tolerant tree species in the canopy of an old-growth coastal temperate rain forest of Chiloé Island, Chile. Journal of Biogeography 35: 1674–1687.
38. Turner IM (2001) The ecology of trees in the tropical rain forest: Cambridge University Press.
39. Chen H, Felker S, Sun S (2010) Allometry of within-fruit reproductive allocation in subtropical dicot woody species. American Journal of Botany 97: 611–619.
40. Zhong Z (1988) Ecological study on evergreen broadleaved forest: Chongqing Southwest China Normal University Press (in Chinese with English abstracts). Pp. 696.
41. Xiong J, Guo S, Pan T (2005) The plant index of Mt. Jinyun Chongqing: Southwest China Normal University Press (in Chinese).
42. Zeng J, Liu Y (1995) An floristic analysis for the xylophyta in Sichuan. Journal of Southwest China Normal University (Natural Science Edition; in Chinese with English abstracts) 20: 686–692.
43. Liu Y (1988) Vegetation summary of Jinyun Mountain Natural Reserve. In: Zhong Z, editor. Ecology study of evergreen broad-leaved forest. Chongqing: Southwest China Normal University Press (in Chinese with English abstracts).
44. Da L, Yang Y, Song Y (2003) Population structure and regeneration types of dominant species in an evergreen broadleaved forest in Tiantong National Forest Park, Zhejiang Province, Eastern China. Acta Phytoecological Sinica 28: 376–384 (in Chinese with English abstract).
45. Sakai A, Ohsawa T, Ohsawa M (1995) Adaptive significance of sprouting of *Euptelea polyandra*, a deciduous tree growing on steep slopes with shallow soil. Journal of Plant Research 108: 377–386.
46. Del Tredici P (2001) Sprouting in temperate trees: a morphological and ecological review. Botanical Review 67: 121–140.
47. Chen H, Niklas KJ, Yang D, Sun S (2009) The effect of twig architecture and seed number on seed size variation in subtropical woody species. New Phytologist 183: 1212–1221.
48. Warton DI, Wright IJ, Falster DS, Westoby M (2006) Bivariate line-fitting methods for allometry. Biological Reviews 81: 259–291.
49. Warton DI, Weber NC (2002) Common slope tests for bivariate errors-in-variables models. Biometrical Journal 44: 161.
50. Falster DS, Warton DI, Wright IJ (2006) SMATR: standardised major axis tests and routines ver 2. 0 ed. Available: http://www.bio.mq.edu.au/ecology/SMATR/Accessed 2014 Aug 1.
51. Webb CO, Donoghue MJ (2005) Phylomatic: tree assembly for applied phylogenetics. Molecular Ecology Notes 5: 181–183; Available: http://phylodiversity.net/phylomatic/version 3 Accessed 2014 Aug 1.
52. Martins EP (2004) COMPARE: Computer programs for the statistical analysis of comparative data, 4.6b. Computer program and documentation distributed by the author; Available: http://compare.bio.indiana.edu Accessed 2014 Aug 1
53. Kruger L, Midgley J, Cowling R (1997) Resprouters vs reseeders in South African forest trees; a model based on forest canopy height. Functional Ecology 11: 101–105.
54. Lepš J, Šmilauer P (2003) Multivariate analysis of ecological data using CANOCO: Cambridge university press.
55. Amarasekare P (2003) Competitive coexistence in spatially structured environments: a synthesis. Ecology Letters 6: 1109–1122.
56. Wang X, Kent M, Fang X (2007) Evergreen broad-leaved forest in Eastern China: its ecology and conservation and the importance of resprouting in forest restoration. Forest Ecology and Management 245: 76–87.
57. Venable DL (2007) Bet hedging in a guild of desert annuals. Ecology 88: 1086–1090.
58. Bullock JM (2000) Gaps and seedling colonization. In: Fenner M, editor. Seeds: the ecology of regeneration in plant communities: CABI Publishing. pp. 375–395.
59. Starrfelt J, Kokko H (2012) Bet-hedging – a triple trade-off between means, variances and correlations. Biological Reviews 87: 742–755.
60. Knox KJE, Clarke PJ (2005) Nutrient availability induces contrasting allocation and starch formation in resprouting and obligate seeding shrubs. Functional Ecology 19: 690–698.
61. Miyawaki A (1999) Creative ecology: restoration of native forests by native trees. Plant Biotechnology 16: 15–25.

19

Community Structure of Skipper Butterflies (Lepidoptera, Hesperiidae) along Elevational Gradients in Brazilian Atlantic Forest Reflects Vegetation Type Rather than Altitude

Eduardo Carneiro[1]*, Olaf Hermann Hendrik Mielke[1], Mirna Martins Casagrande[1], Konrad Fiedler[2]

1 Laboratório de Estudos de Lepidoptera Neotropical, Zoology Department, UFPR. Curitiba, Paraná, Brasil, **2** Division of Tropical Ecology & Animal Biodiversity, University of Vienna, Vienna, Austria

Abstract

Species turnover across elevational gradients has matured into an important paradigm of community ecology. Here, we tested whether ecological and phylogenetic structure of skipper butterfly assemblages is more strongly structured according to altitude or vegetation type along three elevation gradients of moderate extent in Serra do Mar, Southern Brazil. Skippers were surveyed along three different mountain transects, and data on altitude and vegetation type of every collection site were recorded. NMDS ordination plots were used to assess community turnover and the influence of phylogenetic distance between species on apparent community patterns. Ordinations based on ecological similarity (Bray-Curtis index) were compared to those based on phylogenetic distance measures (MPD and MNTD) derived from a supertree. In the absence of a well-resolved phylogeny, various branch length transformation methods were applied together with four different null models, aiming to assess if results were confounded by low-resolution trees. Species composition as well as phylogenetic community structure of skipper butterflies were more prominently related to vegetation type instead of altitude per se. Phylogenetic distances reflected spatial community patterns less clearly than species composition, but revealed a more distinct fauna of monocot feeders associated with grassland habitats, implying that historical factors have played a fundamental role in shaping species composition across elevation gradients. Phylogenetic structure of community turned out to be a relevant additional tool which was even superior to identify faunal contrasts between forest and grassland habitats related to deep evolutionary splits. Since endemic skippers tend to occur in grassland habitats in the Serra do Mar, inclusion of phylogenetic diversity may also be important for conservation decisions.

Editor: M. Alex Smith, University of Guelph, Canada

Funding: Funding was provided by CNPq and CAPES to EC as a doctoral scholarship. OHHM and MMC are granted with "Produtividade de Pesquisa"; see (http://www.cnpq.br/web/guest/apresentacao13). The funders had no role in study design, data collection and analysis, decision to publish, or preparation of the manuscript.

Competing Interests: The authors have declared that no competing interests exist.

* Email: carneiroeduardo@hotmail.com

Introduction

Since the early observations by Forster [1] and von Humboldt [2], species turnover across altitudinal gradients has matured into an important paradigm of community ecology. High mountain chains are usually selected for analyses, mainly because their wide elevational ranges increase the power of detecting patterns in community structure (see [3,4]). Geologically older mountain chains whose peaks are considerably lower as a consequence of long exposition to erosion forces have received less attention, since effects of elevation on species diversity or species composition might not be as evident. In such places, historical events may play an additional fundamental role in assemblage turnover in addition to altitude. More recently, the integration of phylogenetic structure into community ecology has yielded important insights as to how historical factors influence community structure [5].

The Serra do Mar is located near the south-eastern coast of Brazil and represents a good example of an old mountain chain in the southern hemisphere, where animal and plant communities might have been structured through both ecological and historical processes. The start of its rising processes has been dated at about 90 mya, soon after the splitting of Gondwana into its two biggest daughter continents, viz. South America and Africa [6]. Despite its moderate altitudinal amplitude today (peaks generally between 800–1500 m), distinct climatic and pedological shifts occur along these rather short elevational gradients, together with a high endemism rate in certain plant taxa such as Bromeliaceae, Orchidaceae and Gesneriaceae [6,7].

Environmental dimensions change in a predictable manner with increasing altitude in the Serra do Mar. Mean annual temperature decreases by an average of 0.5–0.6 K per 100 m altitude, while soil depth decreases and wind intensity increases [8]. Correlated to those abiotic factors and modulated through superimposed effects of anthropogenic land-use, four different vegetation types can be observed on these mountains: Atlantic rain forest, early succession flora, cloud forest, and grasslands [7]. While Atlantic rain forest prevails in the lowlands (up to 1100 or 1400 m, depending on the mountain), grasslands are generally restricted to altitudes above

1300 m. These grasslands are relict vegetation ecosystems which had been far more widespread in south and southeast Brazil during dry cold periods of the late Quaternary [9]. Since then, this vegetation has become increasingly restricted to mountain tops, due to forest expansion, with many animal species related to this kind of ecosystem being included in red lists [10,11]. Early successional flora is not only found after human interference, but also due to natural disturbance events such as forest fires or landslides [7,8].

Communities of host-specific herbivores, whose life histories are intrinsically related to the vegetation of their habitats, may either respond directly to abiotic environmental gradients, or alternatively their species turnover may rather track biotic changes in vegetation types [12,13]. The challenge remains to disentangle whether abiotic (climate, area size) or biotic conditions (vegetation) are more important as drivers of their community composition and species turnover along elevational gradients [14,15], and how phylogenetic history contributes to understanding contrasting patterns between taxa [5,16].

Various groups of nocturnal Lepidoptera have frequently served as targets to study elevational diversity patterns in the tropics [17–19], yet surprisingly few quantitative butterfly studies do exist from elevational gradients on tropical mountains [20,21]. Especially, the few studies addressing phylogenetic structure in butterfly communities along elevational gradients are extra-tropical [22,23], where faunas are distinctly less diverse (but see [24] for a recent example on a rich Neotropical moth assemblage). Skipper butterflies (Lepidoptera, Hesperioidea) might also have the potential to address elevational patterns and underlying processes, but thus far no specific case study on this somehow elusive family exists from mountain ranges anywhere in the tropics.

Recent studies revealed contrasting patterns of phylogenetic community structure along environmental gradients, i.e. different assemblages respond to environmental or biotic factors depending on their specific requirements, evolutionary history and biogeography [25–28]. Skippers might have particular potential to reveal how vegetation types and altitude influence phylogenetic composition along altitudinal gradients, since major skipper lineages are conservative and contrasting in relation to their larval food plant affiliations [29]. While Hesperiinae larvae feed exclusively on monocotyledonous plants, Pyrginae larvae are bound to various families of dicotyledonous plants, mainly Fabaceae, Malvaceae and Rutaceae [29–31]. In addition, skippers are the only group of butterflies commonly observed throughout all sections of altitudinal gradients in south Brazil (pers. observ.), and they comprise far

Table 1. Two different sets of sample unit delimitations used to analyze skipper assemblages in the Serra do Mar (Brazil)[1].

Samples 1 (m)	Mountain	Veg. Type	Samples 2 (m)	Mountain	Veg. Type	Altitude Class
1000–1100	Anhangava	FOR	998–1060	Anhangava	FOR	low
1100–1200	Anhangava	FOR+ESV	1061–1122	Anhangava	FOR	low
1200–1300	Anhangava	ESV+GRA	1123–1206	Anhangava	ESV	medium
1300–1400	Anhangava	GRA	1207–1289	Anhangava	ESV	medium
1400–1500	Anhangava	GRA	1290–1364	Anhangava	GRA	medium*
900–1000	Araçatuba	FOR+ESV	1365–1440	Anhangava	GRA	high
1000–1100	Araçatuba	ESV	912–938	Araçatuba	FOR	low
1100–1200	Araçatuba	GRA	939–1019	Araçatuba	ESV	low
1200–1300	Araçatuba	GRA	1020–1099	Araçatuba	ESV	low
1300–1400	Araçatuba	GRA	1100–1175	Araçatuba	GRA	low
1400–1500	Araçatuba	GRA	1176–1250	Araçatuba	GRA	medium
1500–1600	Araçatuba	GRA	1251–1325	Araçatuba	GRA	medium
1600–1700	Araçatuba	GRA	1326–1400	Araçatuba	GRA	medium
900–1000	Caratuva	FOR	1401–1475	Araçatuba	GRA	high
1000–1100	Caratuva	FOR+ESV	1476–1550	Araçatuba	GRA	high*
1100–1200	Caratuva	ESV	1551–1625	Araçatuba	GRA	high
1200–1300	Caratuva	ESV	1625–1682	Araçatuba	GRA	high
1300–1400	Caratuva	FOR	980–1031	Caratuva	FOR	low
1400–1500	Caratuva	FOR+GRA	1032–1083	Caratuva	FOR	low
1800–1900	Caratuva	GRA	1084–1158	Caratuva	ESV	low
			1159–1233	Caratuva	ESV	medium
			1234–1306	Caratuva	ESV	medium
			1307–1362	Caratuva	FOR	medium
			1363–1418	Caratuva	FOR	medium
			1419–1488	Caratuva	GRA	high
			1800–1860	Caratuva	GRA	high*

[1]Samples 1: delimited only by altitude; Samples 2: delimited by vegetation type and altitude. Each location is assigned to mountains, elevational belts and vegetation types. Note that the delimitation by altitude plus vegetation increases the number of sample units. Abbreviations: FOR: forest; ESV: early successional vegetation; GRA: grassland.

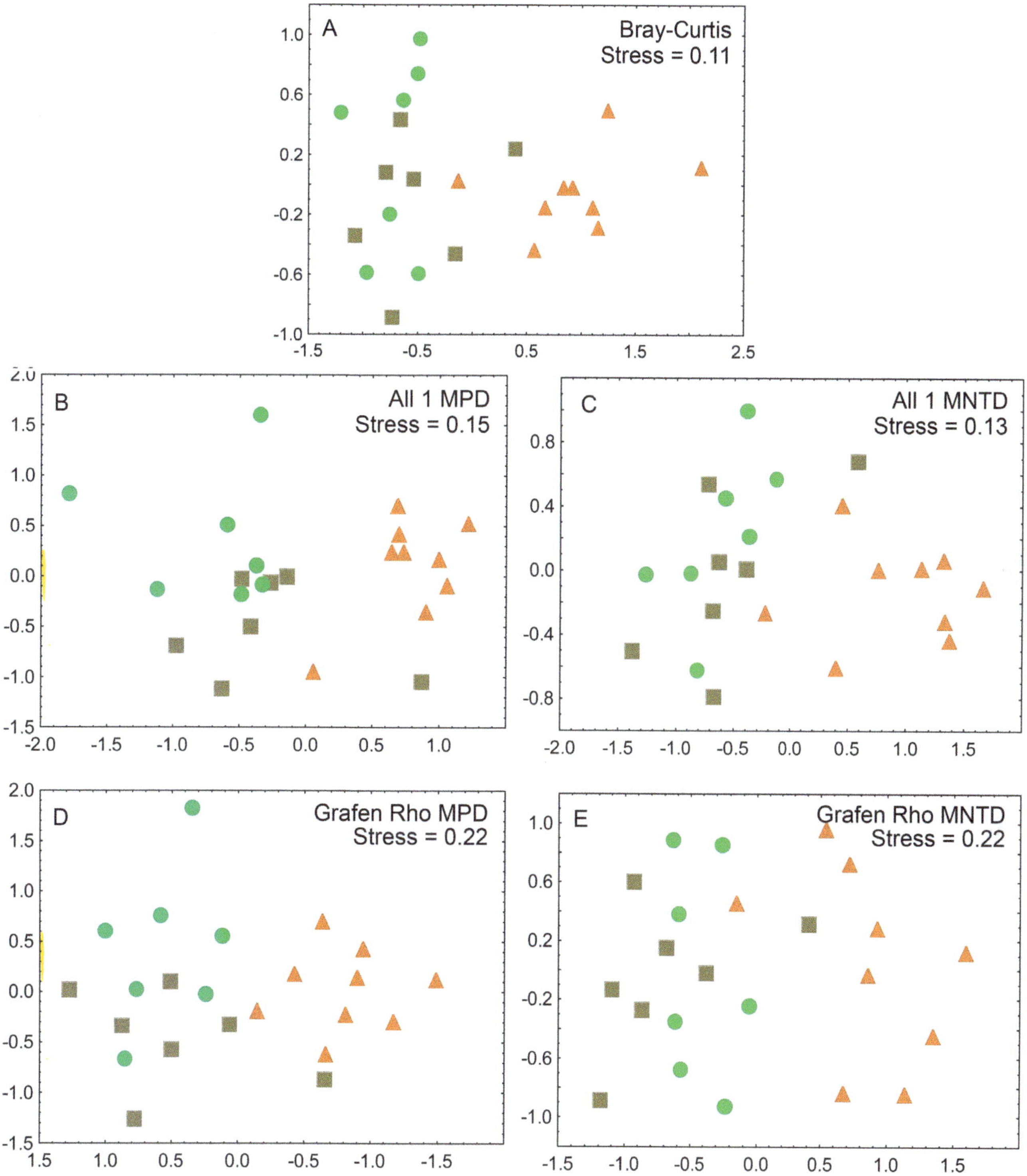

Figure 1. NMDS ordination plots of Hesperiidae assemblages along elevational gradients in Serra do Mar, Brazil. Ordination patterns were assessed based on Bray-Curtis similarities of species lists (A), and compared to two phylogenetic distance indexes (MPD and MNTD) using equal branch lengths (All 1) and Grafen's Rho transformation method (charts B–E). Samples are scored according to altitude and vegetation types. Assemblages are basically ordered along the first axis from low (left) to high elevations (right). Symbols: green circles (forest), brown squares (early successional vegetation), orange triangles (grassland). Stress values indicate goodness of fit of two-dimensional representations to the underlying distance matrices.

larger numbers of grassland endemics when compared to Papilionoidea families in the region [32].

The present study aimed to elucidate the structure of Hesperiidae assemblages along altitudinal gradients of moderate extension, using ecological and phylogenetic measures of community similarity, on three different mountains, in relation to different vegetation types. Specifically, the following hypotheses were tested: 1. Assemblages of skippers are structured according to vegetation type as well as altitude; 2. Phylogenetic similarity is more informative for community responses to altitude and vegetation than ecological similarity (e.g. species composition); 3. Vegetation has a stronger influence on skipper assemblages than altitude; 4. Subfamilies of skippers show discordant community patterns because they differ in life-history traits governing their habitat preferences.

Methods

Study location and sampling methods

Three mountains located in the Serra do Mar of Paraná state, Brazil were sampled in this study: Anhangava (25°23′30″S; 49°00′15″W), Araçatuba (25°54′07″S; 48°59′37″W) and Caratuva mountain (25°13′30″S; 48°51′40″W). The three locations sampled are embedded in protected areas regulated by the state environmental agency (IAP/PR). For location details see [33]. Capture of specimens and their transport to the laboratory for subsequent identification were permitted by licences n° 59.08 (IAP/PR) and 14.595-1 (IBAMA/Sisbio). Hesperiidae specimens were captured with insect nets during up- and down-walking of transects. Each transect was walked 11 times from 2009 to 2011, between 9:00 and 16:00 h. All recorded specimens were immediately labeled according to the elevation (measured to the nearest 10 m using a Garmin 60Cx GPS device) and vegetation type of their sampling locality. In all, 1578 records of 155 species make up the data on which all analyses are based. No endangered species were recorded in this study.

On the mountains in Serra do Mar up to four different vegetation types are present above 900 m: montane forest, cloud forest, early successional vegetation, and grassland. Montane forest refers to a well-developed, tall grown (up to 10 m), vertically stratified Atlantic Rainforest located throughout the slopes of Serra do Mar. Cloud forest stands are more dense, lower in growth (tree height from 3–7 m) with less well defined strata. Successional vegetation and grasslands both lack a canopy stratum. Succession vegetation is dominated by bracken fern *Pteridium aquilinum* (Dennstaedtiaceae), and it is often located where fire or human impact have occurred recently [8]. Grassland sites present a varied floral composition, dominated by several species of Poaceae and Cyperaceae, but occasionally with scattered low trees and shrubs [7].

Although these vegetation types are in part related to altitude (e.g. grasslands are located on mountain tops), their ranges also vary according to a mountain's relief, soil depth, or biogeographical history [7]. This peculiarity allowed us to examine whether vegetation or altitude *per se* plays the more important role on community differentiation. Therefore, skipper samples were delimited and analyzed in two different ways: first, the whole altitudinal range on each mountain was divided into belts of 100 m elevational extent, without considering changes of vegetation types within; second, samples were again delimited by altitude, but in addition the prevalent vegetation type was superimposed to delimit sample sites (Table 1). As a consequence, these latter operational units were not equally sized according to elevational bands, but varied from 50 to 75 m extension each. Because only two specimens of skippers were collected in cloud forest, this vegetation type was excluded from all analyses.

Ecological and phylogenetic community structure analyses

NMDS ordinations were used to search for both ecological and phylogenetic structure in skipper assemblages along elevational gradients. This methodology enables the recognition of spatial gradients across communities through comparisons of pairwise similarities, or distances, between all samples [34]. Brehm & Fiedler [35] evaluated different ordination methods for identifying elevational gradients with incompletely sampled communities and concluded that different techniques performed quite similarly. Furthermore, NMDS has the advantage of fixing *a priori* the number of dimensions to be considered for analysis, and to be grounded on rank statistics which renders this ordination method very robust [36,37].

In a first series of NMDS explorations, based on Bray-Curtis matrix similarities, it was assessed whether inclusion, or exclusion, of hilltopping species (i.e. where adult butterflies aggregate at mountain tops for mate location [38]), or the segregation of samples only by altitude (100 m belts), or by altitude plus vegetation type (50–75 m belts), would affect ordination patterns, as already shown for species richness patterns [33]. Since both these factors indeed influenced the ordination of assemblages (Fig. S1), in the subsequent main series of analyses all hilltopping species were omitted and sample sites were classified according to elevation plus vegetation type. Species that exhibit hilltopping behavior are listed in Carneiro et al. [33]. Because hilltopping species were quite numerous, two samples, one from the 1800–1900 m band on Caratuva and the 1476–1550 m band on Araçatuba, became too small and therefore had to be excluded

Table 2. Spearman rank correlation coefficients *r* (plus associated *p*-values) between altitude of sample sites and the site scores along the two ordination axes extracted from NMDS ordinations[1].

	NMDS Axis 1		NMDS Axis 2	
	r	*p*	*r*	*p*
Bray-Curtis	**0.67**	**0.001**	0.31	0.148
MPD All1	**0.54**	**0.008**	0.07	0.758
MPD Grafen's Rho	**0.53**	**0.009**	0.06	0.792
MNTD All1	**0.54**	**0.008**	**0.60**	**0.002**
MNTD Grafen's Rho	**0.49**	**0.019**	**0.49**	**0.017**

[1]Different sets of Hesperiidae assemblages were considered with different measures of species or phylogenetic composition, sampled along altitudinal gradients in Serra do Mar, Paraná, Brazil. 'All 1' refers to equal branch lengths assigned to the tree topology while 'Grafen's Rho' refers to Grafen's branch length transformation method [45]. Correlations that remain significant after applying a table-wide false discovery rate approach are printed in bold face.

Table 3. Pairwise Spearman rank correlation coefficients *r* (plus associated *p*-values) for the similarity matrices extracted by ecological and phylogenetic metrics[1].

	Bray-Curtis		All 1 MPD		All 1 MNTD		Grafen Rho MPD	
Bray-Curtis	–	–						
All 1 MPD	0.509	<0.001	–	–				
All 1 MNTD	0.849	<0.001	0.397	0.001	–	–		
Grafen Rho MPD	0.608	<0.001	0.787	0.001	0.498	<0.001	–	–
Grafen Rho MNTD	0.877	<0.001	0.357	0.001	0.957	<0.001	0.522	<0.001

[1]Ecological ordination was measured with Bray-Curtis similarity matrix, while four phylogeny-based distance matrices (MPD and MNTD) were calculated using two branch-length options each. 'All 1' refers to equal branch lengths assigned to the tree topology while 'Grafen's Rho' refers to Grafen's Rho branch length transformation method [45]. Correlations that remain significant after applying a table-wide false discovery rate approach are printed in bold face.

altogether. As shown before [33], neither vegetation type nor altitude influenced the efficiency of sampling skippers. Therefore, differences in sample coverage are not expected to affect our results.

The Bray-Curtis similarity index was used to measure ecological similarities of assemblages. Complementarily, the incidence-based Chao-Soerensen index [39] was also calculated, but this did not yield any deviant patters (data not shown). Additionally, two indexes were calculated to assess phylogenetic distances between assemblage samples: the Mean Pairwise Distance (MPD) and the Mean Nearest Neighbor Distance (MNTD) [5]. Both indices were compared because they provide different perspectives of phylogenetic similarities, as an overall pattern of relatedness (MPD) or as how closely related species can be (MNTD) [40]. Additionally, MNTD is more affected by changes at the terminal branches of a phylogeny, whereas MPD is more sensitive to changes at the basis of a phylogeny [40].

To calculate measurements of phylogenetic structure, a phylogenetic hypothesis of the sampled taxa is obviously required. Skipper phylogeny is still very imperfectly resolved [29,41]. Based on the latest phylogenetic approach [29], a tentative community supertree was constructed. Topology of high rank taxa was recovered after [29]. Groups (G) and subgroups (SG) stated by Evans [42] were maintained only when they did not conflict with the current tree topology [29]. Species were clumped according to their respective genera (Fig. S2). When genera or species sampled in our study had not been included in the phylogenetic analyses, they were replaced by their closest tribal affiliates, using expert taxonomical arrangement as best possible surrogate of phylogenetic support [40,43]. Because branch length estimates were not available, and even node ages and fossil calibrations are still largely questioned in butterfly systematics [44], equal branch lengths were arbitrarily assigned to the tree topology (which is the same approach as the nodal distance method; [40]). We then compared these results with branch lengths using Grafen's Rho transformation (r = 0.5) [45] (Fig. S2). Additional branch length transformation methods were also calculated (e.g. Pagel's and Nee's methods: [46,47]), but results did not systematically differ from those presented below (data not shown).

To ascertain that phylogenetic structure of communities obtained from the supertree and branch transformation techniques were different from what would be expected at random, we correlated the empirical phylogenetic distance matrices with matrices produced by four distinct null models (means from 1000 randomizations) through Spearman matrix rank correlation tests (procedure RELATE in PRIMER: [48]). These null models included the shuffling of species labels across the phylogeny (null 0), randomization of species from the sample pool (null 1), randomization of species from the phylogeny pool (null 2), or by swapping versions of sample/species matrix (see [49,50] for details). The choice of an adequate null model is relevant, since their different assumptions might lead to contrasting results [51]. Because our phylogenetic trees do not contain branch length information based on genetic differences, we opted for challenging the results obtained through all these null models and then interpret results.

To test whether taxonomic scale may be informative to community similarities between different altitude and vegetation types, the analyses was first performed for the whole family Hesperiidae, and then separately for communities composed only of Pyrginae+Pyrrhopyginae species (i.e. dicot feeders, hereafter called Pyrginae); or only of Hesperiinae (i.e. monocot feeders). *Urbanus teleus* (Hübner, 1821) was included in the Hesperiinae analyses, since this species has confirmed host-plant records only in

Table 4. Spearman matrix rank correlation coefficients *r* (999 permutations) between phylogenetic distance matrices MPD (mean pairwise distance) and MNTD (mean nearest neighbor distance and four types of null models: 0–3 [49,50][1].

	Model 0		Model 1		Model 2		Model 3	
	r	*p*	*r*	*p*	*r*	*p*	*r*	*p*
Hesperiidae								
MPD								
All 1	**0.41**	**0.001**	0.05	0.725	0.02	0.580	0.22	0.030
Grafen Rho	0.06	0.239	0.05	0.303	0.05	0.300	0.04	0.648
MNTD								
All 1	**0.83**	**0.001**	**0.48**	**0.001**	**0.48**	**0.001**	**0.50**	**0.001**
Grafen Rho	**0.17**	**0.013**	**0.27**	**0.003**	**0.27**	**0.001**	**0.37**	**0.001**
Hesperiinae								
MPD								
All 1	**0.60**	**0.001**	0.00	0.524	0.06	0.782	−0.23	0.996
Grafen Rho	0.02	0.376	0.05	0.248	−0.02	0.671	−0.09	0.935
MNTD								
All 1	**0.69**	**0.001**	**0.33**	**0.001**	**0.33**	**0.001**	**0.33**	**0.001**
Grafen Rho	**0.28**	**0.001**	**0.18**	**0.003**	**0.28**	**0.001**	**0.20**	**0.002**
Pyrginae								
MPD								
All 1	**0.28**	**0.013**	0.11	0.209	−0.08	0.753	**0.36**	**0.023**
Grafen Rho	0.18	0.081	−0.32	0.973	−0.05	0.658	−0.07	0.665
MNTD								
All 1	**0.44**	**0.001**	**0.24**	**0.018**	0.18	0.097	**0.31**	**0.001**
Grafen Rho	0.19	0.091	0.18	0.092	0.14	0.153	0.17	0.097

[1]Two different branch length transformation methods were applied with the mean of 999 randomly generated matrices, according to four null models assumptions (see Methods section). 'All 1': equal branch lengths assigned to the tree topology; 'Grafen's Rho': Grafen's branch length transformation method [45]. Significant correlations here indicate that observed sample matrices are not substantially different from random expectations. Null model tests were applied to the three taxa Hesperiidae, Hesperiinae and Pyrginae. Correlations that remain significant after applying a table-wide false discovery rate approach are printed in bold face.

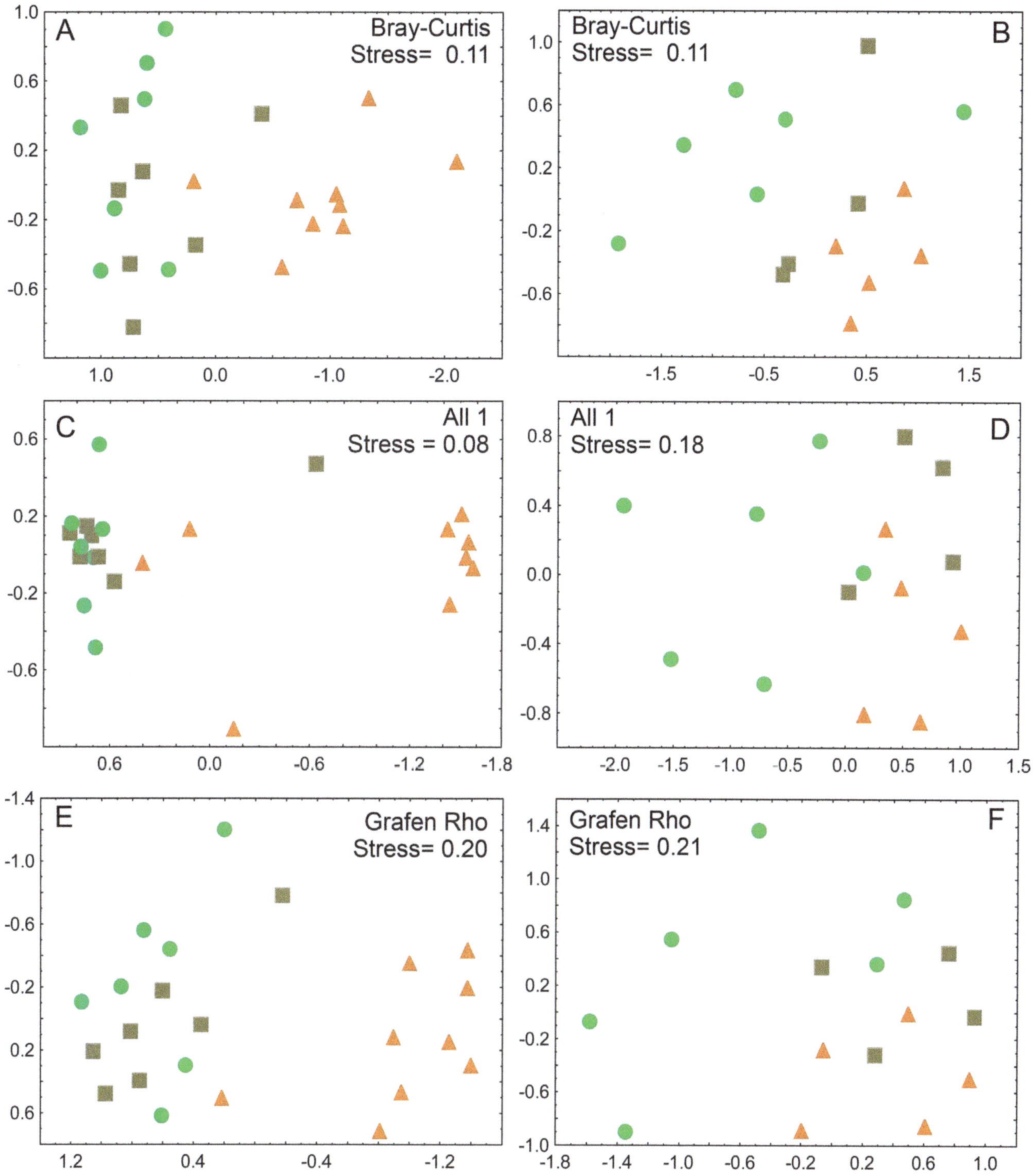

Figure 2. NMDS ordination plots of Hesperiinae (left panels, with monocot-feeding larvae) and Pyrginae (right panels, with larvae feeding on various dicot families) assemblages along elevational gradients in Serra do Mar, Paraná, Brazil. Ordinations are based on ecological (Bray-Curtis: A, B) and phylogenetic (MPD) similarity indexes. Two branch length transformations were applied to obtain MPD: 'All 1' (all branches set to unity: C, D) and Grafen's Rho model (E, F). Samples are partitioned according to altitude and vegetation types. Symbols: green circles (forest), brown squares (early successional vegetation), orange triangles (grassland). Stress values indicate goodness of fit of two-dimensional representations to the underlying distance matrices.

Poaceae [29], disregarding a probably misleading record from Fabaceae [31]. Pyrginae and Hesperiinae similarity matrices were thereafter compared across the same common sampling sites by Spearman matrix rank correlation, aiming to test whether taxa with such contrasting hostplant associations differ in their ecological and phylogenetic similarity patterns in response to

Table 5. Two-way ANOSIM results (*R* statistics and associated *p*-values), evaluating the effects of vegetation type and altitude on ecological similarities (Bray-Curtis) and phylogenetic distances (MPD) of skipper assemblages[1].

	Vegetation		Altitude	
	R^2	p	R^2	p
Hesperiidae				
Bray-Curtis	**0.47**	**0.004**	**0.10**	**0.001**
All1	0.19	0.13	0.13	0.81
Grafen Rho	**0.35**	**0.006**	0.01	0.5
Hesperiinae				
Bray-Curtis	**0.46**	**0.005**	0.05	0.328
All1	**0.49**	**0.004**	0.13	0.18
Grafen Rho	**0.39**	**0.013**	0.08	0.279
Pyrginae				
Bray-Curtis	**0.64**	**0.003**	0.76	0.03
All1	**0.08**	**0.298**	0.39	0.852
Grafen Rho	0.06	0.507	0.40	0.856

[1]Analyses were performed at three different taxonomic levels (entire family Hesperiidae, and two major subfamilies), and for two different branch length transformation methods: 'All 1': equal branch lengths assigned to the tree topology; 'Grafen's Rho': Grafen's branch length transformation method [45]. Results printed in bold remained significant after applying a false discovery rate approach.

elevation and vegetation change. Vegetation and altitude were classified as factors and tested in two-way ANOSIM analyses, where R statistics values served as a measure of effect size. Vegetation was divided into forest, early successional vegetation, and grassland, while the elevational categories included in this analyses were 'low' (900 m–1150 m), 'medium' (1150 m–1400 m), and 'high' (1400 m–1650 m; see Table 1).

Bray-Curtis similarity indexes, NMDS ordinations, RELATE tests and two-way ANOSIM were calculated with PRIMER 6.1.13 [48]. Phylogenetic trees were drawn and branch lengths put in and transformed using Mesquite 2.72 [52], including the PDAP Package [53]. Phylogenetic community distances were calculated using COMDIST and COMDISTNT functions available in PHYLOCOM 4.2 [54]. To avoid spurious significance resulting from multiple tests, a "False Discovery Rate" approach [55] was taken [56]. All test results passing adjusted criteria were assigned with an asterisk (*).

Results

Ecological and phylogenetic community structure at family level

The total skipper fauna collected along all transects comprised of 1578 specimens representing 155 species. Spatial resolution revealed by NMDS ordinations uncovered a nearly even gradient-like pattern mainly represented by the first ordination axis, along which communities were ordered from high to low altitudes (Fig. 1a, 'Bray-Curtis' in Table 2). However, assemblages from grassland sites (from elevations between 1100–1650 m elevation) became segregated from all others, whereas skipper assemblages from forest and early successional vegetation tended to be spatially unordered along both axes. Therefore, an elevational gradient was not evident when looking exclusively to points within each vegetation type, indicating that elevation is secondary to vegetation type in shaping species composition.

When skipper assemblages were ordinated by their phylogenetic distances in the MPD model, spatial patterns were remarkably similar, and NMDS representations achieved almost equal goodness-of-fit (Figs. 1b–e). Application of various branch length transformations did not alter the general pattern observed in NMDS ordinations, but two-dimensional representations had poorer fit (i.e. higher stress values and lower correlations of ordination axis scores with altitude) than if assuming equal branch lengths set to unity. Ordinations based on MPD and MNTD measures, respectively, showed similar patterns with regard to group clustering. Axis 2 significantly correlated with altitude in ordinations based on MNTD, but not for those using MPD (Table 2). Spearman rank correlations between the ecological Bray-Curtis similarity matrix and the four phylogeny-based distance matrices (MPD and MNTD with two branch-length options each) were always highly significant ($p \leq 0.001$), but MNTD index showed higher coefficient values with Bray-Curtis ecological index (0.84 to 0.87), compared to MPD (0.50 to 0.60) (Table 3). Strikingly, phylogenetic indexes showed the lowest coefficient values of correlation between each other (0.35 to 0.52).

Differences between MPD and MNTD distance measures were particularly obvious with regard to null-model tests (Table 4). The MNTD matrix correlated strongly with random matrices created under any of the four null models compared. In contrast, correlations between the MPD matrix and null models were only significant (and much weaker so) when all branch lengths were assigned to one, i.e. in null models 0 and 3 (shuffling of phylogenetic terminal labels and swapping species between samples, respectively). Therefore, MNTD results cannot be reliably interpreted as relating to ecological conditions and we restricted further considerations to MPD measures.

Effects of vegetation, altitude and taxon scale: Pyrginae vs. Hesperiinae

When analyzed separately, the two subfamilies Hesperiinae and Pyrginae revealed important differences in their patterns of assemblage similarities (Fig. 2). The NMDS ordination of Hesperiinae rendered a similar configuration as the entire family (Fig. 2a). Grassland skipper assemblages were distinctly set apart, with no clear differentiation between communities associated with forest and early successional vegetation. Pyrginae assemblages, in contrast, revealed a more distinct grouping of forest sites, as opposed to non-forested grassland and early successional sites (Fig. 2b). Ordinations based on phylogenetic distances were qualitatively similar, but tended to show less segregation between vegetation types, especially for the Pyrginae (Figs. 2c–f). These observations are supported by ANOSIM tests. Vegetation type was the main factor responsible for governing Hesperiidae and Hesperiinae assemblages, rather than altitude *per se* (Table 5). *R* values for comparisons based on phylogenetic distance measures were generally lower than those for comparisons of species composition. This was particularly pronounced in the subfamily Pyrginae.

Discussion

Ecological and phylogenetic community structure

Hesperiidae assemblages differed greatly in their species composition across elevational transects in Serra do Mar, even though sampling occurred over a moderate altitudinal range of 500–700 m extension. Although some herbivorous insect taxa are especially sensitive to abiotic gradients related to altitude [57,58], vegetation may exert more direct and distinct effects on species turnover. This was clearly the case with skipper assemblages in our study. Species turnover along elevational gradients has regularly been demonstrated to occur in many groups of plants and animals (for tropical butterflies and moths, e.g. [18,21]). Yet those studies often required far more extensive altitudinal ranges to uncover species turnover patterns.

In unconstrained ordinations, ecological species composition patterns were strikingly similar to those inferred from phylogenetic distances such as MPD and MNTD. In contrast to the number of studies investigating ecological species turnover along elevational gradients, fewer studies thus far attempted to address phylogenetic turnover in community compositions. This is due to the lack of robust phylogenetic hypotheses for most invertebrate groups, especially in tropical biota. In our case, the finding that MNTD distance patterns were not different from random expectations, opposite to MPD, might reflect the inaccuracy of our phylogenetic tree, in which terminal branches are less well established (presence of many polytomies) than is the resolution among higher level groups [29,41]. Skipper phylogeny has been addressed only recently, and molecular information from many genera, especially of Neotropical origin, is still completely lacking. Nevertheless, MPD data allowed us to reveal almost the same spatial pattern as abundance-weighted species turnover. This corroborates that phylogenetic distance information can be relevant also in groups with substantial uncertainty about their phylogenetic relationships [40,59].

Even though different null models, as expected, showed divergent results when applied to skippers [49,51], a non-random general pattern with regard to phylogenetic community distances was clearly observed. In general, various branch length transformation methods did not massively affect the outcome of ordination analyses, but the assumption of equal branch lengths reduced the information content to levels of random relationships between samples. Therefore, trees without genetically founded branch length estimates can still furnish consistent results [40], once different indexes are used and compared to a variety of null models to verify *a priori* whether phylogenetic inaccuracy may have obscured ecologically relevant aspects of phylogenetic community structure.

Effects of vegetation, altitude and taxon scale: Pyrginae vs. Hesperiinae

While numerous studies have elucidated how elevational gradients influence community similarities, only a few specifically addressed the relevance of vegetation types in relation to merely abiotic gradients on animal assemblages [12,60–62]. This integration, however, is essential, since high altitude habitats often represent different vegetation types which not necessarily are concordant to climatic change with elevation alone [14]. In cases where vegetational and abiotic dimensions were related to butterfly or moth assemblages, altitude (and climate) usually emerged as the single best predictor of faunal composition or richness instead of habitat type [12,60,62]. Accordingly, vegetation type at most emerged as modulating patterns of elevational faunal change [61,63,64].

In contrast, vegetation type played the major role with regard to community differentiation in skipper butterflies of Serra do Mar, instead of altitude *per se*. Along elevational transects of moderate extent, factors beyond mere laps rate may gain higher relative importance, as temperature differences are less prominent than on high mountain ranges like the Andes or Himalaya [4]. Accordingly, the presence of distinct vegetation types should attain higher weight in elevation gradients studies, addressing not only a novel ecological constraint (e.g. humidity and solar incidence are clearly contrasting between grasslands and forests), but also evolutionary dimensions, such as insect-plant interactions.

Life history traits are particularly related to the diversification of Hesperiidae. The two major subfamilies are rather conservative in larval food plant affiliations with either monocot or dicot plant families [29]. Hesperiinae (exclusive monocot feeders) are a more recent lineage than all other dicot feeders, with almost twice the global species richness of Pyrginae [29]. Hence diversification rate in this group must be distinctly higher, in contrast to the relatively low diversification rate estimated for the entire family [65]. Another unrelated group of monocot feeding butterflies (viz. Satyrinae) displays a similar pattern of speciation, and the expansion of grassland habitats around the world has been linked to this "explosive diversification" [66]. The high number of Hesperiinae species endemic to grasslands ecosystems in south Brazil reinforces this statement.

Hesperiinae assemblages of grassland sites were more distinctly clustered when analyses were based on phylogenetic distances rather than abundance-weighted species lists. This may hint towards a historical relationship with grassland ecosystems through evolutionary time. In other words, skipper species in this highly distinct vegetation type represent lineages with unique evolutionary history [16]. Consequently, environmental filtering caused by altitudinal climate shifts seems to play only a secondary role in structuring those assemblages.

Implications for Conservation

Conservation of natural grasslands ecosystems is of transcontinental concern [67]. Atlantic forests are biodiversity hot-spots not only because of high species richness of organisms, but also because of high numbers of endemics [68]. Because mountain grassland patches in the Serra do Mar are inserted amongst this forest landscape, endemic organisms of grasslands are routinely counted to this ecosystem, especially because fine scale distribution of grassland patches is hardly represented in distribution maps. Grassland ecosystems embedded in Atlantic Forest are phylogenetically linked to unique skipper assemblages, in analogy to high-altitude grassland butterflies in the European Alps [23]. The reduction of grassland ecosystems in Atlantic Forest is historically related to climate fluctuations, but nowadays these ecosystems are also threatened by occasional anthropogenic fires, invasive exotic grasses and extensive tourism [69]. The present findings highlight the need of conserving these high altitude habitats also from the perspective of a unique skipper butterfly fauna. Although difficult to identify, Neotropical skippers comprise rich assemblages (compared for example to frugivorous butterflies), and are suitable sensitive biological indicators in Atlantic Forest [68]. Hence, further taxonomic and ecological studies into this family are desired, since for most skipper species we still lack information on hostplant associations, phylogenetic relationships, and geographical distribution records. Even, quite a number of unknown species remain to be described.

Supporting Information

Figure S1 Explorative NMDS ordination plots of Hesperiidae assemblages along elevational gradients in Serra do Mar, Brazil. Ordination patterns were first assessed based on Bray-Curtis similarities, with samples partitioned into 100 m altitudinal bands (a), into 100 m altitudinal bands, but excluding hilltopping species (b), and partitioned according to altitude and vegetation types (excluding hilltopping species) c). Arrows indicate mountain summit samples.

Figure S2 Phylogenetic relationships of Hesperiidae (Insecta, Lepidoptera) species recorded in Serra do Mar, Paraná, Brazil. Topology of high rank taxa was recovered after Warren et al. [29]. Groups (G) and subgroups (SG) stated by Evans [42] were maintained only when not conflicting with the topology published in Warren et al. [29]. Species were clumped according to its respectively genera. As no branch lengths are still available for skipper phylogeny, equal branch lengths (above) and Grafen's Rho transformation (below) were arbitrarily assigned to quantify phylogenetic differences between species.

Acknowledgments

The authors are thankful for the critical review of earlier manuscript versions by André V. L. Freitas (UNICAMP), Emygdio L. A. M. Filho (UFPR), Gilson R. P. Moreira (UFRGS), Maurício O. Moura (UFPR) and two anonymous reviewers.

Author Contributions

Conceived and designed the experiments: EC OHHM MMC KF. Performed the experiments: EC. Analyzed the data: EC KF. Contributed reagents/materials/analysis tools: EC OHHM MMC KF. Wrote the paper: EC OHHM MMC KF.

References

1. Forster JR (1778) Observations made during a voyage round the world on physical geography, natural history, and ethic philosophy. London: G. Robinson. 676 p. Available: http://www.biodiversitylibrary.org/item/106982.
2. von Humboldt A (1849) Aspects of nature, in different lands and different climates; with scientific elucidations. Philadelphia: Lea and Blanchard. 480 p. Available: http://www.biodiversitylibrary.org/item/62179.
3. Rahbek C (2005) The role of spatial scale and the perception of large-scale species-richness patterns. Ecol Lett 8: 224–239. doi:10.1111/j.1461-0248.2004.00701.x.
4. Nogués-Bravo D, Araújo MB, Romdal T, Rahbek C (2008) Scale effects and human impact on the elevational species richness gradients. Nature 453: 216–219. doi:10.1038/nature06812.
5. Webb CO, Ackerly DD, McPeek MA, Donoghue MJ (2002) Phylogenies and community ecology. Annu Rev Ecol Syst 33: 475–505. doi:10.1146/annurev.ecolsys.33.010802.150448.
6. Almeida FFM de, Brito Neves BB de, Dal Ré Carneiro C (2000) The origin and evolution of the South American Platform. Earth-Sci Rev 50: 77–111. doi:10.1016/S0012-8252(99)00072-0.
7. Roderjan CV, Galvão F, Kuniyoshi YS (2002) As unidades fitogeográficas do estado do Paraná. Ciênc E Ambiente 26: 693–712.
8. Roderjan CV, Struminsky E (1992) Caracterização e proposta de manejo da Serra da Baitaca - Quatro Barras. Curitiba: FUPEF/FBPN. 121 p.
9. Behling H (2002) South and southeast Brazilian grasslands during Late Quaternary times: a synthesis. Palaeogeogr Palaeoclimatol Palaeoecol 177: 19–27. doi:10.1016/S0031-0182(01)00349-2.
10. Mikich SB, Bernils RS (2004) Livro vermelho da fauna ameaçada no estado do Paraná. Curitiba: Instituto Ambiental do Paraná. 763 p.
11. Machado AMB, Drummond GM, Paglia AP (2008) Livro vermelho da fauna brasileira ameaçada de extinção. Brasilia: MMA/Fundação Biodiversitas. 160 p.
12. Gutiérrez D (1997) Importance of historical factors on species richness and composition of butterfly assemblages (Lepidoptera: Rhopalocera) in a northern Iberian mountain range. J Biogeogr 24: 77–88. doi:10.1111/j.1365-2699.1997.tb00052.x.
13. Axmacher JC, Brehm G, Hemp A, Tünte H, Lyaruu HVM, et al. (2009) Determinants of diversity in afrotropical herbivorous insects (Lepidoptera: Geometridae): plant diversity, vegetation structure or abiotic factors? J Biogeogr 36: 337–349. doi:10.1111/j.1365-2699.2008.01997.x.
14. Lomolino MV (2001) Elevation gradients of species-density: historical and prospective views. Glob Ecol Biogeogr 10: 3–13. doi:10.1046/j.1466-822x.2001.00229.x.
15. Machac A, Janda M, Dunn RR, Sanders NJ (2011) Elevational gradients in phylogenetic structure of ant communities reveal the interplay of biotic and abiotic constraints on diversity. Ecography 34: 364–371. doi:10.1111/j.1600-0587.2010.06629.x.
16. Graham CH, Fine PVA (2008) Phylogenetic beta diversity: linking ecological and evolutionary processes across space in time. Ecol Lett 11: 1265–1277. doi:10.1111/j.1461-0248.2008.01256.x.
17. Novotny V, Weiblen GD (2005) From communities to continents: beta diversity of herbivorous insects. Ann Zool Fenn 42: 463–475.
18. Beck J, Chey VK (2007) Beta-diversity of geometrid moths from northern Borneo: effects of habitat, time and space. J Anim Ecol 76: 230–237. doi:10.1111/j.1365-2656.2006.01189.x.
19. Muñoz A, Amarillo-Suárez Á (2010) Altitudinal variation in diversity of Arctiidae and Saturniidae (Lepidoptera) in a Colombian cloud forest. Rev Colomb Entomol 36: 292–299.
20. Pyrcz TW, Wojtusiak J (2002) The vertical distribution of pronophiline butterflies (Nymphalidae, Satyrinae) along an elevational transect in Monte Zerpa (Cordillera de Mérida, Venezuela) with remarks on their diversity and parapatric distribution. Glob Ecol Biogeogr 11: 211–221. doi:10.1046/j.1466-822X.2002.00285.x.
21. Pyrcz TW, Wojtusiak J, Garlacz R (2009) Diversity and distribution patterns of Pronophilina butterflies (Lepidoptera: Nymphalidae: Satyrinae) along an altitudinal transect in north-western Ecuador. Neotrop Entomol 38: 716–726. doi:10.1590/S1519-566X2009000600003.
22. Hawkins BA (2010) Multiregional comparison of the ecological and phylogenetic structure of butterfly species richness gradients. J Biogeogr 37: 647–656. doi:10.1111/j.1365-2699.2009.02250.x.
23. Pellissier L, Ndiribe C, Dubuis A, Pradervand J-N, Salamin N, et al. (2013) Turnover of plant lineages shapes herbivore phylogenetic beta diversity along ecological gradients. Ecol Lett 16: 600–608. doi:10.1111/ele.12083.
24. Brehm G, Strutzenberger P, Fiedler K (2013) Phylogenetic diversity of geometrid moths decreases with elevation in the tropical Andes. Ecography 36: 1247–1253. doi:10.1111/j.1600-0587.2013.00030.x.
25. Cavender-Bares J, Keen A, Miles B (2006) Phylogenetic structure of Floridian plant communities depends on taxonomic and spatial scale. Ecology 87: 109–122.
26. Swenson NG, Enquist BJ, Pither J, Thompson J, Zimmerman JK (2006) The problem and promise of scale dependency in community phylogenetics. Ecology 87: 2418–2424.
27. Cadotte MW, Borer ET, Seabloom EW, Cavender-Bares J, Harpole WS, et al. (2010) Phylogenetic patterns differ for native and exotic plant communities across a richness gradient in Northern California. Divers Distrib 16: 892–901. doi:10.1111/j.1472-4642.2010.00700.x.
28. Hoiss B, Krauss J, Potts SG, Roberts S, Steffan-Dewenter I (2012) Altitude acts as an environmental filter on phylogenetic composition, traits and diversity in bee communities. Proc R Soc B Biol Sci 279: 4447–4456. doi:10.1098/rspb.2012.1581.
29. Warren AD, Ogawa JR, Brower AVZ (2009) Revised classification of the family Hesperiidae (Lepidoptera: Hesperioidea) based on combined molecular and morphological data. Syst Entomol 34: 467–523. doi:10.1111/j.1365-3113.2008.00463.x.
30. Beccaloni G, Hall SK, Viloria AL, Robinson GS (2008) Catalogue of the hostplants of the neotropical butterflies = Catálogo de las plantas huésped de las mariposas neotropicales. Zaragoza: Sociedad Entomológica Aragonesa.
31. Robinson GS, Ackerly PR, Kitching IJ, Beccaloni GW, Hernández LM (2010) A database of the World's Lepidopteran hostplants. Available: http://www.nhm.ac.uk/hosts. Accessed 18 August 2010.

32. Dolibaina DR, Mielke OHH, Casagrande MM (2011) Butterflies (Papilionoidea and Hesperioidea) from Guarapuava and vicinity, Paraná, Brazil: an inventory based on records of 63 years. Biota Neotropica 11: 341–354. doi:10.1590/S1676-06032011000100031.
33. Carneiro E, Mielke OHH, Casagrande MM, Fiedler K (2014) Skipper Richness (Hesperiidae) along elevational gradients in Brazilian Atlantic Forest. Neotrop Entomol 43: 27–38. doi:10.1007/s13744-013-0175-8.
34. Legendre P, Legendre L (2013) Numerical ecology. Amsterdam; Boston: Elsevier.
35. Brehm G, Fiedler K (2004) Ordinating tropical moth ensembles from an elevational gradient: a comparison of common methods. J Trop Ecol 20: 165–172. doi:10.1017/S0266467403001184.
36. Minchin PR (1987) An evaluation of the relative robustness of techniques for ecological ordination. Vegetatio 69: 89–107. doi:10.1007/BF00038690.
37. Clarke KR (1993) Non-parametric multivariate analyses of changes in community structure. Aust J Ecol 18: 117–143. doi:10.1111/j.1442-9993.1993.tb00438.x.
38. Pe'er G, Saltz D, Münkemüller T, Matsinos YG, Thulke H-H (2013) Simple rules for complex landscapes: the case of hilltopping movements and topography. Oikos 122: 1483–1495. doi:10.1111/j.1600-0706.2013.00198.x.
39. Chao A, Chazdon RL, Colwell RK, Shen T-J (2005) A new statistical approach for assessing similarity of species composition with incidence and abundance data. Ecol Lett 8: 148–159. doi:10.1111/j.1461-0248.2004.00707.x.
40. Webb (2000) Exploring the phylogenetic structure of ecological communities: an example for rain forest trees. Am Nat 156: 145–155. doi:10.1086/303378.
41. Warren AD, Ogawa JR, Brower AVZ (2008) Phylogenetic relationships of subfamilies and circumscription of tribes in the family Hesperiidae (Lepidoptera: Hesperioidea). Cladistics 24: 642–676. doi:10.1111/j.1096-0031.2008.00218.x.
42. Evans WH (1955) A catalogue of the American Hesperiidae: indicating the classification and nomenclature adopted in the British Museum (Natural History). London: British Museum (Natural History). 552 p.
43. Sanderson MJ, Purvis A, Henze C (1998) Phylogenetic supertrees: Assembling the trees of life. Trends Ecol Evol 13: 105–109. doi:10.1016/S0169-5347(97)01242-1.
44. Wahlberg N, Wheat CW, Peña C (2013) Timing and patterns in the taxonomic diversification of Lepidoptera (butterflies and moths). PLoS ONE 8: e80875. doi:10.1371/journal.pone.0080875.
45. Grafen A (1989) The phylogenetic regression. Philos Trans R Soc Lond B Biol Sci 326: 119–157. doi:10.1098/rstb.1989.0106.
46. Pagel MD (1992) A method for the analysis of comparative data. J Theor Biol 156: 431–442. doi:10.1016/S0022-5193(05)80637-X.
47. Purvis A (1995) A composite estimate of primate phylogeny. Philos Trans R Soc Lond B Biol Sci 348: 405–421. doi:10.1098/rstb.1995.0078.
48. Clarke KR, Warwick RM (2009) PRIMER for Windows. Primer-E Ltd.
49. Gotelli N (2000) Null model analysis of species co-occurrence patterns. Ecology 81: 2606–2621. doi:10.2307/177478.
50. Webb CO, Ackerly DD, Kembel SW (2008) Phylocom: software for the analysis of phylogenetic community structure and trait evolution. Bioinformatics 24: 2098–2100. doi:10.1093/bioinformatics/btn358.
51. Kembel SW, Hubbell SP (2006) The phylogenetic structure of a Neotropical forest tree community. Ecology 87: 86–99. doi:10.1890/0012-9658(2006)87[86:TPSOAN]2.0.CO;2.
52. Maddison WP, Maddison DR (2001) Mesquite, a modular system for evolutionary analysis, Available: http://mesquiteproject.org.
53. Midford PE, Garland Jr P, Maddison WP (2009) PDAP:PDTREE package for Mesquite. Available: http://mesquiteproject.org/pdap_mesquite/.
54. Webb CO, Ackerly DD, Kembel SW (2011) Phylocom: software for the analysis of phylogenetic community structure and trait evolution. Available: http://www.phylodiversity.net/phylocom.
55. Benjamini Y, Hochberg Y (1995) Controlling the false discovery rate: A practical and powerful approach to multiple testing. J R Stat Soc Ser B Methodol 57: 289–300. doi:10.2307/2346101.
56. Pike N (2011) Using false discovery rates for multiple comparisons in ecology and evolution. Methods Ecol Evol 2: 278–282. doi:10.1111/j.2041-210X.2010.00061.x.
57. Whittaker RH (1952) A study of summer foliage insect communities in the Great Smoky Mountains. Ecol Monogr 22: 1–44. doi:10.2307/1948527.
58. Hodkinson ID (2005) Terrestrial insects along elevation gradients: species and community responses to altitude. Biol Rev 80: 489–513. doi:10.1017/S1464793105006767.
59. Swenson NG (2009) Phylogenetic resolution and quantifying the phylogenetic diversity and dispersion of communities. PLoS ONE 4: e4390. doi:10.1371/journal.pone.0004390.
60. Storch D, Konvicka M, Benes J, Martinková J, Gaston KJ (2003) Distribution patterns in butterflies and birds of the Czech Republic: separating effects of habitat and geographical position. J Biogeogr 30: 1195–1205. doi:10.1046/j.1365-2699.2003.00917.x.
61. Axmacher JC, Fiedler K (2008) Habitat type modifies geometry of elevational diversity gradients in geometrid moths (Lepidoptera Geometridae) on Mt Kilimanjaro, Tanzania. Trop Zool 21: 243–251.
62. Illán JG, Gutiérrez D, Wilson RJ (2010) Fine-scale determinants of butterfly species richness and composition in a mountain region. J Biogeogr 37: 1706–1720. doi:10.1111/j.1365-2699.2010.02314.x.
63. Lien VV, Yuan D (2003) The differences of butterfly (Lepidoptera, Papilionoidea) communities in habitats with various degrees of disturbance and altitudes in tropical forests of Vietnam. Biodivers Conserv 12: 1099–1111. doi:10.1023/A:1023038923000.
64. Pellissier L, Pradervand J-N, Pottier J, Dubuis A, Maiorano L, et al. (2012) Climate-based empirical models show biased predictions of butterfly communities along environmental gradients. Ecography 35: 684–692. doi:10.1111/j.1600-0587.2011.07047.x.
65. Heikkilä M, Kaila L, Mutanen M, Peña C, Wahlberg N (2011) Cretaceous origin and repeated tertiary diversification of the redefined butterflies. Proc R Soc B Biol Sci 279: 1093–1099. doi:10.1098/rspb.2011.1430.
66. Peña C, Wahlberg N, Weingartner E, Kodandaramaiah U, Nylin S, et al. (2006) Higher level phylogeny of Satyrinae butterflies (Lepidoptera: Nymphalidae) based on DNA sequence data. Mol Phylogenet Evol 40: 29–49. doi:10.1016/j.ympev.2006.02.007.
67. Suttie JM, Reynolds SG, Batello C (2005) Grasslands of the World. Food & Agriculture Org. 548 p.
68. Brown KS, Freitas AVL (2000) Atlantic forest butterflies: indicators for landscape conservation. Biotropica 32: 934–956. doi:10.1111/j.1744-7429.2000.tb00631.x.
69. Martinelli G (2007) Mountain biodiversity in Brazil. Braz J Bot 30: 587–597. doi:10.1590/S0100-84042007000400005.

Rapid Losses of Surface Elevation following Tree Girdling and Cutting in Tropical Mangroves

Joseph Kipkorir Sigi Lang'at[1,2], James G. Kairo[1], Maurizio Mencuccini[3,7], Steven Bouillon[4], Martin W. Skov[5], Susan Waldron[6], Mark Huxham[2]*

1 Kenya Marine and Fisheries Research Institute, Mombasa, Kenya, **2** School of Life, Sport and Social Sciences, Edinburgh Napier University, Edinburgh, United Kingdom, **3** School of Geosciences, University of Edinburgh, Crew Building, Edinburgh, United Kingdom, **4** Department of Earth and Environmental Sciences, Katholieke Universiteit Leuven, Leuven, Belgium, **5** School of Ocean Sciences, Bangor University, Menai Bridge, Anglesey, United Kingdom, **6** School of Geographical and Earth Sciences, University of Glasgow, Glasgow, South Lanarkshire, United Kingdom, **7** Centre for Ecological Research and Forestry Applications, Universitat Autònoma de Barcelona, Barcelona, Spain

Abstract

The importance of mangrove forests in carbon sequestration and coastal protection has been widely acknowledged. Large-scale damage of these forests, caused by hurricanes or clear felling, can enhance vulnerability to erosion, subsidence and rapid carbon losses. However, it is unclear how small-scale logging might impact on mangrove functions and services. We experimentally investigated the impact of small-scale tree removal on surface elevation and carbon dynamics in a mangrove forest at Gazi bay, Kenya. The trees in five plots of a *Rhizophora mucronata* (Lam.) forest were first girdled and then cut. Another set of five plots at the same site served as controls. Treatment induced significant, rapid subsidence (−32.1±8.4 mm yr^{-1} compared with surface elevation changes of +4.2±1.4 mm yr^{-1} in controls). Subsidence in treated plots was likely due to collapse and decomposition of dying roots and sediment compaction as evidenced from increased sediment bulk density. Sediment effluxes of CO_2 and CH_4 increased significantly, especially their heterotrophic component, suggesting enhanced organic matter decomposition. Estimates of total excess fluxes from treated compared with control plots were 25.3±7.4 tCO_2 ha^{-1} yr^{-1} (using surface carbon efflux) and 35.6±76.9 tCO_2 ha^{-1} yr^{-1} (using surface elevation losses and sediment properties). Whilst such losses might not be permanent (provided cut areas recover), observed rapid subsidence and enhanced decomposition of soil sediment organic matter caused by small-scale harvesting offers important lessons for mangrove management. In particular mangrove managers need to carefully consider the trade-offs between extracting mangrove wood and losing other mangrove services, particularly shoreline stabilization, coastal protection and carbon storage.

Editor: Vanesa Magar, Centro de Investigacion Cientifica y Educacion Superior de Ensenada, Mexico

Funding: This work was funded by grants from the UK Natural Environment Research Council (Grant No. NE/G009589/1), Earthwatch Institute and AVIVA Ltd, London, UK. The funders had no role in study design, data collection and analysis, decision to publish, or preparation of the manuscript, with the exception of volunteers from Earthwatch Institute who helped in field work.

Competing Interests: The authors can confirm that they received funding from the charitable wing of Aviva Ltd. This does not alter the authors' adherence to PLOS ONE policies on sharing data and materials.

* Email: m.huxham@napier.ac.uk

Introduction

Mangrove forests are highly productive systems and often allocate a large proportion of their energy budget to root production [1–3]. Because of the presence of aerial roots, mangroves trap allochthonous organic matter in sediment, with carbon (C) sequestration rates exceeding those of terrestrial tropical forests by a factor of ~6 [4]. Unlike terrestrial forest soils, mangrove sediments do not attain C saturation because of continued sediment accumulation and vertical accretion [5] and hence the size of the C store continues to increase over time [6]. Anoxia, low levels of nutrients and the high lignin content of the roots result in slow decomposition of below-ground organic matter [7–9] and the accumulation of large reserves of peat and C-rich sediments [10–12]. Mangroves are thus amongst the most carbon dense of all forests, with C stocks sometimes exceeding 1000 tonnes C ha^{-1} [12–14], and hence play an important role in global carbon storage [15–17].

As a result of continued vertical accretion and below-ground root growth, surface elevation of mangroves increases over time at rates of up to 4.8 mm yr^{-1} e.g. [5]. These increases in surface elevation are important in allowing mangrove recovery after natural disturbances and are considered essential for many mangroves to survive projected sea level rise of 1.7 to 3.3 mm yr^{-1} this century [18,19].

Human disturbances such as wood harvesting and clearing threaten to impair these important ecological processes. Trends in mangrove loss are alarming, with an estimated 30–50% of forests lost over the past half century [20,21]. Although rates of loss may be declining [22] they remain high, typically 0.7–3% y^{-1}, partly because of high levels of poverty and dense human populations along many tropical coasts [13]. While a recent estimate of the impact of such losses on the total mangrove carbon sink suggested that mangrove destruction could contribute up to 10% of the annual GHG emissions from land use change [12], little understanding exists of the impact of mangrove harvesting on surface elevation dynamics. Reductions in surface elevation after harvesting may be caused by a combination of processes. Following harvesting, root growth and expansion stops, whilst decomposition of dead roots and old organic matter may be accelerated by higher temperatures in the exposed substrate and increased sediment oxidation. Sediment erosion may also increase and the lack of aerial roots may prevent continued accumulation of allochthonous sediment. Additionally, the aerenchymatous tissues of the dying roots may shrink leading to increased bulk density but lower elevation. Finally, leaching of dissolved inorganic and organic C and lateral transport to the sea may occur [2,3,17,23,24].

Few studies have directly measured the impacts of tree removal on below-ground carbon storage and surface elevation. Two studies report on the effects of hurricanes [25,26] and one on total deforestation [27]. Under such extreme conditions carbon losses can be large with resulting 'peat collapse' and coastal erosion. The impacts of smaller scale tree loss are even less well known, although work from Micronesia after non-experimental tree felling [28] and Florida following lightning strikes [29] shows that surface elevation losses might be rapid. Whilst healthy forests are likely to be resilient and show recovery, those under anthropogenic pressure may experience longer term change. Mangroves of the Western Indian Ocean (WIO) region experience small-scale but widespread degradation from indiscriminate harvesting [30,31]. A recent assessment of mangrove decline in Kenya has indicated an annual loss of 0.7%, which underestimates the anthropogenic stressors since it records only total canopy removal [32]. Here we document the first controlled experiment testing the impacts of small-scale cutting, the most common form of mangrove exploitation in the WIO region, on sediment carbon losses and surface elevation. Increasing attention is being paid to avoiding deforestation by Reducing Emissions from Deforestation and Degradation and conservation of forest ecosystems (REDD+) schemes and to managing forests for a range of services, not only wood production. It is essential therefore that the impacts of forest management scenarios on key ecosystem services such as coastal protection and carbon sequestration are analysed to understand possible trade-offs between these and extractive uses.

Mangrove forests of the WIO region are utilized by the local communities for construction and fuelwood [33]. In Kenya mangroves are the only natural forests currently licensed by the Kenya Forest Service (KFS) for wood harvesting [30,34]. In combination with widespread illegal harvesting this has left many mangrove areas either degraded or completely denuded of vegetation [30,35]. Since mangroves meet ~70% of the wood requirements of the coastal population [36], there is always a ready market for mangrove poles, especially in major coastal towns.

We used a controlled experiment to explore the impacts of tree harvesting on: a) surface elevation dynamics; b) the main processes affecting changes in surface elevation i.e., sediment accretion, sediment properties such as bulk density, %C and sediment moisture, and organic matter decomposition; and c) sediment surface C efflux in a natural mangrove forest, here at Gazi bay, Kenya. The C flux from forest floors comes from root (autotrophic) respiration and sediment/soil organic matter decomposition (heterotrophic respiration) [37]. The relative contributions of these sources have not been distinguished in mangroves; such partitioning is useful in understanding carbon stocks and flows in forests. Girdling trees stops the flow of photosynthates to the root system and thereby stops metabolic activities in the roots whilst maintaining the tree canopy. Hence it can be used to separate the components of soil respiration, since CO_2 emitted from the sediment shortly after girdling is assumed to be primarily from organic matter decomposition e.g., [38–40]. Here we also report on the first time this approach has been used for mangroves.

Results

Surface elevation and sediment accretion

At the conclusion of monitoring after 760 days the control plots had gained mean surface elevation of +11.1±10.5 mm, at a mean rate of +4.2±1.4 mm yr^{-1}, while the treatment plots showed a subsidence of −51.3±12.0 mm, at a mean rate of −32.1±8.4 mm yr^{-1} (Figure 1). For a period of 110 days after setting up the horizon marker, both control and treated plots experienced similar trends in sediment accretion, ranging from 5.5 to 8.0 mm (controls, mean 6.4±1.4 mm) and 2.5 to 11.7 mm (treated, mean 5.2±4.6 mm). Disturbance of the horizon marker by crab activities in the control plots did not allow further monitoring beyond four months after set up.

Sediment Surface Carbon Fluxes

Approximately 30 days after girdling, CO_2-C emissions in treated plots increased and remained higher than in controls throughout the girdled period (Table 1 and Figure 2). For the first two months after cutting, CO_2-C emissions in the treated plots were similar to controls, but then increased again for three months before dropping to levels similar to the controls by the end of the sampling period (3.2±0.9 vs. 3.9±1.8 gCO_2–C m^{-2} d^{-1}, respectively; Figure 2). At ~30 days after girdling, the $\delta^{13}C$ signature of the sediment respired CO_2 from the treated plots was significantly more ^{13}C-depleted than in the controls (Figure 3). Methane emissions were highly variable and increased in the treated plots during the girdled period only (Figure 2). Mean emissions of both CO_2–C and CH_4–C were significantly higher in treated than control plots during the girdled period (Table 1). Throughout the treatment period, the mean sediment temperature in the treated plots was higher than the control plots by 0.9 to 5.8°C.

Separation of sediment autotrophic and heterotrophic respiration components

Girdling was not effective in separating the components of sediment respiration since the girdled trees began to lose leaves

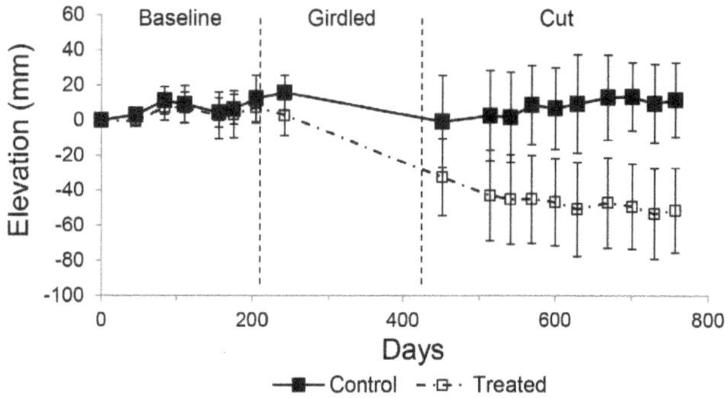

Figure 1. Trends in surface elevation change in control and treated sites in *R. mucronata* forest at Gazi bay, Kenya. Error bars are 95% CI. Vertical broken lines indicate periods when trees were girdled and cut in the treatment plots. Baseline, girdled and cut periods ran from March 2009 to October 2009 (205 days), December 2009 to May 2010 (189 days) and May 2010 to April 2011 (343 days), respectively. The controls and the treatment consisted of five replicates each.

Table 1. Nested design ANOVA for carbon fluxes in *R. mucronata* secondary forest, with mean data for each of six chambers per plot nested within treatment; the data for CO_2 were log-transformed.

Period	Variable	Source of Variation	DF	MS	F	P
Girdled	CO_2	Temperature	1	0.017	0.420	0.518
		Burrows	1	0.017	0.45	0.506
		Treatment	1	0.152	4.73	0.036
		Plot (Treatment)	8	0.025	0.64	0.736
		Error	48	0.039		
	CH_4	Temperature	1	46.140	2.4	0.128
		Burrows	1	2.660	0.14	0.712
		Treatment	1	745.800	20.04	0.000
		Plot (Treatment)	8	55.700	2.9	0.01
		Error	48	19.230		
Clear-cut	CO_2	Temperature	1	0.005	0.16	0.695
		Burrows	1	0.044	1.26	0.267
		Treatment	1	0.001	0.02	0.897
		Plot (Treatment)	8	0.028	0.81	0.6
		Error	48	0.035		
	CH_4	Temperature	1	24.758	3.43	0.07
		Burrows	1	5.73	0.79	0.378
		Treatment	1	1.458	0.19	0.667
		Plot (Treatment)	8	10.834	1.5	0.182
		Error	48	7.225		

and die within one month of girdling; this was much faster than recorded for terrestrial trees e.g., [40]. Therefore, an indirect regression method of estimating autotrophic respiration was used [38]. In each of the control plots, live root biomass was determined after the final sampling in April 2011 (530 days after treatment) by excavations made beneath the pre-marked positions of each chamber. The regression of CO_2 fluxes, measured at the final sampling for each control plot chamber, against the live root biomass and sediment surface temperature adjacent to each chamber, was significant:

Multiple regression: $\ln R_S = -3.093 + 0.0002$*live root biomass + 0.127*temperature; $R^2 = 0.37$; $P = 0.044$, 0.011 and 0.012 for the constant, live root biomass and sediment temperature, respectively.

Heterotrophic respiration (R_H) for final sample values was calculated by applying the equation above, while setting the value of root biomass to zero [38]. Autotrophic respiration (R_A), obtained by subtracting the values for R_H from the measured total sediment respiration (R_S), contributed a mean (±95% CI) of 41.5±11.8% to R_S at the final sampling. Across the entire sampling period, R_A contributed an average of 40.5±7.0% to R_S, which was not significantly different from that obtained at the final sampling (t-test, $t = -0.16$, $P = 0.874$). The partitioning of the components of total R_S in the control plots allowed a comparison of the treated plot CO_2 fluxes (where autotrophic R was zero by definition) and control plot R_H. Sediment respiration of the treated plots was higher than control plots R_H throughout the treatment period (during both girdling and cutting) by 0.6 to 3.7 g CO_2–C m^{-2} d^{-1} (Figure 2).

Estimated C losses from the sediment

Combining data on elevation changes with sediment characteristics provided one estimate of total carbon losses of 35.7±76.9 t C ha^{-1} from the treated compared to control plots.

The net belowground carbon losses due to treatment estimated from sediment surface flux data amounted to 14.2±10.3 tCO_2 ha^{-1} (mean rate of 9.8±7.1 t CO_2 ha^{-1} yr^{-1}) over a period of 530 days after treatment. When only R_H in control plots was considered, the net additional C loss amounted to 36.7±10.7 t CO_2 ha^{-1}, with a mean rate of 25.3±7.4 t CO_2 ha^{-1} yr^{-1}.

Root Decomposition

After 267 days root-bags in the treated plots had lost significantly more mass than those in the controls, with rates of 0.19±0.02 and 0.16±0.03% dry weight loss day^{-1}, respectively (t-test, $t = -2.06$, $P = 0.049$). The trend in decomposition rate in the treated plots indicated that it was increasing with time, while the rate in the control plots remained constant between 156 and 267 days.

Belowground Biochemical Characteristics

The plant roots had a similar $\delta^{13}C$ signature to that of the sediment carbon (Table 2). Treated plots showed significant reductions in % C and sediment moisture and significant increases in bulk density. The sediment carbon stocks to a depth of 1 m ranged from 414.1 to 610.7 t C ha^{-1} for controls and 457.3 to 586.3 t C ha^{-1} for treated plots. Overall the control plots tended to have higher mean C stocks than treated plots, with mean (±95% CI) of 524.1±88.8 vs. 488.4±68.6 tC ha^{-1}, respectively, but the difference was not significant due to the very large small-scale variability in sediment properties.

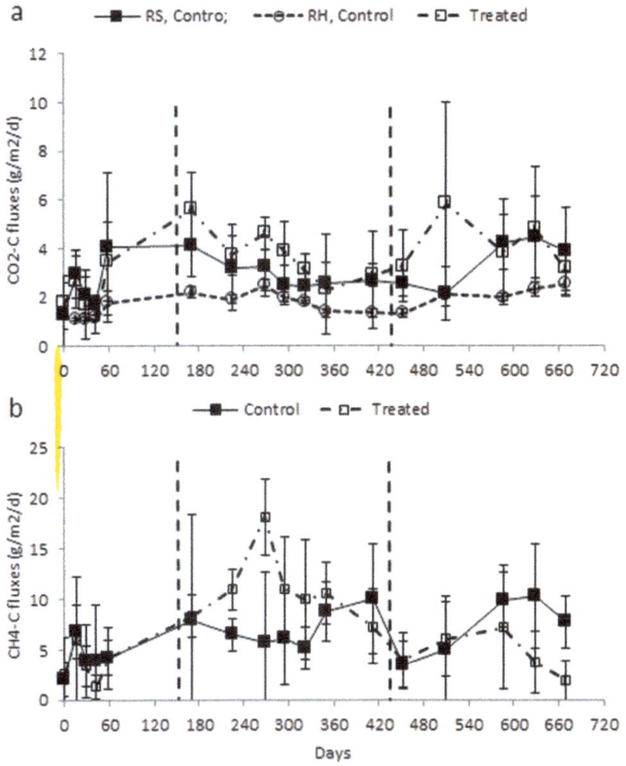

Figure 2. Mean (±95% CI) Carbon emissions. a) CO_2 fluxes of total sediment respiration (R_S) (solid line with filled squares), heterotrophic respiration (R_H) (dashed line with open circles) in control plots and CO_2 fluxes from treated plots (broken line with open squares) and b) CH_4 emissions in *R. mucronata* forest at Gazi bay Kenya. Vertical broken lines indicate periods when trees were girdled and cut in the treatment plots. Sampling for baseline, girdled and cut periods were done from June 2009 to August 2009 (84 days), December 2009 to May 2010 (189 days) and May 2010 to April 2011 (343 days), respectively. The controls and the treatment consisted of five replicates each.

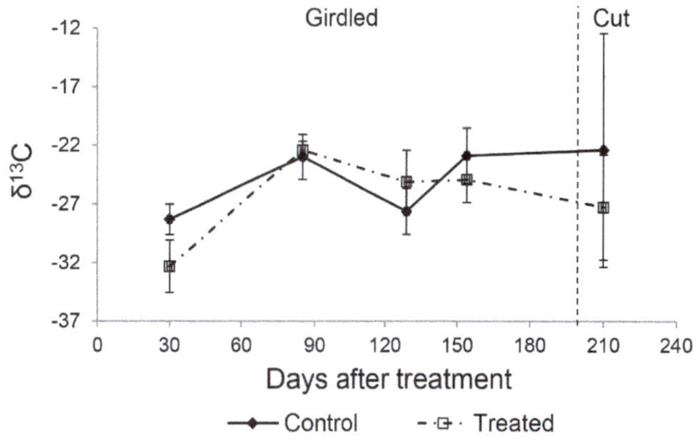

Figure 3. Trends in $\delta^{13}C$ of sediment respired CO_2 in control and treated sites in *R. mucronata* forest at Gazi bay, Kenya. Values are means ±95% CI. Vertical broken line indicates when the trees were clear-cut in treatment plots. The controls and the treatment consisted of five replicates each.

Discussion

Tree death induced rapid and significant subsidence. This was likely due to rapid decomposition of fine roots exacerbated by the absence of new root growth, as shown by the significant difference in root bag decomposition. In addition, sediment compaction and the collapse of arenchymatic tissues due to consolidation of air spaces (as shown by the increase in bulk density), and the significant loss of sediment moisture caused by the treatment, may also have contributed. Rates of sediment accretion did not differ significantly between control and treated plots hence the subsidence was not due to enhanced erosion. The rate of subsidence was surprisingly high given the relatively small scale of the treatments; massive hurricane damage, which led to 'peat-collapse' in Honduran mangroves, caused elevation losses only around ~0.3 as fast as those recorded here (−11 compared with −32 mm yr^{-1} [25]).

Interestingly, the C loss of around 25.3±7.4 t CO_2 ha^{-1} yr^{-1} reported here due to small-scale cutting was similar to that reported for mangrove forests impacted by large-scale clearing (29 tCO_2 ha^{-1} yr^{-1}) [27], and that inferred from peat collapse due to hurricane damage (15 t CO_2 ha^{-1} yr^{-1}) [25]. Much of the C loss occurred within the first year after treatment (mean ±SD? rates of 13.22±9.71 and 7.86±6.77 t CO_2 ha^{-1} yr^{-1} during the girdled and cut periodsrespectively) and by ~1.5 years the C losses induced by treatment began to drop. A similar pattern was recorded in clear-cut mangroves in Belize in which the C emissions in disturbed areas declined with time [27]. However, there was evidence in our study that decomposition of sediment organic carbon (SOC) - not only newly-killed root material - was enhanced by treatment, and that rates of SOC decomposition might be increasing with time. Buried root bags recorded significantly higher rates of decomposition in treated plots (0.19±0.02 vs. 0.16±0.02% dry weight loss day^{-1}, respectively) with most of the difference occurring after the first set of root bags were retrieved (150 days after burial). This was probably due to enhanced sediment surface temperatures in the cut plots due to canopy removal, since the treated plots experienced increases in sediment surface temperatures of 0.8 to 5.9°C compared to the control plots. Therefore, these results highlight the potential impact of physico-chemical changes on C losses in cut forests, which are separate from and additional to the losses from root death per se.

Although our estimates suggest high rates of below-ground C loss caused by tree death (similar to those seen following much larger impacts such as hurricanes) they are likely to represent an underestimate, since surface fluxes of CO_2 cannot account for below-ground, lateral flows of carbon in dissolved inorganic carbon (DIC) [2,3,24].

CH_4 emissions were significantly enhanced during the girdled period, possibly because of the addition of easily fermentable substrates [41–43] from dying roots. However, emissions in both treatment periods (girdled and cut respectively 0.7–1.5 and 0.2–0.9 mmol m^{-2} d^{-1}), did not exceed those reported for a number of pristine mangrove forests worldwide, 0.01–5.0 mmol m^{-2} d^{-1} [40]. Sediment-respired CO_2 collected shortly after girdling showed significant ^{13}C-depletion (−32.3 ‰; Fig. 3). Such depleted signatures are unlikely to arise from respiration of existing organic matter alone, as here $\delta^{13}C$ would be similar to the control. Rather, as coincident with increased CH_4 emissions, the most parsimonious explanation is that oxidation of methane comprised a component of the CO_2 efflux. Assuming a typical $\delta^{13}C$ value for methane in marine environments of −60 ‰ [44], the measured ^{13}C-depletion would represent an additional 12% CO_2 contribution from methane oxidation.

Overall, our experiment showed that small-scale harvesting for wood production as is typical in the countries around the Western

Table 2. Belowground roots and sediment biochemical characteristics in control and cut plots in *R.mucronata* forest at Gazi bay, Kenya.

Variable	Control	Cut
% OC_R	37.2±1.7	-
$\delta^{13}C_R$	−27.0±0.3	-
Sediment moisture content (%)	46.5±4.0	42.0±5.7
Bulk Density (g cm^{-3})	0.84±0.08	0.88±0.10
Sediment C concentration (g C cm^{-3})	0.052±0.008	0.052±0.008
% N	0.38±0.04	0.36±0.11
TOC/TN	18.59±0.8	17.6±0.7
$\delta^{13}C_S$	−27.2±0.2	−27.4±0.1
Sediment C stocks* (t C ha^{-1})	524.1±62.7	488.4±48.4

Values are means ±95% CI, OC_R = organic carbon content of mangrove roots, $\delta^{13}C_R$ and $\delta^{13}C_S$ = carbon isotopic value of mangrove roots and sediment organic matter, respectively, and TOC and TN = total carbon and nitrogen content of the sediment organic matter, respectively.
*Sediment C stocks to 1 m depth.

Indian Ocean caused significant sediment subsidence, which was caused primarily by sediment compaction, loss of sediment moisture and increased organic matter decomposition (both aerobic and anaerobic). Study of lightning damage in Florida mangroves, creating forest gaps larger than our experimental plots, showed even more subsidence (of up to 61 mm), but also suggests that undisturbed forests can recover [29]. Such evidence should not induce complacency since pressures on mangroves in many areas are increasing. Sea level is projected to rise this century [18], with current projections varying between 1.7 and 3.3 mm yr^{-1} [19]. Employing the lower estimate of sea level rise of 1.7 mm yr^{-1}, and the conservative assumption that our cumulative surface elevation loss of 62.4 mm will not increase further in the future, the impact of the loss we observed is equivalent to increasing sea level rise for ~37 years. Hence our study raises serious concerns that the combination of effects caused by global change and small-scale forest harvesting on mangrove surface elevation may be highly damaging to these sensitive ecosystems.

Conclusions

Kenyan mangroves are being lost at a rate of 0.7% cover per year [45], mainly because of the demands of the large and growing population for wood fuel and timber [34,36].

A new approach towards the sustainable management of mangrove forests uses payments for ecosystem services (PES) schemes such as the proposed REDD+ programme [46]. Such schemes should allow combining income from provisioning and regulating ecosystem services (such as timber and coastal protection or carbon sequestration), but only if the trade-offs between them are understood and managed. The present work demonstrates the susceptibility of mangroves to rapid subsidence (with consequent enhanced vulnerability to sea level rise and erosion) and at least short-term carbon loss following relatively small-scale and controlled canopy removal. With this understanding, management regimes aiming to conserve carbon stocks and promote climate resilience should be wary of clear-cutting, particularly in areas that may be exposed to erosion, and should emphasise instead selective cutting, rapid replacement of the lost canopy and the maintenance of un-cut buffer strips on seaward fringes to avoid the risks of erosion.

Materials and Methods

Study site

The study was carried out at Gazi Bay (4° 25′ S and 4° 27′ S; 39° 50′ E and 39° 50′ E), ~55 km south of Mombasa, Kenya. Gazi Bay is a creek system with a total area of 615 ha mangrove forest [47], dominated by *Rhizophora mucronata* (Lam), *Ceriops tagal* (Perr.) C. B. Robinson and *Avicennia marina* (Forsk.) Vierh. The mean annual precipitation of Gazi bay ranges from 1000–1600 mm [47]. The bay receives freshwater from two semi-permanent rivers: Kidogoweni to the north, which discharges in to the Kidogoweni creek, and Mkurumuji river to the south, discharging to the mouth of the bay. The forest is government-owned and permission to use the site was granted by the Kenya Marine and Fisheries Research Institute; our experiments did not affect endangered or protected species, and the individual featured in the accompanying image gave written informed consent (as outlined in the PLOS consent form) to publish this image.

Experimental Design

Five pairs of 12 m×12 m plots were established in March 2009 at high water mark (height above mean sea-level: 2.91±0.01 m) within a *Rhizophora mucronata* dominated forest (specific location: 4°24′32″S 39°31′23″E). In October 2009, five plots (one from each pair) were randomly allocated to 'treatment' and all the trees within them were girdled at ~20 cm above the highest prop root. Girdling is a method that has been used in a number of terrestrial forests to estimate the contribution of root respiration to total sediment gas flux. The rationale is to prevent the flow of carbohydrates from the tree canopy to their roots (thus stopping root respiration) whilst leaving the above-ground components relatively undisturbed; trees may retain foliage for many months after girdling [39,40]. The other plots in the five pairs served as the controls. In May 2010, all the trees in the girdled plots were cut at ~20 cm above the highest stilt roots and all the debris, excepting small fragments, was removed. The treated plots were allowed to stabilize after disturbance for approximately three weeks; thereafter sampling was resumed. During each treatment operation saplings and seedlings were cut down. Hence the experiment consisted of three sampling periods: a) baseline (pre-treatment) (June 2009 to August 2009, 84 days), b) girdled period (December

2009 to May 2010, 159 days) and c) cut period (May 2010 to April 2011, 343 days).

Surface elevation and sediment accretion

Surface elevation dynamics were monitored using surface elevation stations, consisting of two stainless steel rods (6 mm by 1 m) and a horizon marker (kaolin) set up in a 20 cm×20 cm quadrat in each plot. The rods were installed leaving a height of 20 cm above the ground and at opposite corners of the quadrat such that measurements would be made diagonally across it. Height measurements vertically from the ground surface to the heights of the rods were made at seven clearly marked points along a wooden board placed across the rods. Subsequent measurements were made at the same points along the wooden board. All data were averaged to give a single measurement per plot per time. Sediment accretion was determined from measurements of height above the horizon marker. In each quadrat, at least four sediment blocks of 2 cm×2 cm were carefully removed with a sharp knife, the height of sediment above the horizon marker was noted and the block was then carefully replaced in its original position. This approach allows the separation of total surface elevation/ subsidence (which depends on both the accretion or erosion of new sediments and on below-ground processes such as root growth and expansion) from accretion/erosion [25].

Samples for sediment physico-chemical analysis were taken from each plot during February 2010 and August 2012. A sediment core was taken in the centre of the plot using a plastic corer (diameter 6 cm, length 3 m) in February 2010 and again in August 2012. Subsamples of this large core were taken with a small stainless steel corer (diameter 3 cm, length 5 cm) at depths of 0, 2, 4, 8, 10, 20, 30 and 40 cm (February 2010) and depths of 0, 10, 20, 30, 50 and 100 cm (August 2012). To minimize compression of the sediment the coring was done in a series of stages according to the depth profiles for sub-sampling. Sediment samples were oven-dried at 80°C to constant dry weight and bulk density was determined at six depths down to 1 m. The oven-dried samples were transferred to the Department of Earth and Environmental Sciences, KU Leuven, Belgium for analysis of % OC (organic carbon), % N and $\delta^{13}C$ of OC. The concentrations of OC, total N, and $\delta^{13}C$ values of sediment OC were measured on subsamples weighed into Ag cups, acidified with dilute HCl to remove inorganic C, and analysed with a Thermo Flash HT elemental analyser coupled to a Thermo Delta V Advantage IRMS (Conflo IV interface). Data were calibrated with IAEA-C6, and internally calibrated acetanilide and leucine. From the sediment bulk density and OC content, the carbon density and hence the sediment carbon stocks down to 100 cm were derived for the control and treated plots.

Gas fluxes and stable carbon isotope signatures

Gas flux (CO_2, CH_4) samples were collected at approximately monthly intervals at low tide during spring tides using six chambers per plot; some sampling times that were missed due to loss or damage to bags and other equipment. Each chamber was inserted ~5 cm in to the sediment, occupying an area of 0.064 m^2 with an internal volume of 0.011 m^3. The samples from each chamber were taken 20 minutes after closure. Using a 60 ml syringe, at least 240 ml of gas were transferred from each chamber to labelled airtight gas-bags (Cali-5-bond gas bags, Calibrated Instruments Inc. USA). A gas sample of ambient concentration was taken from each chamber before closure; ambient air concentration samples for each plot were collected in one gas bag. Linearity checks were performed by repeatedly sampling the chamber gas for periods of about 60 minutes. They showed that a linear approximation over a 20 minutes period resulted in ~15% underestimation of the slope of gas concentration increase over time. This systematic downward bias was not corrected for. Sediment surface temperature measurements were made beside each chamber with a temperature probe inserted to ~1 cm in to the sediment. The number of crab burrows within the area enclosed by the chamber was noted. The positions of the chambers were marked for subsequent sampling; chambers were always returned to the same sampling positions within plots. Samples for $\delta^{13}C$ analysis of CO_2 were transferred from the chambers to 12 ml pre-evacuated exetainers (Labcoexetainer, Labco Ltd., High Wycombe, UK). Gas flux samples were analysed at the Institute of Atmospheric and Environmental Sciences, University of Edinburgh, UK. For CO_2, the samples were analysed by gas chromatography (GC) using a Perkin Elmer Model 310 with a thermal conductivity detector (TCD). Concentrations of CH_4 were measured using GC (Hewlett Packard 5890 GC, Hewlett Packard Ltd, Stockport, Cheshire, UK) equipped with a flame ionisation detector (FID) and a digital integrator. The $\delta^{13}C$ analysis samples were transferred to the Department of Earth and Environmental Sciences, KU Leuven, Belgium and analysed using a Sercon 20–20 isotope ratio mass spectrometer (IRMS) interfaced with a cryofocussing unit.

In May 2011, at the end of the experiment, all roots beneath each chamber in the control plots were excavated to a depth of 60 cm, washed of sediments and separated in to live root and dead plant materials. Dead roots were differentiated based on the loss of structural integrity, colour and signs of decomposition [5]. The samples were oven-dried at 80°C before weighing.

Decomposition in Root-bags

In September 2010, live roots were excavated within the *R. mucronata* forest contiguous to the experimental plots. Nylon mesh (1 mm) bags each containing ~30 g fresh roots were buried to ~20 cm depth at six random points within each plot. Three bags were retrieved from each plot 156 days after burial, whilst the other three were retrieved 267 days after burial. The contents of each root-bag were rinsed and oven-dried at 80°C for 24 h before weighing. The rate of root decay (% weight loss day^{-1}) was calculated as the % weight loss divided by the number of days buried, using wet-dry weight conversion factors derived from representative samples of live roots, oven-dried at 80°C until constant dry weight.

Statistical analysis

The data for CO_2 were log-transformed and the analysis for each gas was executed using MINITAB 14 software package. Initial analyses included time in repeated measures models, however there were multiple significant interactions preventing legitimate conclusions and so data were separated into the three experimental periods. The gas flux values for each chamber were pooled across each period for the controls and treated plots, giving six single values per period for each plot. For each period, nested ANOVA was carried out for each gas, with variation among plots (i.e. 6 chambers) nested within treatment and sediment surface temperature and crab burrows as covariates. Estimates for respired $\delta^{13}C$ were derived from the Miller-Tans mixing model combined with geometric regression [48,49]. Kayler et al., [49] found that the combination of geometric regression and Miller-Tans mixing model gave the most accurate and precise estimate of $\delta^{13}C_S$ (S = sediment respired CO_2). The gas mixing models are based on the conservation of mass given as [48]:

$$\delta_{obs}[CO_2]_{obs} = \delta_{bg}[CO_2]_{bg} + \delta_s[CO_2]_s$$

This equation describes the gas observed (obs) as coming from two sources: background atmosphere (bg) and source of respiration (s), where δ refers to the isotopic value of each component. Details of the Miller-Tans mixing model combined with geometric regression are discussed by Kayler *et al.*, [49]. The mean $\delta^{13}C$ of the respired CO_2 for each plot was analysed using two-sample t-tests.

To examine the autotrophic contribution to sediment fluxes, stepwise multiple regressions (forward and backward elimination) were performed, with the final CO_2 fluxes measured in each control chamber in April 2011 as the dependent variable and the live root biomass, sediment surface temperature and crab burrows for each chamber as the independent variables. The equation takes the form of $y = k{+}a{*}roots\ {+}b{*}temperature + c{*}burrows$; where k = constant, a, b and c are the coefficients of the estimators. The number of crab burrows was not significant, and hence this term was omitted from the equation. The significant factors were used in estimating the autotrophic respiration from the final CO_2 flux data. First, the heterotrophic respiration (R_H) was calculated as the value of the '*y*' when live root biomass = 0, i.e. $R_H = k{+}\ b{*}temperature$. Then the autotrophic respiration (R_A) was obtained as the difference between total sediment respiration (R_S) and the heterotrophic component (R_H) and expressed as a percentage of R_S (i.e. $\%R_A = (R_S{-}R_H)/R_H{*}100$). To estimate the contribution of R_A across the entire sampling period, the equation was applied to the CO_2 flux data, together with the sediment surface temperature for each control chamber at each sampling time. The mean R_A contribution across the sampling period was then compared with that obtained from the final sampling time.

The total additional C emissions due to treatment (treatment-induced C emissions) were estimated as the area under the response curve based on the trapezoidal rule [50]:

$$Ce = \sum_{i=l}^{n} \frac{(m_i + m_i + 1)}{2} t_i$$

where C_e = treatment induced C emissions, n = number of measurements, m = individual measurements and t = time difference between any two consecutive measurements.

The difference in the rates of root decomposition in the control and treated plots was tested using a two-sample t-test. Sediment carbon concentration (g C cm^{-3}) was calculated as the product of bulk density (BD) and % organic C of the sediment. Thereafter, the sediment carbon stocks down to a depth of 100 cm for each treatment were calculated as the product of carbon concentration and the depth and expressed as t C ha^{-1}: $C_S = C_C * (100{+}E_c)$, where, C_S = sediment C stocks, C_C = C concentration and E_c = elevation change. Since the control plots gained 1.1 cm and the treated plots lost 5.1 cm in surface elevation (*see* Figure 2), the depth for each treatment was adjusted to reflect these changes, i.e. $100{+}E_c = 101.1$ and 94.9 cm for control and treated plots, respectively.

Acknowledgments

Our gratitude goes to our field assistants Laitani Suleiman and Tom Kisiengo Peter, KMFRI staff at Gazi village, the Gazi community and to the many Earthwatch International volunteers who assisted in the field. We would also like to thank Robert Howard (School of GeoSciences, University of Edinburgh, UK) for assisting in laboratory analysis of CO_2 and CH_4 gases. Ken Krauss gave helpful advice on an early draft.

Author Contributions

Conceived and designed the experiments: JL JK MM SB MS SW MH. Performed the experiments: JL JK MS MM MH. Analyzed the data: JL MM SB SW MH. Contributed reagents/materials/analysis tools: SW MM SB. Wrote the paper: JL JK MM MS SB SW MH.

References

1. Twilley RR, Chen RH, Hargis T (1992) Carbon sinks in mangroves and their implications to carbon budget of tropical coastal ecosystems. Water Air Soil Poll 64: 265–288.
2. Bouillon S, Borges A, Castañeda-Moya E, Diele K, Dittmar T, et al. (2008) Mangrove production and carbon sinks: A revision of global budget estimates. Global Biogeochem Cycles 22: GB2013, doi:2010.1029/2007GB003052.
3. Kristensen E, Bouillon S, Dittmar T, Marchand C (2008) Organic carbon dynamics in mangrove ecosystems: A review. Aquat Bot 89: 201–219.
4. Breithaupt JL, Smoak JM, Smith TJ III, Sanders CJ, Hoare A (2012) Organic carbon burial rates in mangrove sediments: Strengthening the global budget. Global Biogeochem Cycles: doi:10.1029/2012GB004375, in press.
5. McKee KL, Cahoon DR, Feller I (2007) Caribbean mangroves adjust to rising sea level through biotic controls on change in soil elevation. Glob Ecol Biogeogr 16: 545–556.
6. Chmura GL, Anisfeld SC, Cahoon DR, Lynch JC (2003) Global carbon sequestration in tidal, saline wetland soils. Global Biogeochem Cycles, 11: 1111–11120.
7. Middleton BA, McKee KL (2001) Degradation of mangrove tissues and implications for peat formation in Belizean island forests. J Ecol 89: 818–828.
8. Gleason SM, Ewel KC (2002) Organic matter dynamics on the forest floor of a Micronesian mangrove forest: An investigation of species composition shifts. Biotropica, 34: 190–198.
9. Huxham M, Lang'at J, Tamooh F, Kennedy H, Mencuccini M, et al. (2010) Decomposition of mangrove roots: Effects of location, nutrients, species identity and mix in a Kenyan forest. Estuar Coast Shelf Sci 88: 135–142.
10. Golley F, Odum HT, Wilson RF (1962) The structure and metabolism of a Puerto Rican red mangrove forest in May. Ecology, 43: 9–19.
11. Fujimoto K, Imayal A, Tabuchi R, Kuramoto S, Utsugi H, et al. (1999) Belowground carbon storage of Micronesian mangrove forests. Ecol Res 14: 409–413.
12. Donato DC, Kauffman J, Murdiyarso D, Kurnianto S, Stidham M, et al. (2011) Mangroves among the most carbon-rich forests in the tropics. Nat Geosci 4: 293–297.
13. Alongi DM (2012) Carbon sequestration in mangrove forests. Carbon Manage 3: 313–322.
14. Trumper K, Bertzky M, Dickson B, van der Heijden G, Jenkins M, et al. (2009) The Natural Fix? The role of ecosystems in climate mitigation. United Nations Environment Programme, UNEPWCMC (Cambridge, UK). 68 pp.
15. Laffoley DA, Grimsditch G (2009) The management of natural coastal carbon sinks. IUCN. 53 pp.
16. Nellemann C, Corcoran E, Duarte CM, Valdés L, De Young C, et al. (2009)Blue Carbon. United Nations Environment Programme. 80 p.
17. McLeod E, Chmuira GL, Bouillon S, Salm R, Björk M, et al. (2011) A blueprint for blue carbon: toward an improved understanding of the role of vegetated coastal habitats in sequestering CO_2. Front Ecol Environ 9: 552–560.
18. IPCC (2007) Climate Change 2007: Synthesis Report. Contribution of Working Groups I, II and III to the Fourth Assessment Report of the Intergovernmental Panel on Climate Change. IPCC (Geneva, Switzerland). 104 p.
19. Nicholls RJ, Cazenave A (2010) Sea-level rise and its impact on coastal zones. Science, 328: 1517–1520.
20. Valiela I, Bowen JL, York JK (2001) Mangrove forests: one of the world's threatened major tropical environments. BioScience, 51: 807–815.
21. Valiela I, Kinney E, Culbertson J, Peacock E, Smith S (2009) Global losses of mangroves and salt marshes, In Duarte, C. M. [ed.], Global Loss of Coastal Habitats: Rates, Causes and Consequences. Fundacion BBVA. pp. 107–133.
22. FAO (Food and Agriculture Organization) (2007) The World's Mangroves, 1980–2005: A Thematic Study in the Framework of the Global Forest Resources Assessment 2005. Forestry Paper 153, FAO, Rome. ix+77 p.
23. Couwenberg J, Dommain R, Joosten H (2010) Greenhouse gas fluxes from tropical peatlands in south-east Asia. Global Change Biol 16: 1715–1732.
24. Alongi DM, de Carvalho N, Amaral A, Costa A, Trott L, et al. (2012) Uncoupled surface and below-ground soil respiration in mangroves: implications for estimates of dissolved inorganic carbon export. Biogeochemistry 109: 151–162.

25. Cahoon DR, Hensel P, Rybczyk J, McKee K, Proffitt E, et al. (2003) Mass tree mortality leads to mangrove peat collapse at Bay Islands, Honduras after Hurricane Mitch. J Ecol 91: 1093–1105.
26. Barr JG, Engel V, Smith TJ, Fuentes JD (2012) Hurricane disturbance and recovery of energy balance, CO_2 fluxes and canopy structure in a mangrove forest of the Florida Everglades. Agric Fore Meteor 153: 54–66.
27. Lovelock CE, Ruess RW, Feller IC (2011) CO_2 efflux from cleared mangrove peat. PLoS ONE, 6: e21279.
28. Krauss K, Cahoon D, Allen J, Ewel K, Lynch J, et al. (2010) Surface elevation change and susceptibility of different mangrove zones to sea-level rise on Pacific High Islands of Micronesia. Ecosystems 13: 129–143.
29. Whelan K (2005) The successional dynamics of lightning-initiated canopy gaps in the mangrove forests of shark river, Everglades National Park, USA. PhD thesis, Florida International University.
30. Kirui KB, Kairo JG, Bosire J, Viergever KM, Rudra S, et al. (2012) Mapping of mangrove forest land cover change along the Kenya coastline using Landsat imagery. Ocean Coastal Manage: doi:10.1016/j.ocecoaman.2011.1012.1004.
31. Abuodha P, Kairo JG (2001) Human-induced stresses on mangrove swamps along Kenya coast. Hydrobiologia 458: 255–265.
32. Dahdouh-Guebas F, Van Pottelbergh I, Kairo JG, Cannicci S, Koedam N (2004) Human-impacted mangroves in Gazi (Kenya): predicting future vegetation based on retrospective remote sensing, social surveys, and distribution of trees. Mar Ecol Prog Ser 272: 77–92.
33. Taylor M, Ravilious C, Green EP (2003)Mangroves of East Africa. UNEP World Conservation Monitoring Centre, (Cambridge, UK). 25 p.
34. Dahdouh-Guebas F, Mathenge C, Kairo JG, Koedam N (2000) Utilization of mangrove wood products around Mida Creek (Kenya) amongst subsistence and commercial users. Econ Bot, 54: 513–527.
35. Bosire JO, Dahdouh-Guebas F, Kairo JG, Koedam N (2003) Colonization of non-planted mangrove species into restored mangrove stands in Gazi Bay, Kenya. Aquat Bot 76: 267–279.
36. Wass P (1995) Kenya's Indigenous Forests: Status, Management and Conservation. IUCN. Xii+250 pp. 252-8317-0292-8315.
37. Alongi DM (2009) The Energetics of Mangrove Forests. (Springer, Netherlands). 216 p.
38. Hanson PJ, Edwards NT, Garten CT, Andrews JA (2000Separating root and soil microbial contributions to soil respiration: A review of methods and observations. Biogeochemistry, 48: 115–146.
39. Hogberg P, Nordgren A, Buchmann N, Taylor AFS, Ekblad A, et al. (2001) Large-scale forest girdling shows that current photosynthesis drives soil respiration. Nature, 411: 789–792.
40. Andersen C, Nikolov I, Nikolova P, Matyssek R, Häberle K-H (2005) Estimating "autotrophic" belowground respiration in spruce and beech forests: decreases following girdling. Europ J Fore Res 124: 155–163.
41. Blodau C (2002) Carbon cycling in peatlands - A review of processes and controls. Environ Rev 10: 111–134.
42. Goreau TJ, de Mello WZ (2007) Mininmizing net greenhouse gas sources from mangrove and wetland soils, In Tateda, Y. [ed.], Greenhouse Gas and Carbon Balances in Mangrove Coastal Ecosystems. pp. 239–248.
43. Kristensen E (2007) Carbon balance in mangrove sediments: The driving processes and their controls, In Tateda, Y. [ed.], Greenhouse gas and carbon balances in mangrove coastal ecosystems. Gendai Tosho. pp. 61–78.
44. Reeburgh WS (2007) Oceanic methane biogeochemistry. Chem Rev 107: 486–513.
45. Kirui KB, Kairo J, Bosire J, Viergever K, Rudra S, et al. (2012) Mapping of mangrove forest land cover change along the Kenya coastline using Landsat imagery. Ocean Coast Manage: doi:10.1016/j.ocecoaman.2011.1012.1004.
46. Locatelli B, Brockhaus M, Buck A, Thompson I (2010) Forests and adaptation to climate change: challenges and opportunities, In Mery G, et al. [eds.], Forests and Society-Responding to Global Drivers of Change. International Union of Forest Research Organizations (IUFRO). pp. 21–42.
47. UNEP (United Nations Environment Program) (1998) Eastern Africa Atlas of Coastal Resources 1: Kenya. (EAF-14) UNEP, 119 p.
48. Miller JB, Tans PP (2003) Calculating isotopic fractionation from atmospheric measurements at various scales. Tellus B, 55: 207–214.
49. Kayler Z, Ganio L, Hauck M, Pypker T, Sulzman EW, et al. (2010) Bias and uncertainty of $\delta^{13}CO_2$ isotopic mixing models. Oecologia, 163: 227–234.
50. Cerone P, Dragomir SS (2000) Trapezoidal-type rules from an inequalities point of view, In Anastassiou, G. [ed.], Handbook of Analytic-Computational Methods in Applied Mathematics. CRC Press. pp. 65–134.

21

Patch Size and Isolation Predict Plant Species Density in a Naturally Fragmented Forest

Miguel A. Munguía-Rosas*, Salvador Montiel

Departamento de Ecología Humana, Centro de Investigación y de Estudios Avanzados del Instituto Politécnico Nacional (CINVESTAV), Mérida, Yucatán, México

Abstract

Studies of the effects of patch size and isolation on plant species density have yielded contrasting results. However, much of the available evidence comes from relatively recent anthropogenic forest fragments which have not reached equilibrium between extinction and immigration. This is a critical issue because the theory clearly states that only when equilibrium has been reached can the number of species be accurately predicted by habitat size and isolation. Therefore, species density could be better predicted by patch size and isolation in an ecosystem that has been fragmented for a very long time. We tested whether patch area, isolation and other spatial variables explain variation among forest patches in plant species density in an ecosystem where the forest has been naturally fragmented for long periods of time on a geological scale. Our main predictions were that plant species density will be positively correlated with patch size, and negatively correlated with isolation (distance to the nearest patch, connectivity, and distance to the continuous forest). We surveyed the vascular flora (except lianas and epiphytes) of 19 forest patches using five belt transects (50×4 m each) per patch (area sampled per patch = 0.1 ha). As predicted, plant species density was positively associated (logarithmically) with patch size and negatively associated (linearly) with patch isolation (distance to the nearest patch). Other spatial variables such as patch elevation and perimeter, did not explain among-patch variability in plant species density. The power of patch area and isolation as predictors of plant species density was moderate (together they explain 43% of the variation), however, a larger sample size may improve the explanatory power of these variables. Patch size and isolation may be suitable predictors of long-term plant species density in terrestrial ecosystems that are naturally and anthropogenically fragmented.

Editor: Eric Gordon Lamb, University of Saskatchewan, Canada

Funding: This study was funded by Consejo NacionaL de Ciencia y Tecnología (project CB-2012-177680), funds to MAM-R. The funder had no role in study design, data collection and analysis, decision to publish, or preparation of the manuscript

Competing Interests: The authors have declared that no competing interests exist.

* Email: munguiarma@mda.cinvestav.mx

Introduction

MacArthur and Wilson's Equilibrium Theory of Island Biogeography (ETIB) postulates that the number of species in oceanic islands can be predicted by island size and isolation. This is firstly because of sample area effects (in any region, larger sample areas will contain more species than smaller samples) and secondly, but more importantly, because of the interplay between immigration and extinction that leads, over time, to a size and isolation-dependent equilibrium number of species (species relaxation) [1,2]. Therefore, the main testable predictions of this theory are that the larger and less isolated the island, the higher the species number at which it should reach equilibrium. The ETIB reached paradigmatic status in biogeography and ecology, and also had an enormous impact on conservation biology [3]. The ETIB has also provided a theoretical framework for understanding habitat fragmentation and making predictions about its effect on biodiversity [3,4,5,6,7].

Since the publication of original monograph of the ETIB in 1967 [2], the authors themselves have suggested that the principles and predictions of the ETIB may apply not only to oceanic islands but also to terrestrial ecosystems that are naturally and anthropogenically fragmented ([2], pp 3–4). However, the current literature is divided about the effects of patch size and isolation on species richness and other measures of biodiversity, such as species density [8,9,10], in terrestrial ecosystems [7,11,12]. In fact, quantitative reviews have shown that patch area and isolation alone are poor predictors of species occupancy in terrestrial forest fragments [13,14]. While the low predictive power of patch area and isolation for biodiversity could be due to several confounding factors associated with anthropogenic disturbance (i.e., patches differ in habitat matrix, elevation, human activity, etc. [15,16,17]), experimental studies in which these factors are controlled have also shown a remarkable lack of consistency in their results, especially with regard to number of species relative to patch size [18,19,20]. A basic condition of the ETIB is that it is only when colonization and emigration rates have reached equilibrium that species number can be accurately predicted by habitat size and isolation [1,2]. Owing to the long life cycle of some plants [21], it may take hundreds or thousands of years after fragmentation to reach a new equilibrium in forest patches [22,23]. Studies addressing the effect of patch and landscape traits on plant biodiversity in anthropogenic forest fragments usually work with forests that have been fragmented for decades or just a few hundred years [e.g., 9, 24]. It is therefore possible that most of the studies of anthropogenic forest fragments have failed to record an effect of patch size and isolation on biodiversity because most of these studies have been conducted in recently formed forest patches, where a new equilibrium has not yet been reached. Therefore, the effects of patch size and isolation on the biodiversity of terrestrial fragmented ecosystems can be more effectively assessed in forests that have been fragmented for a long time.

Naturally fragmented forests offer an excellent opportunity for testing the effects of patch size, isolation and other spatial variables on biodiversity because they have been fragmented for thousands or millions of years [25,26], and consequently can be assumed to have reached equilibrium. The main goal of this study was to assess whether plant species density is explained by patch size and isolation in a naturally fragmented forest on the Yucatan Peninsula. We used species density (number of species in equal-sized samples) as a response variable, as it is a measure of biodiversity that is less influenced by sample area effects than species richness [27,28]. This forest is characterized by the presence of several forest patches (locally known as Petenes, singular Petén) that are roughly circular, variable in size [25,29,30,31] and grow on Quaternary geological formations ca. 1.7 My old [26,32]. Scattered near the coast in a wetland matrix [29,30,31], the Petenes are landscape units that aptly reflect the habitat patch concept as their spatial boundaries naturally contain or delimit populations and communities of plants [12,25]. Therefore, the study area allowed us to test the effect of patch area and isolation while controlling for important confounding factors such as patch shape and matrix. In addition to the effects of patch size and isolation we tested the effect of elevation and patch perimeter. Patch elevation is negatively related to the level of salt water during the rainy season and this affects plant distribution [31,33], while patch perimeter is positively correlated with edge effects [34]. We predicted that plant species density would be positively correlated with patch size and negatively correlated with patch isolation. Patch perimeter would be negatively correlated with plant species density and elevation, positively correlated.

Materials and Methods

Study area

The study area is the Petenes-Celestún-El Palmar biological corridor (19° 53′–21° 11′ N, 90°28′–90°17′W) located along the northwest coast of the Yucatan Peninsula, which has an area of about 240,000 ha [35] (Fig. 1). The weather is tropical subhumid with summer rains, precipitation is 1000–1200 mm y^{-1} and mean temperature 26.1–27.8°C [36]. Our study focused on naturally formed forest patches of semi-evergreen tropical forest, sometimes mixed with tall mangrove species [30,31]. These patches are more abundant in a narrow belt (ca.10 km wide) beside the sea and their abundance decreases toward the mainland where the forest becomes continuous (Fig. 1). Most of these forest patches are roughly circular [30], and those that are amorphous are far less frequent [36]. The forest patches grow on Quaternary geological formations that are approximately 1.7 My old [26,32], and are characterized by taller, more diverse vegetation relative to that of the matrix [30]. It is believed that the presence of these patches is explained by their higher elevation relative to the matrix and the permanent supply of fresh water from one or more sink holes [30]. Tree species such as *Manilkara zapota*, *Metopium brownei*, *Bursera simaruba*, *Laguncularia recemosa* and *Avicennia germinans* dominate the canopy; *Bravaisia tubiflora* and *Sabal yapa* dominate the understorey [26,31,37]. The matrix surrounding the forest patches is dominated by short mangrove species (*Rizophora mangle*, *Conocarpus erectus*), sedges (*Eleocharis cellulose*, *Cladium jamaicense*) and cattails (*Typha dominguensis*) [31,37].

Sampling

From January 2013 to May 2014 we recorded vascular plants in 19 forest patches using belt transects (50×4 m). It was not possible to select these forest patches randomly owing to insurmountable access difficulties. Epiphytes and lianas were not recorded during the vegetation survey owing to the difficulties associated with assessing their presence and abundance (forest canopy height: 26 m [36]). We recorded woody plants with a girth greater than 5 cm (dbh>1.6 cm) and non woody plants taller than 20 cm in five transects per patch. We did not record small seedlings (dbh≤ 1.6 cm) because while propagules do arrive in the forest patches, some species cannot establish because of the water level and increased salinity during the rainy season [33]; recording seedlings might have led to an overestimation of real species density. Following the advice of previous studies [9,27,28,38,39], the area sampled was kept constant in all 19 patches (total sampled area per patch 0.1 ha) to reduce sample area effects. In each patch, the first transect was placed using a random point and the remaining transects were placed systematically: 20 m apart and in a previously defined direction. For all plants, dbh and life form were recorded *in situ*. Plants were identified with the help of field guides [40,41] and expert advice. Unidentified species were morphotyped (N = 2).

The leaves, flower or fruit of plants of uncertain identity were taken to the laboratory for later identification. Biological material was collected under a permit issued by the Mexican Ministry of the Environment (reg. SEMARNAT 31D8B-00780/1303) and field work was conducted with the permission of the authorities of the protected areas (reg. F00.9.DRBRC.002/13 for Ria Celestún Nature Reserve and reg. F00.9.DRBLP.04/13 for Los Petenes Nature Reserve).

Spatial configuration

For each patch sampled we measured its area (ha), perimeter (km) and the elevation at its centre (m a.s.l.). As measures of isolation we used distance to the nearest patch (edge to edge in km), distance to the continuous forest (edge to edge in km) and a connectivity index. The latter was an area-based index weighted by distance, calculated as follows:

$$\text{Connectivity} = \sum_{j}^{N} \frac{A_j}{Dist_j},$$

where A is the area of the focal patch (ha) and Dist is the edge to edge distance (km) between the focal patch and another patch in the surroundings. To calculate this index we used the three patches closest to each sampled patch (Fig. 1). For this index, patch connectivity increases as patch isolation decreases (all values >0). All spatial variables were obtained from digital cartography available in Google Earth Pro 7.2. We chose the most recent images available (2009–2011) and verified that the images represented the current landscape in the field. Spatial variables for all 19 forest patches are available in Appendix S1.

Data analysis

To assess inventory completeness in the patches sampled, we compared the number of species observed in samples to the number of species predicted by three nonparametric estimators: Chao1, ACE and Bootstrap [42]. To minimize potential over- or underestimation of species number by any estimator we used the mean of species predicted by the three estimators as a reference [26]. As previous studies suggest [43,44], we considered an inventory completeness ≥80% as representative; therefore, only

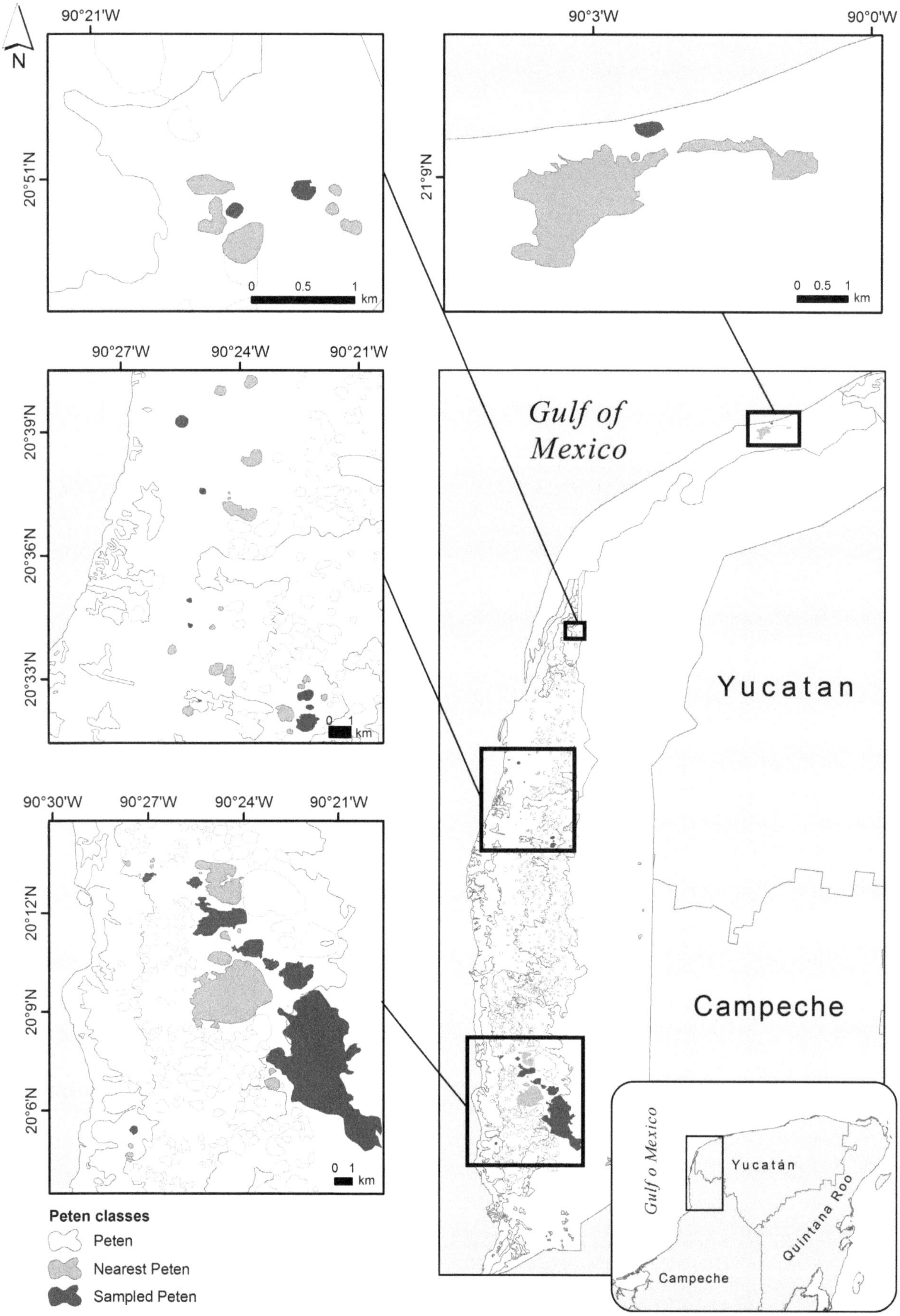
90°21'W
N
20°51'N
0 0.5 1
km
90°3'W
90°0'W
21°9'N
0 0.5 1
km
90°27'W
90°24'W
90°21'W
20°39'N
20°36'N
20°33'N
0 1
km
Gulf of
Mexico
Yucatan
Campeche
90°30'W
90°27'W
90°24'W
90°21'W
20°12'N
20°9'N
20°6'N
0 1
km
Gulf o Mexico
Yucatán
Quintana Roo
Campeche
Peten classes
Peten
Nearest Peten
Sampled Peten

Figure 1. Study area map. Forest patches sampled are in black, and gray patches were used to calculate a connectivity index or the distance to the nearest patch. The white area is the terrestrial portion of the Petenes-Celestún-El Palmar biological corridor. The small rectangle in the inset at the bottom right indicates the position of the study area on the Yucatan Peninsula. All bars represent 1 km.

forest patches that met this criterion were included in the analyses (Table 1).

We used plant species density (number of species in 0.1 ha) as the response variable in a model where the predictors were: patch area (ha) on a logarithmic scale, distance to the nearest patch (km), edge to edge distance from the focal patch to the continuous forest (km), patch connectivity index (see "Spatial configuration" above), patch central elevation (m a.s.l.) and patch perimeter (km). This model was fitted to a generalized linear model with a Poisson error distribution and log link function [45]. To determine the minimal adequate model, the complete model was simplified using the AIC criterion and intermediate models were compared using ANOVAs [45] (Table 2). Examination of the residuals of the minimal adequate model indicated a good fit and no evidence of overdispersion (dispersion parameter = 1).

Results

We recorded 55 different plant species in the 19 forest patches sampled (Appendix S2). Species density per forest patch was 6 to 20 species per 0.1 ha (mean = 14±1.44 species; hereafter mean ±1 standard error) (Table 1). The most abundant species in all of the forest patches we sampled was the shrub *Bravaisia tubiflora* (Acanthaceae). The representativeness of our inventory relative to the mean number of species predicted by the three estimators was, in all cases, equal to or greater than 80% (range = 80%–100%; Table 1).

Forest patch size averages 180.15 ha (range: 1.65–2472.82 ha) and average patch perimeter is 4.14 km (range: 0.51–35.21 km), patch altitude is 8.58 m a.s.l. (range: 3–18 m a. s. l.), focal patch to the continuous forest distance is 14.89 km (range: 3.82–89.98 km), focal patch to the nearest patch distance is 3.8 km (range: 0.03–1.62 km) and the connectivity index is 1794.35 (range: 10.85–12030.91) (Appendix S1).

For the model proposed to explain variation in plant species density, of the six variables initially in the complete model, only two (patch size and distance to the nearest patch) were retained in the minimal adequate model (Table 2). The minimal adequate model had the lowest AIC and did not differ statistically from the complete model ($\chi_4 = 3.11$, $P = 0.54$). The minimal adequate model explained 43% of the total deviance (Table 2). Both of the variables in the minimal adequate model –patch size ($\chi_1 = 19.68$, $P = 0.02$) and distance to the nearest patch ($\chi_1 = 14.78$, $P = 0.03$) – significantly affected plant species density. The relationship between plant species density and patch size is positive and logarithmic ($y = 1.125 \log x + 10.501$; Fig. 2A), while the relationship between plant species density and the distance to the nearest patch is negative and linear ($y = -5.7668x + 16.123$; Fig. 2B). Patch size and distance to the nearest patch explained 23% and 20% of the deviance of plant species density, respectively.

Table 1. Number of species observed in 0.1 ha samples (S) and predicted number of plant species using three nonparametric estimators (Chao1, ACE and Bootstrap) for 19 forest patches on the Yucatan Peninsula.

Patch	S	Chao1	ACE	Bootstrap	Average	Completeness (%)
1	6	6±1.3	7±0.11	7±0.1	6.7±0.3	90
2	9	10±1.4	10±1.7	10±0.1	10±0	90
3	16	16±0.5	16±21	17±0.1	16.3±0.3	98
4	17	20±2.3	19±1.9	19±1.5	19.3±0.3	88
5	19	19±1.1	20±2.2	22±2.1	20.3±0.9	93
6	16	16±1.9	17±1.2	18±1.9	17±0.6	94
7	19	26±2.4	22±2.1	22±1.7	23.3±1.3	81
8	13	13±1.3	14±1.8	14±0.9	13.7±0.3	95
9	16	16±1.8	18±21	18±1.5	17.3±0.7	92
10	18	20±5.3	23±2.4	20±2.1	21±1	86
11	15	15±3.7	16±1.9	17±1.7	16±0.6	94
12	10	13±2	14±1.8	11±0.9	12.6±0.8	80
13	20	22±5.3	23±2.4	23±2.1	22.7±0.3	88
14	11	12±1.4	13±1.4	12±1.2	12.3±0.3	89
15	11	11±3.7	12±1.6	12±1.3	11.7±0.3	94
16	8	8±0.1	8±1.41	9±0.7	8.3±0.3	96
17	10	10±1.4	10±1.4	10±0.5	10±0	100
18	20	26±2.3	24±2.4	23±3.6	24±1.1	83
19	13	13±1.4	13±1.6	13±0.7	13±0	100

Average is the mean number of species predicted by the three estimators. The percent completeness of our inventory relative to the average predicted number of species is shown (Completeness). In all cases, errors represent one standard error of the mean.

Table 2. Log linear models proposed to explain the variation in plant species density (S) in 19 forest patches on the Yucatan Peninsula.

Model	Model description	D^2	AIC
1	log S = log Size+D. nearest patch+D. continuous forest+Connectivity+Elevation+Perimeter	0.53	110
2	log S = log Size+D. nearest patch+D. continuous forest+Connectivity+Elevation	0.53	108
3	log S = log Size+D. nearest patch+D. continuous forest+Connectivity	0.50	107
4	log S = log Size+D. nearest patch+D. continuous forest	0.47	106
5	log S = log Size+D. nearest patch	0.43	105

Model 1 represents the complete model with six explanatory variables: patch size (Size, on a logarithmic scale), distance to the nearest patch (D. nearest patch), distance to continuous forest (D. continuous forest), patch connectivity index (Connectivity), patch elevation (Elevation) and patch perimeter (Perimeter). Model 5 represents the minimal adequate model and models 2–4 are intermediate steps during model simplification. The Akaike information criterion (AIC) and the proportion of explained deviance for each model (D^2) are also shown.

Discussion

Although previous studies have obtained contrasting results for the effects of patch size and isolation on species density [e.g. 8, 9, 10], our results clearly show that patch isolation and patch size predict plant species density in our study area on the Yucatán Peninsula. Unlike those of previous studies our study system has been fragmented for a very long time and important confounding factors such as habitat matrix and shape were kept relatively constant in this study. Therefore, we suggest that patch size and isolation are important determinants of species density in fragmented forests, although in some previous studies confounding effects and, more importantly, little elapsed time since fragmentation (i.e. no equilibrium has been reached) have obscured their relevance.

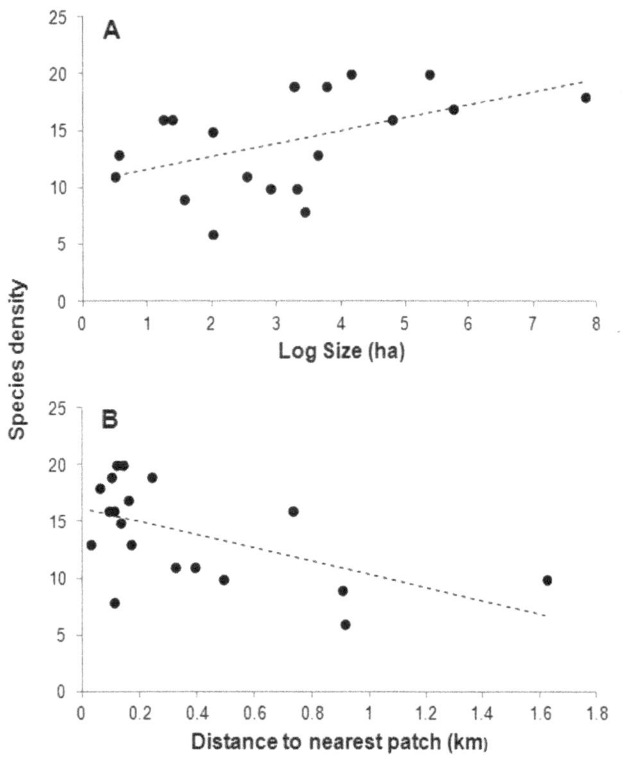

Figure 2. Relationship between plant species density (number of species in 0.1 ha) and patch size (Size) (A) and between plant species density and the distance to the nearest patch (B) for a group of 19 forest patches on the Yucatan Peninsula. Regression lines for A and B are also shown. The scale on the "x" axis for A is logarithmic.

In addition to patch size and isolation, we included other variables in our analysis that were relevant to our study system: patch elevation (negatively correlated with the level of saltwater, which most of the plants in forest patches cannot tolerate; [33]) and patch perimeter (positively correlated with the length of edge in contact with the harsh habitat matrix). However, only patch size and isolation (measured as distance to the nearest patch) significantly explained the variability in species density among forest patches. We believe that patch elevation was not significant because it varies little among patches (3–18 m a.s.l.). Additionally, the internal fresh water supply from the sinkholes that are typically found in petenes [32] may counteract the influence of saltwater from the matrix. Perimeter was probably not a good predictor of species density because most of the patches are roughly circular in shape and therefore variation in area is a better predictor of species density.

In continental islands isolation is clearly the distance to the mainland [1,2], but it seems that this does not apply to terrestrial ecosystems. In our study system we took distance to the continuous forest to be analogous to the distance to the mainland in oceanic islands. However, this variable did not explain the variation in plant density among patches. We suggest that this is because, unlike oceanic islands where the mainland is the primary species pool for colonization, in most forest fragments immigration occurs predominantly from habitats in the vicinity of the patch, rather than from a common mainland [12]. This notion is reinforced by the fact that the distance to the nearest patch significantly explained variation in species density among patches in our study and in other previous studies [46]. Another metric of patch isolation which was not significant in our analysis was a measure of habitat amount weighted by distance to the focal patch (connectivity index). Therefore, our results do not support the suggestion made in a recent study [12] that habitat amount is a better predictor of the number of species in a focal patch than the distance to the nearest patch is.

The most remarkable result of our study was that patch area and isolation significantly predicted species density. However, we must recognize that the predictive value of these variables is moderate (43%; patch area = 23%, patch isolation = 20%). In a previous study [13], the authors evaluated the predictive power of species-area equations and found that these equations usually explain less than a half (49%) of the variation in species number.

Based on this, the authors suggested that predictions by these models are unreliable. We find their position very conservative. Møller and Jennions [47] addressed the question of how much variance can be explained by ecologists. They examined several quantitative reviews of empirical studies and determined that, on average, ecologists can only account for a small portion of variance in their studies (2.5–5.4%) owing to the randomness and noise typical of ecological systems. Therefore, for the results obtained in our study, we think that values of 20 and 23% explained variance are actually quite good. Previous research has suggested that several confounding factors may reduce or nullify the effects of patch size and isolation on species number. For instance, Cook [48] showed that diversity patterns conform better to the prediction of ETIB when matrix species are removed from the patch samples, suggesting that the habitat matrix is creating a confounding effect. However, the results of experimental studies where several confounding effects were controlled also produced inconsistent results with respect to the effects of patch size and isolation on species number [19], suggesting that the time a forest has been fragmented could be even more important than environmental factors in determining the predictive capacity of these variables. Our study system has been fragmented for a period on a scale of geological rather than ecological time and other potential confounding factors were either considered in the model (patch elevation and perimeter) or kept constant (habitat matrix, shape, human disturbance). Therefore, we think that sample size and not being able to choose the forest patches randomly limited the explanatory power of the model.

Another explanation of the contrasting results of previous studies that evaluated the effect of patch size and isolation on biodiversity may lie in how biodiversity was measured. For example, in her review of the effect of habitat fragmentation on biodiversity, Fahrig [7] examined studies where biodiversity was measured as species abundance, species density, species richness, species incidence, genetic diversity, species interactions, species extinction and species turnover. With this kind of variation in the way biodiversity has been measured it is not surprising to find contrasting results and a vague general pattern. In our study we selected the number of species in equal-sized samples (species density) as the response variable. Some authors have preferred species density (obtained from equal-sized samples) over species richness to evaluate species-area relationships [9,27,28,38,39] because this way patch area is not confounded with the sampled area, as occurs when greater sampling effort is invested in larger patches relative to smaller patches.

The most frequently cited mechanisms to explain the association between species richness and habitat size are sample area effects [49], habitat heterogeneity [50], and immigration-extinction balance [1,2]. However, in our study some of these mechanisms may not apply because we used species density as a response variable instead species richness. Sample area effects cannot account for the species density-area relationship found in this study because the area sampled was kept constant. As habitat heterogeneity is usually correlated with area [1,2], habitat heterogeneity is not likely to be the underlying mechanism of this species density-area relationship either. Previous research has suggested that, as occurs with species richness [1,2], variation in species density can be partially explained by demographic processes such as colonization and extinction [51]. Therefore, we suggest that the most likely mechanism underlying the association between species density and patch area or isolation is a size/isolation dependant equilibrium between colonization and extinction; however, we also recognize that further work is needed to unequivocally identify the mechanism.

In conclusion, species density in fragmented terrestrial ecosystems can be predicted by patch area and isolation. The fact that the study area has been fragmented for a very long period of time may have contributed to this result as this ecosystem has very probably reached equilibrium. We encourage other researchers to test for the effect of patch size and isolation on species number in other naturally fragmented forests to assess whether it has greater predictive power in terrestrial ecosystems that have been fragmented for geological periods of time and are likely to be in equilibrium.

Acknowledgments

A Rojas, SG Jurado-Dzib, CR Mezeta-Cob and A Rivas-Arellano provided invaluable assistance in the field. B Magaña elaborated the map in Fig. 1. L Salinas-Peba, L Arias and the staff from the "U Najil Tikin Xiw" herbarium (CICY) helped with plant identification. B Delfosse revised the English.

Author Contributions

Conceived and designed the experiments: MAMR SM. Performed the experiments: MAMR SM. Analyzed the data: MAMR. Contributed reagents/materials/analysis tools: MAMR SM. Contributed to the writing of the manuscript: MAMR SM.

References

1. MacArthur RH, Wilson EO (1963) An equilibrium theory of insular zoogeography. Evolution 17: 373–387.
2. MacArthur RH, Wilson EO (1967) The Theory of Island Biogeography. New Jersey: Princeton University Press. 203p.
3. Haila Y (2002) A conceptual genealogy of fragmentation research: from island biogeography to landscape ecology. Ecol Appl 12: 321–334.
4. Janzen DH (1968) Host plants as islands in evolutionary and contemporary time. Am Nat 102: 592–595.
5. Henney LR (2000) Dynamic equilibrium: a long term, large scale perspective on the equilibrium model of island biogeography. Global Ecol Biogeogr 9: 59–74.
6. Laurance WF (2008) Theory meets reality: how habitat fragmentation research has transcended island biogeographic theory. Biol Conserv 141: 1731–1744.
7. Fahrig L (2003) Effects of habitat fragmentation on biodiversity. Annu Rev Ecol Evol Syst 34: 487–515.
8. Wettstein W, Schmid B (1999) Conservation of arthropod diversity in montane wetlands: Effect of altitude, habitat quality and habitat fragmentation on butterflies and grasshoppers. J Appl Ecol 36: 363–373.
9. Arroyo-Rodríguez V, Pineda E, Escobar F, Benítez-Malvido J (2008) Values of small patches in the conservation of plant-species diversity in highly fragmented rain forest. Conserv Biol 23: 729–739.
10. Farmilo BJ, Melbourne BA, Camac JS, Morgan JW (2014) Changes in plant species density in an experimentally fragmented forest landscape: Are the effects scale-dependent? Austral Ecol 39: 416–423.
11. Gilbert FS (1980) The equilibrium theory of island biogeography: fact or fiction? J Biogeogr 7: 209–235.
12. Fahrig L (2013) Rethinking patch size and isolation effects: the habitat amount hypothesis. J Biogeogr 40: 1649–1663.
13. Boecklen WJ, Gotelli NJ (1984) Island biogeographic theory and conservation practice: species-area or specious-area relationships? Biol Conserv 29: 63–80.

14. Prugh LR, Hodges KE, Sinclair ARE, Brashares JS (2008) Effect of habitat area and isolation on fragmented animal populations. Proc Natl Acad Sci USA 105: 20770–20775.
15. Walter HS (2004) The mismeasure of islands: implication for biogeographical theory and the conservation of nature. J Biogeogr 31: 177–197.
16. Ewers RM, Bidham RK (2006) Confounding factors in the detection of species responses to habitat fragmentation. Biol Rev Camb Philos Soc 81: 117–142.
17. Hu G, Gu J, Feeley KJ, Xu G, Yu M (2013) The effects of landscape variables on the species-area relationship during late-stage habitat fragmentation. Plos One 7: e43894.
18. Robinson GR, Holt RD, Gaines MS, Hamburg SP, Johnson ML, et al. (1992). Diverse and contrasting effects of habitat fragmentation. Science 257: 524–526.
19. Debinski DM, Holt RD (2000) A survey and overview of habitat fragmentation experiments. Conserv Biol 14: 342–355.
20. Collins CD, Holt RD, Foster BL (2009) Patch size effects on plant species decline in an experimentally fragmented landscape. Ecology 90: 2577–2588.
21. Martínez-Ramos M, Alvarez-Buylla ER (1998) How old are tropical rain forest trees? Trends Plant Sci 3: 400–405.
22. Helm A, Hanski I, Pärtel M (2006) Slow response of plant species richness to habitat loss and forest fragmentation. Ecol Lett 9: 72–77.
23. Johansson V, Snäll T, Ranius T (2013) Estimates of connectivity reveal non-equilibrium epiphyte occurrence patterns almost 180 years after habitat decline. Oecologia 172: 607–615.
24. Santos B, Arroyo-Rodríguez V, Moreno CE, Tabarelli M (2010) Edge-related loss of tree phylogenetic diversity in the severely fragmented Brazilian Atlantic Forest. Plos One 5: e12625.
25. Mas JF, Correa J (2000) Analysis of landscape fragmentation in the "Los Petenes" protected area, Campeche, Mexico. Investigaciones Geográficas 43: 42–59.
26. Munguía-Rosas MA, Jurado-Dzib SG, Mezeta-Cob C, Montiel S, Rojas A, et al. (2014) Continuous forest has greater taxonomic, functional and phylogenetic plant diversity than an adjacent naturally fragmented forest. J Trop Ecol 30: 323–333.
27. Yamahura Y, Kawahara T, Iida S, Ozaki K (2008) Relative importance of the area and shape to the diversity of multiple taxa. Conserv Biol 22: 1513–1522.
28. Rosati L, Fipaldini M, Marignani M, Blasi C (2010) Effect of forest fragmentation on vascular plant diversity in a Mediterranean forest archipelago. Plant Biosyst 144: 38–46.
29. Montiel S, Estrada A, León P (2006) Bat assemblages in a naturally fragmented ecosystem in the Yucatan Peninsula, Mexico: species richness, diversity and spatio-temporal dynamics. J Trop Ecol 22: 267–276.
30. Barrera A (1982) Los petenes del noroeste de Yucatán: su exploración ecológica en perspectiva. Biótica 2: 163–169.
31. Rico-Gray V (1982) Estudio de la vegetación de la zona costera inundable del noreste del estado de Campeche, México: Los Petenes. Biótica 7: 171–188.
32. CONANP-SEMARNAT (2006) Programa de conservación y manejo de la Reserva de la Biosfera Los Petenes. Mexico City: CONANP. 206 p.
33. Rico-Gray V, Palacios-Rios M (1996) Salinidad y nivel del agua como factores en la distribución de la vegetación en la ciénega del NW de Campeche, México. Acta Bot Mex 34: 53–61.
34. Murcia C (1995) Edge effects in fragmented forest: implications for conservation. Trends Ecol Evol 10: 58–62.
35. Costa-Lugo E, Parra DA, Andrade-Hernández M, Castillo-Tzab D, Chablé-Santos J, Durán R, Espadas C, et al. (2010). Plan de conservación de la ecoregión Petenes-Celestún-Palmar. Mérida, Yucatán: Pronatura. 177 p.
36. Durán R (1987) Descripción y análisis de la estructura y composición de la vegetación de los petenes del noroeste de Campeche, México. Biótica 12: 191–198.
37. Durán R (1987). Lista florística de la región de los petenes Campeche, México. Biótica 12: 199–208.
38. Arroyo-Rodríguez V, Cavender-Bares J, Escobar F, Melo F, Tabarelli M, et al. (2012) Maintenance of tree phylogenetic diversity in a highly fragmented rainforest. J Ecol 100: 702–711.
39. Arroyo-Rodríguez V, Rös M, Escobar F, Melo F, Santos B, Tabarelli M, et al. (2013) Plant β diversity in fragmented rain forest: Testing floristic homogenization and differentiation hypotheses. J Ecol 101: 1499–1458.
40. Pennington TD, Sarukhán J (2005) Arboles tropicales de México: manual para la identificación de las principales especies. Mexico City: UNAM-Fondo de Cultura Económica. 523 pp.
41. Brokaw N, Bonilla N, Knapp S, MacVean A, Ortíz JJ, et al. (2011). Arboles del mundo maya. Mérida, Yucatán: Natural History Museum. 263 pp.
42. Gotelli NJ, Colwell RK (2011) Estimating species richness. In Magurran AE, McGill BJ, editors. Biological diversity: frontiers in measurement and assessment. New York: Oxford University Press. 39–54.
43. Castillo-Campo G, Halffter G, Moreno C (2008) Primary and secondary vegetation patches as contributors to floristic diversity in a tropical deciduous forest landscape. Biodivers Consev 17: 1701–1714.
44. Barragán F, Moreno C, Escobar F, Halffter G, Navarrete D (2011) Negative impacts of human land use on dung beetle functional diversity. PLoS ONE 6: e17976.
45. Crawley MJ (2013) The R book. 2nd ed. Chichester: John Wiley and Sons Ltd. 1051 p.
46. Prugh LR (2009) An evaluation of patch connectivity measures. Ecol Appl 19: 1300–1310.
47. Møller AP, Jennions MD (2002) How much variance can be explained by ecologists and evolutionary biologists? Oecologia 132: 492–500.
48. Cook WM, Kurt TL, Foster BL, Holt RD (2002) Island theory, matrix effects and species richness patterns in habitat fragments. Ecol Lett 6: 619–623.
49. Connor EF, McCoy ED (1979) The statistics and biology of species-area relationship. Am Nat 113: 791–833.
50. Tews J, Brose U, Grimm V, Tielbörger K, Wichmann MC, et al. (2004) Animal species diversity driven by habitat heterogeneity/diversity: the importance of keystone structures. J Biogeogr 31: 79–92.
51. Grace JB (1999) The factor controlling species density in herbaceous plant communities: an assessment. Perspect Plant Ecol Evol Syst 2: 1–28.

22

Phylogenetic Impoverishment of Amazonian Tree Communities in an Experimentally Fragmented Forest Landscape

Bráulio A. Santos[1]*, Marcelo Tabarelli[2], Felipe P. L. Melo[2], José L. C. Camargo[3], Ana Andrade[3], Susan G. Laurance[4], William F. Laurance[4]

1 Departamento de Sistemática e Ecologia, Universidade Federal da Paraíba, João Pessoa, Paraíba, 58051-900, Brazil, **2** Departamento de Botânica, Universidade Federal de Pernambuco, Recife, Pernambuco, 50670-901, Brazil, **3** Biological Dynamics of Forest Fragment Project, National Institute for Amazonian Research (INPA), C.P. 478, Manaus, Amazonas, 69011-970, Brazil, **4** Centre for Tropical Environmental and Sustainability Science (TESS) and School of Marine and Tropical Biology, James Cook University, Cairns, Queensland, 4878, Australia

Abstract

Amazonian rainforests sustain some of the richest tree communities on Earth, but their ecological and evolutionary responses to human threats remain poorly known. We used one of the largest experimental datasets currently available on tree dynamics in fragmented tropical forests and a recent phylogeny of angiosperms to test whether tree communities have lost phylogenetic diversity since their isolation about two decades previously. Our findings revealed an overall trend toward phylogenetic impoverishment across the experimentally fragmented landscape, irrespective of whether tree communities were in 1-ha, 10-ha, or 100-ha forest fragments, near forest edges, or in continuous forest. The magnitude of the phylogenetic diversity loss was low (<2% relative to before-fragmentation values) but widespread throughout the study landscape, occurring in 32 of 40 1-ha plots. Consistent with this loss in phylogenetic diversity, we observed a significant decrease of 50% in phylogenetic dispersion since forest isolation, irrespective of plot location. Analyses based on tree genera that have significantly increased (28 genera) or declined (31 genera) in abundance and basal area in the landscape revealed that increasing genera are more phylogenetically related than decreasing ones. Also, the loss of phylogenetic diversity was greater in tree communities where increasing genera proliferated and decreasing genera reduced their importance values, suggesting that this taxonomic replacement is partially underlying the phylogenetic impoverishment at the landscape scale. This finding has clear implications for the current debate about the role human-modified landscapes play in sustaining biodiversity persistence and key ecosystem services, such as carbon storage. Although the generalization of our findings to other fragmented tropical forests is uncertain, it could negatively affect ecosystem productivity and stability and have broader impacts on coevolved organisms.

Editor: Xiao-Yong Chen, East China Normal University, China

Funding: The Instituto Nacional de Pesquisas da Amazônia (INPA), Smithsonian Institution, US National Science Foundation, Conselho Nacional de Desenvolvimento Científico e Tecnológico (CNPq), NASA-LBA program, US-AID, Mellon Foundation, Blue Moon Fund, and Marisla Foundation generously supported the Biological Dynamics of Forest Fragment Project. BAS, FPLM and MT are thankful to CNPq for postdoctoral fellowship and research grants. The funders had no role in study design, data collection and analysis, decision to publish, or preparation of the manuscript.

Competing Interests: The authors have declared that no competing interests exist.

* Email: braulio@dse.ufpb.br

Introduction

The current biodiversity crisis driven by human-induced species loss is likely to drastically alter the tree of life [1,2], but this is still poorly understood due to the historical gap between ecology and phylogenetics [3]. Fortunately, this gap is shrinking as evolutionary relationships among species can be incorporated into ecological studies [4–6]. Phylogenetic information, besides providing an additional measure of biodiversity beyond traditional measures of species richness and diversity, is useful to infer community assembly processes and ecosystem stability [7–9].

In the context of conservation, community phylogenetics is a powerful tool to understand how biological communities respond to human disturbances of varying type, intensity, and frequency [10]. Important advances in this field are helping us to understand the effects of fires on community assembly [11–13], the processes involved in community organization during forest regeneration [14,15], and the outcome of species loss and gain on community phylogenetic diversity [16,17]. However, we are still far from drawing general conclusions about the organization and persistence of biodiversity in our rapidly changing world [18].

Several studies have revealed shifts in the taxonomic and functional profile of tropical tree communities in response to habitat loss and fragmentation [19–22]. Compared to continuous forest, tree communities near forest edges and in small forest fragments (hereafter edge-affected habitats), may support only a small fraction of (1) emergent and understory tree species; (2) heavy-wooded, shade-tolerant and slow-growing trees; (3) large-seeded trees dispersed by medium to large-bodied frugivores; and (4) tree species pollinated by specialized vectors and those bearing supra-annual reproduction (see Tabarelli et al. [23] and references therein). Long-term tree monitoring in Central Amazonia reveals that edge-affected habitats are hyperdynamic systems characterized by increased rates of tree recruitment and mortality, especially in the first few decades following edge creation [24]. This hyperdynamism is guided by the remarkable recruitment of fast-growing, early-to-mid successional trees that replace slow-growing late-successional ones [19,25,26], which are most probably killed by increased wind turbulence or desiccation near fragment edges [27,28]. Similarly, in the Atlantic forest of northeast Brazil, a small number of native, fast-growing successional species proliferate from forest edges to entire landscapes [21,29], leading to a sort of regional biotic homogenization [23,30].

However, important questions remain before we can properly predict the destiny of tropical tree communities in human-modified landscapes. Two of them are addressed here: (1) how much evolutionary diversity is being lost in edge-affected habitats along with the taxonomic and functional impoverishment? (2) what is the role of human-induced biotic homogenization on the phylogenetic diversity of remaining communities? A recent study on secondary forests of Costa Rica [31] demonstrated that disturbance-favored early-successional tree species are more phylogenetically related than expected by chance, whereas late-successional species tend to be less related than random expectations drawn from the same regional phylogeny. If these trends hold for Amazonian tree species, we could expect an increase in phylogenetic clustering or a decrease in phylogenetic evenness in the edge-affected habitats, where fast-growing trees are thriving and replacing disturbance-sensitive ones.

In this paper we used one of the largest datasets currently available on tree dynamics in a 1000-km^2 experimentally fragmented Amazonian forest landscape and a recent phylogeny of angiosperms to assess whether tree communities have lost phylogenetic diversity and become phylogenetically clustered since their isolation two decades previously. First, we examined the effects of forest site (1-ha, 10-ha, and 100-ha fragments and continuous forest), time since forest isolation, and distance to nearest forest edge on richness of tree genera, phylogenetic community diversity and structure. Then, we evaluated potential relationships between changes in phylogenetic diversity and the proliferation or declining of 59 tree genera that have significantly changed in importance values since forest isolation. Finally, we estimated the phylogenetic distribution of declining and increasing tree genera and discuss the implications of our results for the ecology and conservation of tropical tree communities in human-modified landscapes.

Methods

Study landscape

The Biological Dynamics of Forest Fragments Project is a 1,000-km^2 experimental landscape in Central Amazonia. Within this landscape, nine forest fragments ranging from 1 to 100 ha in area were isolated from nearby intact forest during the early 1980s by clearing and burning the intervening vegetation to create cattle pastures. Some of the pastures have been abandoned and now support secondary forests. Detailed descriptions of the project, including its study design, fragment histories, the matrices of modified vegetation surrounding fragments, and the methods used for censusing and identifying trees are provided elsewhere [24].

Before fragment isolation, permanent 1-ha forest plots were established within each fragment and in eight comparable sites in nearby intact forest. The present study incorporates tree demography data from 40 1-ha plots distributed throughout the entire landscape, either along forest edges (plot center <100 m from the nearest edge) or in 'intact' forest interiors (170–3,000 m from the edge). After an initial, exhaustive inventory of tree communities, each plot was resampled after fragmentation at typical intervals of 4–6 years to assess tree mortality, growth, and the recruitment of new trees. Altogether, the fates of nearly 32,000 trees (dbh> 10 cm) were followed for periods of up to 18 years (mean 14.7 years) [19]. All the tree specimens from the different censuses were identified together by collecting a voucher specimen for virtually every single tree. In the herbarium, all the voucher specimens were pooled into families (the initial family identification was made in the field) and then each specimen was identified to species or genus/morphospecies level.

In this study we used the same genus-level dataset presented in Laurance et al. [19,32], which were mainly interested in describing broader taxonomic and ecological responses of tree communities to human disturbance in 40 1-ha BDFFP plots. We are aware that examining phylogenetic community structure at the genus level has shortcomings when compared with species-level analyses, as we cannot assume that shifts in tree genera are a proxy for shifts in tree species. However, in the absence of robust species-level phylogeny, the genus-level approach provides reliable information about lineages and can shed light on the structure of hyperdiverse communities such as ours. This approach has been used in other studies, including ant [33] and bacteria communities [34], and contributed considerably for the development of phylogenetic community ecology.

Shifts in phylogenetic community structure and diversity

To evaluate shifts in tree communities we first listed all genera observed in the 40 1-ha plots. Altogether, 267 genera belonging to 62 families were recorded. This genera list was then assembled into a phylogeny (i.e. the regional phylogeny) using the PHYLOMATIC function of Phylocom 4.2 [35] and the maximally resolved supertree of angiosperms R20100701, available for free in the software. The regional phylogeny had branch lengths estimated with the BLADJ algorithm. For this, we used the node ages provided by Bell et al. [36] and further corrected for inconsistencies in syntax and nomenclature of internal nodes and the regional phylogeny following procedures described in Gastauer and Meira-Neto [37]. The resulting time-calibrated regional phylogeny was used in the subsequent analyses (see Appendix S1).

To examine shifts in phylogenetic community structure and diversity within the landscape, we calculated two intra-sample phylogenetic metrics for each community: mean phylogenetic distance (MPD) and net related index (NRI) (see Webb et al. [35] for a complete description of the metrics and Table S1 for raw data of tree plots). We used the argument '-a' of COMSTRUCT function in Phylocom 4.2 to take into account the importance value of each taxon, estimated by the combination of basal area and relative abundance as described in Laurance et al. [19]. To be conservative, we avoided using metrics more sensitive to phylogenetic resolution at the tips of the phylogeny, such as MNTD and NTI, because the evolutionary relationships of many tropical genera within families are not resolved yet.

We treated MPD as the metric of phylogenetic diversity and NRI as the metric of phylogenetic structure. MPD was expressed in million years and represented the observed phylogenetic diversity of a given community. NRI, the measure of 'standarzided effect size' of phylogenetic community structure, was calculated comparing the observed MPD to 999 null communities generated by null model 0 of Phylocom 4.2. This model shuffles species labels across the entire phylogeny [35] and is more appropriate for temporal analyses than other null models (e.g. Phylocom model 2) because it maintains plot abundance distribution, plot species richness, occupancy rates, and conserves the spatial contagion of species (see Norden et al. [31], p.S74, for more details on the use of this null model for temporal analyses).

We then made before vs. after comparisons to assess whether average levels of tree-wide phylogenetic structure changed over time across forest sites. Positive trends of NRI indicate decrease in phylogenetic evenness or increase in phylogenetic clustering since forest isolation, whereas negative trends indicate the opposite. Because communities with similar values of NRI can differ in terms of absolute evolutionary diversity, we made similar comparisons using the observed values of MPD to estimate potential shifts in community phylogenetic diversity.

Increasing and decreasing tree genera

Two previous studies [19,32] identified 28 tree genera that are significantly increasing and 31 that are significantly decreasing in importance value throughout the landscape, totaling 59 genera that respond positively or negatively to human disturbance in demographic terms (see Table S2). The bulk of increasing genera is composed by fast-growing, small-seeded species typical of early to mid successional stages; 15 are thriving in edge-affected habitats [19] and 13 in the continuous forest [32]. The group of decreasing genera is mostly represented by slow-growing, large-seeded species typical of old-growth forests; 18 are declining in edge-affected habitats, six are reducing in the continuous forests and seven are declining in both habitats [19,32].

To assess the relationship between the shifts in phylogenetic diversity and the proliferation or declining of these particular genera, we calculated for each tree community the relative change in MPD and mean importance value of the 59 tree genera that have significantly responded to human disturbance. The relative changes in MPD and mean importance value of genera were calculated as the difference between the final and the initial value, divided by the initial value. Negative relative changes in MPD indicate that tree community loses phylogenetic diversity over time, while positive changes indicate the opposite. If shifts in phylogenetic diversity were influenced by the proliferation of increasing genera, we would expect positive relationship between the gain of increasing genera and the loss of phylogenetic diversity. A similar positive relationship could arise if the negative shifts in MPD were affected by the loss of decreasing genera.

Phylogenetic distribution of increasing and decreasing genera

To assess the phylogenetic distribution of declining and increasing tree genera, we estimated the degree of phylogenetic relationships for each group using NRI, as adopted elsewhere [31]. Both groups of genera, one containing 28 increasing genera and other 31 decreasing genera, were assumed to be a community drawn from the 267-genera regional pool, resulting in a value of NRI for group. According to the findings of Norden et al. [31], we expected that the group of increasing genera was more phylogenetically related than the group of decreasing ones (i.e. $NRI_{increasing} > NRI_{decreasing}$).

Statistical analyses

We used linear mixed models to test for the effects of forest site, time since forest isolation and distance to the nearest forest edge on the richness of tree genera and phylogenetic community metrics. We set forest site (four levels: 1-ha [$n = 4$], 10-ha [$n = 9$], 100-ha forest fragments [$n = 10$] and continuous forest [$n = 17$]), time since forest isolation (first and last census), edge distance and the interaction between forest site×time as fixed effects. We included this interaction term to assess whether the magnitude of the shifts in phylogenetic metrics was greater in the smaller forest fragments (1-ha and 10-ha), where tree communities have changed suddenly since forest isolation [19]. Forest plot ($N = 40$), the subject in which tree censuses were done, was set as the random effect. We adopted the residual maximum likelihood method to separate the variance of fixed and random effects [38]. We used Pearson product-moment correlations to assess the relationship between the relative changes in MPD and mean importance value of decreasing and increasing genera. All analyses were performed in JMP 8 (SAS Institute Inc.).

Results

Genera richness decreased significantly from 116.2±1.2 genera/ha (mean ± SE) before fragmentation to 113.5±1.4 genera/ha after fragmentation (time effect in Table 1), but this reduction occurred regardless of whether plot was in forest fragments or continuous forest (interaction term not significant; Table 1). We observed a similar trend for MPD, which decreased slightly (<2%) but significantly over time and irrespective to plot location (Table 1; Fig. 1). Surprisingly, shifts in phylogenetic structure and diversity were not related with plot distance to the nearest edge (Table 1).

Thirty-two out of 40 communities (80%) experienced a decrease in MPD over the period of tree monitoring, as indicated by negative rates of change in MPD (Fig. 2). In these impoverishing tree communities, the loss of phylogenetic diversity was positively correlated with the increase in importance value of increasing genera ($r = 0.38$; $P < 0.05$) and marginally correlated with the reduction in importance value of decreasing genera ($r = 0.34$; $P = 0.059$), suggesting that both the proliferation of increasing genera and the reduction of decreasing genera are partially underlying the loss of phylogenetic diversity at the study landscape.

Consistent with the reduction in genera richness and MPD, NRI increased after forest isolation, resulting in a 50%-decrease in tree-wide phylogenetic evenness of tree communities throughout the landscape ($NRI_{initial} = -0.33$ vs. $NRI_{final} = -0.17$). Furthermore, as expected, the group of 28 tree genera thriving in the landscape was more phylogenetically clustered than that of 31 declining genera ($NRI_{increasing} = 1.56$ vs. $NRI_{decreasing} = 0.09$).

Discussion

In the last decades, the scientific community has invested immense efforts to anticipate the fate of tropical biodiversity as it becomes increasingly threatened by human-disturbances, such as habitat fragmentation, agricultural expansion and global warming [39,40]. Much has been learnt from either theoretical models or empirical studies, at least at the species and community level, but the scarcity of long-term data and lack of reliable evolutionary information still limit our ability to understand the extent to which human-modified landscapes can operate as a repository for the tropical evolutionary heritage (see references [41–45] for contrasting views).

The present study bridges part of this knowledge gap by documenting shifts in the phylogenetic structure and diversity of tree communities inhabiting multiple sites in a human-modified landscape (see also Arroyo-Rodríguez et al. [46]). Despite the genus-level analyses and the great variation in the phylogenetic metrics across plots and sites, tree communities experienced significant reduction in tree genera richness, loss of phylogenetic

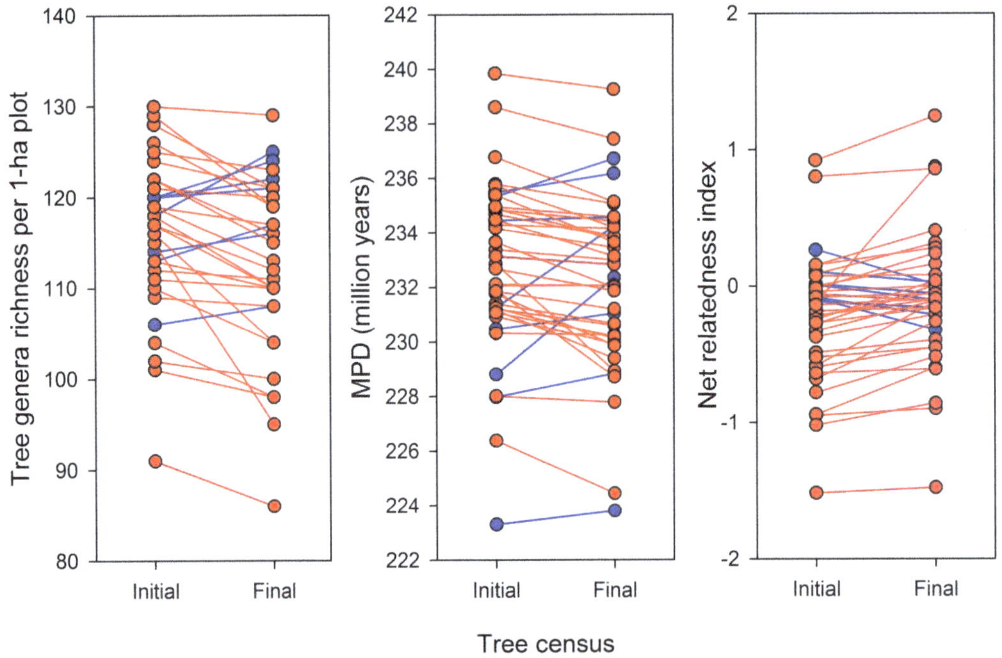

Figure 1. Tree genera richness and phylogenetic metrics of 40 1-ha tree communities in forest fragments and continuous forest in a 1000-km^2 forest landscape of Central Amazonia, Brazil. Initial tree census refers to the first tree inventory carried out before forest isolation in early 1980's; final tree census refers to the last census available in our dataset (on average 14 years after the first census, see Methods). Red lines indicate loss of tree genera and phylogenetic diversity over time.

diversity and reduction in phylogenetic evenness over time throughout the entire landscape. Such impoverishment was correlated with demographic shifts of particular genera, suggesting that the replacement of less related, slow-growing tree taxa by more related, fast-growing ones is partially underlying the erosion of evolutionary heritage at the landscape scale. Considering the long lifespan usually exhibited by tropical tree species, potentially exceeding several centuries [47], it is alarming that such shifts have already emerged within an experimental landscape exposed to habitat loss and fragmentation in the early 1980s.

Identifying the causes for the demographic and phylogenetic shifts we documented in this human-modified landscape is complex, especially across the supposed 'intact' continuous forest plots located from 170–3000 m from the nearest forest edge. A

Table 1. Fixed effects of linear mixed models fitted for repeated measures of tree community attributes in a 1000-km^2 forest landscape of Central Amazonia, Brazil.

Model terms	**DF**	***F*-ratio**	***P*-value**	**Model Adj R^2(%)**
Genera richness				
Site	3,35	2.04	0.126	91.9
Time	1,36	4.69	**0.037**	
Site × Time	3,36	1.76	0.172	
Edge distance	1,35	3.18	0.083	
MPD				
Site	3,35	2.46	0.079	94.2
Time	1,36	9.48	**0.004**	
Site × Time	3,36	1.39	0.261	
Edge distance	1,35	0.23	0.629	
NRI				
Site	3,35	3.59	0.023	93.3
Time	1,36	17.52	**<0.001**	
Site × Time	3,36	0.85	0.476	
Edge distance	1,35	0.14	0.711	

Site is represented by four levels: continuous forest, 1-ha, 10-ha, and 100-ha forest fragments. $N = 40$ tree communities, 1-ha each.

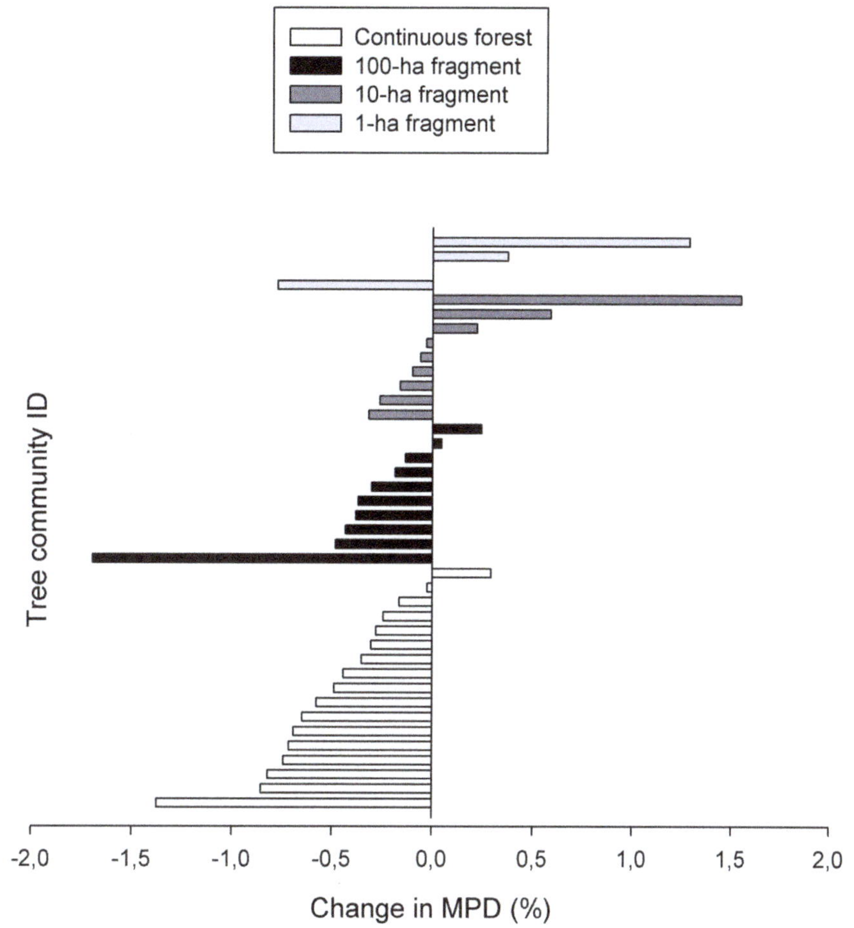

Figure 2. Relative change in mean phylogenetic distance across 40 tree communities in a 1000-km² forest landscape of Central Amazonia, Brazil. Negative values indicate loss of phylogenetic diversity over time.

possibility is that these forest plots could be in a state of disequilibrium recovering from a large-scale past disturbance, leading to time-dependent shifts in community membership and phylogenetic community structure [14,15,31]. Major forest fires has been recognized as the only natural disturbance likely to operate at this scale [32], but charcoal and phytolith data suggest that the study area has been continuously forested for at least 4,500 years [48]. Another possibility is that old-growth forests would be responding to multi-decadal changes in rainfall, which affects forest productivity, species composition and possibly phylogenetic diversity. However, no trends in Central Amazonia rainfall were evident for the 1984–1999 period [32] or across preceding decades [49]. What has been documented so far is that our continuous forest plots are experiencing elevated mortality and turnover rates [50], leading to a significant decline in plot species richness [51] in parallel with the proliferation of fast-growing species [32], which also occur across all forest habitats [19,26]. It has been hypothesized that the elevated mortality and recruitment are a consequence of rising atmospheric CO_2 concentrations, which might stimulate plant growth and competition [32].

Independent findings in the Amazon and Brazilian Atlantic forests have provided evidence for a nonrandom taxonomic and functional impoverishment of tree communities inhabiting edge-affect habitats from local to regional scales [19–21,25], largely due to shifts in physical conditions imposed by the creation of forest edges [25,52]. Such a floristic/functional drift towards communities dominated by few disturbance-adapted fast-growing species ultimately results in increasing levels of biotic homogenization [23,30] and loss of phylogenetic diversity along forest edges [16]. Although the phylogenetic shifts we documented in the study landscape were not related with plot distance to the nearest forest edge, it is not surprising that (1) tree communities experience phylogenetic diversity loss, clustering and possibly homogenization across multiple habitats and spatial scales, and (2) to some extent the evolutionary heritage of sensitive tropical forest biota will not be retained across human-modified landscapes [24].

Nonetheless, we cannot definitively discard the hypothesis that our findings are simply the outcome of stochastic variation in forest dynamics within the study landscape. This possibility is supported by the 17 reference plots in the continuous forest, which also showed significant temporal changes in genera richness and phylogenetic structure and diversity. We would need even longer tree monitoring, possibly over a century, to properly identify the causes of the taxonomic and ecological shifts that ultimately lead to the phylogenetic shifts at that spatial scale. Also, the long-term magnitude of tree community responses to habitat loss and fragmentation remains uncertain, precluding efforts to scale-up conclusions for other tropical forests. It has been proposed that taxonomic, functional and phylogenetic impoverishment of tree communities across multiple spatial scales are more likely to occur among those biota that have evolved and diversified in the presence of less intensive natural or human-related disturbance regimes, such as in Central Amazonia and the Brazilian Atlantic forests [53–55]. In contrast, forests that evolved under intense volcanic activity, frequent hurricanes, and extreme climates, such as some Mexican and Costa Rican forests, support a higher proportion of disturbance-tolerant species than do South America forests, and thus may be less susceptible to human disturbances [56–58]. Although the generality of the phylogenetic impoverishment we detected is unclear, it could negatively affect ecosystem productivity and stability and have broader impacts on coevolved organisms [8,9]. Further studies should examine the generality of both patterns and processes proposed here.

Acknowledgments

We thank the many taxonomic specialists and field technicians who have collected and identified plant material during the study. We are grateful to Víctor Arroyo-Rodríguez and an anonymous reviewer for valuable comments. This is publication number 652 in the BDFFP technical series.

Author Contributions

Conceived and designed the experiments: BAS WFL. Performed the experiments: JLCC AA SGL. Analyzed the data: BAS MT FPLM. Contributed reagents/materials/analysis tools: BAS. Wrote the paper: BAS MT FPLM JLCC SGL WFL.

References

1. Purvis A, Agapow P, Gittleman JL, Mace GM (2000) Nonrandom extinction and the loss of evolutionary history. Science 288: 328–330.
2. Vamosi JC, Wilson JRU (2008) Nonrandom extinction leads to elevated loss of angiosperm evolutionary history. Ecol Lett 11: 1047–1053.
3. Webb CO, Ackerly DD, McPeek MA, Donoghue MJ (2002) Phylogenies and community ecology. Annu Rev Ecol Syst 33: 475–505.
4. Vamosi SM, Heard SB, Vamosi C, Webb CO (2009) Emerging patterns in the comparative analysis of phylogenetic community structure. Mol Ecol 18: 572–592.
5. Pausas JG, Verdu M (2010) The jungle of methods for evaluating phenotypic and phylogenetic structure of communities. Bioscience 60: 614–625
6. Cavender-Bares J, Ackerly DD, Kozak KH (2012) Integrating ecology and phylogenetics: the footprint of history in modern-day communities. Ecology 93(Supplement): S1–S3.
7. Magurran AE (2004) Measuring biological diversity. Oxford: Blackwell Science. 264 p.
8. Cavender-Bares J, Kozak KH, Fine PVA, Kembel SW (2009) The merging of community ecology and phylogenetic biology. Ecol Lett 12: 693–715.
9. Cadotte MW, Dinnage R, Tilman D (2012) Phylogenetic diversity promotes ecosystem stability. Ecology 93(Supplement): S223–S233.
10. Helmus MR, Keller W, Paterson MJ, Yan ND, Cannon CH, et al (2010) Communities contain closely related species during ecosystem perturbance. Ecol Lett 13: 162–174.
11. Verdu M, Pausas JG (2007) Fire drives phylogenetic clustering in Mediterranean Basin woody plant communities. J Ecol 95: 1316–1323.
12. Cavender-Barres J, Reich PB (2012) Shocks to the system: community assembly of the oak savanna in a 40-yearfire frequency experiment. Ecology 93(Supple-(Supplement): S52–S69.
13. Cianciaruso MV, Silva IA, Batalha MA, Gaston KJ, Petchey OW (2012) The influence of fire on phylogenetic and functional structure of woody savannas: Moving from species to individuals. Perspect Plant Ecol Evol Syst 14: 205–216.
14. Letcher SG (2010) Phylogenetic structure of angiosperm communities during tropical forest succession. Proc R Soc B 277: 97–104.
15. Letcher SG, Chazdon RL, Andrade ACS, Bongers F, Breugelf M, et al (2012) Phylogenetic community structure during succession: Evidence from three Neotropical forest sites. Perspect Plant Ecol Evol Syst 14: 79–87.
16. Santos BA, Arroyo-Rodríguez V, Moreno CE, Tabarelli M (2010) Edge-related loss of tree phylogenetic diversity in the severely fragmented Brazilian Atlantic forest. Plos One 5: e12625. doi:101371/journalpone0012625
17. Cadotte MW, Strauss SY (2011) Phylogenetic patterns of colonization and extinction in experimentally assembled plant communities. Plos One 6: e19363. doi:101371/journalpone0019363
18. Winter M, Devictor V, Schweiger O (2013) Phylogenetic diversity and nature conservation: where are we? Trends Ecol Evol 28: 199–204.
19. Laurance WF, Nascimento HEM, Laurance SGW, Andrade A, Ribeiro JELS, et al (2006) Rapid decay of tree-community composition in Amazonian forest fragments. Proc Natl Acad Sci USA 103: 19010–19014.
20. Michalski F, Nishi I, Peres CA (2007) Disturbance-mediated drift in tree functional groups in Amazonian forest fragments. Biotropica 39: 691–701.
21. Santos BA, Peres CA, Oliveira MA, Grillo A, Alves-Costa CP, et al (2008) Drastic erosion in functional attributes of tree communities in Atlantic forest fragments of northeastern Brazil. Biol Conserv 141: 249–260.
22. Lopes AV, Girão LC, Santos BA, Peres CA, Tabarelli M (2009) Long-term erosion of tree reproductive trait diversity in edge-dominated Atlantic forest fragments. Biol Conserv 142: 1154–1165.
23. Tabarelli M, Peres CA, Melo FPL (2012) The 'few winners and many losers' paradigm revisited: Emerging prospects for tropical forest biodiversity. Biol Conserv 155: 136–140.
24. Laurance WF, Camargo JLC, Luizão RCC, Laurance SGW, Pimm SL, et al (2011) The fate of Amazonian forest fragments: A 32-year investigation. Biol Conserv 144: 56–67.
25. Laurance WF, Nascimento HEM, Laurance SGW, Andrade A, Fearnside PM, et al (2006) Rain forest fragmentation and the proliferation of successional trees. Ecology 87: 469–482.
26. Santos GGA, Santos BA, Nascimento HEM, Tabarelli M (2012) Contrasting demographic structure of short- and long-lived pioneer tree species on Amazonian forest edges. Biotropica 44: 771–778.
27. D'Ângelo SA, Andrade ACS, Laurance SGW, Laurance WF, Mesquita RCG (2004) Inferred causes of tree mortality in fragmented and intact Amazonian forests. J Trop Ecol 20: 243–246.
28. Briant G, Gond V, Laurance SGW (2010) Habitat fragmentation and the desiccation of forest canopies: A case study from eastern Amazonia. Biol Conserv 143: 2763–2769.
29. Tabarelli M, Aguiar AV, Girão LC, Peres CA, Lopes AV (2010) Effects of pioneer tree species hyperabundance on forest fragments in northeastern Brazil. Conserv Biol 24: 1654–1663.
30. Lôbo D, Leão T, Melo FPL, Santos AMM, Tabarelli M (2011) Forest fragmentation drives Atlantic Forest of northeastern Brazil to biotic homogenization. Divers Distrib 17: 287–296.
31. Norden N, Letcher SG, Boukili V, Swenson NG, Chazdon R (2012) Demographic drivers of successional changes in phylogenetic structure across life-history stages in plant communities. Ecology 93 (Supplement): S70–S82.
32. Laurance WF, Oliveira AA, Laurance SGW, Condit R, Nascimento HEM, et al (2004) Pervasive alteration of tree communities in undisturbed Amazonian forests. Nature 428: 171–175.
33. Lessard J-P, Fordyce JA, Gotelli NJ, Sanders NJ (2009) Invasive ants alter the phylogenetic structure of ant communities. Ecology 90: 2664–2669.
34. Youssef NH, Ashlock-Savage KN, Elshahed MS (2012) Phylogenetic diversities and community structure of members of the extremely halophilic Archaea (order Halobacteriales) in multiple saline sediment habitats. Appl Environ Microbiol 78: 1332–1344.
35. Webb CO, Ackerly DD, Kembel SW (2008) Phylocom: software for the analysis of phylogenetic community structure and trait evolution. Bioinformatics 24: 2098–2100.
36. Bell CD, Soltis DE, Soltis PS (2010) The age and diversification of the angiosperms re-revisited. Am J Bot 97: 1296–1303.
37. Gastauer M, Meira-Neto JAA (2013) Avoiding inaccuracies in tree calibration and phylogenetic community analysis using Phylocom 4.2. Ecol Inform 15: 85–90.
38. Grafen A, Hails R (2002) Modern Statistics for the Life Sciences. Oxford: Oxford University Press. 368 p.
39. Laurance WF, Peres CA (2006) Emerging threats to tropical forests. Chicago: University of Chicago Press. 520 p.
40. Laurance WF, Sayer J, Cassman KG (2014) Agricultural expansion and its impacts on tropical nature. Trends Ecol Evol 29: 107–116.

41. Wright SJ, Muller-Landau N (2006) The future of tropical forest species. Biotropica 38: 287–301.
42. Laurance WF (2007) Have we overstated the tropical biodiversity crisis? Trends Ecol Evol 22: 65–70.
43. Tabarelli M, Lopes AV, Peres CA (2008) Edge-effects drive tropical forest fragments towards an early-successional system. Biotropica 40: 657–661.
44. Chazdon RL, Peres CA, Dent D, Sheil D, Lugo AE, et al (2009) The potential for species conservation in tropical secondary forests. Conserv Biol 23: 1406–1417.
45. Melo FPL, Arroyo-Rodríguez V, Fahrig L, Martínez-Ramos, Tabarelli M (2013) On the hope for biodiversity-friendly tropical landscapes. Trends Ecol Evol 28: 461–468.
46. Arroyo-Rodríguez V, Cavender-Bares J, Escobar F, Melo FPL, Tabarelli M, et al (2012) Maintenance of tree phylogenetic diversity in a highly fragmented rain forest. J Ecol 100: 702–711.
47. Martínez-Ramos M, Alvarez-Buylla ER (1998) How old are tropical rain forest trees? Trends Plant Sci 3: 400–405.
48. Piperno DR, Becker P (1996) Vegetation history of a site in the Central Amazon Basin derived from phytolith and charcoal records from natural soils. Quat Int 45: 202–209.
49. Mahli Y, Wright J (2004) Spatial patterns and recent trends in the climate of tropical rainforest regions. Philos Trans R Soc Lond B Biol Sci 359: 311–329.
50. Laurance SGW, Nascimento HEM, Laurance WF, Andrade A, Fearnside P, et al (2009) Rapid change in Amazonian forest dynamics: effects of climate change? J Veg Sci 20: 323–333.
51. Laurance SGW, Andrade A, Laurance WF (2010) Unanticipated effects of stand dynamism on Amazonian tree diversity. Biotropica 42: 429–434.
52. Oliveira MA, Santos AMM, Tabarelli M (2008) Profound impoverishment of the large-tree stand in a hyper-fragmented landscape of the Atlantic forest. For Ecol Manage 256: 1910–1917.
53. Barlow J, Gardner TA, Lees AC, Parry L, Peres CA (2012) How pristine are tropical forests? An ecological perspective on the pre-Columbian human footprint in Amazonia and implications for contemporary conservation. Biol Conserv 151: 45–49.
54. Arroyo-Rodríguez V, Rös M, Escobar F, Melo FPL, Santos BA, et al (2013) Plant β-diversity in fragmented rain forests: testing floristic homogenization and differentiation hypotheses. J Ecol 101: 1449–1458.
55. Arroyo-Rodríguez V, Pineda E, Escobar F, Benítez-Malvido J (2009) Value of small patches in the conservation of plant-species diversity in highly fragmented rainforest. Conserv Biol 23: 729–739.
56. Borouncle C, Finegan B (2011) Tree regeneration and understory woody plants show diverse responses to forest–pasture edges in Costa Rica. Biotropica 43: 562–571.
57. Rös M, Escobar F, Halffter G (2011) How dung beetles respond to a human-modified variegated landscape in Mexican cloud forest: a study of biodiversity integrating ecological and biogeographical perspectives. Divers Distrib 18: 377–389.
58. Hernández-Ruedas MA, Arroyo-Rodríguez V, Meave JA, Martínez-Ramos M, Ibarra-Manríquez G, et al. (2014) Conserving Tropical Tree Diversity and Forest Structure: The Value of Small Rainforest Patches in Moderately-Managed Landscapes. Plos One 9: e98931. doi:10.1371/journal.pone.0098931

23

The Soil Biota Composition along a Progressive Succession of Secondary Vegetation in a Karst Area

Jie Zhao[1,2]*, Shengping Li[1,2,3]*, Xunyang He[1,2], Lu Liu[1,2,3], Kelin Wang[1,2]*

1 Key Laboratory of Agro-ecological Processes in Subtropical Region, Institute of Subtropical Agriculture, Chinese Academy of Sciences, Changsha, Hunan, China, 2 Huanjiang Observation and Research Station for Karst Ecosystems, Chinese Academy of Sciences, Huanjiang, Guangxi, China, 3 Graduate University of Chinese Academy of Sciences, Beijing, China

Abstract

Karst ecosystems are fragile and are in many regions degraded by anthropogenic activities. Current management of degraded karst areas focuses on aboveground vegetation succession or recovery and aims at establishing a forest ecosystem. Whether progressive succession of vegetation in karst areas is accompanied by establishment of soil biota is poorly understood. In the present study, soil microbial and nematode communities, as well as soil physico-chemical properties were studied along a progressive succession of secondary vegetation (from grassland to shrubland to forest) in a karst area in southwest China. Microbial biomass, nematode density, ratio of fungal to bacterial biomass, nematode structure index, and nematode enrichment index decreased with the secondary succession in the plant community. Overall, the results indicated a pattern of declines in soil biota abundance and food web complexity that was associated with a decrease in soil pH and a decrease in soil organic carbon content with the progressive secondary succession of the plant community. Our findings suggest that soil biota amendment is necessary during karst ecosystem restoration and establishment and management of grasslands may be feasible in karst areas.

Editor: Dafeng Hui, Tennessee State University, United States of America

Funding: This study was supported by the Chinese Academy of Sciences Action Plan for the Development of Western China (KZCX2-XB3-10), the National Key Technology Research and Development Program of the Ministry of Science and Technology of China (2010BAE00739-02) given to Kelin Wang. The study was also supported by a Western Light Program from CAS and a NSFC program (31300448) given to Jie Zhao. The funders had no role in study design, data collection and analysis, decision to publish, or preparation of the manuscript.

Competing Interests: The authors have declared that no competing interests exist.

* Email: kelin@isa.ac.cn

* These authors contributed equally to this work.

Introduction

Karst landscapes account for about 12% of the Earth's land surface [1]. Ecosystems in karst areas are fragile and suffering extreme degradation in many regions of the world, such as southwest China, Southeast Asia (e.g., Vietnam and Thailand), and Central America (e.g., Mexico, Puerto Rico, Dominican Republic, and Cuba) [2–7]. Karst areas in southern and central Europe (e.g., Austria, Bosnia, Herzegovina, Croatia, Germany, Italy, Russia, and Spain [not all land cover of these countries are karst but there are karst areas in them]) also are facing significant land degradations [3]. One important example of this degradation is "karst rocky desertification", which is characterized in part by shallow soils and exposed bedrock [1]. Karst rocky desertification results largely from anthropogenic activities such as irrational cultivation and extensive logging. Since the 1980s, karst rocky desertification has been recognized as a critical factor limiting the social and economic development of the karst region in southwest China [8–9].

The Chinese central and local governments have promoted the ecological restoration of karst areas in southwest China, and as a consequence many farmlands and plantations have been abandoned and allowed to recover naturally. Several studies have documented the succession of plant communities in the karst region in southwest China. For example, Zeng et al. (2007) identified the following six plant communities occurring during succession following the natural revegetation of areas suffering from rocky desertification: sparse grass, grass, shrub, liana-shrub, deciduous broad-leaved forest, and mixed evergreen and deciduous broad-leaved forest [8]. An earlier study had also identified six vegetation types: grass, grass-shrub, shrub, shrub-tree, tree forest, and climax mixed evergreen and deciduous broad-leaved forest [10]. Although the classification of the plant communities is somewhat different in these two studies, the succession from grass to shrub to forest is considered typical in this area [11–12].

The restoration of degraded karst areas in southwest China is currently evaluated based on the degree of forest vegetation establishment but the restoration of ecosystem services will also depend on the recovery of other organisms including soil organisms. Whether a secondary vegetation succession can promote the increment of biomass and abundance of soil biota in degraded karst areas is largely unknown, and this lack of knowledge limits our ability to understand and predict the functioning of such ecosystems.

In studying the succession of plant communities and soil biota, researchers frequently use space-for-time substitution, i.e., chronosequences [13]. In addition, researchers recognize two general types of succession, which are primary and secondary. Secondary succession is widespread and occurs at a high rate relative to primary succession [14]. In the past two decades, a number of studies have explored the relationship between secondary vegetation succession and soil biota. For example, vegetation succession can influence the composition of soil biota through resource and habitat changes [15–17], and the soil biota can affect the vegetation succession by selective feeding and by altering nutrient availability [18–19]. Few studies, however, have explored the changes in soil biota (especially nematodes) during secondary succession from grassland to other ecosystem types, and these studies have yielded inconsistent results concerning the relationship between vegetation succession and soil biota succession. For example, soil microbial biomass increased during the secondary succession of abandoned fields ranging in age from 1 to 60 years in east-central Minnesota [17] but declined across a secondary vegetation succession from short grassland to tall grassland to shrub-land in northwest France [20]. In addition, soil microbial biomass did not differ between early and late stages in a secondary succession from shrubland to forest in New Zealand [16].

In this study, we determined whether the progressive succession of the plant community is accompanied by the establishments of soil microbial and nematode communities in karst shallow soils; by progressive succession, we mean that the community becomes more complex, with increased species richness, community biomass, and food web trophic links. We explored the compositions of soil microbial and nematode communities along a secondary vegetation succession from grassland to shrubland to forest in the karst region in southwest China, where ecosystems have been severely disrupted and where the soils are very shallow. We test the hypothesis that progressive succession of the plant community should improve the soil habitat (because of increases in soil organic matter and nutrients) and thereby support the establishments of the soil microbial and nematode communities.

Materials and Methods

No specific permits were required for the described field studies. No specific permissions were required for these locations/activities. The location is not privately owned or protected in any way and the field studies did not involve endangered or protected species.

Site description

This study was carried out at the Huanjiang Observation and Research Station for Karst Ecosystems (107°51′–108°43′E, 24°44′–25°33′N), Chinese Academy of Sciences (CAS), Guangxi Province, China. The climate is subtropical monsoon with a distinct wet (from April to September) and dry season (from October to March). The mean annual temperature and precipitation are 18.5°C and 1,389 mm, respectively. The soil developed from a dolostone base and is calcareous.

Three plant communities (or ecosystems) were identified on the hillsides around the Huanjiang Observation and Research Station: grassland, shrubland, and forest. The main criteria used in selecting and designating these vegetation types were as follows: the plant community composition should be different, the disturbance history of the site should be known (based on information obtained from local farmers), and the environmental conditions (soil origin, altitude, slope orientation, and slope gradient) should be as similar as possible. All three vegetation types were situated in the middle of south-facing slopes with gradients ranging from 30 to 50°. The altitude ranged from 280 to 464 m. The grasslands were developed from managed forests

Table 1. Soil physico-chemical properties at two depths in a grassland, shrubland, and forest in the degraded karst region of southwest China.

Property	Depth (cm)	Vegetation type					
		Grassland		Shrubland		Forest	
SWC (%)	0–5	47.3 (1.4)		54.9 (4.4)		44.7 (8.3)	
	5–10	43.2 (2.9)		45.8 (2.6)		40.1 (6.3)	
pH_{WATER}	0–5	8.17 (0.02)	a[a]	7.53 (0.32)	b	6.89 (0.42)	c
	5–10	8.20 (0.03)		7.60 (0.31)		6.85 (0.33)	
TN (g/kg)	0–5	4.79 (0.22)		4.29 (0.20)		5.25 (0.79)	
	5–10	4.33 (0.07)		3.56 (0.14)		4.29 (0.73)	
SOC (g/kg)	0–5	96.9 (5.0)	a	67.8 (9.6)	b	56.5 (8.8)	b
	5–10	87.9 (6.1)		50.6 (13.6)		39.4 (6.9)	
C:N	0–5	20.4 (1.9)	a	16.1 (3.1)	a	10.8 (0.5)	b
	5–10	20.4 (1.7)		14.5 (4.3)		9.2 (0.5)	

[a]Means for a property (averaged over both soil depths) followed by different letters are significantly different ($p<0.05$) according to the LSD test.
SWC (%), pH, TN (g kg^{-1} dry soil), SOC (g kg^{-1} dry soil), and C:N stand for soil water content, soil pH, soil total nitrogen, soil organic carbon, and ratio of soil organic carbon to total nitrogen, respectively.

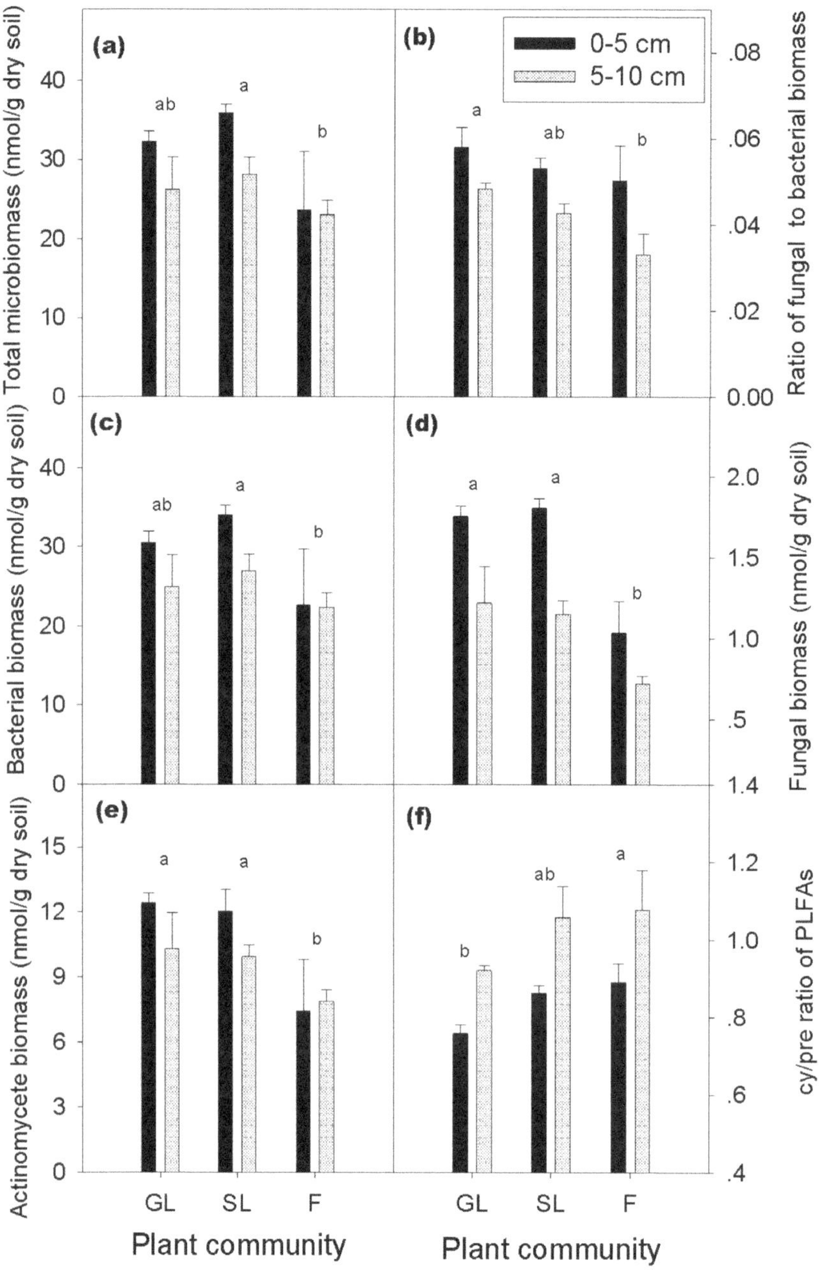

Figure 1. Characteristics of microbial communities (as determined by PLFA analysis) in the grassland (GL), shrubland (SL), and forest (F) at 0–5 and 5–10 cm soil depths. (a) Microbial biomass; (b) Ratio of fungal biomass to bacterial biomass; (c) Bacterial biomass; (d) Fungal biomass; (e) Actinomycetes biomass; and (f) cy/pre ratio. Bars indicate standard errors of means. Within each vegetation type, values with the same letters are not significantly different ($p>0.05$) according to the LSD test.

which had been abandoned for 9 to 10 years, and wildfire had occurred three times since all of the trees had been harvested and replaced by grasses. The dominant plant in the grassland was *Miscanthus floridulu*. The shrublands were developed from managed forests which had been abandoned for 25 to 30 years, at which time all of the trees were removed (i.e., all of the wood was harvested). The common shrub species were *Loropetalum chinensis*, *Litsea coreana* lvl. var. *Sinensis*, and *Pittosporum tonkinense Gagnep*. The forest sites had previously been managed forests and had been abandoned 50 to 55 earlier. The dominant overstory species in the forest was *Cyclobalanopsis glauca*; the common understory species included *Litsea coreana* levl. var. *Sinensis* and *Loropetalum chinensis*.

In 2011, the three identified plant communities at nine sites were selected, among which each plant community including three sites (replicates). A 10×10 m plot was established at each of the grassland and shrubland site, while a 20×30 m plot was established at each of the forest site. Because of anthropogenic disturbances in this area over the past decades, no primary forest ecosystem on calcareous soil that developed from a dolomite base existed near the research station. Therefore, no primary forest ecosystems were included in the present study.

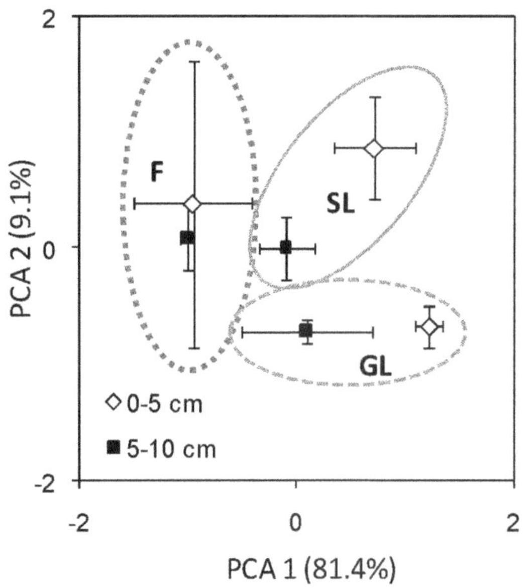

Figure 2. Principal component analysis (PCA) of microbial PLFA biomarkers of the grassland (GL), shrubland (SL), and forest (F) at 0–5 and 5–10 cm soil depth.

Soil sampling and analysis

Soil was sampled in September 2012. Soil cores (2.5 cm in diameter, 5 cm in length) were taken at 0–5 cm and 5–10 cm depths from 15 random locations within each of the nine plots. The 15 cores of the same depth from each plot were combined to form one composite sample, giving three replicate samples for each combination of vegetation type and depth. The surface litter was carefully removed before soil cores were collected. Samples were brought to the laboratory in insulated boxes. Then, the soil was sieved to pass through a 2 mm sieve to remove the roots. The soil water content, microbial and nematode communities were determined on a fresh subsample. Another subsample was thereafter air-dried and used for determination of soil pH, soil organic C and total nitrogen.

Soil pH was determined in 1:2.5 (w/v) soil: water suspensions, and soil water content (SWC %, g of water per 100 g dry soil) was measured by oven-drying the soil for 48 h at 105°C. Soil organic C (SOC, g kg^{-1} dry soil) was determined by dichromate oxidation, and total N (TN, g kg^{-1} dry soil) was measured with an ultraviolet spectrophotometer after Kjeldahl digestion [21].

Phospholipid fatty acids (PLFA) were extracted from 8 g fresh soil and were analyzed according to Bossio and Scow (1998) [22]. Concentrations of each PLFA were calculated relative to 19:0 internal standard concentrations. Bacterial biomass was considered to be represented by 10 PLFAs (i15:0, a15:0, 15:0, i16:0, 16:1ω7, i17:0, a17:0, 17:0, cy17:0, and cy19:0), fungal biomass was considered to be represented by the PLFA 18:2ω6,9, and actinomycete biomass was considered to be represented by three PLFAs (10 Me 16:0, 10 Me 17:0, and 10 Me 18:0) [23–25]. Microbial biomass was considered to be represented by the 10 bacterial PLFAs and the one fungal PLFA. Other PLFAs such as 16:1ω5, 18:1ω7, and 18:1ω9 were also used to analyze the soil microbial community.

Nematodes were extracted from 50 g of fresh soil using the Baermann funnel method [26]. After fixation in a 4% formalin solution, nematodes were counted with a differential interference contrast (DIC) microscope (ECLIPSE 80i, Nikon), and the first 100 individuals encountered were identified to genus. All nematodes were identified to genus when the sample contained fewer than 100 individuals. Nematodes were assigned to main trophic groups (bacterivores, fungivores, herbivores, omnivores, and predators) [27] and colonizer-persister guilds [28].

The microbial data were used to calculate a ratio of fungal biomass to bacterial biomass (F:B) and a cy/pre ratio of PLFAs [(cy17:0+cy19:0)/(16:1ω7c+18:1ω7c)]. The nematode data were used to calculate an enrichment index (EI), a structure index (SI), a maturity index (MI), and a plant-parasitic index (PPI) for each sample [29–30].

Data analysis

Two-way ANOVAs were used to examine the main and interaction effects of plant community and soil depth on dependent variables describing soil physico-chemical properties and soil microbial and nematode communities. Soil physico-chemical properties and microbial community composition were also analyzed by transforming the data to their principal components (PCA) and analyzing these using ANOVAs [31–32]. Data were transformed (natural log, square root, or rank) to meet assumptions of normality and homogeneity of variance. Statistical significance was determined at $p<0.05$. ANOVAs and PCAs were performed using SPSS software (SPSS Inc., Chicago, IL). LSD was used to test differences among treatment means. Tamhane's T2 was used to test differences among treatments when variances of transformed data were unequal. Redundancy analysis (RDA) was used to determine the relationship between soil physico-chemical properties and microbial and nematode community indices, microbial biomass, or nematode abundance. Soil layer was converted to a nominal variable (0, 1) and was treated as a covariable [33–35]. Monte Carlo permutation tests were used to compute statistical significance. RDA was performed with CANOCO 4.5 software [33].

Results

Soil physico-chemical properities

Soil pH was highest in the grassland, lowest in the forest, and intermediate in the shrubland (Table 1). Soil organic carbon (SOC) was greater in the grassland than in the shrubland and forest. Soil C/N was greater in the grassland and shrubland than in the forest. Soil water content (SWC) and total nitrogen (TN) content, however, did not differ among the plant communities (Table 1).

Soil microbial community

Microbial biomass and bacterial biomass were higher in the shrubland than in the forest (Fig. 1a and c). Fungal biomass and actinomycete biomass were higher in the grassland and shrubland than in the forest (Fig. 1d and e). The ratio of fungal to bacterial biomass (F:B) decreased gradually from grassland to shrubland to forest and was significantly higher in the grassland than in the forest (Fig. 1b). The cy/pre ratio of PLFAs increased from grassland to shrubland to forest (Fig. 1f) and was significantly higher in the forest than in the grassland ($p<0.05$) and tended to be higher in the shrubland than in the grassland ($p=0.058$). In addition, ANOVA results of the transformed principal components of the soil microbial community showed that the microbial community structure of the forest differed from that of the grassland and shrubland ($p<0.01$) (Fig. 2). Moreover, the soil microbial community structure tended to differ between the grassland and shrubland ($p=0.069$) (Fig. 2).

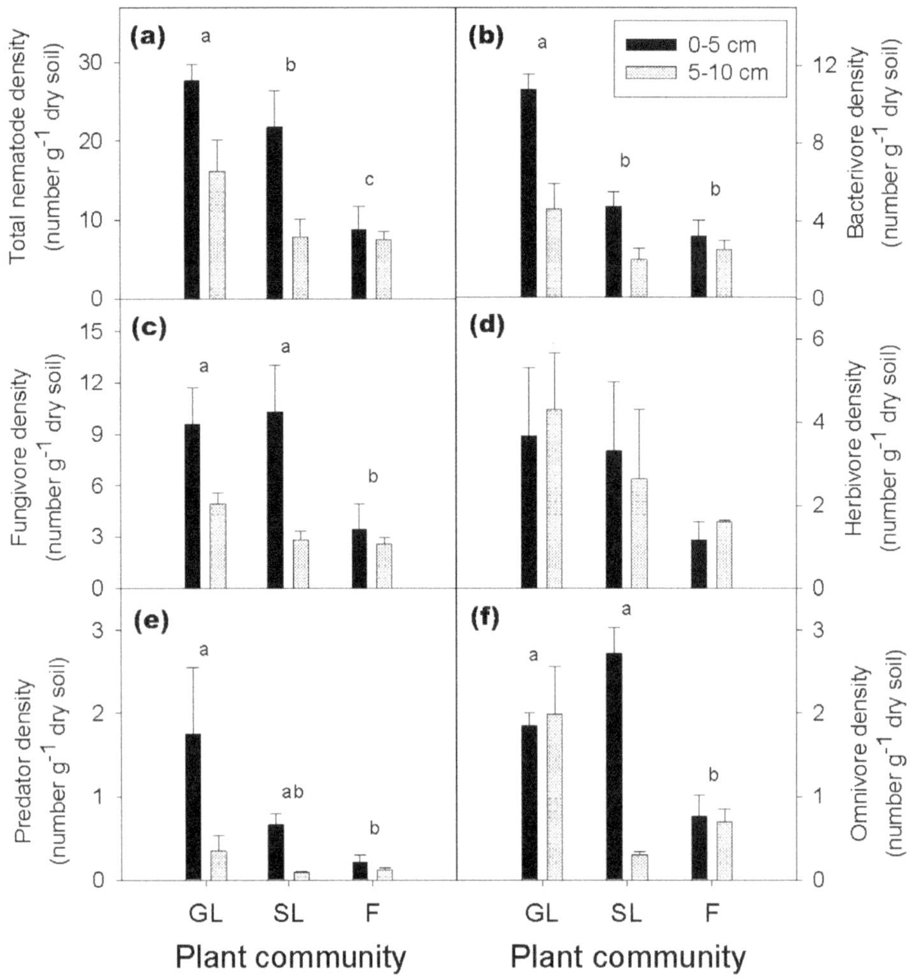

Figure 3. Densities of total nematodes and different nematode trophic groups in the grassland (GL), shrubland (SL), and forest (F) at 0–5 and 5–10 cm soil depth. (a) Total nematodes; (b) Bacterivores; (c) Fungivores; (d) Herbivores; (e) Predators; and (f) Omnivores. Bars indicate standard errors of means. Within each panel, values (averaged across both depths for each plant community) with the same or no letters are not significantly different ($p>0.05$) according to the LSD test.

Soil nematode community

Fungivores and bacterivores were the most abundant trophic groups (Fig. 3b and c). Total nematode density declined gradually from grassland to shrubland to forest at both 0–5 and 5–10 cm soil depths (Fig. 3a). Bacterivore density was greater in the grassland than in the shrubland and forest (Fig. 3b). Fungivore density and omnivore density were greater in the grassland and shrubland than in the forest (Fig. 3c and f); omnivore density in the shrubland, however, was very low at 5–10 cm depth (Fig. 3f). Predator density declined gradually from grassland to shrubland and forest and was significantly greater in the grassland than in the forest ($p = 0.023$) (Fig. 3e). Herbivore density declined gradually from grassland to shrubland to forest but did not significantly differ among the three ecosystems (Fig. 3d).

The weighted soil nematode fauna analysis showed that the nematode communities of all three ecosystems at both 0–5 and 5–10 cm soil depths projected onto quadrat III of the nematode fauna plot (Fig. 4a). The composition of the soil nematode community differed among the three ecosystems, and the value of the nematode structure index (SI) was greater in the grassland than in the forest ($p = 0.003$) (Fig. 4b). The SI value of the shrubland was intermediate to that of the grassland and forest, and tended to differ from that of the grassland ($p = 0.092$) and of the forest ($p = 0.086$). The nematode enrichment index (EI) was not significantly influenced by ecosystem type.

Relationships between soil physico-chemical properties and soil biota

The first two canonical axes explained about 71.6%, 51.9%, and 62.1% of the total variance of microbial and nematode community indices (Fig. 5a), microbial community biomass (Fig. 5b), and nematode community density (Fig. 5c), respectively. The microbial and nematode community indices were related to soil water content ($p = 0.012$, $F = 7.99$), SOC ($p = 0.004$, $F = 7.46$), and soil pH ($p = 0.040$, $F = 3.44$) (Fig. 5a). For each index, SI and MI were mainly and positively correlated with SOC; EI and F:B were mainly and positively correlated with pH; cy/pre was mainly and negatively correlated with pH; and PPI was mainly and positively correlated with TN (Fig. 5a). The microbial community biomass tended to be correlated with soil pH ($p = 0.052$, $F = 4.02$) (Fig. 5b). Total microbial biomass, fungal biomass, bacterial biomass, and actinomycetes biomass were mainly and negatively correlated with TN (Fig. 5b). Nematode community density was significantly related to SOC ($p = 0.010$, $F = 8.19$) (Fig. 5c). Total

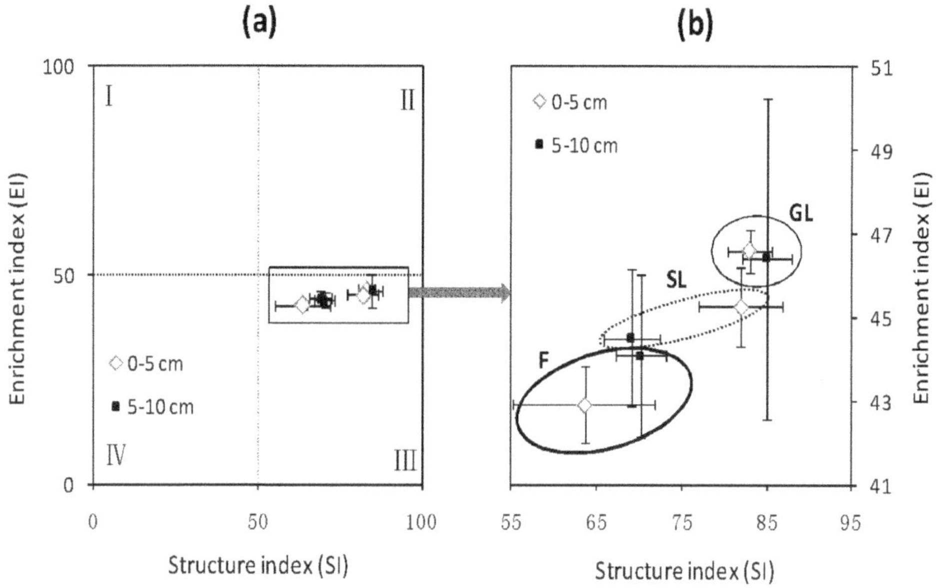

Figure 4. Weighted soil nematode faunal analysis at 0–5 depth and 5–10 cm depth in three vegetation types (GL, grassland; SL, shrubland; and F, forest). (a) Bi-plot with four quadrats; (b) Magnified quadrate-area of bi-plot A. Bars indicate the standard errors of means.

nematode density, omnivore density, and predator density were mainly and positively correlated with pH and SOC; fungivore density was mainly and positively correlated with soil water content and C/N; and herbivore density and bacterivore density were positively correlated with TN, pH, and SOC (Fig. 5c).

Discussion

Succession of soil microbial communities

In this study of a degraded karst region, the progressive secondary succession in vegetation was generally accompanied by a decline of the soil microbial community in that microbial biomass and other microbial variables decreased as the plant community succeeded from grassland to shrubland to forest. Consistent with our results, microbial biomass declined along a secondary succession from grassland to early stages of forest in calcareous soils in northwestern France [20]. In contrast, a previous study reported that succession from grassland to forest increased the soil microbial biomass and biodiversity in karst calcareous soils in southwest China [36]. Another study reported that bacterial phylogenetic diversity was higher but that bacterial metabolic diversity was lower in a grassland than in a forest in karst calcareous soils in southwest China [11]. Conversely, soil microbial biomass decreased with intensity of land degradation from forest to shrubland and grassland in karst calcareous soils in southwest China [37]. In these previous studies in karst areas in southwest China, the responses of soil microbial communities (mainly total microbial biomass) to vegetation succession could be ascribed to the shifts in soil organic carbon, i.e., microbial biomass was positively correlated with soil organic carbon. Although this was also the case in this study, soil organic carbon declined with plant succession. It has been well established that soil organic carbon is usually greater in grasslands than in forests [20,38], and land use changes from grassland to forest have decreased soil carbon stocks worldwide [39]. It is also possible that the relationship between microbial communities and soil organic carbon differs among ecosystems. For instance, the relationship between the F/B ratio and soil organic carbon was positive in agricultural soils but negative in prairie soils at Fermilab, Batavia, in Illinois [40].

In the present study, the increase in the cy/pre ratio with secondary succession might indicate that the soil habitat was becoming more harsh with plant succession. Many previous studies have documented that increasing cy/pre ratios is associated with stress and starvation (e.g., anaerobic conditions, nutrient stress, and water stress) [22,41–42]. Moreover, soil microbial communities in the current study were primarily influenced by soil pH, and the cy/pre ratio was negatively correlated with soil pH. Therefore, soil acidification and the deterioration of the soil habitat occurring with secondary vegetation succession might be the primary environmental factors affecting soil microbial communities. Previous studies had demonstrated that soil acidification significantly reduced the soil microbial biomasses and/or activities and soil nematode abundances [43–45]. In addition, many studies have reported that succession from grassland to forest decreased the soil pH [15,46–48]. Soil pH in the present study was not only negatively associated with the cy/pre ratio but was also positively associated with the F:B ratio. In agreement with our results, the F:B ratio increased slightly with increasing pH as reported by [49]. However, the relationship between the F:B ratio and soil pH is inconsistent [24,50]. Although soil total nitrogen was not influenced by vegetation type, total microbial biomass, bacterial biomass, fungal biomass, and actinomycetes biomass tended to be negatively correlated with soil TN. The reasons for this unexpected result are unclear.

Succession of soil nematode communities

Like soil microbial biomass, soil nematode abundance showed declined patterns with respect to the progressive successional patterns of the plant communities in this study. Few studies have explored the soil nematode communities associated with plant community succession from grassland to forest or from forest to grassland. One study reported that soil nematode densities and biodiversity (i.e., plant feeders and non-plant feeders) increased with secondary succession from savanna to forest in the tropical semi-arid zone of West Africa (Senegal) [51]. Although many

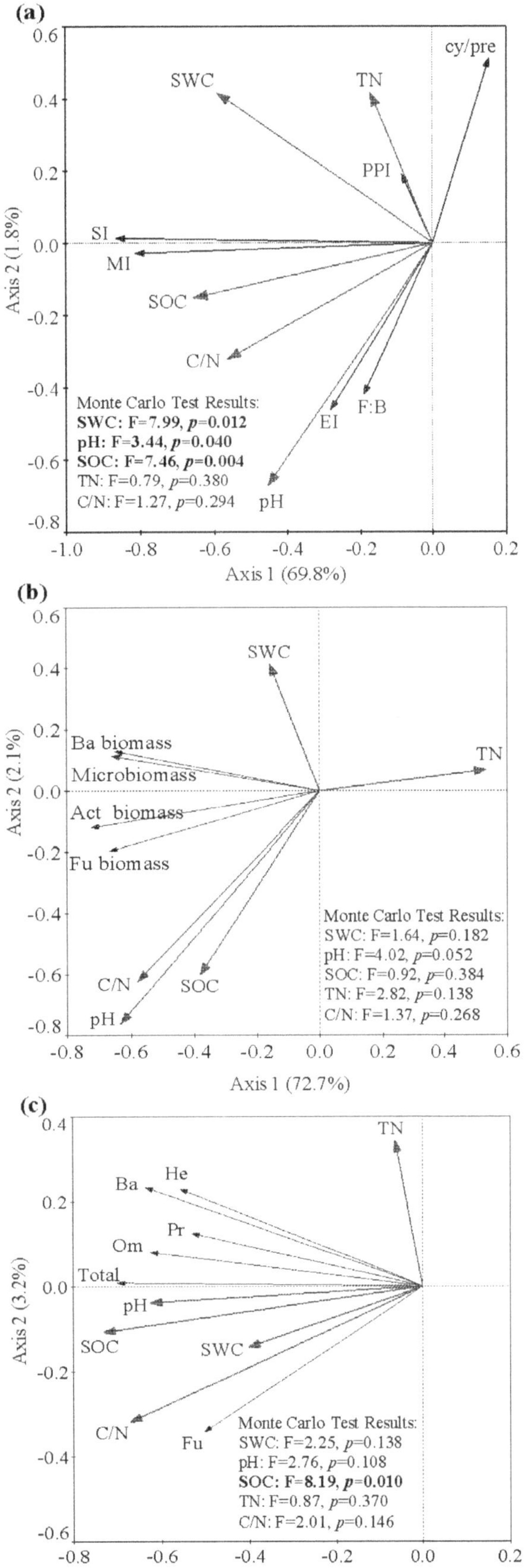

Figure 5. Redundancy analysis (RDA) of microbial and nematode community indices (a), microbial community biomass (b), and nematode community density (c). Ordination diagrams present species scores and environmental factor scores (vectors). Environmental factors include soil water content (SWC), soil pH, soil total nitrogen (TN), soil organic carbon (SOC), and ratio of soil organic carbon to total nitrogen (C:N). Microbial and nematode community indices include ratio of fungal biomass to bacterial biomass (F:B), cy/pre ratio of PLFAs (cy/pre), nematode maturity index (MI), plant parasite index (PPI), structure index (SI), and enrichment index (EI). Variables of microbial community biomass include total microbial biomass (Microbiomass), fungal biomass (Fu biomass), bacterial biomass (Ba biomass), and actinomycetes biomass (Act biomass). Variables of nematode community density include densities of total nematodes (Total), bacterivores (Ba), fungivores (Fu), herbivores (He), omnivores (Om), and predators (Pr).

previous studies have compared the compositions of soil nematode communities of grasslands, shrublands, and forests, it is unclear how inferences based on ecosystem type can be applied to ecosystem succession. For example, Wasilewska (1979) reported that densities of herbivorous, omnivorous, carnivorous, and total nematodes and number of nematode genera were higher in grassland than in forest ecosystems but that densities of bacterivorous and fungivorous nematodes were similar in grassland and forest ecosystems [52]. In another study, land-use change from grassland to shrub heathland suppressed densities of nematode trophic groups (i.e., herbivores, bacterivores, fungivores, omnivores, and predators) in northwest Europe [47].

The weighted soil nematode fauna analysis showed that the nematode communities of all three ecosystems projected onto quadrat III in the nematode fauna plot, which demonstrated that the structures of the soil food webs of all three ecosystems were relatively similar [29]. However, the values of the nematode structure index (SI) decreased from grassland to shrubland to forest ecosystems, which demonstrated that the complexity of the soil food webs decreased in the order grassland > shrubland > forest. In addition, soil organic carbon was the main factor associated with total nematode density and the density of each trophic group. In agreement with our results, many studies have reported that soil organic carbon affects nematode communities [53–55]. Soil nematodes are also affected by spatial heterogeneity, plant community composition, and the soil environment [56]. Although the nematode plant-parasitic index (PPI) was positively correlated with soil total nitrogen, and although the enrichment index (EI) was positively correlated with pH, soil total nitrogen and EI did not significantly change with the secondary vegetation succession. Therefore, soil nematode communities might be more affected by the main soil resource (i.e., soil organic carbon) than by pH or other components of the soil habitat.

Potential implications for land conservation and vegetation recovery in karst areas

The aim of traditional ecosystem restoration in degraded karst areas has usually been to restore forest ecosystems [4–5,8,10]. Our study clearly shows that ecosystem restoration in degraded karst areas might not necessarily be favored by secondary succession to forests. Based on the results of the current study, soil carbon storage and soil biodiversity are greater in grasslands and shrublands than in forests in degraded karst areas. The low soil organic carbon content and biota biomass associated with the forest might be explained in part by karst soils, which are generally very shallow and which can provide only low levels of nutrients and water [57]. The limitations of nutrient and water supply can be exacerbated when the live plant biomass increases during succession [58]. In addition, because, the live biomass is greater for forests than for grasslands or shrublands, natural and/or anthropogenic disruptions (e.g., disease and pest outbreak, drought, harvest, and contamination) may cause greater material loss from forests than from grasslands or shrublands. Consequently, natural and/or anthropogenic disruptions may be more damaging in forests than in grassland or shrubland ecosystems. Therefore, soil carbon and biota ought to be amended in degraded karst areas during vegetation succession or vegetation recovery.

Karst areas of southwest China, Southeast Asian, and Central America have large human populations [1–2]. The challenge is to provide the basic needs of these populations while protecting the fragile karst ecosystems. Agricultural development is the cause of soil erosion and rocky desertification in karst areas [1,12]. As noted in the previous paragraph, disruptions of karst forest ecosystems might be especially destructive. Thus, the cultivation of crops and forestry does not seem to be suitable for maintaining karst ecosystems. In contrast, herbivorous animal husbandry in karst areas of southwest China has provided economic benefits to local farmers while reducing soil erosion [9]. In addition, the perennial C4 grass *Miscanthus* (i.e., the dominant genus in the grassland of the current study) has produced substantial biomass crops worldwide [59]. Another perennial C4 grass, hybrid napiergrass (*Pennisetum hydridum*), has also been widely cultivated and is mainly used for raising livestock in karst areas in southwest China and in other subtropical and tropical areas [60].

Most restoration studies have focused on soil nutrient recovery rather than on soil biota recovery, and many previous studies have used a "space-for-time substitution approach" to explore the soil nutrient status during succession from grassland to forest in karst areas. Most of these studies have reported that soil nutrients accumulate with secondary succession [11,36–37] but that was not the case in the current study. This difference might be explained by the differences in the grasslands. In contrast to grasslands in the earlier studies, the studied grassland was much older and had experienced repeated wildfires.

Conclusions

Contrary to our hypothesis, the progressive succession of vegetation occurring in karst sites in southwest China was accompanied by declines in soil biota abundances and food web complexity. The declines of the soil microbial and nematode communities might be explained by changes in soil habitat (i.e., reduced pH) and resource availability (i.e., reduced soil organic carbon). Because sustainable ecosystem management should integrate aboveground and belowground ecosystem properties and functions [61–62], our results indicate that soil biota amendment is necessary in karst areas during ecosystem restoration and management and establishment of grasslands in degraded karst areas are feasible. This finding provides a scientific basis for the development of herbivorous animal husbandry in karst areas that are under population pressure.

Acknowledgments

We thank Bruce Jaffee for his help in preparing the manuscript. We also thank the editor and two anonymous reviewers for their comments.

Author Contributions

Conceived and designed the experiments: JZ XH KW. Performed the experiments: JZ SL LL. Analyzed the data: JZ SL. Contributed reagents/materials/analysis tools: XH LL. Wrote the paper: JZ SL KW.

References

1. Yuan D (2002) Geology and Geohydrology of Karst and its Relevance to Society; Paris. UNESCO. pp. 15–18.
2. Tuyet D (2001) Characteristics of karst ecosystems of Vietnam and their vulnerability to human impact. Acta Geologica Sinica 75: 325–329.
3. Parise M, De Waele J, Gutierrez F (2009) Current perspectives on the environmental impacts and hazards in karst. Environmental Geology 58: 235–237.
4. Rivera LW, Aide TM (1998) Forest recovery in the karst region of Puerto Rico. Forest Ecology and Management 108: 63–75.
5. Rivera LW, Zimmerman JK, Aide TM (2000) Forest recovery in abandoned agricultural lands in a karst region of the Dominican Republic. Plant Ecology 148: 115–125.
6. Querejeta JI, Estrada-Medina H, Allen MF, Jiménez-Osornio JJ (2007) Water source partitioning among trees growing on shallow karst soils in a seasonally dry tropical climate. Oecologia 152: 26–36.
7. Latinne A, Waengsothorn S, Herbreteau V, Michaux J (2011) Thai limestone karsts: an impending biodiversity crisis. The 1st Environment Asia International Conference. Bangkok, Thailand. pp. 176–187.
8. Zeng F, Peng W, Song T, Wang K, Wu H, et al. (2007) Changes in vegetation after 22 years' natural restoration in the Karst disturbed area in northwestern Guangxi, China. Acta Ecologica Sinica 27: 5110–5119.
9. Wang K, Su Y, Zeng F, Chen H, Xiao R (2008) Ecological process and vegetation restoration in karst region of southwest China. Research of Agricultural Modernization 29: 641–645 (in Chinese with English abstract).
10. Yu L, Zhu S, Wei L, Chen Z, Xiong Z (1998) Study on the natural restoration process of degraded Karst communities–Successional sere. Journal of Mountain Agriculture and Biology 2: 71–77. In Chinese with English abstract.
11. Xiong Y, Xia H, Li Za, Cai Xa, Fu S (2008) Impacts of litter and understory removal on soil properties in a subtropical *Acacia mangium* plantation in China. Plant and Soil 304: 179–188.
12. Qi X, Wang K, Zhang C (2013) Effectiveness of ecological restoration projects in a karst region of southwest China assessed using vegetation succession mapping. Ecological Engineering 54: 245–253.
13. Walker LR, Wardle DA, Bardgett RD, Clarkson BD (2010) The use of chronosequences in studies of ecological succession and soil development. Journal of Ecology 98: 725–736.
14. Walker LR (1999) Ecosystems of disturbed ground. Amsterdam: Elsevier.
15. Schipper LA, Degens BP, Sparling GP, Duncan LC (2001) Changes in microbial heterotrophic diversity along five plant successional sequences. Soil Biology and Biochemistry 33: 2093–2103.
16. Wardle D (1993) Changes in the microbial biomass and metabolic quotient during leaf litter succession in some New Zealand forest and scrubland ecosystems. Functional Ecology 7: 346–355.
17. Zak D, Grigal D, Gleeson S, Tilman D (1990) Carbon and nitrogen cycling during old-field succession: Constraints on plant and microbial biomass. Biogeochemistry 11: 111–129.
18. van der Heijden MGA, Bardgett RD, van Straalen NM (2008) The unseen majority: soil microbes as drivers of plant diversity and productivity in terrestrial ecosystems. Ecology Letters 11: 296–310.
19. De Deyn GB, Raaijmakers CE, Zoomer HR, Berg MP, de Ruiter PC, et al. (2003) Soil invertebrate fauna enhances grassland succession and diversity. Nature 422: 711–713.
20. Chabrerie O, Laval K, Puget P, Desaire S, Alard D (2003) Relationship between plant and soil microbial communities along a successional gradient in a chalk grassland in north-western France. Applied Soil Ecology 24: 43–56.
21. Liu G (1996) Analysis of soil physical and chemical properties and description of soil profiles. Beijing: China Standard.
22. Bossio DA, Scow KM (1998) Impacts of carbon and flooding on soil microbial communities: Phospholipid fatty acid profiles and substrate utilization patterns. Microbial Ecology 35: 265–278.
23. Joergensen RG, Wichern F (2008) Quantitative assessment of the fungal contribution to microbial tissue in soil. Soil Biology and Biochemistry 40: 2977–2991.
24. Frostegård A, Bååth E (1996) The use of phospholipid fatty acid analysis to estimate bacterial and fungal biomass in soil. Biology and Fertility of Soils 22: 59–65.
25. Frostegård Å, Tunlid A, Bååth E (2011) Use and misuse of PLFA measurements in soils. Soil Biology and Biochemistry 43: 1621–1625.
26. Barker KR, Carter CC, Sasser JN (1985) An advanced treatise on Meloidogyne. Volume II: Methodology. Raleigh: United States Agency for International Development. vi + 223 pp. p.
27. Yeates GW, Bongers T, Degoede RGM, Freckman DW, Georgieva SS (1993) Feeding habits in soil nematode families and genera–an outline for soil ecologists. Journal of Nematology 25: 315–331.
28. Bongers T, Bongers M (1998) Functional diversity of nematodes. Applied Soil Ecology 10: 239–251.

29. Ferris H, Bongers T, de Goede RGM (2001) A framework for soil food web diagnostics: extension of the nematode faunal analysis concept. Applied Soil Ecology 18: 13–29.
30. Neher DA (2001) Role of nematodes in soil health and their use as indicators. Journal of Nematology 33: 161–168.
31. Zhao J, Wan S, Li Z, Shao Y, Xu G, et al. (2012) *Dicranopteris*-dominated understory as major driver of intensive forest ecosystem in humid subtropical and tropical region. Soil Biology and Biochemistry 49: 78–87.
32. Zhao J, Wan S, Fu S, Wang X, Wang M, et al. (2013) Effects of understory removal and nitrogen fertilization on soil microbial communities in *Eucalyptus* plantations. Forest Ecology and Management 310: 80–86.
33. Lepš J, Šmilauer P (2003) Multivariate analysis of ecological data using CANOCO. New York: Cambridge University Press. xii + 269 pp. p.
34. Zhao J, Neher D (2013) Soil nematode genera that predict specific types of disturbance. Applied Soil Ecology 64: 135–141.
35. Zhao J, Shao Y, Wang X, Neher DA, Xu G, et al. (2013) Sentinel soil invertebrate taxa as bioindicators for forest management practices. Ecological Indicators 24: 236–239.
36. Zhu H, He X, Wang K, Su Y, Wu J (2012) Interactions of vegetation succession, soil bio-chemical properties and microbial communities in a Karst ecosystem. European Journal of Soil Biology 51: 1–7.
37. Zhang P, Li L, Pan G, Ren J (2006) Soil quality changes in land degradation as indicated by soil chemical, biochemical and microbiological properties in a karst area of southwest Guizhou, China. Environmental Geology 51: 609–619.
38. Evrendilek F, Celik I, Kilic S (2004) Changes in soil organic carbon and other physical soil properties along adjacent Mediterranean forest, grassland, and cropland ecosystems in Turkey. Journal of Arid Environments 59: 743–752.
39. Guo LB, Gifford RM (2002) Soil carbon stocks and land use change: a meta analysis. Global Change Biology 8: 345–360.
40. Allison VJ, Miller RM, Jastrow JD, Matamala R, Zak DR (2005) Changes in soil microbial community structure in a tallgrass prairie chronosequence. Soil Science Society of America Journal 69: 1412–1421.
41. Schmitt A, Glaser B (2011) Organic matter dynamics in a temperate forest soil following enhanced drying. Soil Biology and Biochemistry 43: 478–489.
42. Kieft TL, Ringelberg DB, White DC (1994) Changes in ester-linked phospholipid fatty acid profiles of subsurface bacteria during starvation and desiccation in a porous medium. Applied and Environmental Microbiology 60: 3292–3299.
43. Räty M, Huhta V (2003) Earthworms and pH affect communities of nematodes and enchytraeids in forest soil. Biology and Fertility of Soils 38: 52–58.
44. Chen D, Lan Z, Bai X, Grace JB, Bai Y (2013) Evidence that acidification-induced declines in plant diversity and productivity are mediated by changes in below-ground communities and soil properties in a semi-arid steppe. Journal of Ecology.
45. Rousk J, Brookes PC, Baath E (2010) The microbial PLFA composition as affected by pH in an arable soil. Soil Biology & Biochemistry 42: 516–520.
46. Dmowska E, Ilieva-Makulec K (2006) Secondary succession of nematodes in power plant ash dumps reclaimed by covering with turf. European Journal of Soil Biology 42: S164–S170.
47. Holtkamp R, Kardol P, van der Wal A, Dekker SC, van der Putten WH, et al. (2008) Soil food web structure during ecosystem development after land abandonment. Applied Soil Ecology 39: 23–34.
48. Robertson GP, Vitousek PM (1981) Nitrification potentials in primary and secondary succession. Ecology: 376–386.
49. Bååth E, Anderson T-H (2003) Comparison of soil fungal/bacterial ratios in a pH gradient using physiological and PLFA-based techniques. Soil Biology and Biochemistry 35: 955–963.
50. Rousk J, Brookes PC, Bååth E (2009) Contrasting soil pH effects on fungal and bacterial growth suggest functional redundancy in carbon mineralization. Applied and Environmental Microbiology 75: 1589–1596.
51. Pate E, Ndiaye-Faye N, Thioulouse J, Villenave C, Bongers T, et al. (2000) Successional trends in the characteristics of soil nematode communities in cropped and fallow lands in Senegal (Sonkorong). Applied Soil Ecology 14: 5–15.
52. Wasilewska L (1979) The structure and function of soil nematode communities in natural ecosystems and agrocenoses. Polish Ecological Studies 5: 97–145.
53. Yeates G, Bongers T (1999) Nematode diversity in agroecosystems. Agriculture, Ecosystems & Environment 74: 113–135.
54. Liang W, Lou Y, Li Q, Zhong S, Zhang X, et al. (2009) Nematode faunal response to long-term application of nitrogen fertilizer and organic manure in Northeast China. Soil Biology and Biochemistry 41: 883–890.
55. Freckman D (1988) Bacterivorous nematodes and organic-matter decomposition. Agriculture, Ecosystems & Environment 24: 195–217.
56. Viketoft M (2013) Determinants of small-scale spatial patterns: Importance of space, plants and abiotics for soil nematodes. Soil Biology and Biochemistry 62: 92–98.
57. Zhang X, Wang K (2009) Ponderation on the shortage of mineral nutrients in the soil-vegetation ecosystem in carbonate rock-distributed mountain regions in Southwest China. Earth and Environment 37: 337–341 (in Chinese with English abstract).
58. Guo K, Liu C, Dong M (2011) Ecological adaptation of plants and control of rocky-desertification on karst region of Southwest China. Chinese Journal of Plant Ecology 35: 991–999 (in Chinese with English abstract).
59. Matlaga DP, Davis AS (2013) Minimizing invasive potential of Miscanthus × giganteus grown for bioenergy: identifying demographic thresholds for population growth and spread. Journal of Applied Ecology 50: 479–487.
60. Zhao J, Zhang W, Wang K, Song T, Du H (2014) Responses of the soil nematode community to management of hybrid napiergrass: The trade-off between positive and negative effects. Applied Soil Ecology 74: 134–144.
61. Bardgett RD, Bowman WD, Kaufmann R, Schmidt SK (2005) A temporal approach to linking aboveground and belowground ecology. Trends in Ecology & Evolution 20: 634–641.
62. Smith RS, Shiel RS, Bardgett RD, Millward D, Corkhill P, et al. (2003) Soil microbial community, fertility, vegetation and diversity as targets in the restoration management of a meadow grassland. Journal of Applied Ecology 40: 51–64.

The Impact of Selective-Logging and Forest Clearance for Oil Palm on Fungal Communities in Borneo

Dorsaf Kerfahi[1,2], Binu M. Tripathi[1], Junghoon Lee[2], David P. Edwards[3]*, Jonathan M. Adams[1]*

1 Department of Biological Sciences, Seoul National University, Seoul, Republic of Korea, **2** School of Chemical and Biological Engineering, Interdisciplinary Program of Bioengineering, Seoul National University, Seoul, Republic of Korea, **3** Department of Animal and Plant Sciences, University of Sheffield, Sheffield, United Kingdom

Abstract

Tropical forests are being rapidly altered by logging, and cleared for agriculture. Understanding the effects of these land use changes on soil fungi, which play vital roles in the soil ecosystem functioning and services, is a major conservation frontier. Using 454-pyrosequencing of the ITS1 region of extracted soil DNA, we compared communities of soil fungi between unlogged, once-logged, and twice-logged rainforest, and areas cleared for oil palm, in Sabah, Malaysia. Overall fungal community composition differed significantly between forest and oil palm plantation. The OTU richness and Chao 1 were higher in forest, compared to oil palm plantation. As a proportion of total reads, Basidiomycota were more abundant in forest soil, compared to oil palm plantation soil. The turnover of fungal OTUs across space, true β-diversity, was also higher in forest than oil palm plantation. Ectomycorrhizal (EcM) fungal abundance was significantly different between land uses, with highest relative abundance (out of total fungal reads) observed in unlogged forest soil, lower abundance in logged forest, and lowest in oil palm. In their entirety, these results indicate a pervasive effect of conversion to oil palm on fungal community structure. Such wholesale changes in fungal communities might impact the long-term sustainability of oil palm agriculture. Logging also has more subtle long term effects, on relative abundance of EcM fungi, which might affect tree recruitment and nutrient cycling. However, in general the logged forest retains most of the diversity and community composition of unlogged forest.

Editor: Andrew Hector, University of Oxford, United Kingdom

Funding: This project was funded by a National Research Foundation grant (NRF-2013-031400), Ministry of Education, Science and Technology, South Korea. The funders had no role in study design, data collection and analysis, decision to publish, or preparation of the manuscript.

Competing Interests: Dr. Jonathan M. Adams is a PLOS ONE Editorial Board member. This does not alter the authors' adherence to PLOS ONE Editorial policies and criteria.

* Email: david.edwards@sheffield.ac.uk (DPE); foundinkualalumpur@yahoo.com (JMA)

Introduction

Tropical forests are one of the world's most important reservoirs for biodiversity [1]. They contain an exceptional concentration of the world's species, but are being reduced in area faster than any other ecosystem [2]. Roughly half of the world's natural extent of tropical forest has been logged or converted to different land uses [3]. Some 403 million hectares of tropical rainforests have been included in timber estates and slated for selective logging [4], and between 2000 and 2010, approximately 13 million hectares of forest within the tropics were cleared for agricultural activities [5,6], including oil palm plantations [7].

Anthropogenic disturbances in tropical forests are causing a dramatic decline in global biodiversity, and in associated biological processes that maintain the productivity and sustainability of ecosystems [8]. Several studies have shown that selective logging does not drastically impact the overall species richness and diversity of tropical forest [9–12], however, it has been shown that the impact of selective logging could be anything between fairly mild and severe depending on the intensity of logging [13] with changes in the composition of species, as forest-interior specialists decline and edge-tolerant, gap specialists increase in abundance [14,15]. In contrast, the conversion of both primary and logged forest to agricultural land uses has been to shown to have a far greater negative impact on biodiversity than does logging. Conversion to agriculture results in a major reduction in biodiversity, again across a host of animal and plant taxa [16]. The conversion of primary and logged forest to agricultural plantations also results in a substantial decrease in the functional diversity of tropical ecosystems, with implications for the provision of ecosystem functions, whereas logging has lesser impacts on these metrics [17,18]. As well as affecting plants and larger animals, land use change also affects the soil biota. Land use change affects soil pH, carbon and nutrient content [19,20], causing shifts in soil microbial communities [21–23].

Fungi constitute one of the most diverse and dominant groups of organisms in soil, and they play important ecological roles in the ecosystem as decomposers, pathogens and plant mutualists [24,25]. Understanding the structure and diversity of soil fungal communities is fundamental to the understanding of their function in the ecosystem and their impact on plant communities [26]. However, while minimal work has been done in the tropics to assess the effect of land use changes on soil bacterial communities [23,27], until recently relatively little was known about the impacts of tropical land use change on soil fungal communities. Various studies have suggested that forest clearance to tree plantations or agricultural crops shifts soil fungal communities, linked to strong changes in soil properties [22,28]. However, previous studies of forest clearance to other forms of agriculture on forest fungal communities have mostly been limited to techniques that give relatively low taxonomic resolution (i.e. T-RFLP and PCR-DGGE). Nevertheless, a recent study by McGuire et al. [29], which used high throughput sequencing to analyze soil fungal communities in Southeast Asian tropical forests in west Malaysia, showed that conversion of primary forest to oil palm plantations alters fungal community composition and function, whereas primary and logged forests were more similar in composition and nutrient cycling potential. However, there is a need for further studies to understand the impacts of logging cycles on soil fungal communities and of the conversion of logged forest to agriculture, since logged forests now dominate the tropics [3] and are much more likely to be converted to agriculture than primary forests [2,6].

In this study, we also focused on the rainforests of the Sundaland region, of Southeast Asia, but some 1,800 km away in east Malaysia (Borneo). Across the Sundaland region, the primary forest has been subject to differing degrees of logging intensity. Much of the region's forest (about 50%) has never been logged, while many areas have been subject to one or two logging cycles [30]. Also, oil palm is one of the most rapidly expanding crops in this region. This provides an opportunity to study the effect of different intensities of logging on the soil fungal community and also to evaluate if conversion of forest to oil palm plantation has a stronger impact on soil fungi than logging, as is the case for numerous macroscopic taxa. Our objective here was to understand whether land use change has an impact on the structure and diversity of fungal communities in the Yayasan Sabah (YS) logging concession in Malaysian, Borneo. We compared the fungal communities in forests with different logging histories (unlogged, once-logged and twice-logged), and oil palm plantations. We examined whether there are differences in α and â-diversity, as well as community composition. These results may provide important information for soil management policies, and estimation of ecological impact of land use change in this region.

Materials and Methods

Study area

The study area is located within the Yayasan Sabah (YS) logging concession and contiguous oil palm plantation areas, in Sabah, Malaysian Borneo (4°58′ N, 117°48′ E). The forests in this area are naturally dominated by valuable timber tree species belonging to the family Dipterocarpaceae [31]. Due to logging for the wood industry and clearance for palm oil plantations, the area of forest in Borneo - as elsewhere in the tropics - has been dramatically reduced in recent decades [32].

Fieldwork was conducted in the Ulu Segama-Malua Forest Reserve (US-MFR) and adjacent oil palm estates in Sabah, Borneo. Some areas of forest were logged between 1970 and 1990, and some of these were then re-logged between 2000 and 2007. During the first logging rotation, approximately 113 m^3 per hectare (range 73 m^3 to 166 m^3) of commercially valuable trees > 0.6 m diameter were extracted. During the second logging rotation, an additional 31 m^3 per hectare (range 15 m^3 to 72 m^3) of timber were removed [10,31]. Selectively logged forest in the US-MFR is adjacent to the 45,200 ha Danum Valley Conservation Area (DVCA) and Palum Tambun Watershed Reserve, containing large areas of unlogged forest [10,16]. Oil palm plantations are situated to the north and south of the US-MFR, with mature palms of 20 to 30 years old, planted at a density of 100 trees per hectare.

Soil sampling and DNA extraction

From September to October 2012, twenty-four transects each of 200 m in length were located across four different land uses: unlogged (primary) forests, once-logged and twice-logged forests, and oil palm plantations, with six transects per habitat. Within each habitat, distances between transects ranged from 500 m to 65 km, whereas across habitats, distances between transects ranged from 1 km to 67 km. From each transect, at 50 m intervals, approximately 50 g from the top 5 cm of soil (excluding the leaf litter layer) was taken in a sterile plastic bag using a trowel, giving five samples of soil per transect. The trowels were thoroughly cleaned with ethanol between successive transect sampling. All soil samples were then sieved (2 mm) in laboratory to homogenize the sample and stored at −20°C until DNA extraction [10]. Twenty-four soil DNA extractions (one for each transect) were performed using 0.3 g of soil, with the Power Soil DNA extraction kit (MO BIO Laboratories, Carlsbad, CA, USA) following the directions described by the manufacturer.

PCR amplification and pyrosequencing

Fungal DNA was amplified using ITS primers targeting the internal transcribed spacer (ITS) region 1 and 2. Forward primers comprised the 454 Fusion Primer A-adaptor, a specific multiplex identifier (MID) barcode, and the ITS1F primer (5′-CTTGGTCATTTAGAGGAAGTAA-3′) [33], while the reverse primer was composed of the B-adapter and ITS4 primer (5′-TCCTCCGCTTATTGATATGC-3′) [34]. Polymerase chain reactions (PCR) were performed in 50 μl reactions using the following temperature program: 95°C for 10 min s; 30 cycles of 95°C for 30 s, 55°C for 30 s, 72°C for 30 s; and 72°C for 7 min. The PCR products were purified using the QIAquick PCR purification kit (Qiagen) and quantified using PicoGreen (Invitrogen) spectrofluorometrically (TBS 380, Turner Biosystems, Inc. Sunnyvale, CA, USA). 50 ng of purified PCR product for each sample were combined in a single tube and sent to Macrogen Inc. (Seoul, Korea) for sequencing using 454/Roche GS FLX Titanium Instrument (Roche, NJ, USA).

Sequence processing

Initial quality filtering and denoising were performed following the 454 SOP in the mothur pipeline [35]. The ITS1 region was verified and extracted using the ITS1 extractor for fungal ITS sequences [36]. Putative chimeric sequences were detected and removed via the Chimera Uchime algorithm contained within mother [37]. Operational taxonomic units (OTUs) were assigned using the QIIME implementation of UCLUST [38], with a threshold of 97% pairwise identity. OTUs were classified taxonomically using the classify command in mothur at 80% Naïve Bayesian bootstrap cutoff with 1000 iterations against the UNITE database [39]. Ectomycorrhizal (EcM) fungi were determined by matching taxonomy assignments with established EcM lineages as determined by recent phylogenetic and stable isotope data [40]. The 454 sequence run has been deposited in the NCBI Sequence Read Archive under accession number SRP041467.

Statistical analysis

To correct for differences in number of reads, which can bias diversity estimates, all samples were rarified to 3,347 reads per sample. To test for effects of land use types on the OTU richness

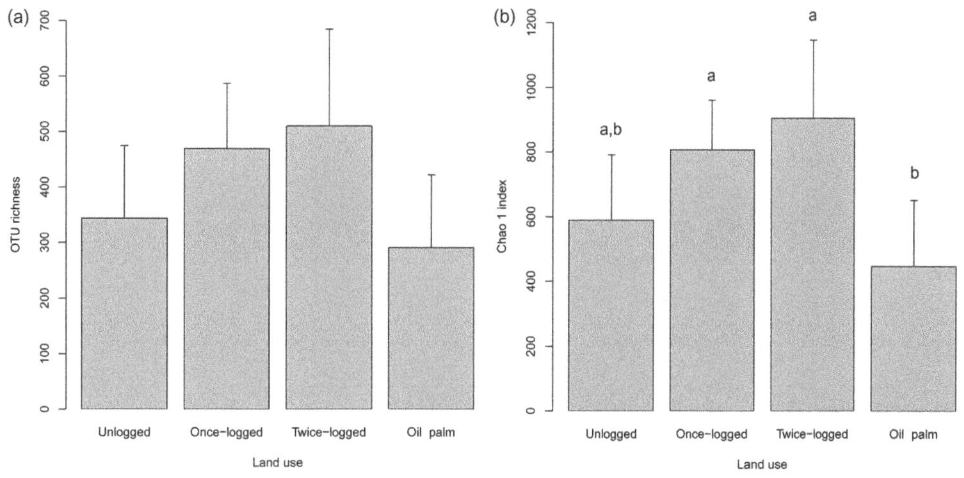

Figure 1. Diversity indices of the fungal community across different land uses in Sabah, Malaysian Borneo. (a) OTU richness and (b) Chao1 index. Pairwise comparisons are shown; different letters denote significant differences between groups at P<0.05.

and diversity indices, we used a linear model (LM) for normal data or generalized linear model (GLM) for non-normal data, considering land use as the major factor. We used the same procedure to test whether relative abundance of the most abundant phyla differed among different land use types. We also assessed the effect of land use on the relative abundance at the order and genus levels within those phyla that showed significant differences due to land use. Post-hoc Tukey tests were used for pairwise comparisons. When neither a linear nor a generalized linear model fitted the data, we used a Kruskall-Wallis test to assess the effect of land use on the relative abundance of fungal taxa, with the Bonferroni correction to assess pairwise comparisons.

To test whether species composition results may have been influenced by pseudoreplication within study sites, we used a Mantel test (Mantel Nonparametric Test Calculator 2.0) [41] to compare transect matrices of fungal compositional to geographic

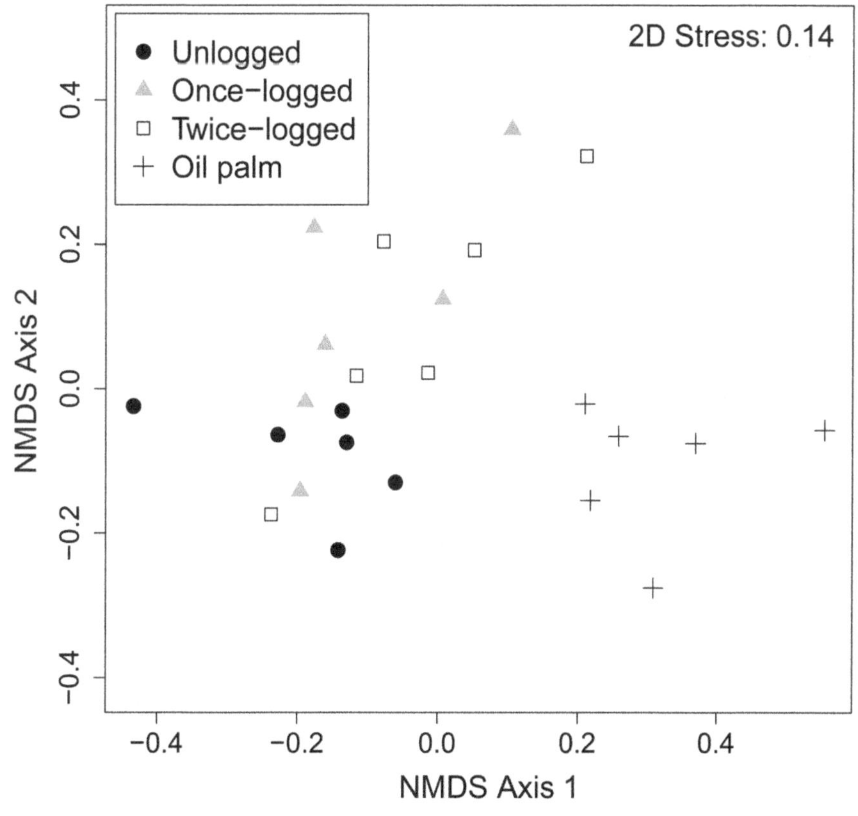

Figure 2. Non-metric multidimensional scaling (NMDS) ordination showing clustering of fungal communities among different land uses in Sabah, Malaysian Borneo.

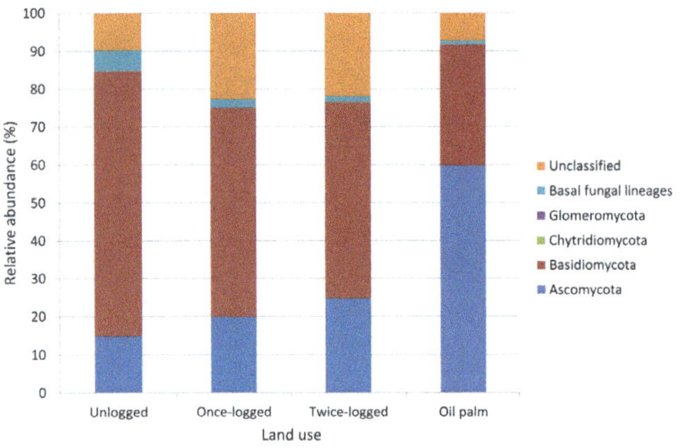

Figure 3. Relative abundance of dominant fungal phyla among different land uses in Sabah, Malaysian Borneo.

distance between pairs of transects within a site and between pairs of transects across the entire dataset [42,43].

The OTU-based community similarity was calculated by using the Bray-Curtis index [44]. Non-metric multidimensional scaling (NMDS), to visualize the change in species composition across the land use types, was conducted in Primer-E software (Version 6, Plymouth, UK). Then, we tested the difference among different land use types using an analysis of similarity (ANOSIM).

We measured β-diversity amongst land use types following Anderson et al. [45], which is defined as the variation in community structure without defining a particular gradient or direction. Therefore, we estimated true β-diversity following Whittaker [46] in [47] for every land use type. In addition, we calculated β-diversity as the average distance from each site to the group centroid [45]. The betadisper function in R was used to test if β-diversity shows any difference between land use types. True β-diversity (i.e. $S/\bar{\alpha}$) for each pair of samples within each of the four land uses was estimated by the following equation:

$$\frac{S}{\bar{\alpha}}=\frac{a+b+c}{(2a+b+c)/2}$$

Where S is the total number of OTUs in two samples, $\bar{\alpha}$ is the average number of OTUs for both samples, a is shared OTUs between both samples, b is OTUs found only in sample 1 and c are OTUs found only in sample 2. To compare true β-diversity among land uses, we used a linear model using land use as factor, and sample as random factor to control for pseudoreplication, as every sample is used in more than one comparison within each land use. Post-hoc Tukey tests were used for pairwise comparisons among different land uses.

Results

A total of 114,744 quality sequences were obtained from the 24 soil samples, with coverage ranging from 3,347–6,456 sequences. After rarifying to 3,347 reads per sample, we obtained a total of 80,328 sequences, and of these around 84% sequences were classified up to phylum level with a total of 5,327 OTUs (defined at ≥97% sequence similarity level). Fungal OUT richness (i.e. number of OTUs) was marginally significantly different across land use types ($F_{3,24} = 2.98$, $P = 0.05$; Fig. 1a), with lowest levels of OTU richness observed in oil palm plantations compared to primary and logged forests. Predicted OTU richness calculated using the Chao1 estimator was significantly higher in logged forests than oil palm plantations ($F3,24 = 4.74$, $P = 0.01$; Fig. 1b), whereas Shannon index did not show any variation among land uses ($F3,24 = 1.93$, $P = 0.15$).

The NMDS plots of pairwise Bray–Curtis dissimilarities showed that fungal communities were clustered significantly across land use types (ANOSIM: $R = 0.51$, $P < 0.001$; Fig. 2). Mantel tests showed no effect of distance on the composition of fungal communities across different land use types in Borneo (all $P > 0.07$). The majority of fungal sequences recovered in our study belonged to the Basidiomycota and Ascomycota, with relative abundances of 52% and 29%, respectively (Fig. 3). The basal

Table 1. Comparison of relative abundance of the dominant fungal orders within the phyla Ascomycota and Basidiomycota among land uses[a].

Taxa	F or χ^{2b}	df	P	Pairwise comparisons[c]
Ascomycota				
Helotiales	3.67	3, 24	0.02	Once-logged > oil palm
Hypocreales	5.54	3, 24	0.006	Once-logged/twice-logged/unlogged < oil palm
Pleosporales	6.42	3, 24	0.003	Once-logged/unlogged < oil palm
Basidiomycota				
Agaricales	10.9	3, 24	0.0001	Once-logged/twice-logged/unlogged > oil palm
Russulales	11.3	3, 24	0.0001	Unlogged > Once-logged/twice-logged/oil palm
Sebacinales	11.1*	3	0.01	Unlogged/twice-logged > oil palm
Sporidiobolales	15.1*	3	0.001	Once-logged/twice-logged/unlogged < oil palm
Thelephorales	10.0*	3	0.01	Unlogged/twice-logged > oil palm
Trichosporonales	13.3	3, 24	0.0001	Once-logged/twice-logged/unlogged > oil palm

[a]Only orders for which significant differences were found are shown.
[b]Effect of land use on relative abundance evaluated by linear or generalized linear model or by the Kruskal-Wallis test (*).
[c]Pairwise comparisons by *post hoc* Tukey test for linear/generalized linear models or *P* values Bonferroni-corrected for Kruskal-Wallis. Differences were considered significant at a P value of <0.05.

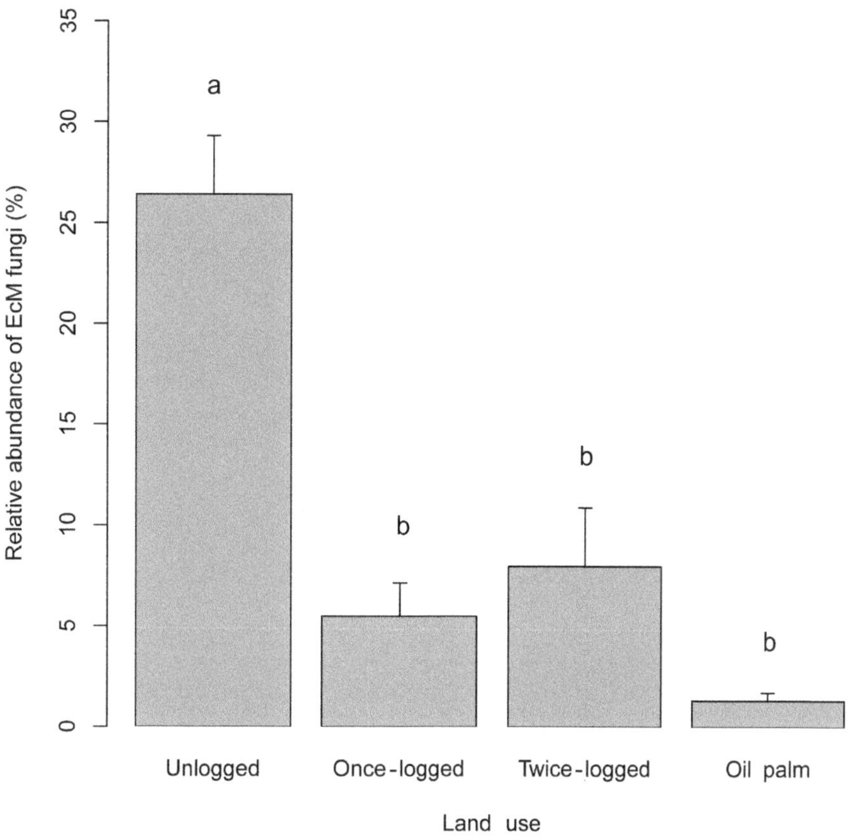

Figure 4. Relative abundance (means ± SD) of ectomycorrhizal (EcM) fungal sequences among different land uses in Sabah, Malaysian Borneo. Pairwise comparisons are shown; different letters denote significant differences between groups at P<0.05.

fungal lineages represented 2%, followed by Glomeromycota and Chytridiomycota with less than 1%, and 15% of the detected sequences were unclassified (Fig. 3). We found a significant change in the relative abundance of the two most dominant phyla: Basidiomycota ($F_{3,24}=4.23$, P = 0.01) and Ascomycota ($F_{3,24}=4.81$, P = 0.01) along different land uses. The abundance of Basidiomycota was greater in the unlogged forest, intermediate in the logged forest, and least in the oil palm. Conversely, the oil palm plantations had greater abundance of Ascomycota compared to forest soils (Fig. 3). At the order level, there were significant differences in the relative abundances of the most dominant orders (Table 1). The results revealed that the relative abundance of *Agaricales*, *Russulales*, *Thelephorales*, *Trichosporonales, Sebacinales* and *Helotiales* were significantly higher in the forest than oil palm plantations (P<0.05; Table 1). However, *Hypocreales*, *Sporidiobolales* and *Pleosporales*, were more abundant in oil palm plantations than unlogged and logged forests (P<0.05; Table 1).

A total of 11,421 sequences belonged to known groups of ectomycorrhizal (EcM) fungi, with the EcM fungi representing around 10% (11,421 sequences) of the total detected fungal sequences, with 180 OTUs. The relative abundance of EcM sequences was significantly different across land use types, with highest and lowest relative abundances observed in primary forest (mean relative abundance = 26%) and oil palm plantations (mean relative abundance = 0.5%), respectively ($\chi^2=18.04$, P<0.001; Fig. 4). From our soil samples, we identified 14 genera belonging to EcM fungi with *Russula* as the most dominant genus (65% of total EcM sequences), followed by *Tomentella*, *Sebacina* and *Lactarius*. The relative abundance of these four dominant EcM genera combined were significantly higher in forest soils compared to oil palm plantations (P<0.05). For *Russula* alone, we also found a difference in abundance between unlogged (greatest abundance), once-logged (less abundant), and twice-logged forest (lowest abundance of *Russula*) (P<0.05 in each case).

The β-diversity, measured as the average distance of all samples to the centroid in each land use type, did not show significant difference among land uses ($F_{3,24}=2.77$, P = 0.06). However, there was a significant effect of forest conversion to oil palm plantations on fungal true β-diversity (i.e. S/α; $F_{3,24}=3.85$, P = 0.01), with oil palm having lowest true β-diversity compared to unlogged and logged forests (Fig. 5). Logging did not produce any significant change in true β-diversity.

Discussion

Our results showed that fungal OTU richness and Chao1 index differed among land uses. Oil palm plantations had lower OTU richness, which can be attributed to the effects of anthropogenic intervention and forest conversion. This contrasts with the findings on bacterial communities of Lee-Cruz et al. [27] in the same study site, where they found that OTU richness and diversity indices did not differ among land uses, and that α-diversity was similar in forests and oil palm plantations.

We found that the structure of fungal communities differed most fundamentally between forests and oil palm plantations. This finding mirrors that of McGuire et al. [29], who found that the community of soil fungi collected across three different land uses in Malaysia differed between oil palm plantations and forests. Changes in fungal community composition in logged forest and

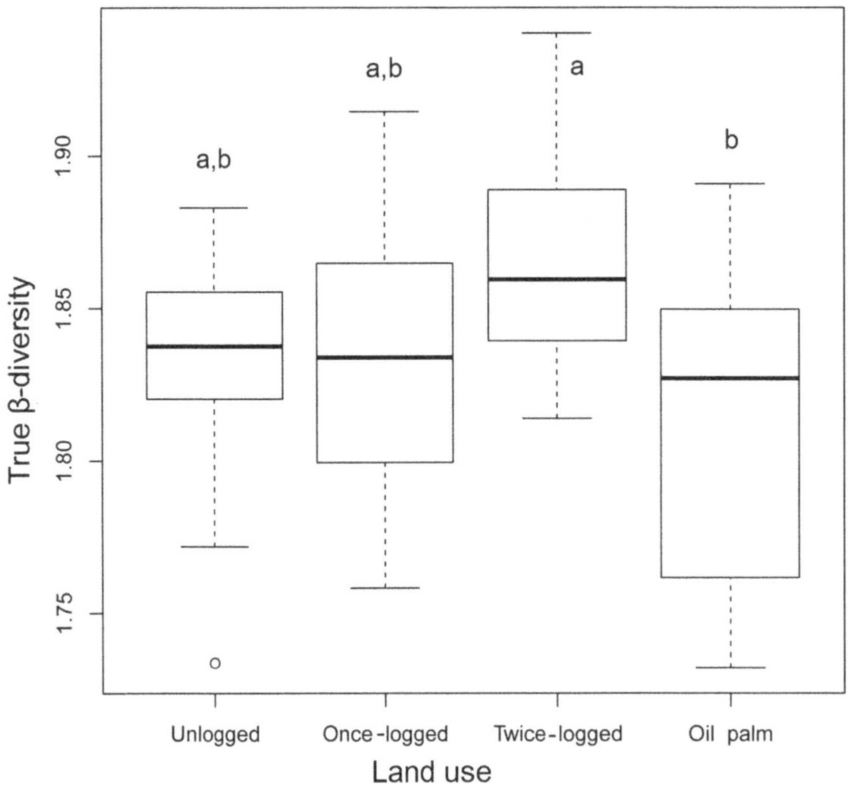

Figure 5. Fungal community true β-diversity (i.e. $S/\bar{\alpha}$) among the four land uses in Sabah, Malaysian Borneo. Boxes show the lower quartile, the median and the upper quartile. Pairwise comparisons are shown; different letters denote significant differences between groups at $P < 0.05$.

former forest areas could directly impact the functioning of soil communities and their ability to provide key ecosystem services, such as decomposition and nutrient recycling [22]. Given that fertilizer prices are predicted to rise dramatically in the coming decades [48], these results suggest that improvements in agricultural methods by establishing diversified farms could be necessary for sustaining vital soil biodiversity and ecosystem-service values.

The abundance pattern of fungal taxa detected here is similar to the soils of the Neotropics and elsewhere, where Basidiomycota and Ascomycota are also the most prevalent groups [49–53]. The most abundant orders of Basidomycota in the forest areas we sampled were *Russulales*, *Agaricales*, *Sporidiobolales*, *Telephorales*, *Trichosporonales* and *Sebacinales*. They all became less abundant in oil palm plantations compared to primary forest. These orders are generally reported to be the most varied and abundant groups of fungi in forests around the world [40,54,55], they have also been characterized as being lignicolous, saprobic or mycorrhizal, associated with litter decomposition, and they are known to degrade plant-derived cellulose [56]. In our forest plots, the ectomycorrhizal orders *Russulales* and *Telephorales* were the most common. This pattern has also been found in forests from the Neotropics and African tropics [57–59].

The lower abundance of Basidiomycota in oil palm soils may be due to the lack of large quantities of coarse woody debris, often derived from roots or branches of forest trees, in oil palm plantations. However, it may be worth investigating whether the change in abundance of Basidiomycota - and change in their overall community structure - with conversion of forest to oil palm may have implications for long term nutrient processing in the oil palm soils. Since Basidiomycota as a group are often able to break down relatively recalcitrant substrates and changes in their abundance and community composition might impede nutrient recycling in oil palm soils. The converse increase in Ascomycota, may be seen in terms of the relatively lignin-poor and nutrient-rich character of most of the organic matter reaching the soil from roots and leaves of oil palm, and from the herbaceous weedy layer that grows under the palms.

We found that both a history of logging, and forest conversion to oil palm plantation, resulted in shifts in EcM fungal communities. The most drastic effects were with forest conversion to oil palm. However, logging history also had detectable effects EcM fungi relative abundance. Ectomycorrhizas are one of the most important widespread types of mycorrhiza in forests of the cool temperate and boreal latitudes [60], and they also form an important group and often the dominant ecologically and economically important minority of Dipterocarpaceae family trees in tropical Asia [61–65].

The lower abundance of EcM fungi in logged forests might be due to a thinner, more incomplete root mat following past logging disturbance. The much lower abundance of EcM fungi in oil palm plantation soils could be partly due to the much lower abundance of potential host roots (e.g. Dipterocarpaceae which are absent from the palm plantations). Such fungi have generally been found to recover slowly from disturbance even when potential host plants are present [66]. Among the detected genera in our study, we found that *Russula*, *Sebacina*, *Lactarius* and *Tomentella* showed significant impact of land use change. *Russula* was relatively the most abundant EcM genus in our samples: this genus has been found to be common on roots of dipterocarp forests, tropical and southern hemisphere angiosperm forests [62,67,68].

This study also investigated the impacts of land use change on the turnover of fungal communities across space (β-diversity), and results suggest that there is a spatial homogenization of fungal communities in oil palm agriculture compared to forest: fungal β-diversity in oil palm plantations was lower than forest. Habitat conversion to agriculture also reduces β-diversity of soil bacteria in Amazonia [69]. In contrast, across the same land use system and sites in Borneo as we studied here, there was actually an increase in β-diversity of bacteria with both logging and conversion to oil palm [27].

There are some caveats that accompany our findings. We worked in only one biogeographic region and on only one form of agriculture. It is thus important to replicate this work in other logging systems and in other key expanding crops, including soya, sugar cane, and cacao. We also did not explicitly demonstrate the impacts of changing fungal composition on soil ecosystem functions and services: this is a major knowledge gap, with critical importance to the development of sustainable logging and, in particular, agricultural systems in the tropics.

In conclusion, the conversion of both primary and logged forest to oil palm drives a change in the overall fungal community, including EcM fungi abundance, and an associated decrease in total fungal community beta-diversity. This finding invites further studies that investigate the long-term implications of such changes for agricultural sustainability. There was a more subtle long-term impact of logging on fungal communities. Most measurable features of the unlogged forest fungal community remained unchanged after logging. However, there were significant changes in EcM fungal abundance due to logging, which could have a pervasive impact since EcM fungi are thought to play a key role in tree growth and community structure. Despite this, the lack of drastic changes in the overall forest fungal community structure following logging strengthens the view that logged forest is not necessarily an irretrievably damaged and drastically altered system, and that protecting it from conversion to oil palm may still have considerable conservation benefits.

Acknowledgments

We thank Yayasan Sabah, Danum Valley Management Committee, Sabah Parks, the state Secretary, Sabah Chief Minister's Departments, and the Sabah Biodiversity Council for permission to conduct research. We also, thank the Royal Society's Southeast Asian Rainforest Research Program (SEARRP) and the Borneo Rainforest Lodge for logistical support and site access.

Author Contributions

Conceived and designed the experiments: DPE JMA. Performed the experiments: DK BMT DPE. Analyzed the data: DK BMT DPE JL. Contributed reagents/materials/analysis tools: DK BMT DPE JMA JL. Contributed to the writing of the manuscript: DK BMT DPE JMA.

References

1. Whitmore TC (1990) An introduction to tropical rain forests. Oxford: Clarendon Press. 226 pp.
2. Asner GP, Rudel TK, Aide TM, Defries R, Emerson R (2009) A contemporary assessment of change in humid tropical forests. Conserv Biol 23: 1386–1395.
3. Wright SJ (2005) Tropical forests in a changing environment. Trends Ecol Evol 20: 553–560.
4. Blaser J (2011) Status of tropical forest management 2011: International Tropical Timber Organization.
5. Food and Agriculture Organization of the United Nations (2010) Global forest resources assessment 2010 : main report. Rome: FAO. 340 p.
6. Hansen MC, Stehman SV, Potapov PV, Loveland TR, Townshend JR, et al. (2008) Humid tropical forest clearing from 2000 to 2005 quantified by using multitemporal and multiresolution remotely sensed data. Proc Natl Acad Sci U S A 105: 9439–9444.
7. Morris RJ (2010) Anthropogenic impacts on tropical forest biodiversity: a network structure and ecosystem functioning perspective. Philos Trans R Soc Lond B Biol Sci 365: 3709–3718.
8. Achard F, Eva HD, Stibig H-J, Mayaux P, Gallego J, et al. (2002) Determination of Deforestation Rates of the World's Humid Tropical Forests. Science 297: 999–1002.
9. Gibson L, Lee TM, Koh LP, Brook BW, Gardner TA, et al. (2011) Primary forests are irreplaceable for sustaining tropical biodiversity. Nature 478: 378–381.
10. Edwards DP, Larsen TH, Docherty TD, Ansell FA, Hsu WW, et al. (2011) Degraded lands worth protecting: the biological importance of Southeast Asia's repeatedly logged forests. Proc Biol Sci 278: 82–90.
11. Berry NJ, Phillips OL, Lewis SL, Hill JK, Edwards DP, et al. (2010) The high value of logged tropical forests: lessons from northern Borneo. Biodiver Conserv 19: 985–997.
12. Woodcock P, Edwards DP, Newton RJ, Vun Khen C, Bottrell SH, et al. (2013) Impacts of intensive logging on the trophic organisation of ant communities in a biodiversity hotspot. PLoS One 8: e60756.
13. Burivalova Z, Şekercioğlu ÇH, Koh LP (2014) Thresholds of Logging Intensity to Maintain Tropical Forest Biodiversity. Curr Biol 24: 1–6.
14. Hamer K, Hill J, Benedick S, Mustaffa N, Sherratt T, et al. (2003) Ecology of butterflies in natural and selectively logged forests of northern Borneo: the importance of habitat heterogeneity. J Appl Ecol 40: 150–162.
15. Cleary DF, Boyle TJ, Setyawati T, Anggraeni CD, Loon EEV, et al. (2007) Bird species and traits associated with logged and unlogged forest in Borneo. Ecol Appl 17: 1184–1197.
16. Edwards DP, Hodgson JA, Hamer KC, Mitchell SL, Ahmad AH, et al. (2010) Wildlife-friendly oil palm plantations fail to protect biodiversity effectively. Conserv Lett 3: 236–242.
17. Edwards F, Edwards D, Larsen T, Hsu W, Benedick S, et al. (2013) Does logging and forest conversion to oil palm agriculture alter functional diversity in a biodiversity hotspot? Anim Conserv 17: 163–173.
18. Baraloto C, Hérault B, Paine C, Massot H, Blanc L, et al. (2012) Contrasting taxonomic and functional responses of a tropical tree community to selective logging. J Appl Ecol 49: 861–870.
19. McGrath DA, Smith CK, Gholz HL, Oliveira FdA (2001) Effects of Land-Use Change on Soil Nutrient Dynamics in Amazônia. Ecosystems 4: 625–645.
20. Murty D, Kirschbaum MUF, McMurtrie RE, McGilvray H (2002) Does conversion of forest to agricultural land change soil carbon and nitrogen? a review of the literature. Global Change Biol 8: 105–123.
21. Cornejo FH, Varela A, Wright SJ (1994) Tropical Forest Litter Decomposition under Seasonal Drought: Nutrient Release, Fungi and Bacteria. Oikos 70: 183–190.
22. Lauber CL, Strickland MS, Bradford MA, Fierer N (2008) The influence of soil properties on the structure of bacterial and fungal communities across land-use types. Soil Biol Biochem 40: 2407–2415.
23. Tripathi B, Kim M, Singh D, Lee-Cruz L, Lai-Hoe A, et al. (2012) Tropical Soil Bacterial Communities in Malaysia: pH Dominates in the Equatorial Tropics Too. Microb Ecol 64: 474–484.
24. Orgiazzi A, Lumini E, Nilsson RH, Girlanda M, Vizzini A, et al. (2012) Unravelling soil fungal communities from different Mediterranean land-use backgrounds. PloS one 7: e34847.
25. Wu TH, Chellemi DO, Martin KJ, Graham JH, Rosskop EN (2007) Discriminating the effects of agricultural land management practices on soil fungal communities. Soil Biol Biochem 39: 1139–1155.
26. Martin F, Cullen D, Hibbett D, Pisabarro A, Spatafora JW, et al. (2011) Sequencing the fungal tree of life. New Phytol 190: 818–821.
27. Lee-Cruz L, Edwards DP, Tripathi BM, Adams JM (2013) Impact of Logging and Forest Conversion to Oil Palm Plantations on Soil Bacterial Communities in Borneo. Appl Environ Microbiol 79: 7290–7297.
28. Lupatini M, Jacques RJS, Antoniolli ZI, Suleiman AKA, Fulthorpe RR, et al. (2013) Land-use change and soil type are drivers of fungal and archaeal communities in the Pampa biome. World J Microbiol Biotechnol 29: 223–233.
29. McGuire KL, D'Angelo H, Brearley FQ, Gedallovich SM, Babar N, et al. (2014) Responses of soil fungi to logging and oil palm agriculture in Southeast Asian tropical forests. Microbial Ecol.
30. Wilcove DS, Giam X, Edwards DP, Fisher B, Koh LP (2013) Navjot's nightmare revisited: logging, agriculture, and biodiversity in Southeast Asia. Trends Ecol Evol 28: 531–540.
31. Fisher B, Edwards DP, Giam XL, Wilcove DS (2011) The high costs of conserving Southeast Asia's lowland rainforests. Front Ecol Environ 9: 329–334.
32. Gibbs HK, Ruesch AS, Achard F, Clayton MK, Holmgren P, et al. (2010) Tropical forests were the primary sources of new agricultural land in the 1980s and 1990s. Proc Natl Acad Sci U S A 107: 16732–16737.
33. Gardes M, Bruns TD (1993) ITS primers with enhanced specificity for basidiomycetes–application to the identification of mycorrhizae and rusts. Mol Ecol 2: 113–118.

34. White TJ, Bruns T, Lee S, Taylor J (1990) Amplification and direct sequencing of fungal ribosomal RNA genes for phylogenetics. PCR protocols: a guide to methods and applications 18: 315–322.
35. Schloss PD, Gevers D, Westcott SL (2011) Reducing the effects of PCR amplification and sequencing artifacts on 16S rRNA-based studies. PLoS One 6: e27310.
36. Nilsson RH, Veldre V, Hartmann M, Unterseher M, Amend A, et al. (2010) An open source software package for automated extraction of ITS1 and ITS2 from fungal ITS sequences for use in high-throughput community assays and molecular ecology. Fungal Ecol 3: 284–287.
37. Edgar RC, Haas BJ, Clemente JC, Quince C, Knight R (2011) UCHIME improves sensitivity and speed of chimera detection. Bioinformatics 27: 2194–2200.
38. Edgar RC (2010) Search and clustering orders of magnitude faster than BLAST. Bioinformatics 26: 2460–2461.
39. Abarenkov K, Nilsson RH, Larsson KH, Alexander IJ, Eberhardt U, et al. (2010) The UNITE database for molecular identification of fungi - recent updates and future perspectives. New Phytol 186: 281–285.
40. Tedersoo L, May T, Smith M (2010) Ectomycorrhizal lifestyle in fungi: global diversity, distribution, and evolution of phylogenetic lineages. Mycorrhiza 20: 217–263.
41. Liedloff A (1999) Mantel nonparametric test calculator. Version 2.0. School of Natural Resource Sciences, Queensland University of Technology, Australia.
42. Ghazoul J (2002) Impact of logging on the richness and diversity of forest butterflies in a tropical dry forest in Thailand. Biodiver Conserv 11: 521–541.
43. Ramage BS, Sheil D, Salim HM, Fletcher C, MUSTAFA NZA, et al. (2013) Pseudoreplication in tropical forests and the resulting effects on biodiversity conservation. Conserv Biol 27: 364–372.
44. Magurran AE (2004) Measuring biological diversity. Maldan, MA: Blackwell Pub. viii, 256 p.
45. Anderson MJ, Crist TO, Chase JM, Vellend M, Inouye BD, et al. (2011) Navigating the multiple meanings of beta diversity: a roadmap for the practicing ecologist. Ecol Lett 14: 19–28.
46. Whittaker RH (1960) Vegetation of the Siskiyou mountains, Oregon and California. Ecol Monogr 30: 279–338.
47. Koleff P, Gaston KJ, Lennon JJ (2003) Measuring beta diversity for presence-absence data. J Anim Ecol 72: 367–382.
48. Piesse J, Thirtle C (2009) Three bubbles and a panic: An explanatory review of recent food commodity price events. Food Policy 34: 119–129.
49. O'Brien HE, Parrent JL, Jackson JA, Moncalvo JM, Vilgalys R (2005) Fungal community analysis by large-scale sequencing of environmental samples. Appl Environ Microbiol 71: 5544–5550.
50. Wubet T, Christ S, Schöning I, Boch S, Gawlich M, et al. (2012) Differences in soil fungal communities between European beech (Fagus sylvatica L.) dominated forests are related to soil and understory vegetation. PloS one 7: e47500.
51. Gomes NC, Fagbola O, Costa R, Rumjanek NG, Buchner A, et al. (2003) Dynamics of fungal communities in bulk and maize rhizosphere soil in the tropics. Appl Environ Microbiol 69: 3758–3766.
52. Oros-Sichler M, Gomes NC, Neuber G, Smalla K (2006) A new semi-nested PCR protocol to amplify large 18S rRNA gene fragments for PCR-DGGE analysis of soil fungal communities. J Microbiol Methods 65: 63–75.
53. Bridge PD, Newsham KK (2009) Soil fungal community composition at Mars Oasis, a southern maritime Antarctic site, assessed by PCR amplification and cloning. Fungal Ecol 2: 66–74.
54. Geml J, Laursen GA, Timling I, McFarland JM, Booth MG, et al. (2009) Molecular phylogenetic biodiversity assessment of arctic and boreal ectomycorrhizal Lactarius Pers. (Russulales; Basidiomycota) in Alaska, based on soil and sporocarp DNA. Mol Ecol 18: 2213–2227.
55. Geml J, Laursen GA, Herriott IC, McFarland JM, Booth MG, et al. (2010) Phylogenetic and ecological analyses of soil and sporocarp DNA sequences reveal high diversity and strong habitat partitioning in the boreal ectomycorrhizal genus Russula (Russulales; Basidiomycota). New Phytol 187: 494–507.
56. Kuramae EE, Hillekens RHE, de Hollander M, van der Heijden MGA, van den Berg M, et al. (2013) Structural and functional variation in soil fungal communities associated with litter bags containing maize leaf. Fems Microbiol Ecol 84: 519–531.
57. Ba AM, Duponnois R, Moyersoen B, Diedhiou AG (2012) Ectomycorrhizal symbiosis of tropical African trees. Mycorrhiza 22: 1–29.
58. Smith ME, Henkel TW, Aime MC, Fremier AK, Vilgalys R (2011) Ectomycorrhizal fungal diversity and community structure on three co-occurring leguminous canopy tree species in a Neotropical rainforest. New Phytol 192: 699–712.
59. Tedersoo L, Sadam A, Zambrano M, Valencia R, Bahram M (2010) Low diversity and high host preference of ectomycorrhizal fungi in Western Amazonia, a neotropical biodiversity hotspot. Isme Journal 4: 465–471.
60. Smith SE, Read DJ (2010) Mycorrhizal symbiosis: Academic press.
61. Brearley FQ (2012) Ectomycorrhizal Associations of the Dipterocarpaceae. Biotropica 44: 637–648.
62. Peay KG, Kennedy PG, Davies SJ, Tan S, Bruns TD (2010) Potential link between plant and fungal distributions in a dipterocarp rainforest: community and phylogenetic structure of tropical ectomycorrhizal fungi across a plant and soil ecotone. New Phytol 185: 529–542.
63. Moyersoen B, Becker P, Alexander I (2001) Are ectomycorrhizas more abundant than arbuscular mycorrhizas in tropical heath forests? New Phytol 150: 591–599.
64. Haug I, Weiß M, Homeier J, Oberwinkler F, Kottke I (2005) Russulaceae and Thelephoraceae form ectomycorrhizas with members of the Nyctaginaceae (Caryophyllales) in the tropical mountain rain forest of southern Ecuador. New Phytol 165: 923–936.
65. Natarajan K, Senthilarasu G, Kumaresan V, Riviere T (2005) Diversity in ectomycorrhizal fungi of a dipterocarp forest in Western Ghats. Curr Sci 88: 1893–1895.
66. Peay KG, Schubert MG, Nguyen NH, Bruns TD (2012) Measuring ectomycorrhizal fungal dispersal: macroecological patterns driven by microscopic propagules. Mol Ecol 21: 4122–4136.
67. Tedersoo L, Jairus T, Horton BM, Abarenkov K, Suvi T, et al. (2008) Strong host preference of ectomycorrhizal fungi in a Tasmanian wet sclerophyll forest as revealed by DNA barcoding and taxon-specific primers. New Phytol 180: 479–490.
68. Riviere T, Diedhiou AG, Diabate M, Senthilarasu G, Natarajan K, et al. (2007) Genetic diversity of ectomycorrhizal Basidiomycetes from African and Indian tropical rain forests. Mycorrhiza 17: 415–428.
69. Rodrigues JL, Pellizari VH, Mueller R, Baek K, Jesus Eda C, et al. (2013) Conversion of the Amazon rainforest to agriculture results in biotic homogenization of soil bacterial communities. Proc Natl Acad Sci U S A 110: 988–993.

Discriminating the Drivers of Edge Effects on Nest Predation: Forest Edges Reduce Capture Rates of Ship Rats (*Rattus rattus*), a Globally Invasive Nest Predator, by Altering Vegetation Structure

Jay Ruffell[1]*, Raphael K. Didham[1,2], Paul Barrett[3], Nic Gorman[3], Rhonda Pike[3], Andrée Hickey-Elliott[3], Karin Sievwright[3], Doug P. Armstrong[3]

1 School of Animal Biology, The University of Western Australia, Perth, Australia, **2** Centre for Environment and Life Sciences, the Commonwealth Scientific and Industrial Research Organisation, Perth, Australia, **3** Wildlife Ecology Group, Te Kura Mātauranga o ngā Taonga ā Papatuanuku, Massey University, Palmerston North, New Zealand

Abstract

Forest edges can strongly affect avian nest success by altering nest predation rates, but this relationship is inconsistent and context dependent. There is a need for researchers to improve the predictability of edge effects on nest predation rates by examining the mechanisms driving their occurrence and variability. In this study, we examined how the capture rates of ship rats, an invasive nest predator responsible for avian declines globally, varied with distance from the forest edge within forest fragments in a pastoral landscape in New Zealand. We hypothesised that forest edges would affect capture rates by altering vegetation structure within fragments, and that the strength of edge effects would depend on whether fragments were grazed by livestock. We measured vegetation structure and rat capture rates at 488 locations ranging from 0–212 m from the forest edge in 15 forest fragments, seven of which were grazed. Contrary to the vast majority of previous studies of edge effects on nest predation, ship rat capture rates increased with increasing distance from the forest edge. For grazed fragments, capture rates were estimated to be 78% lower at the forest edge than 118 m into the forest interior (the farthest distance for grazed fragments). This relationship was similar for ungrazed fragments, with capture rates estimated to be 51% lower at the forest edge than 118 m into the forest interior. A subsequent path analysis suggested that these 'reverse' edge effects were largely or entirely mediated by changes in vegetation structure, implying that edge effects on ship rats can be predicted from the response of vegetation structure to forest edges. We suggest the occurrence, strength, and direction of edge effects on nest predation rates may depend on edge-driven changes in local habitat when the dominant predator is primarily restricted to forest patches.

Editor: John F. Valentine, Dauphin Island Sea Lab, United States of America

Funding: The research was funded by Marsden Fund Grant MAU0707 to DPA and A.R.E Sinclair. RKD was supported during the writing of this manuscript by an Australian Research Council Future Fellowship FT100100040. JR was supported by a Hackett Postgraduate Research Scholarship from the University of Western Australia. The funders had no role in study design, data collection and analysis, decision to publish, or preparation of the manuscript.

Competing Interests: The authors have declared that no competing interests exist.

* Email: ruffell.jay@gmail.com

Introduction

The majority of the world's forests are fragmented by human activities to some extent [1], and the species that occur within them are vulnerable to the impacts of forest fragmentation. Moreover, the extent of forest fragmentation is likely to increase dramatically into the foreseeable future as the human population and its resource requirements grow [2,3]. Understanding the impacts of forest fragmentation on ecological systems is therefore fundamentally important for effective conservation management in many regions, and has become a major focus for conservation research in recent decades [4].

One potential impact of forest fragmentation that has received considerable attention is the effect of forest edges on predation rates of forest birds' nests [5,6]. Nest predation is the major cause of nest failure in birds, and is believed to strongly influence the population dynamics of many species [7,8]. Causes of variability in nest predation rates are therefore of substantial interest to ecologists and conservation managers [6], and there are now many hundreds of studies describing the relationship between nest predation rates and distance from the forest edge.

As a result of this research, forest edges are widely believed to reduce nest success by increasing rates of nest predation [9]. However, although several reviews of the topic have shown that predation rates can indeed be greater at forest edges [10–12], there is a growing body of evidence which suggests that these effects are inconsistent and context dependent [9,10,13–16]. This variability presents a significant issue for predicting and managing the effects of forest fragmentation on birds. The challenge for researchers is to understand the factors that give rise to edge effects, as well as those that moderate their influence, so that the impacts of fragmentation can be predicted and managed effectively [4,5,17]. For this reason, there is a need for researchers to move beyond simply describing edge effects on nest predation rates to identifying the mechanisms underlying their occurrence and strength [5,18,19]. In particular, because edge effects on nest predation are likely to depend to a large degree on the response of predators to fragmentation-induced changes in habitat, a primary focus of research should be the relationships between forest edges, habitat characteristics, and the abundance or activity of dominant nest predators [6,15,20]. This point has largely been overlooked, but might well be the key to understanding the observed variability in edge effects on nest predation [6].

Additionally, because variability in edge effects is likely to be due, at least in part, to differences among predators [6,12,13,16,21], there is an urgent need for studies of edge effects on nest predation in a more representative range of regions [11]. At present, the vast majority of studies have been conducted in North America or Europe, with only a handful in the southern hemisphere [11,13]. Overcoming this geographical bias will be important both for region-specific management and for providing insights into the generality of existing theory which has been derived primarily from North American and European systems.

The ship rat (*Rattus rattus*; also known as the black rat) is a globally invasive nest predator which has been widely implicated in the declines of birds and other taxa in many regions, particularly where prey species are naive to mammalian predators [22,23–30]. In New Zealand, for example, ship rats are thought to be the primary predator of the eggs and chicks of forest birds [31]. However, despite their major impact on the viability of bird populations in many systems, we know of only two studies that have examined the effects of forest edges on ship rats. Delgado et al. [32] examined the effects of forest edges on seed predation rates by ship rats in the Canary Islands. They found that predation rates were higher at the forest edge than in the forest interior, and hypothesised that higher vegetation density at the forest edge may have been responsible. In contrast, Christie et al. [33] found that ship rat capture rates in New Zealand forests were lower at the forest edge, although they did not discuss potential reasons for this finding.

In this study, we examined the effects of forest edges on ship rat capture rates (which we used as a proxy for nest predation risk) in forest fragments in a pastoral landscape in New Zealand. We took a mechanistic approach, examining how local habitat characteristics might give rise to edge effects as well as causing variability in those effects. Based on previous findings that ship rat encounter rates correlate with vegetation structure [34,35] and that vegetation structure is often affected by forest edges [5], we hypothesised that effects of forest edges on rat capture rates would operate via changes in vegetation structure within forest patches. Moreover, because vegetation structure can also be strongly affected by livestock grazing [36,37], we additionally hypothesised that the effects of forest edges on rat capture rates would vary depending on whether forest fragments were grazed by livestock. Such an interaction could arise, for example, if grazing overwhelmed any effects of edge distance on vegetation structure or if grazing itself was concentrated at the forest edge.

Our specific aims were to (1) examine whether ship rat capture rates vary with distance from edge in forest fragments in New Zealand; (2) examine if livestock grazing causes variability in the strength of this relationship; and (3) determine the mechanisms underlying both the occurrence of edge effects and variability in their strength, by testing the hypothesis that edge effects on rat capture rates are mediated by changes in vegetation structure.

Methods

Study design and data collection

The study took place near the town of Benneydale (75° 220′E, 38° 320′S), in the Waikato region of central North Island. The landscape consists primarily of remnants of mature native podocarp-broadleaf forest dominated by tawa (*Beilschmiedia tawa*) in a matrix of pastoral farming. Native forest cover is approximately 23% of the landscape, as measured in ArcGIS (version 10.0) from the New Zealand Land Cover Database in a 10 km radius from the centre of the study area. Remnants in the area contain most of the common native and exotic birds found in forests throughout mainland New Zealand, and many of these native species are believed to be affected by nest predation from ship rats [22]. Most notably, the viability of a North Island robin (*Petroica longipes*) population distributed among these remnants has been shown to depend on a low abundance of ship rats [38].

Rats were sampled in 13 forest patches, ranging from 2–19 ha. Six of these were fenced to exclude livestock, five were grazed, and a further two contained a fenced and an unfenced section. The fenced and unfenced sections of these partly fenced patches were treated as separate patches for our analyses, and hereafter we refer to a sample size of 15 patches. There was no targeted rat control in any of the patches at the time the study began. All patches were on private land, with land access permissions obtained through the Tiroa E & Te Hape B Trusts (http://www.tiroatehape.maori.nz). Specific site locations are provided as GPS coordinates in Supporting Information.

Rats were captured during a kill trapping program which was part of a wider study on the effects of ship rats on North Island robins. Victor snap traps, placed within handmade corflute tunnels and pegged to the ground, were laid out on a 50 m grid throughout the entire area of each patch. These were left in place for 4–5 months prior to trapping to reduce rat neophobia, then baited with peanut butter, set, and checked the following day to see if they caught a rat. Individual patches were trapped in August of either 2008 or 2009, with no replication across years. Trapping was conducted on the same night for all patches within each year. The Massey University Animal Ethics Committee determined that trapping did not require ethics approval, but the traps used pass the humaneness standard for vertebrate control set by New Zealand's National Animal Welfare Advisory Committee [39]. The study did not involve any protected or endangered species.

In total, we deployed 488 traps across our 15 patches, with the number of traps per patch ranging from 5–76 (mean = 35). The location of each trap was recorded with a handheld GPS. We used Google Earth's drawing tools to create polygons that mapped the boundary of each forest patch, then used ArcGIS to measure the distance from each trap to the nearest polygon boundary. We estimate measurement error for patch boundaries to be $\sim +/-$ 10 m, a similar scale to GPS measurement error. Estimated distance of traps from the forest edge ranged from 0–212 m. Patch sizes were similar between grazed and ungrazed patches (Table S1).

We used the binomial data on whether a trap caught a rat over a single night as the response variable in our statistical models (see below). We chose this response variable because of its similarity to the 'rat tracking rate' (RTR) index, which measures the proportion of footprint tracking tunnels tracked by rats (or equivalently, the probability that an individual tunnel will be tracked) over a single night, for tunnels spaced at 50 m and baited with peanut butter. The RTR index is the standard protocol used as a proxy for rat abundance in New Zealand [40], and has been shown to relate to predation risk for several bird species [41–43]. Moreover, it may be a better measure of nest predation rates than artificial nests [34]. As a result, we believe that our response variable was likely to reflect the risk of nest predation from ship rats. The use of snap traps within our tunnels allowed us to identify rats as ship rats (as opposed to Norway rats, *Rattus norvegicus*, which are also occasionally caught in mainland New Zealand forests [31]).

To examine whether effects of edge distance on rat capture rates operated via changes in vegetation structure, we took a number of measurements to quantify the structure of the vegetation in a 15 m radius around each trap: (1) understorey density, scored as either 0 ('sparse'), 1 ('average'), or 2 ('dense'); (2) percent cover of dead tree fern fronds on the ground; (3) percent cover of living vegetation <1.5 m above ground; (4) percent canopy cover; (5) average canopy height, scored as either 0–5 m, 5–15 m, or >15 m; and (6) presence of vines (supplejack, *Ripogonum scandens*). We selected this set of variables because we believed that together they captured the variation in vegetation structure that we observed among our trap sites. Variables 1–5, which were difficult to measure objectively, were measured by two observers and averaged. Vegetation structure data, and all other data used in this study, are provided in Table S4.

Statistical analyses

Net effects of distance from edge and livestock grazing on rat capture probability. Initially, we used binomial generalised linear mixed models ('GLMMs') to examine the net effects of distance from forest edge, livestock grazing, and their interaction on the probability of rat capture (i.e. without examining the hypothesised pathways via altered vegetation structure). GLMMs were fitted using the 'lme4' package [44] in R version 2.13.1 [45]. We used binomial GLMMs because trapping outcome was a binary variable, and because of the hierarchical structure of our data (traps nested within patches). Patch was included as a random factor in these models to account for this nesting.

We developed five candidate models that included all combinations of distance from edge, livestock grazing, and their interaction as predictors of rat capture probability, including an intercept-only null model, then used AIC to estimate which of these models best described variation in rat capture probability. Where there were multiple models that were reasonably well supported by the data (i.e. a ΔAIC of <4 [46]), we used the 'model.avg' function in the R package 'MuMIn' [47] to average parameter estimates across these models. This approach reduces bias in parameter estimates when multiple models are similarly likely [46].

Testing the role of vegetation structure in edge effects and livestock grazing effects. We used path analysis to test our hypothesis that distance from edge, livestock grazing, and their interaction affected rat capture probability via changes in vegetation structure. Path analysis allowed us to test support for this hypothesised causal structure, and to estimate the strength of each relationship while accounting for all of the relationships implied by the causal structure [48]. We used generalised multilevel confirmatory path analysis [49], which allowed us to account for the nesting of traps within forest patches and binomial rat capture data [49].

Generalised multilevel confirmatory path analysis does not allow for reciprocal effects among variables [49]. However, most of our six measured vegetation variables were correlated with one another, and we could not rule out the possibility that they were reciprocally related. For example, a high cover of dead tree fern fronds was likely to reduce the recruitment of seedlings into the understorey, while at the same time a dense understorey was likely to suppress the growth of tree ferns. To avoid the possibility of reciprocal relationships, we took the conservative approach of including only two vegetation variables in our path models: understorey density and presence of vines. These two variables were not correlated with each other but together were correlated with all other vegetation variables.

However, this conservative approach could potentially underestimate the extent to which vegetation structure mediated the effects of forest edges and livestock grazing on rat capture rates, because any variation in vegetation structure that was not captured by understorey density or the presence of vines could be attributed to paths not mediated by these variables. We therefore conducted a sensitivity analysis to examine whether the inclusion of additional measures of vegetation structure would increase the extent to which edge effects were mediated by vegetation structure in our path model (Appendix S1). To do this, we converted our six vegetation variables into four uncorrelated principal components axes, which together explained 83% of variation in vegetation structure, and used these components as the measures of vegetation structure in our path models. We did not use this approach in our main analysis because the inclusion of principal component axes may have reduced our ability to reject incorrectly specified path models (Appendix S1).

In developing the causal structure of our path model, we initially constructed a 'full' version of our hypothesised model which included all possible paths consistent with our hypothesis: (1) distance from edge, livestock grazing, and their interaction all influenced the two vegetation variables, and (2) both vegetation variables affected rat capture probability. We also included direct paths from distance from edge to rat capture probability and from livestock grazing to rat capture probability, to allow for any effects of these variables on rat capture probability that did not operate through our measured vegetation variables (Figure 1a).

We attempted to reduce this full model into a more parsimonious one by dividing it into a series of sub-models, one for each endogenous variable (i.e. a variable with paths leading into it) in the path model. Each of these sub-models was a standard mixed effects model, with the endogenous variable as the response variable and all variables directly 'upstream' of it as its predictors [49,50]. These models included patch as a random factor, and were either linear mixed models ('LMMs'; vegetation variables as response) or binomial GLMMs (rat capture probability as response). LMMs and GLMMs were fitted with the 'nlme' package [51] and 'lme4' package, respectively, in R. We fitted these models with restricted maximum likelihood ('REML'), inspected plots of residuals against fitted values and against each predictor, and where necessary dealt with heterogeneity using nlme's 'varIdent' function [52]. We then attempted to reduce the number of paths in each sub-model by re-fitting them with maximum likelihood and using backwards selection based on AIC [50].

Reduced models were re-fitted with REML, re-validated, and then reassembled into our final, most parsimonious path model [50]. The overall fit of this final path model was then tested using Shipley's [49] d-sep test for generalised multilevel path models. This test involves identifying a set of 'independence claims' (pairs of variables which should be uncorrelated under statistical control if the path model is correct), then measuring whether observed levels of correlation across all independence claims can be explained by random variation [49]. For our final path model, path coefficients were calculated as the estimated slopes of each of the retained variables in the final LMM and GLMM sub-models [49].

Path analyses sometimes calculate path coefficients based on standardised variables to enable measurement of the relative

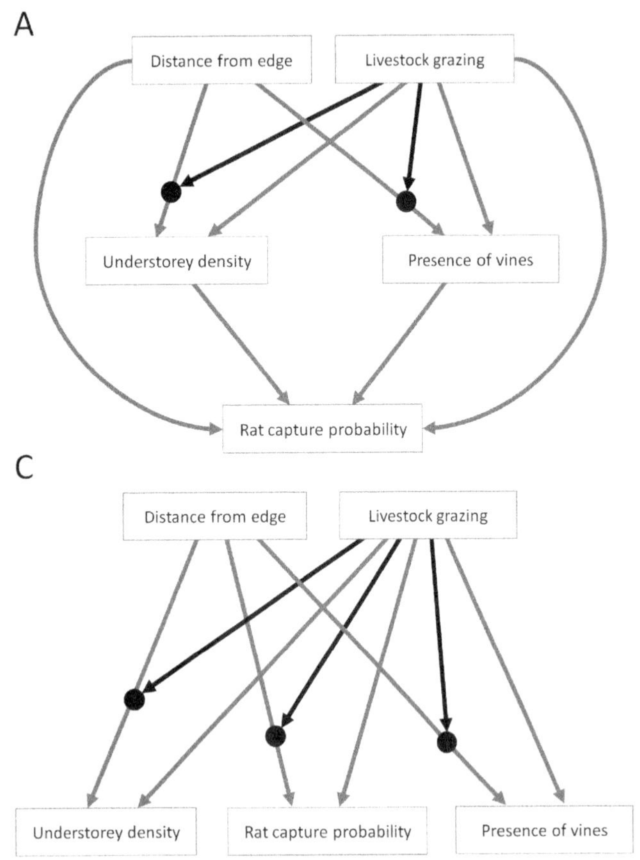

Figure 1. Alternative models of causal relationships between edge distance, grazing, vegetation structure, and rat capture probability. A: model hypothesised in this study, in which distance from forest edge and grazing indirectly influenced rat capture probability by driving changes in vegetation structure, and in which grazing altered the strength of the distance from edge effect; **B**: an alternative 'direct edge effects' model, where livestock grazing effects on rat capture probability were mediated by vegetation structure, but edge effects were not; **C**: an alternative 'direct effects only' model, where distance from forest edge, livestock grazing, and their interaction affected vegetation structure variables and rat capture probability, but where neither edge effects nor livestock grazing effects on rat capture probability were mediated by vegetation structure. Black arrows represent variables which alter the strength of the relationship they feed into (i.e. interaction effects).

strengths of the different pathways (e.g. [50]). We did not take this approach, instead using unstandardised versions of all variables. This was because different path coefficients had different interpretations as a result of different transformations of predictor variables (distance from edge was square root transformed, while other variables were untransformed; see below) and different link functions for response variables (logit link for rat captures, identity link for other variables) used in models.

Alternative causal models. A lack of evidence against a hypothesised path model does not necessarily mean that it is correct, so researchers should also consider alternative models that could plausibly explain relationships among variables [53]. We developed two alternative models that could plausibly explain relationships between distance from edge, livestock grazing, vegetation, and rat capture probability: (1) a 'direct edge effects' model, where livestock grazing effects on rat capture probability were mediated by vegetation structure, but edge effects were not (Figure 1b); and (2) a 'direct effects only' model, where neither edge effects nor livestock grazing effects on rat capture probability were mediated by vegetation structure (Figure 1c). We used the d-sep test [49] to determine whether there was also support for these causal relationships among variables.

Data transformations. Distance from edge was square-root transformed prior to all analyses, for two reasons. First, some of the paths in our models were modelled with LMMs, and the square-root performed best out of a range of transformations at linearising relationships among variables. Second, this transformation forced fitted values of the relationship between distance from edge and rat capture probability to be a decelerating curve, rather than the sigmoidal curve fitted by a standard binomial GLMM on untransformed predictor variables. We considered a decelerating curve to be a more biologically realistic shape for edge effects, because changes are expected to be greatest near the forest edge.

Results

We caught rats in 22% of our traps across all patches (108 rats across 488 traps), all of which were ship rats. Rat capture rates varied widely among patches, with the proportion of traps that caught rats ranging from 0% to 65%.

Net effects of distance from edge and livestock grazing on rat capture probability

In our initial 'net effects' models of the effects of distance from edge and livestock grazing on rat capture probability, AIC suggested the global model was best (i.e an effect of distance from edge, grazing, and their interaction), but there was a similar weight of evidence (Δ AIC <1) for all three models that included a distance effect (i.e. the global model, the distance and grazing main effects model, and the distance-only model). In contrast, there was little evidence to support the livestock grazing-only or intercept-only models (Δ AIC >4) (Table S2). We therefore used a model averaging approach to estimate parameters across the three models that included a distance effect.

Fitted values from these model-averaged parameters showed that rat capture probability clearly increased with distance from edge. In forest fragments grazed by livestock, capture probabilities were estimated to be 78% lower at the forest edge than 118 m into the forest interior (the greatest distance measured in unfenced fragments) (Figure 2). This trend was similar in fragments from which livestock were excluded, with capture probabilities estimated to be 51% lower at the forest edge than 118 m into the forest interior. These ungrazed fragments also tended to have higher rat capture probabilities for a given distance from the forest edge (Figure 2). However, unconditional confidence intervals marginally overlapped zero for the effects of distance, grazing, and their interaction (Table S3). Nonetheless, we believe there is strong support for an effect of edge distance on rat capture rates. First, unconditional confidence intervals are conservative in that they incorporate uncertainty in model structure. Second, only those models which included a distance effect were well-supported as judged by AIC (Table S2), and a distance main effect was significant in each of these models (distance-only model: $p = 0.01$; main effects model: $p = 0.02$; full model: $p = 0.01$).

Mediating effects of vegetation structure on edge effects and livestock grazing effects

Backwards variable selection within sub-models allowed us to simplify our hypothesised path model (i.e. the model in which the effects of forest edges and livestock grazing on rat capture probability were mediated by vegetation structure; Figure 1a). This resulted in the removal of the direct effect of livestock grazing on rat capture probability, as well as the livestock grazing by distance from edge interaction on the presence of vines (Figure 3). This model provided a good fit to the data (d-sep test, $\chi^2 = 8.98$, $df = 8$, $p = 0.344$). In contrast, our two alternative path models which also could have plausibly explained relationships among distance from edge, livestock grazing, vegetation structure and rat capture probability (Figure 1b,c) were both rejected by the d-sep test ($p<0.05$), providing further support for our hypothesised path model.

In line with our causal hypothesis, path coefficients for the final reduced path model (Figure 3) suggested that the effects of distance from edge on rat capture probability were largely mediated by changes in vegetation structure: the understorey became denser and vines became more common with increasing distance from the forest edge, and these in turn were associated with increased rat capture probability. The model also suggested a significant direct effect of distance from edge on rat capture probability which operated independently of our measures of vegetation structure. However, part of this direct effect of distance from edge may have included vegetation-mediated effects which were not captured by our two vegetation structure variables. Indeed, a sensitivity test found that this direct effect weakened and became non-significant when the model included additional measured variation in vegetation structure (Figure S1). Our path model also suggested that livestock grazing had a negative effect on rat capture probability, which was mediated by changes in vegetation structure (Figure 3). Finally, the model suggested that grazing altered the strength of the vegetation-mediated effect of distance from edge on rat capture probability, with the effects of

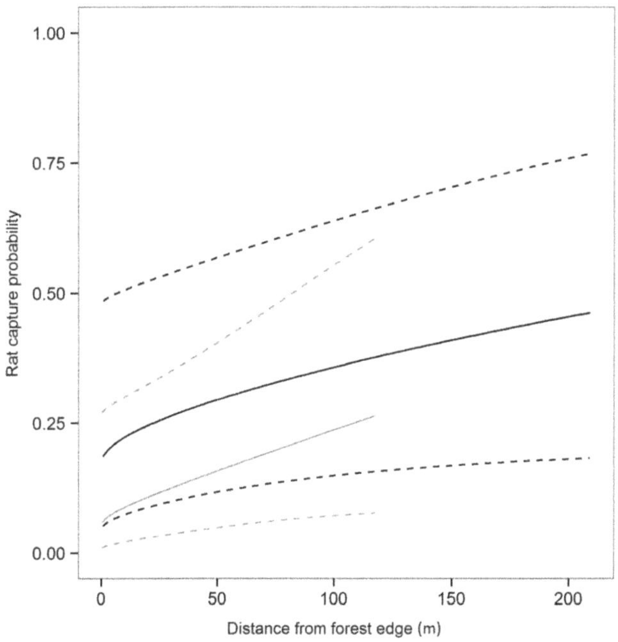

Figure 2. Model-predicted values for the relationship between rat capture probability and distance from forest edge. Values were predicted separately for those patches that were grazed by livestock (grey lines) and those that were not (black lines). Dotted lines show 95% confidence intervals for predicted values. The truncated values for grazed patches reflect the reduced range over which edge distances were measured in these patches.

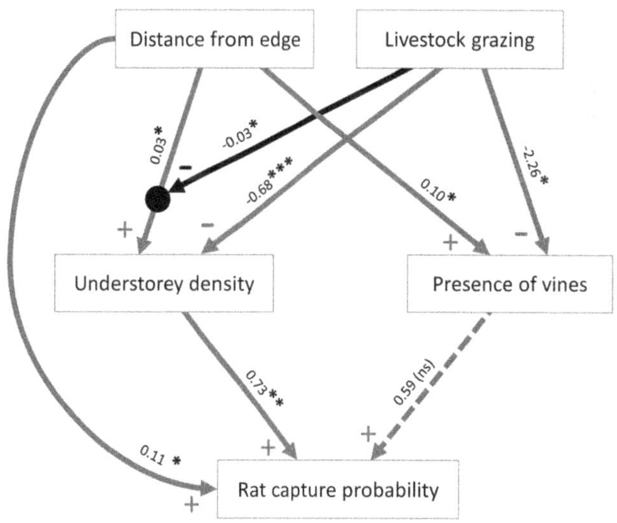

Figure 3. Final path model showing relationships between distance from forest edge, grazing, vegetation structure, and rat capture probability. Plus/minus symbols at each arrow head show a positive/negative effect, and asterisks denote significance of each path: ***<0.001; **<0.01; *<0.05; ns = not significant. Dotted arrows show non-significant paths, and black arrows show variables which alter the strength of the relationship they feed into (i.e. interaction effects). Note that path coefficients are not directly comparable because of different scales of measurement of predictor variables (distance square root transformed; other variables untransformed) and different link functions for response variables (logit link for rat capture probability; identity link for other variables) used in models.

distance from edge on understorey density being weaker in grazed than ungrazed fragments (Figure 3).

Discussion

Predicting and managing the effects of forest edges on nest predation rates requires an understanding of the mechanisms driving their occurrence, but few studies have moved beyond describing spatial variation in nest predation rates to testing hypotheses about the underlying causes [5,6,18,54]. In this study, path analysis allowed us to identify edge-driven changes in vegetation structure as a major driver of edge effects on the capture rates of ship rats, a globally invasive nest predator. Vegetation structure has been found to correlate with nest predation rates previously [55–57], although we know of no studies that have identified change in vegetation structure as a driver of edge effects on nest predation. We also found that the relationships between forest edges, vegetation structure, and ship rats gave rise to 'reverse' edge effects in our study system, in which rat capture rates were higher in the interior of forest fragments than at the forest edge. This finding contrasts with the common assumption that forest edges increase nest predation rates [9], and supports a growing body of evidence which suggests that edge effects on nest predation rates are inconsistent, context dependent, and ultimately unpredictable without an understanding of the mechanisms involved (e.g. [13,15,16,21,56,58]).

The 'within patch' mechanism underpinning edge responses in our study system contrasts with the widely held belief that edge effects on nest predation rates result from large-scale processes that operate beyond individual forest patches (e.g. [6,13,18,58]). In particular, the most frequently cited mechanism for edge effects on nest predation rates is 'matrix spill-over', where generalist predators reach high densities in adjacent matrix land uses and then forage in forest remnants, leading to elevated nest predation at the forest edge [6,13,20,59,60]. Matrix spill-over is likely to cause edge effects wherever dominant nest predators (1) increase in response to matrix land uses and (2) range across both forest and matrix habitats. However, while this has often proven true for predator communities in the temperate northern hemisphere studies that have dominated the literature [16,18] it was unlikely to be the case in our study system, because ship rats typically decrease in response to pastoral land use and remain within forest habitats [31,61].

Instead, ship rats responded to edge-driven changes in habitat within forest patches, resulting in capture rates that were higher in the forest interior than at the forest edge. This response may have occurred via the movement of individuals, since the typical home range length of a ship rat (100–300 m; [35]) would encompass multiple traps, or via population-level changes in local abundance. It may also underlie the results of a previous study which found the same 'reverse' edge effects on ship rat capture rates in New Zealand forests [33]. We expect low matrix spill-over for other matrix types in New Zealand (for example, urban areas, plantation forests, or cropland), since ship rats also decrease in response to these land uses [31,62,63].

We suggest that where dominant nest predators are primarily forest species that do not range across forest and matrix habitats, edge effects on nest predation rates may be more closely tied to changes in habitat within forest patches than to changes in habitat in the wider landscape. In these cases, forest edges may have positive, neutral, or negative effects on nest predation rates, depending on if and how important resources for dominant predators vary with distance from the forest edge. This is in line with suggestions that variability in edge effects on nest predation may largely result from differences in habitat use by different predator communities [6,15,16,21].

Our net effects models suggested that livestock grazing reduced the capture rates of ship rats, and path analysis indicated that this effect was also mediated by vegetation structure. These results support previous findings on the effects of livestock grazing on ship rat capture rates in New Zealand [35]. In contrast, evidence for our hypothesis that livestock grazing altered the strength of edge effects on rat capture rates was equivocal. In support of this hypothesis, our path analysis suggested that grazing moderated the effect of distance from edge on vegetation structure, which in turn influenced rat capture probability. Similarly, our AIC-best net effects model specified that the effect of distance from edge on rat capture probability depended on whether fragments were grazed. However, there was similar support for models in which edge effects operated independently of grazing effects, and model-averaged fitted values suggested that any moderating influence of livestock grazing on the effect of forest edges on rat capture rates was weak.

Although we identified vegetation structure as a major mechanism underlying both edge effects and livestock grazing effects on rat capture rates, vegetation structure was potentially correlated with a large number of other variables that determine habitat quality for ship rats. For example, microclimate, vegetation composition, and densities of ship rat prey (primarily invertebrates; [31]) and ship rat predators (primarily stoats *Mustela erminea*, but also feral cats *Felis catus*, weasels *M. nivalis*, and ferrets *M. putorius*); [31]) may have been correlated with vegetation structure (e.g. [5,64,65]) and were likely to influence habitat quality for ship rats. Moreover, while our path model included understorey density and presence of vines as our measures of vegetation structure, these variables were also correlated with our additional vegetation variables (canopy height, canopy cover, cover of dead tree fern fronds, and cover of living vegetation <1.5 m above ground). As such, ship rats may not have been responding to understorey density and the presence of vines per se, but these variables may instead have been proxies for the true measures of habitat quality that influenced ship rat capture rates.

It is also important to point out that the density of birds' nests might well be correlated with vegetation structure. If so, it is possible that increases in rat density could be balanced by increases in nest density, such that ship rat capture rates become decoupled from per capita (i.e. per-nest) predation rates. Such a decoupling could occur, for example, if rats become satiated at high nest densities and decrease their per capita rate of predation (i.e. a Type II or Type III functional response; [66]). We know of no studies which have found that nest density increases with increasing vegetation structure for any of the native bird species found in our study area, although this relationship has been found for a species which occurs elsewhere in New Zealand (saddleback *Philesturnus carunculatus*; [67]).

Our path analysis suggested that there were additional effects of distance from edge on rat capture probability that were not mediated by understorey density or presence of vines. While this may reflect the fact that forest edges affect rat capture rates in ways unrelated to vegetation structure, it is also likely that this 'direct path' at least partially represented the effects of vegetation structure on rat capture rates that were not captured by understorey density and vines alone. Indeed, in a sensitivity analysis that included all six of our original vegetation variables as an orthogonal set of principle components (to avoid potential reciprocal effects among variables), the direct path between distance from edge and rat capture probability weakened and became non-significant (Figure S1). This suggests that the observed direct path between distance from edge and rat capture probability may be a consequence of the small number of variables we used to characterise vegetation structure.

Our study highlights the value of a mechanistic approach for understanding variability in edge effects on nest predation rates and for predicting how forest edges will affect the abundance or impacts of ship rats elsewhere. Our approach suggested that changes in vegetation structure largely mediated both the occurrence of edge effects on rat capture probability and variability in their strength due to livestock grazing. This implies that edge effects on ship rats can be predicted from the response of vegetation structure to forest edges. Although our measures of vegetation density increased with distance from the forest edge, the reverse is often true [5,68], even within other lowland podocarp-broadleaf forests in New Zealand [64]. This suggests that edge effects on ship rats may reverse in other systems. Support for this suggestion comes from a previous study of edge effects on ship rats in the Canary Islands, where vegetation density was higher at the edge than in the forest interior [69]. As predicted by our path model, ship rat density was also higher at the forest edge, the opposite trend to that found in our study [69]. These results suggest that variability in edge effects on ship rats can be understood and predicted, even to the point of changing sign, based on an understanding of how vegetation structure responds to forest edges. Our results also raise the possibility of reducing the impacts of ship rats by reducing the structural complexity of vegetation within patches, although this strategy would generally be inappropriate in New Zealand given that native forest of high conservation value is often characterised by a structurally complex understorey [35].

Path analysis is a promising approach for understanding the roles of forest edges and local habitat in influencing nest predation rates. However, studies of edge effects by their very nature tend to produce hierarchically structured data (i.e. measurements are taken at multiple distances from edge across several forest patches) and non-normal response variables (e.g. predator capture probability, predator counts, or daily predation rates of real or artificial nests), making analysis with traditional path modelling methods difficult [49]. Recent methodological developments that enable the inclusion of random effects and non-normal response variables in path models, such as generalised multilevel confirmatory path analysis, make path analysis a powerful approach for understanding the mechanisms that underlie edge-driven changes in nest predation rates.

Supporting Information

Figure S1 Final path model showing estimated causal relationships between distance from the forest edge, livestock grazing, vegetation structure (principal components Veg PC1-4) and probability of rat capture. Prior to path analysis vegetation variables were converted to six principal components, but only the first four were related to distance from edge, livestock grazing, or rat capture probability, so Veg PC5–6 were omitted from the analysis. The first four components explained 34%, 22%, 15%, and 12% of the total measured variance in vegetation structure respectively. Plus/minus symbols at each arrow head show a positive/negative effect, and asterisks denote significance of each path: ***<0.001; **<0.01; *<0.05; ns = not significant. Dotted arrows show non-significant paths, and black arrows show variables which alter the strength of the relationship they feed into (i.e. interaction effects). Note that path coefficients are not directly comparable because of different scales of measurement of predictor variables (distance square root transformed; other variables untransformed) and different link functions for response variables (logit link for rat capture probability; identity link for other variables) used in models.

Table S1 Summary statistics for patch size and sampling effort (number of rat traps) per patch for grazed and ungrazed patches.

Table S2 Levels of support for candidate models used to test the effects of distance from forest edge, livestock grazing, and their interaction on rat capture probability. ΔAIC measures change in AIC relative to the best model, while the Akaike weight w_i gives the probability that model i is the most parsimonious in the set. The table shows fixed terms only, but candidate models also included a patch-level random intercept to model non-independence of rat traps within the same forest patch.

Table S3 Parameter estimates and 95% confidence intervals (in parentheses) for effects of distance from edge, livestock grazing, and their interaction on rat capture probability. Multiple candidate models were supported by the data (ΔAIC <4; see Table S1), so parameter estimates and confidence intervals were averaged across these models to account for uncertainty in model structure. Parameter estimates are given on the scale of the logit link used in binomial GLMMs.

Table S4 Data used in the analyses.

Acknowledgments

We thank the Tiroa & Te Hape Trusts and R. Peacocke for permission to conduct the research on their land, and farm managers R. Peacocke, I. Valler, T. Ballantine, R. Walker, and L. Crown for their support. Trapping was conducted with help from postgraduate Wildlife Management students at Massey University. We thank members of the Cross Validated online statistics forum for suggesting phylogenetic PCA as a means of producing uncorrelated principal components of hierarchically structured data, and three anonymous reviewers for comments which improved the manuscript.

Author Contributions

Conceived and designed the experiments: DPA JR RKD. Performed the experiments: DPA PB NG RP AHE KS. Analyzed the data: JR. Wrote the paper: JR DPA RKD PB NG RP AHE KS.

References

1. Bryant D, Nielsen D, Tangley L (1997) The last frontier forests - ecosystems and economies on the edge. Washington D.C.: World Resources Institute.
2. Foley JA, DeFries R, Asner GP, Barford C, Bonan G, et al. (2005) Global consequences of land use. Science 309: 570–574.
3. Day JW, Hall CA, Yáñez-Arancibia A, Pimentel D, Ibáñez Martí C, et al. (2009) Ecology in Times of Scarcity. BioScience 59: 321–331.
4. Ewers RM, Didham RK (2006) Confounding factors in the detection of species responses to habitat fragmentation. Biological Reviews 81: 117–142.
5. Murcia C (1995) Edge effects in fragmented forests: implications for conservation. Trends in Ecology & Evolution 10: 58–62.
6. Lahti D (2009) Why we have been unable to generalize about bird nest predation. Animal Conservation 12: 279–281.

7. Newton I (1998) Population limitation in birds. London: Academic Press.
8. Martin TE (1993) Nest predation and nest sites. BioScience 43: 523–532.
9. Lahti D (2001) The "edge effect on nest predation" hypothesis after 20 years. Biological Conservation 99: 365–374.
10. Andrén H (1995) Effects of landscape composition on predation rates at habitat edges. In: Fahrig L, Merriam G, editors. Mosaic Landscapes and Ecological Processes. London: Chapman & Hall. pp. 225–255.
11. Batary P, Baldi A (2004) Evidence of an edge effect on avian nest success. Conservation Biology 18: 389–400.
12. Paton P (1994) The effect of edge on avian nest success: how strong is the evidence? Conservation Biology 8: 17–26.
13. Chalfoun AD, Thompson FR, Ratnaswamy MJ (2002) Nest predators and fragmentation: a review and meta-analysis. Conservation Biology 16: 306–318.
14. Donovan TM, Jones PW, Annand EM, Thompson III FR (1997) Variation in local-scale edge effects: mechanisms and landscape context. Ecology 78: 2064–2075.
15. Falk KJ, Nol E, Burke DM (2011) Weak effect of edges on avian nesting success in fragmented and forested landscapes in Ontario, Canada. Landscape Ecology 26: 239–251.
16. Cox WA, Thompson III FR, Faaborg J (2012) Landscape forest cover and edge effects on songbird nest predation vary by nest predator. Landscape ecology 27: 659–669.
17. Debinski DM, Holt RD (2000) A survey and overview of habitat fragmentation experiments. Conservation Biology 14: 342–355.
18. Thompson FR, Donovan TM, DeGraaf RM, Faaborg J, Robinson SK (2002) A multi-scale perspective of the effects of forest fragmentation on birds in eastern forests; George TL, Dobkin DS, editors. Camarillo: Cooper Ornitholigical Society. 8–19 p.
19. Tewksbury JJ, Garner L, Garner S, Lloyd J, Saab V, et al. (2006) Tests of landscape influence: nest predation and brood parasitism in fragmented ecosystems. Ecology 87: 759–768.
20. Storch I, Woitke E, Krieger S (2005) Landscape-scale edge effect in predation risk in forest-farmland mosaics of central Europe. Landscape Ecology 20: 927–940.
21. Nour N, Mattysen E, Dhont A (1993) Artificial nest predation and habitat fragmentation: different trends in bird and mammal predators. Ecography 16: 111–116.
22. Innes J, Kelly D, Overton JM, Gillies C (2010) Predation and other factors currently limiting New Zealand forest birds. New Zealand Journal of Ecology 34: 86–114.
23. Moors P, Atkinson I, Sherley G (1992) Reducing the rat threat to island birds. Bird Conservation International 2: 93–114.
24. Jouventin P, Bried J, Micol T (2003) Insular bird populations can be saved from rats: a long-term experimental study of white-chinned petrels *Procellaria aequinoctialis* on Ile de la Possession (Crozet archipelago). Polar Biology 26: 371–378.
25. Monteiro LR, Ramos JA, Furness RW (1996) Past and present status and conservation of the seabirds breeding in the Azores Archipelago. Biological Conservation 78: 319–328.
26. Martin J-L, Thibault J-C, Bretagnolle V (2000) Black rats, island characteristics, and colonial nesting birds in the mediterranean: consequences of an ancient introduction. Conservation Biology 14: 1452–1466.
27. Harris DB, Macdonald DW (2007) Interference competition between introduced black rats and endemic Galápagos rice rats. Ecology 88: 2330–2344.
28. MacFarland CG, Villa J, Toro B (1974) The Galápagos giant tortoises (*Geochelone elephantopus*) Part I: Status of the surviving populations. Biological Conservation 6: 118–133.
29. Pryde MA, O'Donnell CF, Barker RJ (2005) Factors influencing survival and long-term population viability of New Zealand long-tailed bats (*Chalinolobus tuberculatus*): Implications for conservation. Biological Conservation 126: 175–185.
30. Towns DR, Daugherty CH (1994) Patterns of range contractions and extinctions in the New Zealand herpetofauna following human colonisation. New Zealand Journal of Zoology 21: 325–339.
31. Innes J (2001) Advances in New Zealand mammalogy 1990–2000: European rats. Journal of the Royal Society of New Zealand 31: 101–125.
32. Smith AC, Koper N, Francis CM, Fahrig L (2009) Confronting collinearity: comparing methods for disentangling the effects of habitat loss and fragmentation. Landscape Ecology 24: 1271–1285.
33. Christie J, Brown D, Westbrooke I, Murphy EC (2009) Environmental predictors of stoat (*Mustela erminea*) and ship rat (*Rattus rattus*) capture success. DOC Research & Development Series 305. Wellington, New Zealand Department of Conservation.
34. Getzlaff C, Sievwright K, Hickey-Elliot A, Armstrong DP (2013) Predator indices from artificial nests and tracking tunnels: do they tell the same story? New Zealand Journal of Ecology 37: 232–239.
35. Innes J, King CM, Bridgman L, Fitzgerald N, Arnold G, et al. (2010) Effect of grazing on ship rat density in forest fragments of lowland Waikato, New Zealand. New Zealand Journal of Ecology 34: 227–232.
36. Hobbs RJ (2001) Synergisms among habitat fragmentation, livestock grazing, and biotic invasions in southwestern Australia. Conservation Biology 15: 1522–1528.
37. Smale MC, Dodd MB, Burns BR, Power IL (2008) Long-term impacts of grazing on indigenous forest remnants on North Island hill country, New Zealand. New Zealand Journal of Ecology 32: 57.
38. Armstrong DP, Gorman NIC, Pike R, Kreigenhofer B, McArthur N, et al. (2014) Strategic Rat Control for Restoring Populations of Native Species in Forest Fragments. Conservation Biology 28: 713–723.
39. Beausoleil N, Fisher P, Warburton B, Mellor D (2010) How humane are our pest control tools? Part 1. Vertebrate toxic agents and kill traps in mammal species. Christchurch, New Zealand Ministry of Agriculture and Forestry.
40. Gillies C, Williams D (2013) Using tracking tunnels to monitor rodents and mustelids. Version 2.5.2. Hamilton, Department of Conservation.
41. Innes J, Nugent G, Prime K, Spurr EB (2004) Responses of kukupa (*Hemiphaga novaeseelandiae*) and other birds to mammal pest control at Motatau, Northland. New Zealand Journal of Ecology 28: 73–81.
42. Innes J, Hay R, Flux I, Bradfield P, Speed H, et al. (1999) Successful recovery of North Island kokako *Callaeas cinerea wilsoni* populations, by adaptive management. Biological Conservation 87: 201–214.
43. Armstrong DP, Raeburn EH, Lewis RM, Ravine D (2006) Estimating the viability of a reintroduced New Zealand robin population as a function of predator control. Journal of Wildlife Management 70: 1020–1027.
44. Bates D, Maechler M, Bolker B (2011) lme4: Linear mixed-effects models using S4 classes. Version 0.999999-0.
45. R Core Team (2012) R: A Language and Environment for Statistical Computing. Vienna: R Foundation for Statistical Computing.
46. Burnham KP, Anderson DR (2002) Model selection and multi-model inference: a practical information-theoretic approach. New York: Springer Verlag.
47. Barton K (2013) MuMin: multi-model inference. Version 1.9.13.
48. Grace J (2006) Structural equation modeling and natural systems. Cambridge, UK: Cambridge University Press.
49. Shipley B (2009) Confirmatory path analysis in a generalized multilevel context. Ecology 90: 363–368.
50. Laliberté E, Tylianakis JM (2012) Cascading effects of long-term land-use changes on plant traits and ecosystem functioning. Ecology 93: 145–155.
51. Pinheiro J, Bates D, DebRoy S, Sarkar D, Team RC (2013) nlme: Linear and Nonlinear Mixed Effects Models. Version 3.1–111.
52. Zuur AF, Ieno EN, Walker N, Saveliev AA, Smith GM (2009) Mixed effects models and extensions in ecology with R. London: Springer.
53. Grace J (2008) Structural equation modelling for observational studies. Journal of Wildlife Management 72: 14–22.
54. Tewksbury JJ, Levey DJ, Haddad NM, Sargent S, Orrock JL, et al. (2002) Corridors affect plants, animals, and their interactions in fragmented landscapes. Proceedings of the National Academy of Sciences of the United States of America 99: 12923.
55. Major RE, Kendal CE (1996) The contribution of artificial nest experiments to understanding avian reproductive success: a review of methods and conclusions. Ibis 138: 298–307.
56. Spanhove T, Lehouck V, Boets P, Lens L (2009) Forest fragmentation relaxes natural nest predation in an Afromontane forest. Animal Conservation 12: 267–275.
57. Martin JL, Joron M (2003) Nest predation in forest birds: influence of predator type and predator's habitat. Oikos 102: 641–653.
58. Hartley MJ, Hunter ML (1998) A Meta-Analysis of Forest Cover, Edge Effects, and Artificial Nest Predation Rates. Conservation Biology 12: 465–469.
59. Githiru M, Lens L, Cresswell W (2005) Nest predation in a fragmented Afrotropical forest: evidence from natural and artificial nests. Biological Conservation 123: 189–196.
60. Andrén H, Angelstam P (1988) Elevated predation rates as an edge effect in habitat islands: experimental evidence. Ecology 69: 544–547.
61. Dowding JE, Murphy EC (1994) Ecology of ship rats (*Rattus rattus*) in a kauri (*Agathis australis*) forest in Northland, New Zealand New Zealand Journal of Ecology 18: 19–28.
62. Morgan DKJ, Waas JR, Innes J (2009) An inventory of mammalian pests in a New Zealand city. New Zealand Journal of Zoology 36: 23–33.
63. King CM, Innes JG, Flux M, Kimberley MO, Leathwick JR, et al. (1996) Distribution and abundance of small mammals in relation to habitat in Pureora Forest Park. New Zealand Journal of Ecology 20: 215–240.
64. Young A, Mitchell N (1994) Microclimate and vegetation edge effects in a fragmented podocard-broadleaf forest in New Zealand. Biological Conservation 67: 63–72.
65. Humphrey J, Hawes C, Peace A, Ferris-Kaan R, Jukes M (1999) Relationships between insect diversity and habitat characteristics in plantation forests. Forest Ecology and Management 113: 11–21.
66. Holling C (1959) The components of predation as revealed by a study of small-mammal predation of the European pine sawfly. The Canadian Entomologist 91: 293–320.
67. Steffens KE, Seddon PJ, Mathieu R, Jamieson IG (2005) Habitat selection by South Island saddlebacks and Stewart Island robins reintroduced to Ulva Island. New Zealand Journal of Ecology 29: 221–229.
68. Didham RK, Lawton JH (1999) Edge structure determines the magnitude of changes in microclimate and vegetation structure in tropical forest fragments. Biotropica 31: 17–30.
69. Delgado JD, Arévalo JR, Fernández-Palacios JM (2001) Road and topography effects on invasion: edge effects in rat foraging patterns in two oceanic island forests (Tenerife, Canary Islands). Ecography 24: 539–546.

Habitat Degradation and Seasonality Affect Physiological Stress Levels of *Eulemur collaris* in Littoral Forest Fragments

Michela Balestri[1,2], Marta Barresi[2], Marco Campera[1,2], Valentina Serra[2], Jean Baptiste Ramanamanjato[3], Michael Heistermann[4], Giuseppe Donati[1]*

1 Department of Social Sciences, Oxford Brookes University, Oxford, United Kingdom, **2** Department of Biology, University of Pisa, Pisa, Italy, **3** QIT Madagascar Minerals, Rio Tinto, Tolagnaro, Madagascar, **4** Endocrinology Laboratory, German Primate Center, Leibniz Institute for Primate Research, Goettingen, Germany

Abstract

The littoral forest on sandy soil is among the most threatened habitats in Madagascar and, as such, it represents a hot-spot within a conservation hot-spot. Assessing the health of the resident lemur fauna is not only critical for the long-term viability of these populations, but also necessary for the future re-habilitation of this unique habitat. Since the Endangered collared brown lemur, *Eulemur collaris*, is the largest seed disperser of the Malagasy south-eastern littoral forest its survival in this habitat is crucial. In this study we compared fecal glucocorticoid metabolite (fGCM) levels, a measure of physiological stress and potential early indicator of population health, between groups of collared brown lemurs living in a degraded forest fragment and groups occurring in a more preserved area. For this, we analysed 279 fecal samples collected year-round from 4 groups of collared brown lemurs using a validated 11-oxoetiocholanolone enzyme immunoassay and tested if fGCM levels were influenced by reproductive stages, phenological seasons, sex, and habitat degradation. The lemurs living in the degraded forest had significantly higher fGCM levels than those living in the more preserved area. In particular, the highest fGCM levels were found during the mating season in all animals and in females during gestation in the degraded forest. Since mating and gestation are both occurring during the lean season in the littoral forest, these results likely reflect a combination of ecological and reproductive pressures. Our findings provide a clear indication that habitat degradation has additive effects to the challenges found in the natural habitat. Since increased stress hormone output may have long-term negative effects on population health and reproduction, our data emphasize the need for and may add to the development of effective conservation plans for the species.

Editor: Jason M. Kamilar, Midwestern University & Arizona State University, United States of America

Funding: This work received funding from a Rufford Small Grant (Reference number: 8953-2) and Oxford Brookes University (Mentoring Scheme: HRSU-ASQL). The funders had no role in study design, data collection and analysis, decision to publish, or preparation of the manuscript.

Competing Interests: The authors declare that the affiliation of one of them to a commercial company (QIT Madagascar Minerals) does not alter their adherence to PLOS ONE policies on sharing data and materials.

* Email: gdonati@brookes.ac.uk

Introduction

The term "stressor" is used to refer to any internal or external stimuli that perturb homeostasis of living organisms [1], [2]. There are several potential stressors for free-ranging animals. These can be natural, such as adverse climatic factors [3], high predation pressure [4], decrease in food availability [5], social aggression or competition [6], or anthropogenic, such as habitat degradation and fragmentation [7], [8], logging and hunting [9], and noise pollution [10].

In support of behavioural studies, hormonal studies have been recently used to assess animals' adaptability to such challenges and their welfare [11–13]. In vertebrates, stressors, depending on their severity, may cause a physiological response that entails an increase in glucocorticoid (i.e., a class of steroid hormones) secretion from the adrenal cortex [1], [14]. This response is considered adaptive in helping the animal to face critical periods that are threatening to homeostasis [13], [15], [16].

In spite of the positive short-term effects, the action of glucocorticoids may cause severe problems when animals are exposed to long-term (chronic) stressors [14], [17]. Among the most common problems caused by long-term elevated glucocorticoid levels are immune suppression, atrophy of tissue, reproductive suppression, gastric ulcers, and muscle wasting [18–21]. Furthermore, species with slow life histories, such as primates, may be particularly strongly impacted by lost reproductive opportunities resulting from chronic stress [22]. Since chronic stressors can lead to health risks [19], [23], high glucocorticoid levels are often assumed to indicate lower individual fitness or population viability [17]. However, this generalized view is debated, since recent studies suggest a complex relationship between glucocorticoids, stressors, and fitness [5], [24–26]. Despite the associated fitness costs, chronic stress has been proposed to evolve in species where it has an adaptive role in helping to face long-term stressors [13].

Reproductive stages may considerably affect glucocorticoid levels in vertebrates [27]. In the case of primates, physiological stress levels have been shown to increase in males during mating seasons as a consequence of high reproductive competition [28–34], but not in species where mate competition is low [22], [35],

[36]. Females, on the other hand, have higher energetic demands during gestation and lactation and this may also lead to higher physiological stress [34], [37–41]. This seems to hold true despite the fact that, in primates, females have been shown to have lower energetic costs during reproductive stages than other mammals of similar size [42–44]. Furthermore, females are expected to have higher glucocorticoid levels during gestation than during lactation because of (1) placental release of corticotropin-releasing hormone which directly affects both fetal and maternal HPA-axis activity (reviewed by [45]), (2) increased synthesis of cortisol binding corticosteroid-binding globulin, (e.g. [46]) and (3) pregnancy-related increases in levels of estrogens (e.g. [47]).

Habitat disturbance has also been found to be associated with physiological stress in primates [48–51] and in other vertebrate species [52–55]. Anthropogenic disturbance is a widespread phenomenon in Madagascar where a large proportion of the original habitats have been lost [56], [57]. Forest disturbance due to anthropogenic pressure has been shown to reduce food availability and diversity, emphasizing the ecological unpredictability of the island [58]. In fact, Madagascar has been shown to differ from other primate habitats, due to its relatively unpredictable rainfall which leads to irregular fruiting patterns, making these environments challenging especially for frugivores [58], [59]. Despite this, recent studies have highlighted some degree of flexibility in frugivorous lemurs, which demonstrate a level of tolerance to habitat disturbance [60], [61]. Several frugivorous lemurs respond to habitat disturbance by integrating fallback food species and/or by shifting to a more folivorous diet [62], [63]. Additionally, activity patterns and ranging behaviour may be modified in order to maximize resource access or, alternatively, to conserve energy [64], [65].

The littoral forest of South-Eastern Madagascar is one of the most threatened habitats on the island with only a few hundred hectares of fragmented forest left [66], [67]. The Endangered collared brown lemur, *Eulemur collaris*, is the largest frugivorous lemur living in these forests, where it shows high social and ecological flexibility [61], [68]. However, a reduction of food availability and quality [61], [69] and an increase in parasite load [70] in the more disturbed areas indicate that these lemurs may be exposed to high physiological stress which, in turn, may lead to increased health risks. Since *Eulemur collaris* is also the largest seed disperser in the littoral forest [71] its survival in this habitat is crucial. Thus, assessing the impact of habitat disturbance on the stress physiology and welfare of this species is not only important for the long-term viability of the local populations, but also necessary for the future re-habilitation of this unique habitat [72].

In this paper, we examined how reproductive and phenological seasons, habitat degradation, and sex affect the physiological responses of collared brown lemurs. This response was investigated by comparing fecal glucocorticoid metabolite (fGCM) levels [73] between groups living in a degraded fragment (Mandena) and groups living in a more preserved fragment of littoral forest (Sainte Luce) in South-Eastern Madagascar. Fecal samples can be easily collected without disturbing the animal, thereby allowing frequent sampling, even over a long time period [74], thus they can be used as a powerful non-invasive measure of physiological stress levels in free-ranging animals [1], [12], [17], [37].

This study aims to elucidating whether the behavioural and ecological flexibility previously recorded in collared brown lemurs living in littoral forest fragments [61] may be sufficient to compensate for the non-optimal environment or whether the animals show increased signs of physiological stress.

Against this background we predicted higher fGCM levels:

1. In males during the mating season compared to other reproductive stages, because the mating season represents a period with pronounced reproductive competition in many species of primates;
2. In females during gestation and lactation compared to other reproductive stages, because of the expected higher energy demands;
3. During the lean season than during the season of abundance since the former is expected to be a time of food shortage in forest fragments due to low levels of fruit availability;
4. In the more degraded forest during stressful reproductive stages and the lean season, since anthropogenic disturbance may amplify fruit shortage and potentially increase exposure to climatic fluctuations and predators.

Materials and Methods

Ethics Statement

We conducted this study with the authorization of the Commission Tripartite of the Direction des Eaux et Forêts de Madagascar (Autorisation de recherche n.29/11/MEF/SG/DGF/DCB.SAP/SCB du 20/01/11) and this research was ethically approved by the University of Pisa (Animal Care and Use Board). We captured the adult individuals via cages, using banana slices as bait, and we rapidly anesthetized them with Zoletil 100 (5 mg/kg of tiletamine hydrochloride). All animals recovered from anesthesia within 1.5 hours and were subsequently released at the site of capture. The lemurs were followed until regaining full mobility in trees, and there were no injuries as a consequence of the captures.

Study Sites and Subjects

The data were collected in two littoral forest areas from February 2011 to January 2012: Mandena (MAN) and Sainte Luce (STL) in South-East Madagascar. The Conservation Zone of MAN (24°57′S, 47°0′E) consists of two fragments of around 240 ha of degraded littoral forest [75]. The average canopy height in MAN is 8.9±SD 4.4 m [69]. The second study site, the littoral forest of STL (24°46′S, 47°10′E), around 30 km North of Fort Dauphin, is among the most intact littoral ecosystems in Madagascar and contains a very high diversity of vegetation [66]. The study area was located in a 252-ha-fragment of well-preserved littoral forest and swamp, 190 of which are included in the Conservation Zone [75]. The average canopy height in STL is 14.7±SD 4.3 m [69]. Previous botanical analyses illustrate that floristically MAN and STL represent the same habitat although structural differences indicate higher degradation in the former area [69]. Several lemur-focused studies confirm that MAN contains lower quality resources than STL in terms of fruit nutritional values and size of feeding trees [61], [65].

Eulemur collaris is a medium-sized lemur with body mass of 2.15±SD 0.25 kg and body length of 46.1±SD 2.6 cm [61]. These lemurs live in multi-male multi-female groups and show no clear dominance of one sex [61], [76]. In this study we analysed hormonal data collected from all adult individuals of four different groups (n = 22): two groups in MAN (group AB and group C), and two groups in STL (group A and group B).

To ensure continuous observations of the groups, four animals (one for each group) were captured and equipped with radio-collars in order to monitor them via the use of radio-telemetry (Biotrack). Collection of fecal samples began approximately one month after capturing the animals to minimize the risk that fGCM levels were influenced by the capture event itself. Individuals were identified via collars as well as individual characteristics such as age, sex, size, canine length, tail shape, fur colour, and other distinctive traits.

Fecal sample collection and GC analysis

Each habituated group was followed four days (from 6 a.m. to 6 p.m.) per month in order to collect fecal samples and behavioral observations. A total of 279 fecal samples were collected from 22 subjects (mean per individual: 12.7±SE 0.3; range: 2–25). Each individual was sampled every 18.0±SE 1.9 days (range: 8.3–44.3 days). The samples were collected immediately after defecation. Site, group, date, time, and identity of the donor were recorded. Fecal samples were preserved in 10 ml tubes with 96% ethanol and stored at room temperature for 7–12 months before further processing for hormone analysis [77].

We collected 12 additional fecal samples to evaluate possible degradation of fGCM concentrations over one-year storage as reported for other species [78], [79]. For this, each fecal sample was divided into three aliquots and stored in ethanol as described above. Aliquots were kept at the field station at ambient temperature before being processed (see below), thus simulating the conditions under which the study samples were collected and stored. The first aliquot was extracted after 3 months, while the other two aliquots were extracted after 6 and 12 months, respectively, in order to match the longest storage time study samples were stored in ethanol. Fecal extracts were stored at −20°C before the final hormone analysis. The results showed no significant effect of storage duration on fGCM levels (RM ANOVA: Storage effect: $F_{2,22} = 1.44$, $p = 0.258$), a finding in line with what was found in *Propithecus verreauxi* [32] and in *Eulemur rufifrons* [33]. Thus, there was no indication that variation in storage duration biased our hormone data.

Fecal samples for hormone analyses were homogenized in their original ethanolic solvent by mechanical squashing of the fecal pellets with a metal stick. The ethanolic fecal suspension (including a 2-mL ethanol rinse) was decanted into a 50-mL propylene tube, and steroids were extracted by vortexing for 15 min. Following centrifugation at 3000 rpm for 10 min the supernatant was decanted, the exact volume recorded and stored at −20°C until analysis [77]. The remaining fecal pellets were dried in an oven to a constant weight and the dry weight of individual samples was determined in order to compensate for differences in fecal fibers and water content [80].

Fecal extracts were analyzed for immunoreactive 11oxoetiocholanolone (3α,11oxo-CM), a group specific measurement of 5-reduced cortisol metabolites with a 3α,11oxo-structure [81]. The assay has been successfully applied to monitor adrenocortical activity and glucocorticoid output from fecal samples in various primate species (e.g. [81]), including other species of lemurs [32], [82]. It has also been used successfully to monitor physiological stress in the redfronted lemur (*Eulemur rufifrons* [33]), a species closely related to the collared brown lemur. We used reverse-phase high pressure liquid chromatography analysis (HPLC) to characterize the immunoreactive metabolites measured by the 11oxoetiocholanolone EIA. HPLC was carried out as described by Heistermann et al. [81]. To evaluate possible sex differences in 11oxoetiocholanolone immunoreactivity profiles, we performed HPLC on both a male and a female sample. HPLC also allowed us to evaluate whether certain fecal androgens, which could potentially be detected by antibodies raised against cortisol metabolites [81], [83], were co-measured by the 11oxoetiocholanolone EIA.

HPLC analysis indicated that almost all immunoreactivity was detected as distinct peaks between fractions 9 and 31 - positions where cortisol metabolites in our HPLC system elute (Fig. 1) [81]. The similarity between HPLC glucocorticoid immunoreactivity profiles from the collared brown lemur samples and those derived from fecal samples of other primate species [81], including the redfronted lemur [33], strongly suggests that the11oxoetiocholanolone assay is reliable in detecting glucocorticoid output in our study species. In this respect, the presence of only small amounts of immunoreactivity measured after fraction 40 (positions where certain potentially cross-reacting androgen metabolites elute [81]) suggests a low degree of co-measurement of these androgens in our assay (Fig. 1). Furthermore, HPLC profiles were almost identical between the male and female sample in terms of both number and elution position (i.e. characteristic) of metabolites measured, indicating that the 11oxoetiocholanolone assay measures the same immunoreactive compounds in both males and females.

For measurement of 3a,11 oxo-CM levels, fecal extracts were diluted (1:20 to 1:600 depending on concentration) with assay buffer (0.04 M PBS, pH 7.2) and duplicate 50 µl aliquots were measured by microtiterplate EIA along with 50 µl aliquots of reference standard in doubling dilutions over the range of 1.02–125 pg [81]. Briefly, following incubation of the plates overnight at 4°C, the plates were washed three times and incubated with 150 µl streptavidin–peroxidase (HRP) for 30 min in the dark at room temperature after which (following a second washing step) 150 µl of HRP-substrate solution was added to each well. Following substrate incubation (45 min), the enzyme reaction was stopped with 50 µl 2 M H_2SO_4 to each well and absorbance was measured at 450 nm (reference 630 nm) on a plate reader. Sensitivity of the assay was 3 pg. Serial dilutions of fecal extracts from samples of different animals gave displacement curves parallel to that obtained for the standard (*t*-test for difference in slopes between sample dilution curve and standard curve: $t_{11} = 0.669$, $p = 0.517$). Intra-assay coefficients of variation (CV) for low- and high value quality controls were 7.7% (n = 16) and 6.9% (n = 16), respectively. Respective figures for inter-assay CV values were 10.3% (n = 20) and 14.5% (n = 20). All hormone levels reported are expressed as ng/g dry fecal mass.

Data analyses

In order to evaluate the effect of food availability on fGCM levels, we distinguished between a lean season (April–October) and a season of food abundance (November–March). The two seasons were distinguished on the basis of previous multi-annual studies in STL [84] and phenological data collected in MAN during our study period [65].

In order to evaluate the effect of the reproductive stages, we distinguished between four main stages: mating (May to mid-June), gestation (mid-June to September), lactation (October to December), and non-reproductive (January to April) [85].

All 279 fecal samples collected from the 22 adult individuals were used in the analyses. All adult females gave birth during the study period. Statistical comparisons were conducted using a General Linear Mixed Model (GLMM) with reproductive stages (nested in phenological seasons), sites, and sexes as fixed factors, and individuals as a random factor. Both main effects and two-way interaction effects were evaluated in the model. We controlled for the time of the day (morning or afternoon) when each sample was collected by including it in the model as fixed factor, since it has been shown to potentially affect fGCM levels [41], [86]. In fact, the fGCM levels were higher during the afternoon (median: 766.5 ng/g, quartiles: 438.5–1439.8 ng/g, n = 181) than during the morning (median: 600.3 ng/g, quartiles: 424.9–1013.3 ng/g, n = 98) (GLMM, Time of day: $F_{1,265} = 6.14$, $P = 0.014$). We also included in the model the sample weight as a covariate (fixed

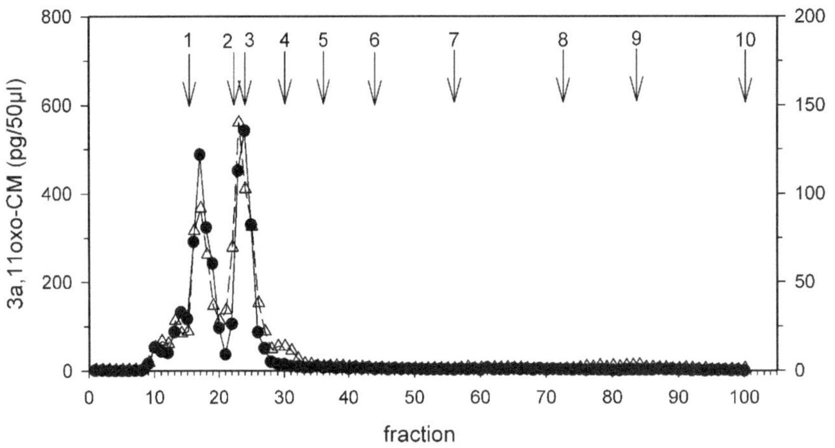

Figure 1. HPLC profiles of *Eulemur collaris*. HPLC profiles of immunoreactivity detected with the 3α,11oxo-CM assay in a male (black circles) and female (white triangles) fecal sample extract. Arrows indicate elution positions of reference standards (1) cortisol (fraction 15), (2) corticosterone (22), (3) 11β-hydroxyetiocholanolone (23/24), (4) 11-oxoetiocholanolone (29), (5) βandostrane-3,11,17-trione (36), (6) testosterone (44), (7) androstendione, dehydroepiandrosterone (55/56), (8) epiandrosterone, 5β-DHT, 5β-androstrane-3β-ol-17-one (72), (9) 5β-androstrane-3α-ol-17-one (82/83), (10) androsterone (100).

effect) to control for the potential effects of the fecal mass on hormone concentrations [86]. The sample weight was negatively correlated with the fGCM levels (GLMM, Weight: $F_{1,265} = 67.15$, $p<0.001$).

We used Duncan's tests as post hoc analyses. We tested for normal distribution of residuals (Kolmogorov-Smirnov test) and equality of variances (Levene's test) as underlying assumptions of the GLMM. Residual values of fGCM levels were not normally distributed and therefore the data were ln transformed. We performed all tests with SPSS 21.0 considering $p<0.05$ as threshold of significance.

Results

The fGCM levels excreted by the 22 adult individuals (Fig. 2) did not differ between males (median: 666.1 ng/g, quartiles: 361.4–1318.5 ng/g, n = 164) and females (median: 740.1 ng/g, quartiles: 487.2–1309.1 ng/g, n = 115) (GLMM, Sex: $F_{1,265} = 1.54$, $p = 0.216$).

The fGCM levels differed between the four reproductive stages (GLMM, Reproductive stage: $F_{3,265} = 2.62$, $p = 0.048$), with a median of 1307.6 ng/g (quartiles: 710.9–2291.2 ng/g, n = 19) during mating, 766.5 ng/g (quartiles: 445.2–1473.1 ng/g, n = 111) during gestation stage, 612.4 ng/g (quartiles: 358.6–1132.4 ng/g, n = 115) during lactation stage, and 638.7 ng/g (quartiles: 474.6–990.3 ng/g, n = 34) during the non-reproductive stage. The fGCM levels during the mating season were higher than during gestation (Duncan: $p = 0.003$), lactation ($p<0.001$), and non-reproductive ($p<0.001$) stages.

The lemurs also had higher fGCM values during the lean season (median: 803.5 ng/g, quartiles: 463.3–1473.8 ng/g, n = 180) than during the season of food abundance (median: 569.3 ng/g, quartiles: 355.0–1002.0 ng/g, n = 99) (GLMM, Phenological season: $F_{1,265} = 12.72$, $p<0.001$).

The fGCM levels were higher in MAN (median: 870.2 ng/g, quartiles: 546.3–1714.9 ng/g, n = 126) than in STL (median: 590.9 ng/g, quartiles: 401.1–1014.6 ng/g, n = 153) (GLMM, Site: $F_{1,265} = 20.27$, $p<0.001$).A different pattern was found between the two sexes during the four reproductive stages (GLMM, Sex* Reproductive stage: $F_{3,265} = 2.95$, $p = 0.033$). In particular, males during the mating season had higher fGCM levels than during the periods corresponding to female gestation (Duncan: $p = 0.004$), lactation ($p<0.001$), and non-reproductive stage ($p<0.001$). Females during lactation had lower fGCM levels than during mating ($p = 0.008$) and gestation ($p = 0.030$). Interaction effects indicated that the two sexes did not show a different pattern between the two sites (GLMM, Sex*Site: $F_{1,265} = 0.10$, $p = 0.755$) and between the two phenological seasons (GLMM, Sex*Phenological season: $F_{1,265} = 0.11$, $p = 0.742$) (Table 1 and Table 2).

Results of pair-wise comparisons of mean differences between sites, reproductive stages, and phenological seasons for ln transformed fGCM values in males and females are shown in Table 3 and Table 4, respectively.

Discussion

Consistent with our predictions, the highest fGCM values were found in females during the gestation period in the degraded forest of MAN and in males during the mating season in both sites. Phenological season also played a role in shaping the fGCM output with higher values exhibited during the lean season, whilst sex had no significant effect. Fecal glucocorticoid metabolite levels were higher during the afternoon rather than during the morning, a characteristic of nocturnal animals which, from a chronobiological perspective, *Eulemur* species belong to [87]. The covariate sample weight also significantly influenced the fGCM output.

As predicted, in males we found higher fGCM levels during the mating season than during the other reproductive stages. Mating season appears to be a stressful period also for male mouse lemurs (*Microcebus murinus* [28]), male ring-tailed lemurs (*Lemur catta* [31], [34]), male sifakas (*Propithecus verreauxi* [32]), and male red-fronted lemurs (*Eulemur rufifrons* [33]). In fact, during this study, collared brown lemur males showed higher aggression rates during the mating season as compared to the other reproductive stages (Serra et al., unpublished data). It is reasonable to assume that the higher fGCM levels found during this time of the year may be related to a general increase of aggression rates and high reproductive competition [88]. Increased aggression rates have been shown to affect fGCM levels in males of other primate species, such as Eastern chimpanzees (*Pan troglodytes* [29]), chacma baboons (*Papio ursinus* [30]), Verreaux's sifakas (*Propithecus verreauxi* [32]) and in males of other vertebrates (wolves,

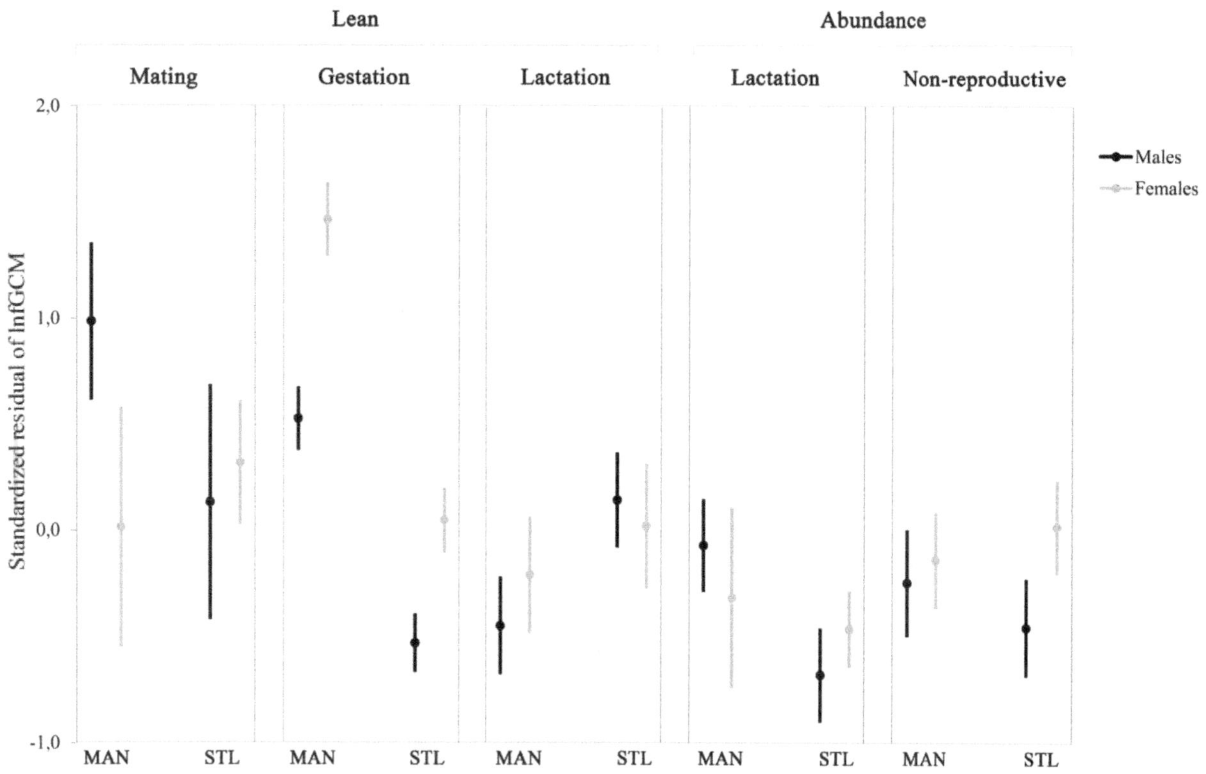

Figure 2. Fecal glucocorticoid metabolite levels of *Eulemur collaris* over the study period. The figure shows standardized residuals of lnfGCM after controlling for the sample weight. MAN: Mandena, STL: Sainte Luce. Lean: May–October 2011, Abundance: February–April 2011 and November 2011–January 2012. Mating: 1st May–15th July, Gestation: 16th July–30th September, Lactation: 1st October–31st December, Non-reproductive: 1st January–30th April. Values are means and standard errors.

Canis lupus, [89]; bison bulls, *Bison bison*, [90]; American alligators, *Alligator mississippiensis*, [91]). Conversely, low physiological stress levels have been observed in species where mate competition is low, such as males of muriquis (*Brachyteles arachnoides*, [35]), tufted capuchin monkeys (*Cebus apella*, [36]), and red-bellied lemurs (*Eulemur rubriventer* [22]), while no differences were found between mating and non-mating seasons in male rhesus macaques (*Macaca mulatta*, [92]).

In our study, we also found high fGCM levels in females during the mating season, suggesting that this period may also be stressful for them. This is in accordance with findings on other primates (e.g. South-american squirrel monkeys, *Saimiri sciureus*, [93], [94]) and mammals (e.g. giant pandas, *Ailuropoda melanoleuca*, [95], [96]) where the GC elevation has been attributed to the effects of ovarian cycling and general anxiousness. However, we must consider that our results indicate a very high variability within the mating season, which may be due to two confounding factors. Firstly, we had a small sample size due to the very short mating season [58], and we may have accidentally included samples belonging to the previous and/or the subsequent reproductive stage. Secondly, individuals may have different levels of physiological stress depending on their dominance status [97].

By far, the highest fGCM levels were found in females in the more degraded site during gestation (see Fig. 2). It is well known that gestation represents a serious challenge for females, as during this period they have increased energetic demands, as previously shown for females of ring-tailed lemurs (*Lemur catta*, [37]), white faced capuchins (*Cebus capucinus*, [40]), and other mammalian species (little brown myotis, *Myotis lucifugus*, [98]; red squirrels, *Tamiasciurus hudsonicus*, [99]). Higher glucocorticoid levels during gestation may not necessarily correspond to higher physiological stress, however, since placental hormones and fetal estrogens also stimulate cortisol production [100]. In the Malagasy littoral forest, the physiological increase in glucocorticoid levels due to gestation may be enhanced by the additional physiological stress due to the concomitant lean season [84]. However, the significantly lower fGCM levels found in the females inhabiting the less disturbed fragment during gestation strongly suggests that the degraded habitat conditions are largely responsible for this effect. Interestingly, during gestation in the degraded forest males had similar fGCM levels as the females within the same site. This supports the idea that habitat degradation and seasonal food availability override the potential effect of female reproductive state on fGCM levels. The fact that pregnant females showed relatively low levels of fGCM in the less disturbed area is also in line with previous studies showing that lemurs minimize maternal energetic investment during gestation [100].

Contrary to other studies, collared brown lemur females did not show high fGCM levels during lactation, a potentially stressful period for females due to the burden of infant carrying and maternal care (ring-tailed lemurs, *Lemur catta*, [34], [38], but see [101]; Assamese macaques, *Macaca assamensis*, [102]; rhesus macaques, *Macaca mulatta*, [39]). The lack of elevated fGCM levels in lactating females of collared brown lemurs may have been caused by the overriding effect of the concomitant increase in food availability. Similar results have been shown in *Lemur catta* at Beza Mahafaly which also showed low physiological stress levels during lactation [37]. In fact, lactation in collared brown lemurs is synchronized with the transition from the lean season to the season of food abundance, when young leaves and ripe fruits increase in

Table 1. Fecal glucocorticoid metabolite levels (ng/g) in males of *Eulemur collaris* over the study period.

	MAT (11)	GES (68)	LAC (68)	NRE (17)	LEA (109)	ABU (55)	Total
MAN	**1856**	**1227**	**673**	**572**	**1015**	**601**	**837**
(82)	1308–3151	683–1822	355–1006	376–842	597–1679	376–1011	500–1440
STL	**1019**	**430**	**844**	**571**	**587**	**486**	**498**
(82)	602–1618	298–590	300–1495	329–990	340–984	282–1002	303–1002
Total	**1564**	**611**	**704**	**572**	**770**	**512**	
	770–3151	412–1314	310–1280	376–842	428–1435	306–1011	

Values are medians and quartiles.
MAN: Mandena (more degraded site); STL: Sainte Luce (more preserved site); MAT: mating; GES: gestation; LAC: lactation; NRE: non reproductive; LEA: lean season; ABU: period of food abundance.

Table 2. Fecal glucocorticoid metabolite levels (ng/g) in females of *Eulemur collaris* over the study period.

	MAT (11)	GES (68)	LAC (68)	NRE (17)	LEA (109)	ABU (55)	Total
MAN	**711**	**1981**	**593**	**822**	**1621**	**561**	**1012**
(82)	676–2291	1621–2683	386–846	397–982	679–2462	386–982	600–1981
STL	**1265**	**713**	**588**	**813**	**729**	**630**	**713**
(82)	1091–1580	505–882	439–1060	566–1168	507–1060	475–931	483–1035
Total	**1178**	**976**	**588**	**822**	**896**	**609**	
	694–1936	596–1844	421–1013	559–1000	551–1621	392–957	

Values are medians and quartiles.
MAN: Mandena (more degraded site); STL: Sainte Luce (more preserved site); MAT: mating; GES: gestation; LAC: lactation; NRE: non reproductive; LEA: lean season; ABU: period of food abundance.

Table 3. *P* values of pair-wise comparisons of mean differences between sites in ln transformed fGCM values across different reproductive stages in males of *Eulemur collaris* (Duncan post-hoc).

Site		MAN	MAN	MAN	MAN	MAN	STL	STL	STL	STL	STL
	Stage	**MAT-L**	**GES-L**	**LAC-L**	**LAC-A**	**NRE-A**	**MAT-L**	**GES-L**	**LAC-L**	**LAC-A**	**NRE-A**
MAN	MAT-L	-	-	-	-	-	-	-	-	-	-
MAN	GES-L	0,12	-	-	-	-	-	-	-	-	-
MAN	LAC-L	**0,00**	0,06	-	-	-	-	-	-	-	-
MAN	LAC-A	**0,01**	0,23	0,46	-	-	-	-	-	-	-
MAN	NRE-A	**0,00**	0,11	0,74	0,65	-	-	-	-	-	-
STL	MAT-L	0,07	0,71	0,12	0,37	0,21	-	-	-	-	-
STL	GES-L	**0,00**	0,02	0,58	0,23	0,41	**0,04**	-	-	-	-
STL	LAC-L	0,10	0,85	0,08	0,29	0,15	0,82	**0,03**	-	-	-
STL	LAC-A	**0,00**	**0,04**	0,84	0,37	0,62	0,09	0,70	0,06	-	-
STL	NRE-A	**0,00**	0,07	0,92	0,50	0,79	0,14	0,55	0,10	0,78	-

Median sample size: 15 (range: 5–36).
MAN: Mandena (more degraded site); STL: Sainte Luce (more preserved site); MAT: mating; GES: gestation; LAC: lactation; NRE: non reproductive; L: lean period; A: period of food abundance.

Table 4. P values of pair-wise comparisons of mean differences between sites in ln transformed fGCM values across different reproductive stages in females of *Eulemur collaris* (Duncan post-hoc).

Site		MAN	MAN	MAN	MAN	MAN	STL	STL	STL	STL	STL
	Stage	**MAT-L**	**GES-L**	**LAC-L**	**LAC-A**	**NRE-A**	**MAT-L**	**GES-L**	**LAC-L**	**LAC-A**	**NRE-A**
MAN	MAT-L	-	-	-	-	-	-	-	-	-	-
MAN	GES-L	**0,07**	-	-	-	-	-	-	-	-	-
MAN	LAC-L	0,15	**0,00**	-	-	-	-	-	-	-	-
MAN	LAC-A	**0,08**	**0,00**	0,73	-	-	-	-	-	-	-
MAN	NRE-A	0,23	**0,00**	0,77	0,55	-	-	-	-	-	-
STL	MAT-L	0,60	0,17	0,06	**0,03**	0,10	-	-	-	-	-
STL	GES-L	0,23	**0,00**	0,78	0,56	0,97	0,10	-	-	-	-
STL	LAC-L	0,64	**0,03**	0,31	0,19	0,42	0,35	0,42	-	-	-
STL	LAC-A	0,11	**0,00**	0,84	0,86	0,65	**0,04**	0,66	0,24	-	-
STL	NRE-A	0,65	**0,03**	0,30	0,19	0,42	0,36	0,42	0,98	0,24	-

Median sample size: 9 (range: 3–25).
MAN: Mandena (more degraded site); STL: Sainte Luce (more preserved site); MAT: mating; GES: gestation; LAC: lactation; NRE: non reproductive; L: lean period; A: period of food abundance.

their availability [56], [84], [103]. Additionally, primates produce some of the most dilute milk of all mammals [42] and, in particular, milk-producing costs for *Eulemur* species are among the lowest amongst primates [104].

As expected, we found that seasonality had a strong effect on fGCM output in collared brown lemurs. This was in the expected direction as we found higher fGCM values during the lean season, when fruit availability was low [65]. In primates physiological stress levels have been repeatedly found to be shaped by fruit availability (olive baboons, *Papio anubis*, [105]; ring-tailed lemurs, *Lemur catta*, [37], [38]; chimpanzees, *Pan troglodytes*, [29]; black howler monkeys, *Alouatta pigra*, [106]; yellow baboons, *Papio cynocephalus*, [107]; Mexican howler monkeys, *Alouatta palliata*, [51]) and periods of presumed nutritional stress (Eastern red colobus, *Procolobus rufomitratus*, [108]). In particular, in MAN, the percentage of tree species with ripe fruits averaged 4.4% during the mating season and the gestation period while it stands at 14.2% during lactation and non-reproductive stages [65]. Thus, low fruit availability seems to have an additive effect on reproductive stages and habitat degradation, and the combination of these three factors is likely to reflect fGCM output in our study species.

This study clearly indicates that levels of fGCM were higher for collared brown lemurs in the degraded forest fragment of MAN when compared to lemurs in the more preserved forest of STL, suggesting a higher level of physiological stress in animals living in disturbed areas. The most likely explanation for this difference may be found in the lower levels of food availability [65] and quality [61] recorded in MAN as compared to STL which may result in increased nutritional stress. Other stressors, such as a higher predation risk in the disturbed, more open MAN forest, may have contributed to the recorded difference. In support of this, the lemurs' primary predator, the fossa (*Cryptoprocta ferox*), was reported several times in MAN but not in STL over the last decade [109].

Previous studies show that collared brown lemurs in the degraded forest shape their ranging, feeding, and activity pattern to cope with a decrease in food abundance [61], [65]. In particular, lemurs in MAN during our study period had larger home ranges and traveled shorter daily distances than lemurs in STL [65]. Our results indicate that, although the collared brown lemurs seem to cope with habitat degradation by changing their behavioral ecology, living in a degraded forest area nevertheless increases physiological stress. This may have an effect on the long-term viability of the population. These effects may include higher vulnerability to diseases, reduced reproduction, and even a higher mortality rate [17]. Our finding of a higher parasite burden for the lemurs in MAN as compared to those in STL [70] is in accordance with this, and may indicate that elevated glucocorticoid levels do pose a health risk to the animals, although the cause-and-effect relationship between these two parameters is not entirely clear.

Our findings are in line with other recent studies which found that habitat degradation and fragmentation are associated with an increase in GC outputs in a variety of primates (Eastern red colobus, *Procolobus rufomitratus*, [108]; black howler monkeys, *Alouatta pigra*, [49]; Yucatan spider monkeys, *Ateles geoffroyi*, [50]; Mexican howler monkyes, *Alouatta palliata*, [51]; but see [9] for effect of logging and hunting on brown spider monkeys, *Ateles hybridus* and red howler monkeys, *Alouatta seniculus*) and other animal species (American redstarts, *Serophaga ruticilla*, [52]; spotted salamanders, *Ambystoma maculatum*, [53]; African savanna elephants, *Loxodonta africana*, [54]; agile antechinus, *Antechinus agilis*, [55]; see also [110] for effects of human disturbance on elks, *Cervus elaphus*, and wolves, *Canis lupus*). Conversely, the only previous study which compared lemurs' fGCs levels in disturbed and undisturbed habitats [22] showed opposite results. In the latter study, red-bellied lemurs (*Eulemur rubriventer*) in the undisturbed forest of Ranomafana showed higher fecal cortisol levels than those in the disturbed habitat during the lean period. This result might be explained by an attenuated response to prolonged stress to reduce costs of continued stress hormone production [14]. In fact, *Eulemur rubriventer* in the disturbed habitat showed higher infant mortality [22]. Conversely, lemurs in MAN have birth and mortality rates similar to those in STL [84] and in other more preserved forests [111]. Thus, lemurs in MAN do not give indications of an attenuated response to habitat degradation, but they do not seem to exhibit any clear negative effects at the population level. Furthermore, *Eulemur rubriventer* in Ranomafana relied on the exotic *Psidium cattleianum* in the disturbed area, and this might have shielded them during lean periods [22]. In fact, exotic fruits are known to sometimes provide a nutritionally higher resource and a longer temporal availability than native species [112]. Contrary to Tecot's study [22], collared brown lemurs do not seem to rely on exotic species in MAN [61] and this may also explain the high fGCM levels detected during the lean season.

In addition to Tecot [22], studies on other animals (bighorn sheep, *Ovis canadiensis* [113]; Canadian grizzly bears, *Ursus arctos*, [114]; African forest elephants, *Loxodonta cyclotis*, [115]; spotted salamanders, *Ambystoma maculatum*, [53]) found higher fGCM in undisturbed habitats. Thus, the hypothalamic–pituitary–adrenal axis response may differ between species [13] and the relationship between GC levels and stressors may be not so clear-cut.

Another possible stressor that may have influenced the higher fGCM levels found in lemurs living in MAN is their proximity to a mining site. The area of MAN in which the study took place is in fact very close to the machinery set up in the region to extract titanium deposits [116]. This may have exposed the lemurs to chronic stressors such as anthropogenic noise and light pollution. In particular, noise pollution may lead to behavioural changes [10] and to an increase in fGCM levels [117]. *Eulemur collaris* is a species known to shape its 24-h activity depending on luminosity [68], [118]. Hence, artificial light pollution may potentially alter the species behaviour. For example, human activities have been shown to alter the activity budget of bighorn sheep [113]. Further studies focusing on these aspects may give a clearer insight on the impact of mining on the area.

Conclusions

By comparing fGCM levels in collared brown lemurs living in a degraded and in a more preserved forest fragment, we found higher fGCM levels occur in those individuals living in the former situation, which appears to be a stressful environment. Higher fGCM levels in disturbed habitats suggest that coping with a harsher environment has a cost of increased physiological stress in these lemurs. This study underlines the importance of physiological investigations to assess population health of threatened species and the potentially detrimental effect of habitat loss on animal welfare. Because of the paucity of studies comparing lemurs living in disturbed environments over the long-term, more research is urgently required to evaluate the consequences of chronic physiological stress on the highly threatened lemur populations.

Acknowledgments

This work was carried out under the collaboration agreement with the Department of Animal Biology and Anthropology of the University of

Antananarivo and QIT Madagascar Minerals (QMM). We thank the Madagascar Institute for the Conservation of Tropical Environments (MICET), the Association of Managers of Forests of Ambatoatsinana (FIMPIA), the Mandena Management Committee (COGEMA), and the Ministère des Eaux et Forets for their collaboration and permission to work in Madagascar. We thank the Costwold Wildilife Park & Gardens for giving us precious help for the validation test. We are grateful to Silvana Borgognini for her support during the early stages of this research and to Rachel Sawyer and Lauren Lansdowne for the language revision of this manuscript. Our gratitude goes to Sergio Tofanelli, Stefania Bertoncini and Giulio Petroni for giving us the opportunity to work in their laboratories. We acknowledge the QMM biodiversity staff, especially Manon Vincelette, Johny Rabenantoandro, Christophe Rambolamanana, Laza Andriamandimbiarisoa, Faly Randriatafika, David Rabehevitra, Sylvio Angelico, Claude Soanary, and the field assistants (Kadofa, Josette, Faris, Philemon, Germain, Crescent) in MAN and STL. We thank Murielle Ravaolahy and Aristide Andrianarimisa for their contribution to this research and Andrea Heistermann for conducting the hormone analyses.

Author Contributions

Conceived and designed the experiments: GD MH. Performed the experiments: M. Barresi M. Balestri MC VS. Analyzed the data: M. Barresi M. Balestri MC VS MH GD. Contributed reagents/materials/ analysis tools: JBR MH GD. Wrote the paper: M. Balestri MC JBR MH GD.

References

1. Creel S (2001) Social dominance and stress hormones. Trends Ecol Evol 16: 491–97.
2. Chrousos GP (2009) Stress and disorders of the stress system. Nat Rev Endocrinol 5: 374–381.
3. Wingfield J (2013) Ecological processes and the ecology of stress: the impacts of abiotic environmental factors. Func Ecol 27: 37–44.
4. Clinchy M, Sheriff MJ, Zanette L (2013) Predator-induced stress and the ecology of fear. Funct Ecol 27: 56–65.
5. Busch DS, Hayward LS (2009) Stress in a conservation context: a discussion of glucocorticoid actions and how levels change with conservation- relevant variables. Biol Conserv 142: 2844–2853.
6. Creel S, Dantzer B, Goymann W, Rubenstein DR (2013) The ecology of stress: effects of the social environment. Funct Ecol 27: 66–80.
7. Wingfield JC, Hunt K, Breuner C, Dunlap K, Fowler GS, et al. (1997) Environmental stress, field endocrinology, and conservation biology. In: Clemmons JR, Buchholz R, editors. Behavioral Approaches to Conservation in the Wild. Cambridge University Press, Cambridge. pp. 95–131.
8. Fahrig L (2003) Effects of habitat fragmentation on biodiversity. Annu Rev Ecol Evol Syst 34: 487–515.
9. Rimbach R, Link A, Heistermann M, Gómez-Posada C, Galvis N, et al. (2013) Effects of logging, hunting, and forest fragment size on physiological stress levels of two sympatric ateline primates in Colombia. Conserv Physiol 1: doi:10.1093/conphys/cot031.
10. Francis CD, Barber JR (2013) A framework for understanding noise impacts on wildlife: an urgent conservation priority. Front Ecol Environ 11: 305–213.
11. Keay JM, Singh J, Gaunt MC, Kaur T (2006) Fecal glucocorticoids and their metabolites as indicators of stress in various mammalian species: a literature review. J Zoo Wildlife Med 37: 234–244.
12. Goymann W (2012) On the use of non-invasive hormone research in uncontrolled, natural environments: the problem with sex, diet, metabolic rate and the individual. Method Ecol Evol 3: 757–765.
13. Boonstra R (2013) Reality as the leading cause of stress: rethinking the impact of chronic stress in nature. Funct Ecol 27: 11–23.
14. Romero LM (2004) Physiological stress in ecology: lessons from biomedical research. Trends Ecol Evol 19: 249–255.
15. Sapolsky RM, Romero LM, Munck AU (2000) How do glucocorticosteroids influence stress responses? Integrating permissive, suppressive, stimulatory and preparative actions. Endocr Rev 21: 55–89.
16. Wingfield JC, Romero LM (2001) Adrenocortical responses to stress and their modulation in free-living vertebrates. In: McEwen BS, Goodman HM, editors. Handbook of Physiology; Section 7: The Endocrine System; Volume IV: Coping With The Environment: Neural and Endocrine Mechanisms. New York: Oxford Univ. Press. pp 211–234.
17. Pride ER (2005) High faecal glucocorticoid levels predict mortality in ring-tailed lemurs (*Lemur catta*). Biol Lett 1: 60–63.
18. Cameron JL (1997) Stress and behaviorally induced reproductive dysfunction in primates. Semin Reprod Endocr 15: 37–45.
19. Sapolsky RM (2002) Endocrinology of the stress response. In Behavioral Endocrinology. In: Becker JB, Breedlove SM, Crews D, MaCarthy MM, editors. Cambridge, MTP Press. pp. 409–450.
20. Heistermann M, Ademmer C, Kaumanns W (2004). Ovarian cycle and effect of social changes on adrenal and ovarian function in *Pygatherix nemaeus*. Int J Primatol 25: 689–708.
21. Cavigelli S, Chaudhry HS (2012) Social status, glucocorticoids, immune function, and health: can animal studies help us understand human socioeconomic-status-related health disparities? Horm Behav 62: 295–313.
22. Tecot S (2013) Variable energetic strategies in disturbed and undisturbed rain forest habitats: fecal cortisol levels in southeastern Madagascar. In: Masters J. Gamba M, Génin F, Tuttle R, editors. Leaping Ahead: Advances in Prosimian Biology (Developments in Primatology: Progress and Prospects). New York: Springer. pp. 185–195.
23. Dhabhar F, McEwen B (2000) Bidirectional effects of stress and glucocorticoid hormones on immune function: possible explanations for paradoxical results. In: Ader R, Felton D, Cohen N, editors. Psychoneuroimmunology. New York: Academic Press. pp. 301–338.
24. Breuner CW, Patterson SH, Hahn TP (2008) In search of relationships between the acute adrenocortical response and fitness. Gen Comp Endocr 157: 288–295.
25. Bonier F, Martin PR, Moore IT, Wingfield JC (2009) Do baseline glucocorticoids predict fitness? Trends Ecol Evol 24: 634–642.
26. Madliger CL, Love OP (2013) The need for a predictive, context-dependent approach to the application of stress hormones in conservation. Conserv Biol doi: 10.1111/cobi.12185.
27. Romero LM (2002) Seasonal changes in plasma glucocorticoid concentrations in free-living vertebrates. Gen Comp Endocrinol 128: 1–24.
28. Schilling A, Perret M, Predine J (1984) Sexual inhibition in a prosimian primate. A pheromone-like effect. J Endocrinol 102: 143–151.
29. Muller M, Wrangham R (2004) Dominance, cortisol and stress in wild chimpanzees (*Pan troglodites schweinfurthii*). Behav Ecol Sociobiol 55: 332–340.
30. Bergman T, Beehner J, Cheney D, Seyfarth R, Whitten P (2005) Correlates of stress in free-ranging male chacma baboons, *Papio hamadryas ursinus*. Anim Behav 70: 703–713.
31. Pride ER (2005) Optimal group size and seasonal stress in ring-tailed lemurs (*Lemur catta*). Behav Ecol 16: 550–560.
32. Fichtel C, Kraus C, Ganswindt A, Heistermann M (2007) Influence of reproductive season and rank on fecal glucocorticoid levels in free-ranging male Verreaux's sifakas (*Propithecus verreauxi*). Horm Behav 51: 640–648.
33. Ostner J, Kappeler P, Heistermann M (2008) Androgen and glucocorticoid levels reflect seasonally occurring social challenges in male redfronted lemurs (*Eulemur fulvus rufus*). Behav Ecol Sociobiol 62: 627–638.
34. Starling AP, Charpentier MJE, Fitzpatrick C, Scordato ES, Drea CM (2010) Seasonality, sociality, and reproduction: long-term stressors of ring-tailed lemurs (*Lemur catta*). Horm Behav 57: 76–85.
35. Strier KB, Ziegler TE, Wittwer DJ (1999) Seasonal and social correlates of fecal testosterone and cortisol levels in wild male muriquis (*Brachyteles arachnoides*). Horm Behav 35: 125–134.
36. Lynch JW, Ziegler TE, Strier KB (2002) Individual and seasonal variation in fecal testosterone and cortisol levels of wild male Tufted Capuchin monkeys, *Cebus apella nigritus*. Horm Behav 41: 275–287.
37. Cavigelli SA (1999) Behavioural patterns associated with faecal cortisol levels in free-ranging female ring-tailed lemurs, *Lemur catta*. Anim Behav 43: 166–179.
38. Pride RE (2005) Foraging success, agonism, and predator alarms: behavioral predictors of cortisol in *Lemur catta*. Int J Primatol 26: 295–319.
39. Hoffman CL, Ayala JE, Mas-Rivera A, Maestripieri D (2010) Effects of reproductive condition and dominance rank on cortisol responsiveness to stress in free-ranging female rhesus macaques. Am J Primatol 72: 559–565.
40. Carnegie S, Fedigan LM, Ziegler TE (2011) Social and environmental factors affecting fecal glucocorticoids in wild, female white-faced capuchins (*Cebus capucinus*). Am J Primatol 869: 861–869.
41. Rimbach R, Heymann EW, Link A, Heistermann M (2013) Validation of anenzyme immunoassay for assessing adrenocortical activity and characterizationof factors that affect fecal glucocorticoid metabolite levels in two New World primates. Gen Comp Endocrinol 191: 13–23.
42. Lee PC (1996) The meanings of weaning: growth, lactation, and life history. Evol Anthropol 5: 87–96.
43. Dufour DL, Sauther ML (2002) Comparative and evolutionary dimensions of the energetics of human pregnancy and lactation. Am J Hum Biol 14: 584–602.
44. Rosetta L, Lee PC, Garcia C (2011) Energetics during reproduction: a doubly labeled water study of lactating baboons. Am J Phys Anthropol 144: 661–668.
45. Coe CL, Lubach GR (2014) Vital and vulnerable functions of the primate placenta critical for infant health and brain development. Front Neuroendocrinol. Available: http://dx.doi.org/10.1016/j.yfrne.2014.03.004.
46. Albrecht E, Pepe G (1998) Secretion and metabolism of steroids in primate mammals during pregnancy. In: Bazer, F. (Ed.), Endocrinology of Pregnancy. Humana Press, New York, pp. 319–351.

47. McLean M, Smith R (1999) Corticotropin-releasing hormone in human pregnancy and parturition. Trends Endocrinol Metab 10: 174–178.
48. Chapman CA, Wasserman MD, Gillespie TR, Speirs ML, Lawes ML, et al. (2006) Do food availability, parasitism, and stress have synergistic effects on Red Colobus populations living in forest fragments? Am J Phys Anthropol 131: 525–34.
49. Martínez-Mota R, Valdespino C, Sánchez-Ramos MA, Serio-Silva JC (2007) Effects of forest fragmentation on the physiological stress response of black howler monkeys. Anim Cons 10: 374–379.
50. Rangel-Negrín A, Alfaro JL, Valdez RA, Romano MC, Serio-Silva JC (2009) Stress in Yucatan spider monkeys: effects of environmental conditions on fecal cortisol levels in wild and captive populations. Anim Conserv 12: 496–502.
51. Dunn JC, Cristóbal-Azkarate J, Schulte-Herbrüggen B, Chavira R, Veà JJ (2013) Travel time predicts fecal glucocorticoid levels in free-ranging howlers (*Alouatta palliata*). Int J Primatol 34: 246–259.
52. Marra PP, Holberton RL (1998) Corticosterone levels as indicators of habitat quality: effects of habitat segregation in a migratory bird during the non-breeding season. Oecologia 116: 284–292.
53. Homan RN, Regosin JV, Rodrigues DM, Reed JM, Windmiller BS, et al. (2003) Impacts of varying habitat quality on the physiological stress of spotted salamanders (*Ambystoma maculatum*). Anim Conserv 6: 11–18.
54. Jachowski DS, Slotow R, Millspaugh JJ (2012) Physiological stress and refuge behavior by african elephants. PLOS One 7(2): e31818. doi:10.1371/journal.pone.0031818.
55. Johnstone CP, Lill A, Reina RD (2012) Does habitat fragmentation cause stress in the agile antechinus? An haematological approach. J Comp Physiol 182: 139–155.
56. Mittermeier R, Louis EE, Richardson M, Schwitzer C, Langrand O et al. (2010) Lemurs of Madagascar. Washington DC: Conservation International. 767 p.
57. Schwitzer C, Mittermeier RA, Johnson SE, Donati G, Irwin M, et al. (2014) Averting lemur extinctions amidst Madagascar's political crisis. Science 343: 842–843.
58. Wright PC (1999) Lemur traits and Madagascar ecology: coping with an island environment. Yearb Phys Anthropol 42: 31–72.
59. Dewar RE, Richard AF (2007) Evolution in the hypervariable environment of Madagascar. P Natl Acad Sci USA 104: 3–27.
60. Gould L, Sussman RW, Sauther ML (1999) Natural disasters and primate populations: the effects of a 2-year drought on a naturally occurring population of ring-tailed lemurs (Lemur catta) in southwestern Madagascar. Int J Primatol 20: 69–84.
61. Donati G, Kesch K, Ndremifidy K, Schmidt SL, Ramanamanjato JB, et al. (2011) Better few than hungry: flexible feeding ecology of collared lemurs *Eulemur collaris* in littoral forest fragments. PLOS One 6 (5): e19807. doi: 10.1371/journal.pone.0019807.
62. Irwin MT (2007) Living in forest fragments reduces group cohesion in Diademed Sifakas (*Propithecus diadema*) in Eastern Madagascar by reducing food patch size. Am J Primatol 69: 434–347.
63. Ratsimbazafy JH, Ramarosandratana HV, Zaonarivelo RJ (2002) How do black-and-white ruffed lemurs still survive in a highly disturbed habitat? Lemur News 7: 7–10.
64. Irwin MT (2008) Diademed sifaka (*Propithecus diadema*) ranging and habitat use in continuous and fragmented forest: higher density but lower viability in fragments? Biotropica 40: 231–240.
65. Campera M, Serra V, Balestri M, Barresi M, Ravaolahy M, et al. (2014) Effects of habitat quality and seasonality on ranging patterns of *Eulemur collaris* in littoral forest fragments. Int J Primatol. doi:10.1007/s10764-014-9780-6.
66. Bollen A, Donati G (2006) Conservation status of the littoral forest of southeastern Madagascar: A review. Oryx 40: 57–66.
67. Consiglio T, Schatz GE, McPherson G, Lowry II PP, Rabenantoandro J, et al. (2006) Deforestation and plant diversity of Madagascar's littoral forests. Conserv Biol 20: 1799–1803.
68. Donati G, Bollen A, Borgognini-Tarli SM, Ganzhorn JU (2007) Feeding over 24-h cycle: Dietary flexibility of cathemeral collared lemurs (*Eulemur collaris*). Behav Ecol Sociobiol 61: 1237–1251.
69. Rabenantoandro J, Randriatafika F, Lowry II PP (2007) Floristic and structural characteristics of remnant littoral forest sites in the Tolagnaro area. In: Ganzhorn JU, Goodman SM, Vincelette M, editors. Biodiversity, ecology and conservation of the littoral ecosystems of South-eastern Madagascar. Washington DC: Smithsonian Institution Press. pp. 65–77.
70. Lazdane K, Broll A, Theisinger O, Bearder SK, Donati G (2014) Susceptibility of collared brown lemurs (*Eulemur collaris*) to gastrointestinal parasite infections in remnant fragments of Malagasy littoral forest. Ital J Zool. doi:10.1080/11250003.2014.915993.
71. Bollen A, Van Elsacker L, Ganzhorn JU (2004) Relations between fruits and disperser assemblages in the littoral forest of south-east Madagascar: a community level approach. J Trop Ecol 20: 599–612.
72. Ganzhorn JU, Fietz J, Rakotovao E, Schwa D, Zinner DN (1999) Lemurs and the regeneration of dry deciduous forest in Madgascar. Conserv Biol 13: 794–804.
73. Lane J (2006) Can non-invasive glucocorticoid measures be used as reliable indicators of stress in animals? Anim Welfare 15: 331–342.
74. Palme R, Rettenbacher S, Touma C, El-Bahr SM, Möstl E (2005) Stress hormones in mammals and birds: Comparative aspects regarding metabolism, excretion and noninvasive measurement in fecal samples. Ann N Y Acad Sci 1040: 162–171.
75. Ganzhorn JU, Andrianasolo T, Andrianjazalahatra T, Donati G, Fietz J (2007) Lemurs in evergreen littoral forest fragments. In: Ganzhorn JU, Goodman SM, Vincelette M, editors. Biodiversity, Ecology and Conservation of the Littoral Ecosystems of South-Eastern Madagascar. Washington DC: Smithsonian Institution Press. pp. 223–236.
76. Pereira ME, Kappeler PM (1997) Divergent systems of agonistic behaviour in lemurid primates. Behaviour 134: 225–274.
77. Kraus C, Heistermann M, Kappeler PM (1999) Physiological suppression of sexual function of subordinate males. A subtle form of intrasexual competition among male sifakas (*Propithecus verreauxi*)? Physiol Behav 66: 855–861.
78. Khan MZ, Altmann J, Isani SS, Yu J (2002) A matter of time: evaluating the storage of fecal samples for steroid analysis. Gen Comp Endocrinol 128: 57–64.
79. Hunt K, Wasser S (2003) Effect of long-term preservation methods on fecal glucocorticoid concentrations of grizzly bear and African elephant. Physiol Biochem Zool 76: 918–928.
80. Wasser SK, Thomas R, Nair PP, Guirdy C, Southers J, et al. (1993) Effect of dietary fibre on steroid measurements in baboons (*Papio cynocephalus cynocephalus*). J Reprod Fertil 97: 569–574.
81. Heistermann M, Palme R, Ganswindt A (2006) Comparison of different enzymeimmunoassays for assessment of adrenocortical activity in primates based on fecal analysis. Am J Primatol 68: 257–273.
82. Hämäläinen A, Heistermann M, Fenosoa ZSA, Kraus C (2014) Evaluating capture stress in wild gray mouse lemurs via repeated fecal sampling: method validation and the influence of prior experience and handling protocols on stress responses. Gen Comp Endocrinol 195: 68–79.
83. Ganswindt A, Palme R, Heistermann M, Borragan S, Hodges JK (2003) Non-invasive assessment of adrenocortical function in the male African elephant (*Loxodonta africana*) and its relation to musth. Gen Comp Endocrinol 134: 156–166.
84. Bollen A, Donati G (2005) Phenology of the littoral forest of Sainte Luce, southeast Madagascar. Biotropica 37: 32–43.
85. Donati G (2002) Activity pattern of *Eulemur fulvus collaris* in relation to environmental variations in the littoral forest of Sainte Luce (Fort-Dauphin, Madagascar). PhD Thesis, University of Pisa.
86. Hodges JK, Heistermann M (2011) Field endocrinology: monitoring hormonal changes in free-ranging primates. In: Setchell, J.M., Curtis, D.J. editors. Field and Laboratory Methods in Primatology: A Practical Guide. Cambridge: University Press. pp. 353–370.
87. Erkert HG, Cramer B (2006) Chronobiological background to cathemerality: circadian rhythms in *Eulemur fulvus albifrons* (Prosimii) and *Aotus azarai boliviensis* (Anthropoidea). Folia Primatol 77: 87–103.
88. Wingfield JC, Hegner RE, Dufty AM, Balle GF (1990) The "Challenge Hypothesis": Theoretical implications for patterns of testosterone secretion, mating systems, and breeding strategies. Am Nat 136: 829–846.
89. Sands JL, Creel S (2004) Social dominance, aggression and faecal glucocorticoid levels in a wild population of wolves, *Canis lupus*. Anim Behav 67: 387–396.
90. Mooring MS, Patton ML, Lance VA, Hall BM, Schaad EW, et al. (2006) Glucocorticoids of bison bulls in relation to social status. Horm Behav 49: 369–375.
91. Hamlin HJ, Lowers RH, Guillette LJ (2011) Seasonal androgen cycles in adult male American alligators (*Alligator mississippiensis*) from a barrier island population. Biol Reprod 85: 1108–1113.
92. Higham JP, Heistermann M, Maestripieri D (2013) The endocrinology of male rhesus macaque social and reproductive status: a test of the challenge and social stress hypotheses. Behav Ecol Sociobiol 67: 19–30.
93. Schiml PA, Mendoza SP, Saltzman W, Lyons DM, Mason WA (1996) Seasonality in squirrel monkeys (*Saimiri sciureus*). Physiol Behav 60: 1105–1113.
94. Schiml PA, Mendoza SP, Saltzman W, Lyons DM, Mason WA (1999) Annual physiological changes in individually housed squirrel monkeys (*Saimiri sciureus*). Am J Primatol 4: 93–103.
95. Lindburg G, Czekalaa NM, Swaisgood R (2001) Hormonal and behavioral relationships during estrus in the giant panda. Zoo Biol 20: 537–543.
96. Owen MA, Czekalaa NM, Swaisgood R, Karen S, Lindburg G (2005) Seasonal and diurnal dynamics glucocorticoids and behavior in giant pandas. Ursus 16: 208–221.
97. Creel S, Sands JL (2003) Is social stress a consequence of subordination or a cost of dominance? In: De Waal FBM, Tyack PL, editors. Animal Social Complexity. Intelligence, Culture, and Individualized Societies. Harvard University Press, Cambridge, Massachusetts, London. pp. 153–169.
98. Reeder DM, Kosteczko NS, Kunz TH, Widmaier EP (2004) Changes in baseline and stress-induced glucocorticoid levels during the active period in free-ranging male and female little brown myotis, *Myotis lucifugus* (Chiroptera: Vespertilionidae). Gen Comp Endoc 136: 260–269.
99. Dantzer B, McAdam AG, Palme R, Fletcher QE, Boutin S, et al. (2010) Fecal cortisol metabolite levels in free-ranging North American red squirrels: Assay validation and the effects of reproductive condition. Gen Comp Endocr 67: 279–286.
100. Tilden C (2008) Low fetal energy deposition rates in lemurs: another energy conservation strategy. In: Fleagle JG, Gilbert CG, editors. Elwyn Simons, a Search for Origins. Springer, New York. pp. 311–318.

101. Gould L, Ziegler TE, Wittwer DJ (2005) Effects of reproductive and social variables on fecal glucocorticoid levels in a sample of adult male ring-tailed lemurs (*Lemur catta*) at the Beza Mahafaly Reserve, Madagascar. Am J Primatol 67: 5–23.
102. Fürtbauer I, Heistermann M, Schülke O, Ostner J (2014) Low female stress hormone levels are predicted by same- or opposite-sex sociality depending on season in wild Assamese macaques. Psychoneuroendocrinology 48: 19–28.
103. Wright PC, Razafindratsita T, Pochron ST, Jernvall J (2005) The key to frugivory in Madagascar. In: Dew J, Boubli H, editors. Tropical Fruits and Frugivores: the Search for Strong Interactors. The Netherlands: Springer Press. pp. 121–138.
104. Tilden CD, Oftedal OT (1997) Milk composition reflects patterns of maternal care in prosimian primates. Am J Primatol 41: 195–211.
105. Sapolsky RM (1986) Endocrine and behavioral correlates of drought in wild olive baboons (*Papio anubis*). Am J Primatol 11: 217–227.
106. Behie AM, Pavelka MS, Chapman CA (2010) Sources of variation in faecal cortisol levels in howler monkeys in Belize. Am J Primatol 72: 600–606.
107. Gesquiere LR, Onyango PO, Alberts SC, Altmann J (2011) Endocrinology of year-round reproduction in a highly seasonal habitat: Environmental variability in testosterone and glucocorticoids in baboon males. Am J Phys Anthropol 144: 169–176.
108. Chapman CA, Saj T, Snaith TV (2007) Temporal dynamics of nutrition, parasitism and stress in colobus monkeys: implications for population regulation and conservation. Am J Phys Anthropol 134: 240–250.
109. Donati G, Ramanamanjato JB, Ravoahangy AM, Vincelette M (2007) Translocation as a conservation measure for a threatened species: the case of *Eulemur collaris* in the Mandena littoral forest, south-eastern Madagascar. In: Ganzhorn JU, Goodman SM, Vincelette M, editors. Biodiversity, Ecology and Conservation of the Littoral Ecosystems of South-Eastern Madagascar. Washington DC: Smithsonian Institution Press. pp. 237–245.
110. Creel S, Fox JE, Hardy A, Sands J, Garrott B, et al. (2002) Snowmobile activity and glucocorticoid stress responses in wolves and elk. Conserv Biol 16: 809–814.
111. Overdorff DJ, Merenlender AM, Talata P, Telo A, Forward ZA (1999) Life history of *Eulemur fulvus rufus* from 1988–1998 in southeastern Madagascar. Am J Phys Anthropol 108: 295–310.
112. Johnson CA, Raubenheimer D, Rothman JM, Clarke D, Swedell L (2013) 30 days in the life: daily nutrient balancing in a wild chacma baboon. PLOS One 8(7): e70383. doi:10.1371/journal.pone.0070383.
113. Sayre RW (1996) Ecology of Bighorn Sheep in Relation to Habitat and Oil. PhD thesis, University of North Dakota.
114. Wasser SK, Davenport B, Ramage ER, Hunt KE, Parker M, et al. (2004) Scat detection dogs in wildlife research and management: application to grizzly and black bears in the Yellowhead Ecosystem, Alberta, Canada. Can J Zool 82: 475–492.
115. Munshi-South J, Tchignoumba L, Brown J, Abbondanza N, Maldonado JE, et al. (2008) Physiological indicators of stress in African forest elephants (*Loxodonta africana cyclotis*) in relation to petroleum operations in Gabon, Central Africa. Divers Distrib 14: 995–1003.
116. Vincelette M, Theberge M, Randrihasipara L (2007) Evaluations of forest cover at regional and local levels in the Tolagnaro region since 1950. In: Ganzhorn JU, Goodman SM, Vincelette M, editors. Biodiversity, Ecology and Conservation of the Littoral Ecosystems of South-Eastern Madagascar. Washington DC: Smithsonian Institution Press. pp. 49–58.
117. Blickley JL, Word KR, Krakauer AH, Phillips JL, Sells SN, et al. (2012) Experimental chronic noise is related to elevated fecal corticosteroid metabolites in lekking male greater sage-grouse (*Centrocercus urophasianus*). PLOS One 7(11): e50462. doi:10.1371/journal.pone.0050462.
118. Donati G, Baldi N, Morelli V, Ganzhorn JU, Borgognini-Tarli SM (2009) Proximate cues and ultimate determinants of brown lemur cathemerality. An Behav 77: 317–325.

PERMISSIONS

All chapters in this book were first published in PLOS ONE, by The Public Library of Science; hereby published with permission under the Creative Commons Attribution License or equivalent. Every chapter published in this book has been scrutinized by our experts. Their significance has been extensively debated. The topics covered herein carry significant findings which will fuel the growth of the discipline. They may even be implemented as practical applications or may be referred to as a beginning point for another development.

The contributors of this book come from diverse backgrounds, making this book a truly international effort. This book will bring forth new frontiers with its revolutionizing research information and detailed analysis of the nascent developments around the world.

We would like to thank all the contributing authors for lending their expertise to make the book truly unique. They have played a crucial role in the development of this book. Without their invaluable contributions this book wouldn't have been possible. They have made vital efforts to compile up to date information on the varied aspects of this subject to make this book a valuable addition to the collection of many professionals and students.

This book was conceptualized with the vision of imparting up-to-date information and advanced data in this field. To ensure the same, a matchless editorial board was set up. Every individual on the board went through rigorous rounds of assessment to prove their worth. After which they invested a large part of their time researching and compiling the most relevant data for our readers.

The editorial board has been involved in producing this book since its inception. They have spent rigorous hours researching and exploring the diverse topics which have resulted in the successful publishing of this book. They have passed on their knowledge of decades through this book. To expedite this challenging task, the publisher supported the team at every step. A small team of assistant editors was also appointed to further simplify the editing procedure and attain best results for the readers.

Apart from the editorial board, the designing team has also invested a significant amount of their time in understanding the subject and creating the most relevant covers. They scrutinized every image to scout for the most suitable representation of the subject and create an appropriate cover for the book.

The publishing team has been an ardent support to the editorial, designing and production team. Their endless efforts to recruit the best for this project, has resulted in the accomplishment of this book. They are a veteran in the field of academics and their pool of knowledge is as vast as their experience in printing. Their expertise and guidance has proved useful at every step. Their uncompromising quality standards have made this book an exceptional effort. Their encouragement from time to time has been an inspiration for everyone.

The publisher and the editorial board hope that this book will prove to be a valuable piece of knowledge for researchers, students, practitioners and scholars across the globe.

LIST OF CONTRIBUTORS

Xuejiao Bai
College of Forestry, Shenyang Agricultural University, Shenyang, Liaoning Province, China
State Key Laboratory of Forest and Soil Ecology, Institute of Applie Ecology, Chinese Academy of Sciences, Shenyang, Liaoning Province, China

Tania Brenes-Arguedas
Centro Cientifico Tropical, San Jose, Costa Rica

Ji Ye, Xugao Wang, Zhanqing Hao and Zuoqiang Yuan
State Key Laboratory of Forest and Soil Ecology, Institute of Applied Ecology, Chinese Academy of Sciences, Shenyang, Liaoning Province, China

Fei Lin, Shuai Shi and Dingliang Xing
State Key Laboratory of Forest and Soil Ecology, Institute of Applied Ecology, Chinese Academy of Sciences, Shenyang, Liaoning Province, China
University of Chinese Academy of Sciences, Beijing, China

João M. B. Carreiras
Tropical Research Institute (IICT), Lisboa, Portugal
National Centre for Earth Observation (NCEO), Centre for Terrestrial Carbon Dynamics (CTCD), University of Sheffield, Sheffield, South Yorkshire, United Kingdom

Joshua Jones
Department of Geography and Earth Sciences, Aberystwyth University, Aberystwyth, Ceredigion, United Kingdom

Richard M. Lucas
Department of Geography and Earth Sciences, Aberystwyth University, Aberystwyth, Ceredigion, United Kingdom
Centre for Ecosystem Science, School of Biological, Earth and Environmental Sciences, University of New South Wales, Sydney, New South Wales, Australia

Cristina Gabriel
Tropical Research Institute (IICT), Lisboa, Portugal

Yamato Tsuji
Primate Research Institute, Kyoto University, Inuyama, Aichi, Japan

Matthew V. DiLeo, Richard M. Bostock and David M. Rizzo
Department of Plant Pathology, University of California Davis, Davis, California, United States of America

Gaolong Zhu
Department of Geography, Minjiang University, Fuzhou, China
International Institute for Earth System Sciences, Nanjing University, Nanjing, China

Weimin Ju
International Institute for Earth System Sciences, Nanjing University, Nanjing, China, 3 Institute for Climate and Global Change Research, Nanjing University, Nanjing, China

J. M. Chen
International Institute for Earth System Sciences, Nanjing University, Nanjing, China
Department of Geography, University of Toronto, Toronto, Ontario, Canada

Yibo Liu
Jiangsu Key Laboratory of Agricultural Meteorology, College of Applied Meteorology, Nanjing University of Information Science and Technology, Nanjing, China

Sheng Xu, Wenduo Xu, Wei Chen, Xingyuan He, Yanqing Huang and Hua Wen
State Key Laboratory of Forest and Soil Ecology, Institute of Applied Ecology, Chinese Academy of Sciences, Shenyang, China

Fang Wang
Smithsonian Conservation Biology Institute, National Zoological Park, Front Royal, Virginia, United States of America
School of Life Sciences, Peking University, Beijing, China

William J. McShea
Smithsonian Conservation Biology Institute, National Zoological Park, Front Royal, Virginia, United States of America

Dajun Wang, Sheng Li, Hao Wang and Zhi Lu
School of Life Sciences, Peking University, Beijing, China

Qing Zhao
Department of Fish, Wildlife, and Conservation Biology, Colorado State University, Fort Collins, Colorado, United States of America

Li-hua Tu, Gang Chen, Yong Peng, Hong-ling Hu, Ting-xing Hu, Jian Zhang and Xian-wei Li
College of Forestry, Sichuan Agricultural University, Ya'an, Sichuan, China

Li Liu
Personnel Department, Sichuan Agricultural University, Ya'an, Sichuan, China

Yi Tang
College of Horticulture, Sichuan Agricultural University, Ya'an, Sichuan, China

Katiane Silva Conceição, Francisco Aparecido Rodrigues and Marinho Gomes de Andrade
Departamento de Matema´tica Aplicada e Estatı´stica, Instituto de Ciências Matemáticas e de Computação, Universidade de São Paulo, São Carlos, São Paulo, Brazil

Werner Ulrich
Department of Animal Ecology, Nicolaus Copernicus University in Toruń, Toruń, Poland

Carlos Alberto Ribeiro Diniz
Department of Statistics, Federal University of São Carlos, São Carlos, Brazil

Pornwiwan Pothasin
Department of Biology, Faculty of Science, Chiang Mai University, Chiang Mai, Thailand

Stephen G. Compton
School of Biology, University of Leeds, Leeds, United Kindgom
Department of Zoology and Entomology, Rhodes University, Grahamstown, South Africa

Prasit Wangpakapattanawong
Department of Biology, Faculty of Science, Chiang Mai University, Chiang Mai, Thailand
World Agroforestry Center c/o Knowledge Support Center for the Greater Mekong Subregion, Faculty of Social Sciences, Chiang Mai University, Chiang Mai, Thailand

Junfeng Niu and Ping Zhao
Key Laboratory of Vegetation Restoration and Management of Degraded Ecosystems, South China Botanical Garden, Chinese Academy of Sciences, Guangzhou, Guangdong, China

Zhaozhong Feng and Xiaoke Wang
State Key Laboratory of Urban and Regional Ecology, Research Center for Eco-Environmental Sciences, Chinese Academy of Sciences, Beijing, China

Weiwei Zhang
Key Laboratory of Black Soil Ecology, Northeast Institute of Geography and Agroecology, Chinese Academy of Sciences, Harbin, Heilongjiang, China

Carlos Iñiguez-Armijos
Departamento de Ciencias Naturales, Universidad Técnica Particular de Loja (UTPL), Sección de Ecología, Loja, Ecuador
Institute for Landscape Ecology and Resources Management (ILR), Research Centre for Bio Systems, Land Use and Nutrition (IFZ), Justus-Liebig-University Giessen, Giessen, Germany

Adrián Leiva
Departamento de Ciencias Naturales, Universidad Técnica Particular de Loja (UTPL), Sección de Ecología, Loja, Ecuador

Hans-Georg Frede and Lutz Breuer
Institute for Landscape Ecology and Resources Management (ILR), Research Centre for Bio Systems, Land Use and Nutrition (IFZ), Justus-Liebig-University Giessen, Giessen, Germany

Henrietta Hampel
Universidad de Cuenca, Quinta Balzaín, Cuenca, Ecuador

Aritz Ruiz-González
Department of Zoology and Animal Cell Biology, University of the Basque Country, UPV/EHU, Vitoria-Gasteiz, Spain
Systematics, Biogeography and Population Dynamics Research Group, Lascaray Research Center, University of the Basque Country, UPV/EHU, Vitoria-Gasteiz, Spain
Conservation Genetics Laboratory, National Institute for Environmental Protection and Research, ISPRA, Ozzano dell'Emilia, Bologna, Italy

Mikel Gurrutxaga
Systematics, Biogeography and Population Dynamics Research Group, Lascaray Research Center, University of the Basque Country, UPV/EHU, Vitoria-Gasteiz, Spain
Department of Geography, University of the Basque Country, UPV/EHU, Vitoria-Gasteiz, Spain

Samuel A. Cushman
U.S. Forest Service, Rocky Mountain Research Station, Flagstaff, AZ, United States of America

María José Madeira and Benjamin J. Go´mez-Moline
Department of Zoology and Animal Cell Biology, University of the Basque Country, UPV/EHU, Vitoria-Gasteiz, Spain

Systematics, Biogeography and Population Dynamics Research Group, Lascaray Research Center, University of the Basque Country, UPV/EHU, Vitoria-Gasteiz, Spain

Ettore Randi
Conservation Genetics Laboratory, National Institute for Environmental Protection and Research, ISPRA, Ozzano dell'Emilia, Bologna, Italy
Department 18/Section of Environmental Engineering, Aalborg University, Aalborg, Denmark

Richard Schuster and Peter Arcese
Department of Forest and Conservation Sciences, University of British Columbia, Vancouver, British Columbia, Canada

Tara G. Martin
Department of Forest and Conservation Sciences, University of British Columbia, Vancouver, British Columbia, Canada
Ecosciences Precinct, CSIRO Ecosystem Sciences, Brisbane, Queensland, Australia

Nerea Abrego
Department of Biological and Environmental Science, University of Jyväskylä, Jyväskylä, Finland
Department of Plant Biology and Ecology, University of the Basque Country (UPV/EHU), Bilbao, Spain

Gonzalo García-Baquero
Department of Plant Biology and Ecology, University of the Basque Country (UPV/EHU), Bilbao, Spain

Panu Halme
Department of Biological and Environmental Science, University of Jyväskylä, Jyväskylä, Finland
Natural History Museum, University of Jyväskylä, Jyväskylä, Finland

Otso Ovaskainen
Department of Biosciences, University of Helsinki, Helsinki, Finland

Isabel Salcedo
Department of Plant Biology and Ecology, University of the Basque Country (UPV/EHU), Bilbao, Spain

James W. Raich
Department of Ecology, Evolution & Organismal Biology, Iowa State University, Ames, Iowa, United States of America

Deborah A. Clark
Department of Biology, University of Missouri-St. Louis, St. Louis, Missouri, United States of America

Luitgard Schwendenmann
School of Environment, The University of Auckland, Auckland, New Zealand

Tana E. Wood
United States Department of Agriculture Forest Service, International Institute of Tropical Forestry, Rio Piedras, Puerto Rico
Fundación Puertorriqueña de Conservación, San Juan, Puerto Rico

Zhang-Ting Huang, Yan Zhang and Yu-Qi Ying
School of Environment and Resources, Zhejiang A&F University, Lin'an, Zhejiang, China

Pei-Kun Jiang
School of Environment and Resources, Zhejiang A&F University, Lin'an, Zhejiang, China
Zhejiang Provincial Key Laboratory of Carbon Cycling in Forest Ecosystems and Carbon Sequestration, Zhejiang A&F University, Lin'an, Zhejiang, China

Scott Xiaochuan Chang
Department of Renewable Resources, University of Alberta, Edmonton, Alberta, Canada

Haiyang Wang
Institute of Landscape Ecology of Montane Horticulture, Southwest University, Chongqing, 400716, China

Hui Feng and Yanru Zhang
Department of Botany, College of Horticulture and Landscape Architecture, Southwest University, Chongqing, 400716, China

Hong Chen
Institute of Landscape Ecology of Montane Horticulture, Southwest University, Chongqing, 400716, China
Department of Botany, College of Horticulture and Landscape Architecture, Southwest University, Chongqing, 400716, China

Eduardo Carneiro, Olaf Hermann Hendrik Mielke and Mirna Martins Casagrande
Laboratório de Estudos de Lepidoptera Neotropical, Zoology Department, UFPR. Curitiba, Paraná, Brasil

Konrad Fiedler
Division of Tropical Ecology & Animal Biodiversity, University of Vienna, Vienna, Austria

Joseph Kipkorir Sigi Lang'at
Kenya Marine and Fisheries Research Institute, Mombasa, Kenya
School of Life, Sport and Social Sciences, Edinburgh Napier University, Edinburgh, United Kingdom

James G. Kairo
Kenya Marine and Fisheries Research Institute, Mombasa, Kenya

Maurizio Mencuccini
School of Geosciences, University of Edinburgh, Crew Building, Edinburgh, United Kingdom
Centre for Ecological Research and Forestry Applications, Universitat Auto`noma de Barcelona, Barcelona, Spain

Steven Bouillon
Department of Earth and Environmental Sciences, Katholieke Universiteit Leuven, Leuven, Belgium

Susan Waldron
School of Geographical and Earth Sciences, University of Glasgow, Glasgow, South Lanarkshire, United Kingdom

Mark Huxham
School of Life, Sport and Social Sciences, Edinburgh Napier University, Edinburgh, United Kingdom

Martin W. Skov
School of Ocean Sciences, Bangor University, Menai Bridge, Anglesey, United Kingdom

Miguel A. Munguía-Rosas and Salvador Montiel
Departamento de Ecología Humana, Centro de Investigación y de Estudios Avanzados del Instituto Politécnico Nacional (CINVESTAV), Mérida, Yucatán, México

Bráulio A. Santos
Departamento de Sistemática e Ecologia, Universidade Federal da Paraíba, João Pessoa, Paraíba, 58051-900, Brazil

Marcelo Tabarelli and Felipe P. L. Melo
Departamento de Botânica, Universidade Federal de Pernambuco, Recife, Pernambuco, 50670-901, Brazil

José L. C. Camargo and Ana Andrade
Biological Dynamics of Forest Fragment Project, National Institute for Amazonian Research (INPA), C.P. 478, Manaus, Amazonas, 69011-970, Brazil

Susan G. Laurance and William F. Laurance
Centre for Tropical Environmental and Sustainability Science (TESS) and School of Marine and Tropical Biology, James Cook University, Cairns, Queensland, 4878, Australia

Jie Zhao, Kelin Wang and Xunyang He
Key Laboratory of Agro-ecological Processes in Subtropical Region, Institute of Subtropical Agriculture, Chinese Academy of Sciences, Changsha, Hunan, China
Huanjiang Observation and Research Station for Karst Ecosystems, Chinese Academy of Sciences, Huanjiang, Guangxi, China

Shengping Li and Lu Liu
Key Laboratory of Agro-ecological Processes in Subtropical Region, Institute of Subtropical Agriculture, Chinese Academy of Sciences, Changsha, Hunan, China
Huanjiang Observation and Research Station for Karst Ecosystems, Chinese Academy of Sciences, Huanjiang, Guangxi, China
Graduate University of Chinese Academy of Sciences, Beijing, China

Binu M. Tripathi and Jonathan M. Adams
Department of Biological Sciences, Seoul National University, Seoul, Republic of Korea

Dorsaf Kerfahi
Department of Biological Sciences, Seoul National University, Seoul, Republic of Korea
School of Chemical and Biological Engineering, Interdisciplinary Program of Bioengineering, Seoul National University, Seoul, Republic of Korea

Junghoon Lee
School of Chemical and Biological Engineering, Interdisciplinary Program of Bioengineering, Seoul National University, Seoul, Republic of Korea

David P. Edwards
Department of Animal and Plant Sciences, University of Sheffield, Sheffield, United Kingdom

Jay Ruffell
School of Animal Biology, The University of Western Australia, Perth, Australia

Raphael K. Didham
School of Animal Biology, The University of Western Australia, Perth, Australia
Centre for Environment and Life Sciences, the Commonwealth Scientific and Industrial Research Organisation, Perth, Australia

Paul Barrett, Nic Gorman, Rhonda Pike, Andrée Hickey-Elliott, Karin Sievwright and Doug P. Armstrong
Wildlife Ecology Group, Te Kura Mātauranga o ngā Taonga ā Papatuanuku, Massey University, Palmerston North, New Zealand

Michela Balestri and Marco Campera
Department of Social Sciences, Oxford Brookes University, Oxford, United Kingdom
Department of Biology, University of Pisa, Pisa, Italy

Marta Barresi and Valentina Serra
Department of Biology, University of Pisa, Pisa, Italy

Jean Baptiste Ramanamanjato
QIT Madagascar Minerals, Rio Tinto, Tolagnaro, Madagascar

Michael Heistermann
Endocrinology Laboratory, German Primate Center, Leibniz Institute for Primate Research, Goettingen, Germany

Giuseppe Donati
Department of Social Sciences, Oxford Brookes University, Oxford, United Kingdom

Index